AGRICULTURAL MECHANICS:
Fundamentals and Applications

AGRICULTURAL MECHANICS:
Fundamentals and Applications

ELMER L. COOPER, Ed.D.

H. Edward Reiley
Consulting Editor

 Delmar Publishers Inc.®

NOTICE TO THE READER

Delmar Staff:
   Administrative editor: Mark W. Huth
   Developmental editor: Marjorie A. Bruce
   Production editor: Ruth Saur
   Art director: Ronald Blackman
   Art coordinator: Tony Canabush
   Design coordinator: John Orozco

For information, address Delmar Publishers Inc.
2 Computer Drive West, Box 15–015
Albany, New York 12212–9985

Printed in the United States of America
Published simultaneously in Canada
by Nelson Canada,
a Division of International Thomson Limited

10 9 8 7 6 5 4 3 2

Library of Congress Cataloging in Publication Data

Cooper, Elmer L.
   Agricultural mechanics.

   Includes index.
   1. Agricultural mechanics.   I. Reiley, H. Edward.
II. Title.
S675.3.C66   1986   631.3   85–20568
ISBN 0–8273–2271–2

CONTENTS

PREFACE

Agricultural Mechanics: Fundamentals and Applications is written for students and individuals needing written materials on basic mechanical skills. It grew out of the need for an easy to read, easy to understand, and highly illustrated text on modern agricultural mechanics for high school and post-secondary programs. It addresses the specific needs of students enrolled in production agriculture, ornamental horticulture, agricultural business, agricultural mechanics, and natural resources programs. The text starts with very basic and general information, such as career opportunities, and then provides competency-based instruction on basic mechanical skills and applications.

Agricultural Mechanics: Fundamentals and Applications is highly illustrated to clarify the written word. It contains instructions on materials such as wood, sheet metal, structural steel, electrical wire, rust preventatives, wood preservatives, and paint and other finishes. It includes procedures utilizing hand tools, portable power tools, and stationary power equipment. The selection and use of nails, screws, bolts, and glue are covered in detail. Processes include woodworking, metalworking, electric and gas welding, heating and cutting, small engine maintenance, electrical wiring, plumbing, and concrete and masonry work.

A competency-based approach is used throughout the text. This makes the text consistent in format, easy to use for individualized instruction, easy to teach from, simple for substitute teachers, and easy to assess student progress. Each unit is part of a section and includes (1) a statement of objective, (2) competencies to be developed, (3) a list of new terms, (4) a materials list, (5) highly illustrated text material, (6) student activities, and (7) a self-evaluation. All new terms are defined in the text as well as in the glossary.

The appendix includes 51 project plans with bills of materials and construction procedures. The projects were carefully selected to match the skills covered in the text. Plans include some projects that have become classics in the field, some that cover targeted enterprises, and some that are new and innovative. The projects were selected because they are used in high school agricultural programs. The appendix also includes 36 tables containing information for estimating, planning, selecting, purchasing, and building in agricultural mechanics. The project plans, tables, glossary and index provide unique reference materials which, for many users, are alone worth the price of the text.

An instructor's guide is also available. The guide provides a key to student self-evaluations, numerous transparency masters, and suggestions for using the text.

About the Author

Dr. Elmer L. Cooper is an associate professor in agricultural education at the University of Maryland. He has taught vocational agriculture in the public schools for seventeen years, including grades eight through twelve and adult classes. Dr. Cooper served as the State Program Specialist in Agriculture/Agribusiness and Renewable Natural Resources in Maryland for a number of years. He also has experience in teacher education at two universities. This extensive background well qualifies Dr. Cooper for selecting, organizing, and writing the contents of *Agricultural Mechanics: Fundamentals and Applications.*

ACKNOWLEDGMENTS

The author wishes to thank the many individuals who provided encouragement and assistance during the preparation of this text. He would particularly like to thank the many vocational agriculture teachers, students, and faculty at North Harford High School, Hereford High School, Linganore High School, Walkersville High School, the Institute of Applied Agriculture and the Department of Agricultural Engineering at the University of Maryland, the Towson State University, and other educational institutions for providing the settings and/or being a subject in the photographs.

The expert contributions by agricultural instructors Steen Westerberg, Vernon Marshall, Charles Cramer, James Ferrant, and Thomas Hawthorne are especially noteworthy. Appreciation is expressed to photographers Frederick Doepkens, Richard Kreh, and Ronald Seibel for their invaluable services.

The early approval and continuous support of the project by Dr. Clifford Nelson, Chairman of the Department of Agricultural and Extension Education, University of Maryland, is appreciated and gratefully acknowledged. In like manner, appreciation is expressed to his wife, Dollye Cooper, for her constant support and assistance.

The continuous encouragement and periodic handling of typing, compilation, and mailing tasks by Cathie Galeano, and typing by Sandy Smith and the typists at TERP Service are gratefully acknowledged. The author would also like to express appreciation to the many schools, colleges, curriculum centers, businesses, and government offices across the nation for supplying photographs and illustrative materials for use in the text.

Section 1
EXPLORING CAREERS IN AGRICULTURAL MECHANICS

Unit 1
Mechanics in the World of Agriculture

Objective

To determine how mechanical skills, concepts, and principles are used in agriculture and related occupations.

Competencies to Be Developed

After studying this unit, you will be able to:
- Define agriculture and agricultural mechanics.
- Define occupation and occupational cluster.
- Describe the role of mechanics and mechanical applications in society.
- Demonstrate knowledge of contributions made by mechanical application to the development of agriculture.
- Name 8 inventors of important agricultural machines.

Terms to Know

agriculture	business
agribusiness	profit
renewable natural	trade
resources	mechanical
occupational cluster	mechanic
occupation	agricultural mechanics
employment	efficiency

Materials List

Pencil
Paper
Encyclopedias

Historically, the word *agriculture* meant to farm or to grow plants or animals. Today, agriculture is defined as those activities concerned with the production of plants and animals, and the related supplies, services, mechanics, products, processing and marketing. Actually, modern agriculture covers so many activities that a simple definition is not possible. The United States Department of Education uses the phrase "agriculture/agribusiness and renewable natural resources" to refer to the broad range of activities associated with agriculture. *Agribusiness* refers to the network of commercial firms that have developed with or stems out of agriculture.

Renewable natural resources are the resources provided by nature that can replace or renew themselves. Examples of such resources are wildlife, trees, and fish. Some occupations in renewable natural resources are game trapper, forester, and waterman. A waterman uses boats and specialized equipment to harvest fish, oysters, and other seafood.

All jobs and types of work in the field of agriculture/agribusiness and renewable natural resources make up an occupational cluster. An *occupational cluster* is a group of related jobs. There are many occupations that are related to agriculture as shown in figure 1–1. Agricultural mechanics is only one of the many careers that fall under the broad term of agriculture.

Figure 1-1. The industry of agriculture provides interesting careers in production, management, science, education, finance, communication, government, conservation, and mechanics.

FRUIT PRODUCTION

PESTICIDES

TURF MANAGEMENT

GARDENING

GREENHOUSE PRODUCTION

NURSERY OPERATIONS

BERRY PRODUCTION

LANDSCAPE DESIGN

FLORICULTURE

Figure 1–1. *Continued*

```
┌─────────────────────────────────┐
│         ONE FARM WORKER         │
│  (JOBS INVOLVING THE PRODUCTION │
│        OF FOOD AND FIBER)       │
└─────────────────────────────────┘
```

```
┌──────────────────────────────────┐   ┌──────────────────────────────────┐
│  TWO JOBS IN AGRICULTURAL         │   │  TWO JOBS IN AGRICULTURAL         │
│  SUPPLIES AND SERVICES (JOBS      │   │  PROCESSING AND DISTRIBUTION      │
│  THAT SUPPORT PRODUCTION          │   │  (JOBS THAT PROCESS AND MARKET    │
│  AGRICULTURE)                     │   │  THE PRODUCTS OF PRODUCTION       │
│                                   │   │  AGRICULTURE)                     │
└──────────────────────────────────┘   └──────────────────────────────────┘
```

Figure 1–2. The industry of agriculture now includes five people for each worker on a farm. There are about four nonfarm agricultural jobs for each farm job.

Agriculture depends on many nonfarm workers for goods and services. For every farm worker in the United States, there are said to be four workers in nonfarm agricultural jobs, figure 1–2. Nonfarm agriculturists provide machinery, equipment, fertilizer, feed, seed, money, research, and government services. Similarly, marketing, transporting, processing, and distributing farm products are done by nonfarm agriculturalists. This network of nonfarm agricultural workers helped the average farmer feed about 80 people in 1985, figure 1–3. On the most productive and

efficient farms, however, one farmer was said to feed more than 200 people.

The term *occupation* means business, employment, or trade. It is the work a person does regularly to earn a living. *Business* generally refers to work done for profit. *Profit* refers to income made from the sale of goods or services after expenses have been taken out. An example of profit is the income made by a local machinery dealer or a local garden center minus expenses.

Employment means work done for pay. Employees are paid by the hour, day, week, month, or year. A small-engines mechanic employed at the local machinery dealer is probably paid by the hour. The sales manager is probably paid an annual salary with bonuses when the sales volume is very high.

Trade refers to specific kinds of work or businesses, especially those that require skilled mechanical work. *Mechanical* means having to do with a machine, mechanism, or machinery. A person who is specifically trained to perform mechanical tasks is a *mechanic*. The mechanic must be skilled in the use of tools and machines. The mechanic must also be able to select appropriate materials, use a variety of processes, and analyze problems. The term *mechanics* is defined as the branch of physics dealing with motion, and the action of forces on bodies or fluids.

The term *agricultural mechanics* is the selection, operation, maintenance, service, selling, and use of power units, machinery, equipment, structures, and utilities used in agriculture. Hundreds of different jobs are available in agricultural mechanics. A total of 99 out of 305 occupational entries in *A Concise Handbook of Occupations* are mechanical in nature.

Figure 1–3. The average American farmer produced enough to feed about 80 people in 1985 (*courtesy of the U.S. Department of Agriculture*).

□ AGRICULTURE—A BASIC INDUSTRY

Agriculture is a very complex industry. The industry produces plant and animal products from which thousands of commodities are made. Since every person and many industries depend upon agriculture, it is said to be a basic industry. Some products of agriculture are food, oils, fiber, lumber, ornamental trees and shrubs, flowers, leather, fertilizers, feed, seed, and more. Basic agricultural products form the raw materials for many items of everyday living.

Fabrics for clothing, curtains, and floor coverings are made from oils such as corn oil, soybean oil, and cottonseed oil. Plastics of all kinds are also made from oils. Products from animals are used to make materials such as glue, leather, and paint. The manufacture of automobiles, furniture, airplanes, radios, stereos, and computers all depend on agriculture for certain raw materials. The construction of homes, boats, and factories all depend on agriculture for lumber and other basic commodities. Most dwellings in America are surrounded by lawn, shrubs, or other plants for beautification. These are also agricultural commodities.

Agriculture is, indeed, a basic industry upon which all people depend. It is the backbone of the American society.

□ THE ROLE OF AGRICULTURAL MECHANICS

Mechanical applications are found throughout agriculture. Some of these applications are shown in figures 1–4 through 1–8. A few examples of jobs involving agricultural mechanics are:

- the engineer who designs tractors and other farm and ranch machines

Figure 1–4. Landscaping includes machinery and mechanical operations that require people in agricultural mechanics to keep machinery repaired and in top operating condition (*courtesy of the U.S. Department of Agriculture*).

Figure 1–5. Agricultural construction is an important segment of agricultural mechanics (*courtesy of the U.S. Department of Agriculture*).

- the electrician who installs greenhouse climate controls
- the soil conservationist who constructs terraces to control erosion
- the hardware store employee who must locate repair parts for agricultural tools and machines

Figure 1–6. Workers in agricultural mechanics are responsible for the design, manufacture, testing, sales, and service of farm buildings, machinery, and equipment (*courtesy of the U.S. Department of Agriculture*).

Figure 1–7. Many jobs in modern agricultural mechanics involve computerized operations (*courtesy of the U.S. Department of Agriculture*).

- the lawn equipment service mechanic who repairs lawn tractors
- the weldor who repairs farm machinery

Processing plants for field crops, livestock, poultry, fruits, and vegetables all have machinery. Such machinery requires designers, engineers, operators, maintenance and repair personnel, and construction workers. Even people with jobs in finance, publications, and communications may need some knowledge of mechanics when their assignments deal with agriculture.

Figure 1–8. Many jobs in agricultural mechanics are found in suburban or urban settings. Here, heat from the sun is trapped in the solar pond on the left and is then used to heat the production greenhouses on the right (*courtesy of the U.S. Department of Agriculture*).

☐ THE INFLUENCE OF MECHANIZATION

At the birth of the United States in 1776, over 90 percent of the American colonists were farmers, yet many of General Washington's troops at Valley Forge died for lack of food and clothing. Today, less than 10 percent of all Americans work in agriculture, yet there are generally food surpluses in America. The ratio of farm workers to nonfarm people in America approximately reversed itself in 200 years. In 1776, the farm to nonfarm ratio was approximately 9 to 1. Today, the ratio of agricultural workers to the remaining population is approximately 1 to 9. Mechanization has played a major role in this rise in production efficiency. *Efficiency* means ability to produce with a minimum waste of time, energy, and materials.

America provided the inventors for many of the world's most important agricultural machines. Cyrus McCormick invented the reaper in 1834 to cut small grain crops. Later, the combine was invented which cut and threshed the grain in the field. Today, one modern combine operator can cut and thresh as much grain in one day as 100 persons could cut and bundle in one day in the 1830s.

Of great significance was Thomas Jefferson's invention of an iron plow to replace the wooden plows of the time. Later, in 1837, a man named John Deere developed a steel plow which permitted plowing of the rich prairie soils of the great American West. In 1793, Eli Whitney invented the cotton gin. The cotton gin removes the cotton seed from the cotton fiber. This paved the way for an expanded cotton and textile industry. In 1850, Edmund W. Quincy invented the mechanical corn picker. Joseph Glidden's development of barbed wire permitted establishment of ranches with definite boundaries. In 1878, Anna Baldwin invented the vacuum milking machine, and in 1904, Benjamin Holt invented the tractor.

Many of the early inventors worked alone or with one or two partners. They all were workers in the area of agricultural mechanics. By the early 1900s, many people worked in factories or operated businesses. The companies that were formed to produce agricultural machinery or process agricultural products turned to invention also. For instance, mechanical cotton pickers were developed in the 1930s by several American companies.

Agriculture has become highly mechanized in the developed countries of the world. For the undeveloped countries, many engineers, teachers, and technicians have sought simple, tough, and reliable small machines to improve agriculture. In such countries, America's highly developed, complex, computerized, and expensive machinery will not do. Most countries do not have people trained for the variety of agricultural mechanics jobs that are needed to support America's agriculture.

Many features, such as rubber tires, have been standard equipment on American farms since the 1930s. Yet, a machine with rubber tires is useless if a tire is damaged and repair services are not available. This is the case in most undeveloped countries in Central and South America, Asia, and Africa. Much of the world cannot compete with American agriculture because the related agricultural products and services are not available to support the farm worker.

□ STUDENT ACTIVITIES

1. Copy the words that are italicized in the unit and write a proper definition for each.
2. Interview a Cooperative Extension Specialist for Agricultural Resources in your county or city. Ask the specialist to describe the different jobs people do in your locality that are regarded as agricultural or agriculturally related.
3. Look up "inventors" or "inventions" in an encyclopedia. Pick out the inventions that relate to agriculture and report your findings to the class.
4. Select three or five classmates to join you in a debate on the role of agriculture in society. One team should support the position that agriculture is the backbone of society. The opposing team should support the notion that it is not.
5. Consider an everyday product such as bread, milk, leather gloves, or a corsage for Mother's Day. Trace the production, processing, and marketing of the item from its source to finished product. List points along the way where agricultural mechanics are used.

□ SELF-EVALUATION

A. Multiple Choice. Select the best answer.
1. The production of plants and animals and the related supplies, services, mechanics, products, processing, and marketing defines
   a. horticulture
   b. renewable natural resources
   c. agricultural mechanics
   d. agriculture
2. Agribusiness is
   a. the same as agricultural mechanics
   b. limited to the sale of agricultural products
   c. business stemming from agriculture
   d. special work done by medical doctors

3. Examples of renewable natural resources are
   a. oil, gas, and coal
   b. fish, trees, and wildlife
   c. rubber, steel, and water
   d. air, soil, and minerals
4. The term occupation means
   a. business
   b. employment
   c. trade
   d. all of these
5. Agricultural mechanics stems mostly from
   a. physics
   b. biology
   c. medicine
   d. horticulture
6. Agricultural products come from
   a. soil and coal
   b. plants and animals
   c. iron ore and aluminum
   d. atomic fuel
7. Products of agriculture include
   a. leather seat covers
   b. paint
   c. flower arrangements
   d. All of these
8. Agricultural mechanics includes the occupation of
   a. garden tractor repairman
   b. automobile mechanic
   c. pile driver
   d. systems analyst
9. Mechanization of agriculture has resulted in
   a. decreased soil production
   b. decreased farm expenses
   c. increased production efficiency
   d. increased numbers of farm workers
10. Cyrus McCormick invented the
    a. steel plow
    b. cotton gin
    c. milking machine
    d. reaper
11. In 1776, the ratio of farm workers to nonfarm people was approximately
    a. 9 to 1
    b. 1 to 1
    c. 4 to 1
    d. 1 to 9
12. Today, the ratio of agricultural workers to the remaining population is approximately
    a. 9 to 1
    b. 1 to 9
    c. 4 to 1
    d. 1 to 1

Unit 2
Career Options in Agricultural Mechanics

Objective

To determine how skills in agricultural mechanics may be used to earn a good living.

Competencies to Be Developed

After studying this unit, you will be able to:
* List the major divisions in the agricultural cluster of occupations.
* Identify occupations in agriculture that require mechanical skills.
* Describe the relationship between mechanical applications and success in certain agricultural occupations.
* Conduct an in-depth study of one or more jobs in agricultural mechanics.
* Establish tentative personal goals for using agricultural mechanics skills.

Terms to Know

off-the-farm agricultural jobs
occupational division
agribusiness and agricultural production

FFA
4-H
Boy Scouts

Materials List

A computerized, microfiche, or other career information system
Occupational Outlook Handbook

Figure 2–1. The career options available to vocational agriculture students provide "many hats" from which to choose (*courtesy of The National Future Farmer*).

Figure 2–2. The serenity of rural areas attracts many people to careers in agriculture (*courtesy of the U.S. Department of Agriculture*).

Many hats to choose from! This is one way to state the career opportunities and options in agriculture. Figure 2–1 shows a young man with many hats stacked on his head and holding many tools. The picture suggests that young men and women who have agricultural skills have many good jobs available to them. The serenity of rural areas also attract many people to careers in agriculture, figure 2–2.

A. Agricultural engineers and soil and water technicians seek better ways to use water wisely.

B. Repair of electronic equipment is a necessary part of modern agriculture.

C. Agriculture provides many jobs for those who prefer to work outdoors.

Figure 2–3. Many occupations are available in the field of agriculture (*courtesy of the U.S. Department of Agriculture*).

Future Farmers of America (*FFA*) members and vocational agriculture students have bright futures. Such students are preparing for an industry with many opportunities, figure 2–3. The advice is simple, "Plan ahead!"

☐ AGRICULTURE IS NUMBER ONE

Agriculture is America's number one employer. Some estimate there are 23 million jobs in the broad field of agriculture. This is nearly 22 percent of America's

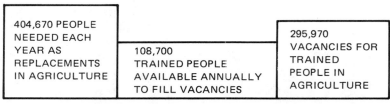

| 404,670 PEOPLE NEEDED EACH YEAR AS REPLACEMENTS IN AGRICULTURE | 108,700 TRAINED PEOPLE AVAILABLE ANNUALLY TO FILL VACANCIES | 295,970 VACANCIES FOR TRAINED PEOPLE IN AGRICULTURE |

SOURCE: U.S. DEPARTMENT OF EDUCATION

Figure 2–4. Employment outlook for agriculture/agribusiness as reported in an interdepartmental study conducted by the United States government in 1979 (*courtesy of the U.S. Department of Education*).

total work force. A large proportion of these jobs are off-the-farm agricultural jobs. *Off-the-farm agricultural jobs* are those jobs requiring agricultural skills, but are not regarded as farming or ranching. The efficiency of this unique mix of farm and nonfarm agricultural jobs permits the average American to use only 14 percent of income on food. This is the lowest of any nation in the world.

The opportunities in agriculture are found at all levels of employment. The first level of employment is laborer; this level requires the least amount of preparation. A few laborer jobs require less than a high school diploma. Additional training is needed for the better paying jobs. Other job levels include semi-skilled, skilled, managerial, and professional. Some professional jobs, such as veterinarian and scientist, require a doctor's degree. The doctor's degree requires seven or more years of college education. It is important to make a career choice as early as possible so the proper training can be obtained. Career goals should be set while in high school. High school courses, work experience, and future schooling can then help prepare for the chosen career. Jobs are available in agricultural occupations at every work level.

Students enrolled in an agricultural course have the edge on others who seek careers in agriculture. Vocational agriculture prepares the student specifically for the world of work. With so many areas inside the broad field of agriculture, choices must be narrowed. Choosing a division within agriculture generally permits more specializing and better job opportunities. An interdepartmental study made by the United States government reports that there are many more jobs available in agriculture than there are qualified people to fill them, figure 2–4. This means an excellent employment outlook for vocational agriculture graduates.

□ AGRICULTURAL DIVISIONS

A number of United States government agencies have worked together to classify occupations. In 1981 the National Center for Educational Statistics published

A Classification of Instructional Programs. This book arranges all occupations into occupational clusters and divisions. All jobs in agriculture, both farm and nonfarm, are in the agricultural cluster. An *occupational division* is a group of occupations or jobs within a cluster that requires similar skills.

Divisions in Agribusiness and Agricultural Production

The *agribusiness and agricultural production* area contains eight divisions. They are

- agricultural business and management
- agricultural mechanics
- agricultural production
- agricultural products and processing
- agricultural services and supplies
- horticulture
- international agriculture
- agribusiness and agricultural production, other.

Figure 2–5. Many people make careers of operating the massive machinery used in agricultural occupations (*courtesy of the U.S. Department of Agriculture*).

Figure 2–6. This solar pond is used to heat nearby greenhouses and provides examples of mechanical applications in agriculture (*courtesy of the U.S. Department of Agriculture*).

Notice that agricultural mechanics is one of the divisions. Farm equipment operator is a part of agricultural mechanics. Many young people find satisfaction in a career operating heavy equipment, figure 2–5. There are many interesting jobs available in agriculture that use mechanical applications. Figure 2–6 shows the construction and use of a solar pond to heat greenhouses. Figure 2–7 shows career classifications in agribusiness, production agriculture, and renewable natural resources.

Divisions in Agricultural Sciences

The agricultural sciences has six divisions. They are

- agricultural sciences, general
- animal sciences
- food sciences
- plant sciences
- soil sciences
- agricultural sciences, other.

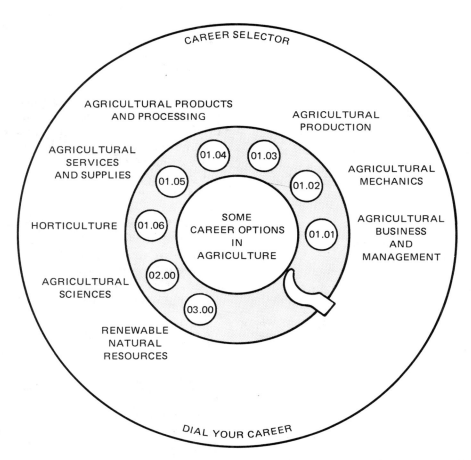

Figure 2–7. Dial a career in agriculture! The numbers on the dial correspond to the 1981 U.S. Department of Education Classification of careers. Note that six items are numbered 01. This means they are all divisions under Agribusiness and Agricultural Production (*courtesy of the U.S. Department of Education*).

Figure 2–8. Scientific application is an important part of agriculture. In this photo, laser technology is being used in a land leveling operation (*courtesy of The National Future Farmer magazine*).

Notice that each division has an "s" on science. This is because there are numerous sciences under each division.

American agriculture is based on scientific knowledge. Production, management, and mechanics all rely on information obtained through the scientific approach. Figure 2–8 shows laser technology being used in a land leveling operation.

Divisions in Renewable Natural Resources

The renewable natural resources area contains seven divisions. They are

- renewable natural resources, general
- conservation and regulation
- fishing and fisheries
- forestry production and processing
- forestry and related sciences
- wildlife management
- renewable natural resources, other.

There are many jobs listed under the renewable natural resources division that require agricultural mechanics.

□ CAREER SELECTION

Agriculture teachers, guidance counselors, and librarians are sources of information regarding careers in the field of agriculture. Figure 2–9 provides a breakdown of the various divisions in agriculture as listed in *A Classification of Instructional Programs*. The publication also contains a description of each program.

There are specific job titles in agricultural mechanics. They are classified under the following categories:

- agricultural mechanics, general
- agricultural electrification, power, and controls
- agricultural mechanics, construction, and maintenance skills
- agricultural power machinery
- agricultural structures, equipment, and facilities
- soil and water mechanical practices
- agricultural mechanics, other

Figure 2–10 shows some specific jobs that are included in these classifications.

Farmers, ranchers, greenhouse operators, pesticide applicators, veterinarians, wildlife officers—all are better at their jobs if they have agricultural mechanics skills. Even those who simply use buildings, equipment, or materials find mechanical skills help them solve problems. As students plan courses in vocational agriculture, they should get as much experience in agricultural mechanics as possible. Such experiences should be stimulating and rewarding. They may occur in the classroom, shop, greenhouse, school farm, at home, in supervised occupational experience programs, or FFA. In FFA, students may participate in agricultural mechanics or tractor operation contests. There are proficiency awards for farm and home safety and agricultural mechanics. Proficiency award winners receive up to $500 depending on the level of the award. Additionally, agricultural mechanics may be used as a speech topic in any organization.

Members of *4-H* conduct many projects using agricultural mechanics skills. Some of these projects are wood science, electricity, tractor safety and maintenance, automotive and small engines. The *Boy Scouts of America* has over 115 merit badge areas of which at least 25 involve agricultural mechanics skills. Some of these are camping, drafting, electricity, energy, engineering, farm mechanics, forestry, gardening, home repairs, landscape architecture, machinery, soil and water conservation, metalwork, plumbing, and woodwork.

□ MEETING THE CHALLENGE

Students having trouble choosing a specific career within agriculture should examine figure 2–11. This may help them to set career goals. Figure 2–12 indicates what other vocational agriculture students plan on doing. Most importantly, students should talk with the teacher or vocational counselor to learn more about an appropriate career. Answering the following

0.1   AGRIBUSINESS AND AGRICULTURAL PRODUCTION

01.01   AGRICULTURAL BUSINESS AND MANAGEMENT
01.0101   Agricultural Business and Management, General
01.0102   Agricultural Business
01.0103   Agricultural Economics
01.0104   Farm and Ranch Management
01.0199   Agricultural Business and Management, Other

01.02   AGRICULTURAL MECHANICS
01.0201   Agricultural Mechanics, General
01.0202   Agricultural Electrification, Power, and Controls
01.0203   Agricultural Mechanics, Construction and Maintenance Skills
01.0204   Agricultural Power Machinery
01.0205   Agricultural Structures, Equipment, Facilities
01.0206   Soil and Water Mechanical Practices
01.0299   Agricultural Mechanics, Other

01.03   AGRICULTURAL PRODUCTION
01.0301   Agricultural Production, General
01.0302   Animal Production
01.0303   Aquaculture
01.0304   Crop Production
01.0305   Game Farm Management
01.0399   Agricultural Production, Other

01.04   AGRICULTURAL PRODUCTS AND PROCESSING
01.0401   Agricultural Products and Processing, General
01.0402   Food Products
01.0403   Nonfood Products
01.0499   Agricultural Products and Processing, Other

01.05   AGRICULTURAL SERVICES AND SUPPLIES
01.0501   Agricultural Services and Supplies, General
01.0502   Agricultural Services
01.0503   Agricultural Supplies Marketing
01.0504   Animal Grooming
01.0505   Animal Training
01.0506   Horseshoeing
01.0599   Agricultural Services and Supplies, Other

01.06   HORTICULTURE
01.0601   Horticulture, General
01.0602   Aboriculture
01.0603   Floriculture
01.0604   Greenhouse Operation and Management
01.0605   Landscaping
01.0606   Nursery Operation and Management
01.0607   Turf Management
01.0699   Horticulture, Other

01.07   INTERNATIONAL AGRICULTURE
01.0701   International Agriculture

01.99   AGRIBUSINESS AND AGRICULTURAL PRODUCTION, OTHER
01.9999   Agribusiness and Agricultural Production, Other

02.   AGRICULTURAL SCIENCES

02.01   AGRICULTURAL SCIENCES, GENERAL
02.0101   Agricultural Sciences, General

02.02   ANIMAL SCIENCES
02.0201   Animal Sciences, General
02.0202   Animal Breeding and Genetics
02.0203   Animal Health
02.0204   Animal Nutrition
02.0205   Animal Physiology
02.0206   Dairy
02.0207   Fisheries Science
02.0208   Livestock
02.0209   Poultry
02.0299   Animal Sciences, Other

Figure 2–9.   Agricultural programs as listed in *A Classification of Instructional Programs,* National Center for Education Statistics, U.S. Department of Education, Washington, DC 20202, 1981 *(courtesy of the U.S. Department of Education).* Figure continues on page 14.

02.03 FOOD SCIENCES
    02.0301 Food Sciences, General
    02.0302 Dairy Processing
    02.0303 Food Distribution
    02.0304 Food Engineering
    02.0305 Food Packaging
    02.0306 Food Technology
    02.0399 Food Sciences, Other

02.04 PLANT SCIENCES
    02.0401 Plant Sciences, General
    02.0402 Agronomy
    02.0403 Horticulture Science
    02.0404 Ornamental Horticulture
    02.0405 Plant Breeding and Genetics
    02.0406 Plant Pathology (Applied)
    02.0407 Plant Physiology
    02.0408 Plant Protection (Pest Management)
    02.0409 Range Management
    02.0410 Turf Management Science
    02.0499 Plant Sciences, Other

02.05 SOIL SCIENCES
    02.0501 Soil Sciences, General
    02.0502 Soil Chemistry
    02.0503 Soil Conservation
    02.0504 Soil Management and Fertility
    02.0505 Soil Microbiology
    02.0506 Soil Physics
    02.0599 Soil Sciences, Other

02.99 AGRICULTURAL SCIENCES, OTHER
    02.9999 Agricultural Sciences, Other

03.   RENEWABLE NATURAL RESOURCES

03.01 RENEWABLE NATURAL RESOURCES, GENERAL
    03.0101 Renewable Natural Resources, General

03.02 CONSERVATION AND REGULATION
    03.0201 Conservation and Regulation

    03.0202 Conservation
    03.0203 Resources Protection and Regulation
    03.0299 Conservation and Regulation, Other

03.03 FISHING AND FISHERIES
    03.0301 Fishing and Fisheries, General
    ------- Commercial Fishing Operation (see 49.0303)
    03.0302 Fisheries
    03.0399 Fishing and Fisheries, Other

03.04 FORESTRY PRODUCTION AND PROCESSING
    03.0401 Forestry Production and Processing, General
    03.0402 Forest Production
    03.0403 Forest Products Utilization
    03.0404 Forest Products Processing Technology
    03.0405 Logging
    03.0406 Pulp and Paper Production
    03.0499 Forestry Production and Processing

03.05 FORESTRY AND RELATED SCIENCES
    03.0501 Forestry and Related Sciences, General
    03.0502 Forestry Science
    03.0503 Forest Biology
    03.0504 Forest Engineering
    03.0505 Forest Hydrology
    03.0506 Forest Management
    03.0507 Forest Mensuration
    03.0508 Urban Forestry
    03.0509 Wood Science
    03.0599 Forestry and Related Sciences, Other

03.06 WILDLIFE MANAGEMENT
    03.0601 Wildlife Management

03.99 RENEWABLE NATURAL RESOURCES, OTHER
    03.9999 Renewable Natural Resources, Other

Figure 2.9. *Continued*

- AGRICULTURAL ELECTRIFICATION, POWER AND CONTROLS
  Electrician
  Electrician assistant
  Safety technician
  Service company representative

- AGRICULTURAL MECHANICS, CONSTRUCTION AND MAINTENANCE SKILLS
  Construction maintenance worker
  Lubrication specialist
  Machinery and equipment repairman
  Shop foreman
  Safety specialist

- AGRICULTURAL POWER MACHINERY
  Equipment deliveryperson
  Equipment mechanic
  Farm equipment operator
  Local dealer
  Machinery setup person
  Parts person
  Salesperson
  Service representative
  Tractor mechanic

- AGRICULTURAL STRUCTURES, EQUIPMENT AND FACILITIES
  Agriculture structure contractor
  Appraiser
  Building maintenance supervisor
  Construction supervisor
  Farmstead planner
  Waste handling specialist

- SOIL AND WATER MECHANICAL PRACTICES
  Agronomist
  Appraiser
  Conservationist
  Conservation technician
  Ecology technician
  Irrigation operator
  Manager
  Soil technician
  Water control manager

Figure 2–10. Some challenging jobs available in agricultural mechanics.

questions may be helpful in gaining insight into a career choice.

- Do you want to feel pride in a job well done?
- Do you want to help others by doing a job for them which they cannot do themselves?
- Do you want to keep farm machinery operating through planting and harvest?
- Do you want to work on engines, build buildings, operate equipment, or install electrical equipment?
- Are you interested in welding, hydraulics, soil management, water management, or use of tools?

Agricultural mechanics is a career that can lead to great specialization. It can provide a good living locally or it can lead to international opportunities. Agricultural mechanics skills are needed; students planning a career in this field need to get prepared. To get prepared, students should

- visit with people who have jobs in agriculture that use mechanical skills
- prepare a list of questions to ask workers about their jobs, figure 2–13 (page 18)
- talk to as many people as possible about agricultural mechanics
- plan to learn every skill possible in school, at home, and on the job
- learn what and why in the classroom
- learn how through shop and laboratory activities

Figure 2–11. Areas of career opportunities in American agriculture (*courtesy of The National Future Farmer magazine and the College of Agriculture, University of Illinois*).

Decisions, Decisions, Decisions . . .

Research conducted by *The National FUTURE FARMER* magazine in 1982 revealed the following statistics about FFA members:

73.6 percent plan to continue education after high school graduation

of those, 43.1 percent plan to attend a four-year institution; 21.6 percent plan to enter a two-year junior college; and 35.3 percent plan to enroll in a vocational technical school

almost half of those FFA members who plan to attend a post-secondary school made their decision to do so as a freshman in high school

Of the entire FFA membership, 96.8 percent said they plan to have a career in either production agriculture or an ag-related field

of those, 54.1 percent, 21.4 percent plan for a career in ag sales or service, 17.6 percent chose natural resources or forestry, nine percent chose ag processing, another nine percent chose a career as a vo-ag instructor, 5.8 wanted horticulture, and a whopping 37.2 marked "other;" when asked to list those "other" careers, at least 100 different ag career occupations were listed, everywhere from ag mechanic to plant genetist.   ●●●

Figure 2–12.   Career goals of FFA members (*courtesy of The National Future Farmer magazine*).

The following is a list of questions which could be used when interviewing people about their occupation. This list is *not* complete and is intended to be used as a guide for developing questions.

1. Why did you pick this job?
2. How did you get started in your occupation?
3. How did you choose your place of training?
4. What educational, training, and other qualifications are there for the job?
5. If you should wish to change jobs, would the training contribute in any way?
6. Do you think this job would have a good future for me?
7. How could I get started in this career?
8. What is the salary range of this occupation?
9. What could a beginning person expect to make?
10. What are the fringe benefits?
11. Do you get paid vacations?
12. Do you have medical insurance?
13. Is there any chance of layoff? If so, how often?
14. What sort of planning does this business have for retirement?
15. What do you or don't you like about your job?
16. What are the advantages?
17. What are the disadvantages?
18. What are the hours and working conditions?
19. Do you ever have to work holidays? If so, which ones?
20. Do you ever work on weekends?
21. Is there a special uniform you must wear, or are you free to wear what you want? Does the company provide the uniform or does the employee?
22. What tools do you need?
23. Do you have to buy your own equipment?

Figure 2–13.   Some questions for career exploration interviews. Figure continues on page 18.

24. What are the physical requirements?
25. What do you do in this occupation?
26. How much traveling is involved?
27. What kinds of people do you work with?
28. Are there chances for advancement?
29. What are your responsibilities?
30. Do you belong to a union?
31. What's a typical day like for you in this job?
32. Is there any on-the-job training?
33. Has there ever been a time when you couldn't stand your job? If so, why and when?
34. Do you have to move if the company does?
35. What work experience did you have before you started to work in this occupation?
36. Who depends on your work? Upon whom do you depend?
37. Are there opportunities for advancement in this job? If so, what are the requirements for advancement?
38. How does your job affect your personal life?
39. What kinds of people do you meet?
40. Do you work mainly with people or things?
41. Do you work a lot with ideas?
42. Does your job offer opportunities to be creative?
43. Are people with your skills usually needed—even when business may be bad?
44. Is your work at all seasonal?
45. Could you briefly describe the personal qualities a person would need to do your job—strength, height, agility, ability to think rapidly, ability to make decisions, ability to deal with other people, etc.?
46. Would you recommend this kind of work for your children?
47. How do you spend your time after work?
48. If you could have any job in the world, what would you like to be?
49. Do you still go to school for special training?
50. When are people promoted? When are people fired?

Figure 2–13. *Continued*

□ **STUDENT ACTIVITIES**

1. Check the *Dictionary of Occupational Titles* for agricultural mechanics job titles that interest you.
2. Ask your school guidance counselor to help you use the career information systems available in your school. Ask about a computerized system and a microfiche system. Use these or other systems to find information on three job titles of your choice in agricultural mechanics.
3. Use the above sources and the *Occupational Outlook Handbook* to prepare a report on the job in agricultural mechanics which interests you the most. Include the following parts in your report: a) job title, b) outlook for the future, c) working conditions, d) personal and educational requirements for success, e) special advantages of the job, and f) hazards and other disadvantages.
4. Consult the yellow pages of your local phone book. Record the names, addresses, and phone numbers of all agencies who hire people in the job of most interest to you.
5. Use the classified ads section of a large newspaper. Mark all jobs mentioned where a knowledge of agricultural mechanics would be helpful.
6. Visit a business which hires people with agricultural mechanics skills. Ask the employer and some workers about the working conditions, travel requirements, employee benefits, skills needed, salary, and other items of interest to you.

□ **SELF-EVALUATION**

A. Multiple Choice. Select the best answer.
1. America's number one employer is
   a. agriculture
   b. chemicals
   c. oil
   d. steel

2. The estimated number of agricultural-related jobs in the United States is
   a. 5 million
   b. 11 million
   c. 23 million
   d. 44 million
3. The first level of employment in agriculture is
   a. skilled
   b. semiskilled
   c. professional
   d. laborer
4. The U.S. Department of Education classifies agricultural jobs by the number(s)
   a. 01
   b. 02
   c. 03
   d. all of these

B. Matching. Match the job classification in column I with the job title in column II.

Column I	Column II
1. Agricultural Electrification, Power and Controls	a. farmstead planner
2. Agricultural Mechanics, Construction and Maintenance Skills	b. electrician assistant
3. Agricultural Power Machinery	c. machinery setup person
4. Agricultural Structures, Equipment and Facilities	d. safety specialist
5. Soil and Water Mechanical Practices	e. ecology technician

Section 2
USING THE AGRICULTURAL MECHANICS SHOP

Unit 3
Shop Orientation and Procedures

Objective

To recognize major work areas and use safe procedures when working in an agricultural mechanics shop.

Competencies to Be Developed

After studying this unit, you will be able to:
- Identify major work areas in an agricultural mechanics shop.
- State school policies regarding shop procedures.
- State general safety precautions regarding the shop.
- Inform parents of shop policies and procedures.
- Sign a shop policy and procedure statement and a physical problems statement.
- Have parents sign a shop policy and procedure statement and a physical problems statement.
- Submit signed policy and procedures statement and physical problems statement to the teacher.

Terms to Know

hands-on experience	supervised occupational
electricity	experience program
scale drawing	policy
plan reading	procedure
farmstead layout	safe
concrete	efficient

Materials List

Tape
Drawing paper
Pencil, eraser, and ruler
Shop policies and procedures statement form
Physical problems statement form

□ THE SHOP AS A PLACE TO LEARN

The agricultural mechanics or ag mechanics shop is a wonderful place to learn. Such shops may be found in schools, on farms, or in other agricultural businesses. Some people have shops in their basement or in special buildings, figure 3–1. Good lighting, adequate electrical power, and grounded electrical outlets are essential for safety and efficiency in all shop areas. A study bench and vise are basics for the shop. Some people have tools and equipment that are easy to move from place to place, figure 3–2.

Sometimes students and teachers refer to the agricultural mechanics shop in the school as the Ag Mech Shop or simply the Ag Shop. Students learn many things that help them now and in the future. Important skills for now include the use of basic hand tools and power equipment, figure 3–3. Most teenagers and young adults find many uses for tools such as repair of bicycles, automobiles, stereos, appliances, garden equipment, and engines. Agricultural students use tools to help them in their agricultural projects and occupational experience programs.

Figure 3–1. Many farm and nonfarm workers have shops in their basements or in a special building.

The modern agricultural mechanics shop has facilities for the serious student to learn skills that are useful for a lifetime. Many adults who are vocational agriculture graduates learned mechanical skills when in high school. They now use these skills at home, in their businesses, and for leisure.

Agricultural Shop Operations

Most agricultural mechanics operations fall under several broad categories:

* selection, care, and correct use of hand tools and power equipment

Figure 3–2. A large tool box can hold frequently used tools, fasteners, and lubricants. The portability of the tool box means that the tools can be taken to the site for repairs to buildings and/or machinery.

Figure 3–3. An agriculture instructor demonstrates the proper use of a radial arm saw.

* woodworking and simple carpentry
* use of sheet metal (metalworking)
* handling and fastening structural steel, including welding
* pipe selection and fastening (pipe fitting)
* rope work such as tying knots and splicing
* repair of machinery
* painting and finishing

Many of these items deal with the use of materials for constructing equipment or buildings used in agriculture. Therefore, a large amount of space in a shop is used for material storage.

Agricultural Power and Machinery

The agricultural mechanics shop must have large open spaces for machinery repairs, figure 3–4. Students typically build projects in the shop. They also bring equipment from their homes, farms, or places of employment and repair it as a way of learning through *hands-on experiences*, figures 3–5 and 3–6. Hands-on-experience simply means learning by doing an operation rather than just reading or talking about it. It is very important, however, that reading, discussion, and demonstrations precede hands-on experiences for most activities.

Agricultural power and machinery experiences in-

Figure 3–4. The agricultural mechanics shop must have large open areas for students to work on large projects.

clude the selection, management, adjustment, operation, maintenance, and repair of:

• gas and diesel engines
• trucks
• tractors
• field machines
• feed handling equipment
• crop storage and handling equipment
• special machines such as those used in horticulture
• other mechanical devices used in agriculture

Agricultural Electrification

American agriculture is a very efficient industry. Because of this efficiency, the average American spends only 14 percent of income on food. American agriculture is more efficient than those of most other coun-

Figure 3–5. Students repair a spreader in the fenced-in area outside an agricultural mechanics shop.

Figure 3–6. An agricultural student completes a tractor tune-up.

tries. This is due in part to the use of energy to operate machines.

Electricity is a form of energy that can produce light, heat, magnetism, and chemical changes. By using magnetism, engineers make electric motors and controls. Also, they create high voltage currents to provide spark for gasoline engines..

The agricultural student can benefit much from a knowledge of electricity and its use. This includes the use of electricity in the home, on the farm, and in all agricultural settings. Knowledge of electricity includes the selection, installation, and maintenance of wiring and electrical equipment, figure 3–7. Many of these skills may be learned in the agricultural mechanics shop.

Agricultural Buildings and Equipment

Many people in agriculture are around buildings and equipment in their work. This is true whether the worker is a crop farmer, nursery operator, greenhouse manager, herdsman, machinery dealer, packing house operator, agricultural banker, or a vocational agriculture teacher. Many agricultural businesses are family businesses. Many of the construction and maintenance procedures on buildings and equipment are done by the owner, family members, or employees, figure 3–8. Therefore, agriculturists should have a thorough knowledge of building construction, and repair and maintenance of equipment.

Figure 3–7. Electric motor maintenance and repair are important skills for agricultural workers (*courtesy of the U.S. Department of Agriculture*).

Scale Drawing and Plan Reading. □ Most agricultural mechanics projects are too large to be drawn actual size on paper. Therefore, drawings must be scaled down (reduced) to make the dimensions proportionately smaller so the project can be planned on paper. *Scale drawing* is object representation on paper and uses a smaller dimension to represent a larger dimension, e.g., ¼ inch on the paper = 1 foot on the object, or ¼″ = 1′. This means that each foot of the actual object is represented by ¼ inch on the drawing. The dimension ratio used must be indicated on the drawing. The actual dimensions of the object are stated on the drawing.

Scale drawings can also be made of a part or object that is too small to show details clearly. The actual dimensions of the part or object are increased on the drawing. A scale drawing that states 2″ = 1″ means that the drawing is twice the size of the actual part or object.

Students build projects from plans in a book, or they design their own projects personally. Either way, they need to know about scale drawing and plan reading.

Plan reading simply means using the scale drawing to build the project from the information given. The ability to do scale drawing and plan reading is basic to all construction, figure 3–9.

Farmstead Layout. □ *Farmstead layout* refers to the efficient arrangement of buildings on a farm. While arrangement has some relationship to construction, students typically study farmstead layout in the classroom rather than in a shop. Mistakes on a plan can be corrected quickly and easily. However, once construction is completed, mistakes are difficult and expensive to correct.

Figure 3–8. Equipment operators frequently are required to maintain and repair the equipment they operate (*courtesy of the U.S. Department of Agriculture*).

Figure 3–9. A plan or blueprint provides the worker with information on materials, dimensions, and details of construction.

Figure 3–10. This student is examining the impeller in a water pump. The next step will be to replace worn parts to restore the pumping capacity of the unit.

Functional Requirements of Buildings. □ Functional skills are needed to plan, install, or repair water systems, irrigation equipment, sprinkler systems, drainage pipe, and sewage systems, figure 3–10. This requires the agriculturalist to be a junior engineer or a jack-of-all-trades to keep operating costs down.

Figure 3–11. After pouring concrete, the worker must smooth the surface, provide construction joints, and finish the edges.

Figure 3–12. The bulldozer is used to construct terraces to control the flow of water on slopes. Proper installation of terraces and other soil and water controls help conserve both water and soil (*courtesy of the U.S. Department of Agriculture*).

Concreting. □ *Concrete* is a mixture of portland cement, water, sand, and course aggregates (stones). Concrete is used in agriculture for patios, landscape features, feeding floors, walks, roads, bridges, holding tanks, and other projects. Concrete can be poured to make objects or structures of nearly any shape. It is hard, durable, and resistant to abuse. Hence, much concrete is used in agriculture, figure 3–11.

Soil and Water Management

Mechanical skills needed for soil and water management are generally developed in the classroom or field, not in a shop. These include land leveling, land measurement, mapping, drainage, irrigation, terracing, contouring, and strip cropping, figure 3–12. However, some of the basic shop skills are useful in learning soil and water management practices.

□ MAJOR WORK AREAS IN THE AGRICULTURAL MECHANICS SHOP

Briefly, the major work areas in a modern agricultural mechanics shop are:

- tool storage
- materials storage
- woodworking
- finishing
- metal working (including welding)
- electricity
- machinery repair
- spray painting

Figure 3–13. Shops with large open areas and marked traffic lanes provide safe working conditions for agricultural students.

In some schools, one or more of these areas are combined, such as tool storage and materials storage. In other schools, these areas are subdivided. For example, metal working may be in three or four areas such as cold metal, hot metal, sheet metal, and welding.

Work areas in the shop are generally not divided by definite barriers. Major walkways are often identified with strips of vinyl tape or painted lines. This helps students identify work areas and keep walkways clear of tools and materials, figure 3–13.

Large open areas should be available inside the agricultural mechanics shop for students to build large projects and repair machinery which they use in their supervised occupational experience programs, figure 3–14.

The term *Supervised Occupational Experience Program* (SOEP) refers to activities that the student does outside the agricultural class or laboratory to develop agricultural skills.

Many school agricultural mechanics shops have fenced-in areas with concrete or blacktopped surfaces. Such areas are located just outside the shop and greatly increase the amount of space available for students to develop valuable shop skills, figure 3–15.

□ POLICIES FOR SAFE PROCEDURES IN THE SHOP

The school shop is similar to well-equipped shops used by contractors, farmers, or hobbyists. Such home or farm shops must be equipped with basic hand tools, power equipment, and materials storage. Most procedures that are recommended for safe operation of a school shop also apply to home shops. Students are encouraged to think in terms of both situations as they study Ag Mech Shop procedures and organization.

Figure 3–14. This farm machinery carrier is being constructed by students for use in their supervised occupational experience program.

Figure 3–15. Large wood and metal projects are frequently built in areas outside the school shop. Typically, the area is secured by a woven wire fence and locked gates during nonschool hours.

Shop Size

The school shop must be big enough for 20 or more students to work at one time. Home and farm shops generally need work areas for only several people at one time. To be safe and efficient, the shop must be large enough to meet reasonable standards for space for each person in the shop. *Safe* means free from harm or danger. *Efficient* means able to produce with a minimum of time, energy, and expense. Even with the recommended amount of space per pupil, learning opportunities and personal safety are decreased without careful management.

Proper Instruction

Students must be properly instructed in the correct and safe way to use each tool. To avoid crowding which creates serious hazards, students must be spread throughout the shop. The instructor generally has students working in different shop areas during the shop period.

Safety Policies and Procedures

Every school shop should have policies and procedures to improve instruction and help assure student

---

SHOP POLICIES AND PROCEDURES

Safe Conduct and Dress. I AGREE TO:
1. Purchase school insurance or bring a note signed by my parent(s) verifying that I have suitable personal accident and medical insurance.
2. Inform my teacher of any allergies or handicaps before using the shop.
3. Occupy my assigned place at the beginning of the period.
4. Wear safety glasses at all times when in the shop, use special goggles as needed for handling chemicals and caustic materials, and wear shields or helmet when conducting hazardous operations.
5. Wear prescribed protective clothing and not wear long neckties, loose sleeves, or other loose clothing.
6. Practice general cleanliness and orderliness at all times.
7. Not wear finger rings when handling molten metal or working on electrical systems.
8. Not loiter in the shop.
9. Not throw objects in the shop.
10. Avoid "horseplay" at all times in the shop.
11. Report all accidents including minor cuts, scratches, and splinters to the instructor promptly.
12. Remain in the shop or classroom at all times except when excused by the instructor.
13. Help with shop cleanup and storage duties until the job is finished.

Safe Machine Use. I AGREE TO:
1. Never operate power equipment unless specifically authorized.
2. Never operate machines unless the guards are in place.
3. Use all tools for the purpose intended and in the approved manner as taught by the instructor.
4. Stand to the side of grinding wheels, buffers, and blades while the machine is gaining speed, and out of the fire path when furnaces and other burners are being started.
5. Remove keys from chucks when adjustments are complete.
6. Fasten all work securely before drilling, milling, sanding, and other such operations.
7. Avoid talking to or otherwise distracting others using machines or doing hazardous activities.
8. Not leave machines unattended, nor repair, oil, clean, or adjust them while they are running.
9. Not force machines beyond the capacity for which they were designed.
10. Report to the teacher all electrical equipment that is not properly grounded or otherwise safe to use.
11. Report to the teacher all tools and machinery in need of repair, and any hazards that I observe.

Safe Materials Handling. I AGREE TO:
1. Obtain the teacher's permission and read the label before using any pesticides or other chemicals.
2. Handle, use and store pesticides and other chemicals properly.
3. Not use gasoline as a cleaner or solvent, and to handle all fuels in the prescribed ways.
4. Not permit water to be poured into acid or molten metal, nor permit grease or oil to come into contact with molten metal.
5. Handle and store all soiled materials, rags, paints, solvents, and flammable materials in the approved manner.
6. Lift or move heavy objects in the approved manner only.

I have read the above rules and discussed them in class with my instructor. I realize they are for my protection and I will do all I can to see that they are enforced. I will observe all precautions given by my instructor or others assigned to supervise my participation in school and school-related activities.

Signed _____ / _____
                  student's signature   date

_____ / _____
                  parent's signature   date

Figure 3–16. Example of a "Shop Policies and Procedures" statement

ALLERGIES AND PHYSICAL PROBLEMS STATEMENT

The agriculture teachers and others responsible for providing treatment for me in case of an emergency should be aware of the following:

1. I am allergic to _____

2. I should not be given the following medication _____

3. I have the following physical or personal problems _____

4. My family physician is _____
5. My physician's phone number is _____
6. My parents' names are _____
7. My parents' phone numbers are _____

Signed _____ / _____
          student's signature   date

_____ / _____
          parent's signature   date

Figure 3–17. Example of an allergies and physical problems statement

safety. A *policy* is a plan of action or a way of management. A *procedure* is a method of doing things or a particular course of action. Figure 3–16 gives an example of a good policies and procedures statement for an agricultural mechanics shop. The instructor may add to or take away from this list according to established requirements in the school. It is suggested that the student discuss every item in the instructor's policy and procedure list in detail with the instructor and parents. Students and parents should sign the statement to indicate complete understanding. The signed statements may then be filed with other student records at the school.

Another important procedure to help protect students in the shop is to fill out and sign an allergies and physical problems statement, as shown in figure 3–17. Parents should sign these statements also to assure that proper medical treatment can be given in case of accident or other emergency.

□ **STUDENT ACTIVITIES**

1. Measure the length, width and other major dimensions of your agricultural mechanics shop. How many square feet are available per pupil in your class?
2. Make (or ask your teacher to provide) a drawing of your shop and any fenced-in area which may be used by students. Label the tool storage, lumber storage, metal storage, woodworking, metalworking, plumbing, electrical, and project areas.
3. List five projects which other vocational agriculture students have constructed in the shop.
4. List five projects which you would like to make in the shop.
5. Write, correctly spell, and define all italicized words in this unit.
6. Refer to figure 3–16. Study each statement and list the number of each statement that is related to safety.
7. Read, discuss with your parents, sign, and obtain your parents' signatures on the policies and procedures statement for your shop. Return the signed document to your teacher.
8. Fill out, sign, and have your parents sign a personal allergies and physical problems statement for you. Return the signed statement to your teacher.

□ **SELF-EVALUATION**

A. Multiple Choice. Select the best answer.
1. The agricultural mechanics shop is also referred to as the
   a. Ag mechanics shop
   b. Ag mech shop
   c. Ag shop
   d. all of these
2. The agricultural mechanics shop is a good place for serious students to learn skills that are useful
   a. now
   b. in the future
   c. both of these
   d. none of these
3. Agricultural mechanics includes
   a. woodworking and carpentry
   b. metalworking and welding
   c. pipe fitting and irrigation
   d. all of these
4. Hands-on experience means
   a. made by hand tools only
   b. a process of learning by doing
   c. a procedure requiring many people to help
   d. a wasteful method of education
5. Large open spaces are needed in agricultural mechanics shops as compared with other shops for
   a. safe operation of stationary power equipment
   b. meeting fire code regulations
   c. storing materials for the school custodians
   d. student project work
6. Safety in the shop depends upon
   a. students
   b. teacher(s)
   c. school shop designers
   d. all of these

7. Safety in the shop depends upon
   a. students staying in assigned areas
   b. students wearing proper clothing
   c. use of safety glasses by all persons in the shop
   d. all of these
8. Which of the following enhances student safety in the shop?
   a. shop cleanliness and orderliness
   b. proper instruction
   c. machines that are kept in adjustment
   d. all of these

B. Matching. Match the definitions in column I with the correct term in column II.

Column I	Column II
1. energy source for agriculture	a. agricultural shop operations
2. construction and repair	b. agricultural power and machinery
3. irrigation and drainage	c. agricultural electrification
4. woodworking, metalworking, welding, and pipefitting	d. agricultural buildings and equipment
5. efficient arrangement of buildings	e. scale drawing
6. portland cement, water, sand and stone	f. farmstead layout
7. selection, management, adjustment, operation, maintenance, and repair	g. concrete
8. object representation on paper	h. soil and water management

Unit 4
Personal Safety in Agricultural Mechanics

Objective

To interpret safety colors and codes, protect the body against injury, and work safely in agricultural mechanics settings.

Competencies to Be Developed

After studying this unit, you will be able to:
* State how to create a safe place to work.
* Recognize hazards in agricultural mechanics.
* List the types of parts and areas identified by various safety colors.
* Describe what each safety color means.
* Select appropriate protective clothing and devices for personal protection.

Terms to Know

safety	noise duration
focal color	decibel
noise intensity	

Materials List

Examples of shop protective clothing and devices

Students sometime grow weary of teachers, parents, and others who constantly remind them to work and play safely. Yet far more people are injured each year than need to be to carry on the work of society.

Most people react to injury with anger at themselves or others who were responsible. This is because many injuries could have been avoided. People are willing to accept severe pain and hardship if they must but they are resentful if they could have escaped the loss by being more careful.

Injury and disability are troublesome problems in society. Not only does the victim suffer, but others suffer as well. Friends of the injured are frightened and inconvenienced. Parents, guardians, spouses, and others frequently must take off work for visits to the doctor's office or hospital. They often forfeit income and must pay for extra expenses. Therefore, it is well worth the time to create safe places to work and to learn safety in everything. *Safety* means freedom from accidents. Some common causes of accidents are indicated in figure 4–1.

Many accidents leave the victim partially or totally disabled. This carries a lifetime of regret. Many accidents are fatal. This results in needless loss of loved ones and bereavement to those left behind. Obviously, the thinking person will make every reasonable effort to work safely. This helps avoid the hardships caused by accidents. Accidents in farm populations involve machinery, drownings, firearms, falls, falling objects, burns and others, figure 4–2.

□ THE SAFE PLACE TO WORK

Work in agricultural mechanics involves extensive contact with tools and machinery. Therefore, workers should be especially aware of the hazards that exist and take special precautions as needed. By taking the

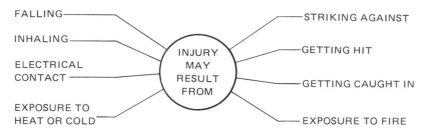

Figure 4–1. Some common causes of accidents

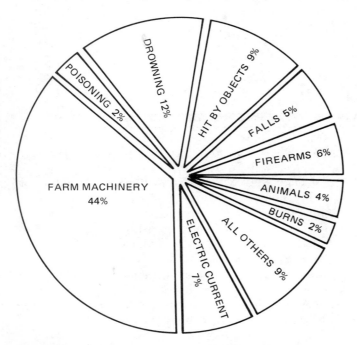

Figure 4–2. Many accidents which happen to farm people are machinery related.

following precautions a safer work place can be created.

- Install all electrical devices according to the National Electrical Code.
- Install all machinery according to the manufacturer's specifications.
- Keep all tools and equipment adjusted or fitted according to specifications.
- Use tools and equipment skillfully.
- Provide proper storage for tools, materials, fuels, chemicals, and waste materials.
- Keep work areas clean and free of tools, materials, grease, and dirt.
- Keep moving parts properly shielded.
- Manage all situations to avoid the likelihood of falling objects.
- Avoid areas where objects may fall.
- Avoid the flight path of objects that could be thrown by machines.
- Protect eyes, face, feet, and other parts of the body with protective clothing and devices.
- Move slowly enough to avoid creating hazards to self and others.
- Read and heed all precautions.

These precautions help to make the work place safe. However, each person should have insurance to cover any personal injury or property damage that may occur.

□ SAFETY COLORS

National organizations have worked together to develop a safety color coding system for shops. The American Society of Agricultural Engineers and the Safety Committee of the American Vocational Association have published such a code. In developing the code, these agencies drew upon materials published by the American National Standards Institute (ANSI), the United States Department of Transportation (DOT), the National Safety Council (NSC), and the Occupational Safety and Health Act (OSHA).

Colors in the coding system are used to

- alert people to danger or hazards
- help people to locate certain objects
- make the shop a pleasant place to work
- promote cleanliness and order
- help people to react quickly to emergencies

Each color or combination of colors conveys a specific message, based upon a standard code. Students need to memorize what each color stands for in order for the safety message to be understood in the shop. Shops must be properly painted for the color coding system to make its contribution to shop safety and efficiency.

The following descriptions show how each safety color is used to convey a safety message and help students work safely in the shop. The colors are listed in figure 4–3 and used as follows:

Red = Danger. Red is used to identify areas or items of danger or emergency such as safety switches and fire equipment.

Orange = Warning. Orange is used to designate machine hazards such as edges and openings. Orange is also used as background for electrical switches, levers, and controls.

Yellow = Caution. Yellow, like the amber traffic light, means be cautious. It is used to identify parts of machines such as wheels, levers, and knobs that control or adjust the machine. Yellow and black stripes are used in combination to mark stairs, protruding objects, and other stationary hazards.

Blue = Information. Blue is used for signs if a warning or caution is intended. Such signs are made of white letters on blue background and carry messages such as "OUT OF ORDER" or "DO NOT OPERATE."

Green = Safety. Safety green is a special shade of green and indicates the presence of safety equipment, safety areas, first aid, and medical practice.

Purple = Radioactivity. Purple is the traditional color used to designate radiation hazards. However, a black and yellow diagonal striped pattern has been

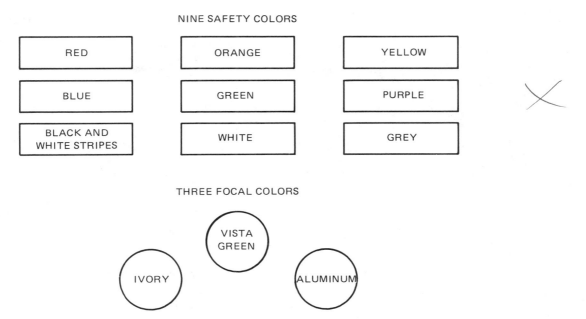

Figure 4–3. The nationally accepted shop safety color coding system uses nine safety colors and three focal colors to designate objects and areas.

designated as the marking for radiation as of 1986.

White. White is used to mark off traffic areas. White arrows indicate the direction of traffic. White lines also mark work areas around objects in the shop. Yellow may be used in place of white to mark areas and lanes.

White and Black Stripes. White and black in alternate stripes or checkers are traffic markings. An example of such use is to mark barricades which stop traffic.

Gray. Gray is used on floors of work areas in the shop. It is a restful color and provides good contrast for other safety colors. It is used to paint body areas of machines and may be used on table tops if painting is desired.

☐ FOCAL COLORS

The nationally accepted shop safety color coding system includes three focal colors. A *focal color* is used to draw attention to large items such as machines, cabinets, and floors. The focal colors provide contrast for the safety colors and create pleasant surroundings for people using the shop. The focal colors are ivory, vista green, and aluminum.

Ivory. Ivory is used to highlight or improve visibility of certain items. These items include tool storage chests, table edges, and freestanding vises and anvils.

Vista Green. Vista green is a special shade of green. It is used to paint bodies of machines, cabinets, and stationary tools such as vises. It is regarded as a pleasing color and contrasts with the safety colors.

Aluminum. Aluminum is used on waste containers such as those for scrap wood, scrap metal, and rags.

Some of these colors are used in combinations with other colors to mark pipes, hoses, and vents. Such markings identify the material in the lines as oxygen, natural gas, compressed air, air at atmospheric pressure, and water.

A properly color-coded shop is attractive and pleasant. The coloring system helps guide the user safely through many enjoyable experiences in agricultural mechanics.

☐ PROTECTIVE CLOTHING AND DEVICES

Most work situations in agricultural mechanics require some type of body protection. The eyes, ears, hands, arms, feet, and legs are easily injured, figure 4–4. The best protection against injury is to prevent the accident from happening. The next best approach is to protect the body where injury may occur. Both lines of defense are needed to minimize injuries.

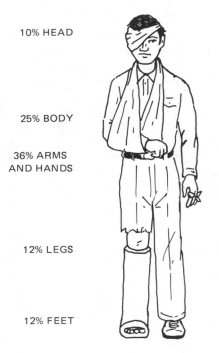

10% HEAD

25% BODY

36% ARMS
AND HANDS

12% LEGS

12% FEET

Figure 4–4. Studies show that any part of the body may be injured in the shop.

Safety Glasses and Face Shields

The face and eyes are regarded as the most critical parts of the body to be protected. This is because the eyes are so easily damaged and the face is easily disfigured. Flying objects striking the head can easily cause blindness or result in death. Acids, caustic chemicals, fertilizers, pesticides, solvents, molten metal, and hot water are all dangerous materials. Those who work around these materials should shield themselves against the possibility of accidental contact with them.

Safety Glasses and Goggles. □ Safety glasses and goggles offer minimum eye protection and are the first line of defense for the eyes, figure 4–5. Glasses and goggles should be the approved type with special impact-resistant lenses and side shields. They should fit the face and be kept clean for proper visibility. Single-piece goggles and cup goggles are used where special eye protection is needed against chemicals, flying objects, or damaging light. Special shaded lenses are needed when welding.

Many states have special laws requiring students to wear safety glasses or cover goggles upon entering and participating in vocational shops in the schools

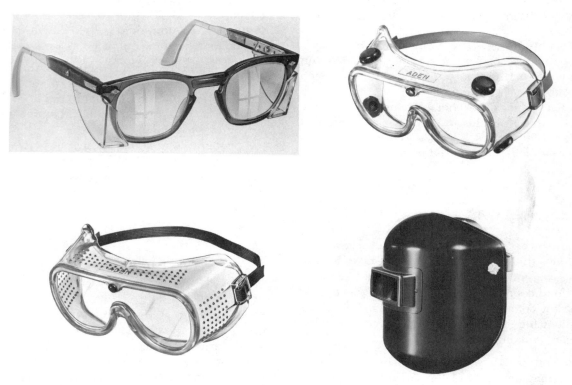

Figure 4–5. Safety glasses, goggles, or face shields should be worn when working around materials, processes, and machines (*courtesy of J. I. Scott Advertising/Marketing*).

Some states have laws requiring an approved face shield when an individual is

- participating in, or exposed to, the immediate vicinity where hot molten metal or solder is being prepared, poured, or used in any form
- participating in, or exposed to, the immediate vicinity where milling, turning, sawing, shaping, grinding, sanding, cutting, or stamping of any solid materials are taking place
- participating in, or exposed to, the immediate vicinity where heat treating, tempering, or kiln firing of any material is taking place.

Approved cup goggles, helmets, or hand shields are required when in the areas of gas welding, electric arc welding, or weld flash exposure.

Students must always wear safety glasses as minimum protection against eye injury. When students are in special hazard areas, special eye and face shields must be used, figure 4–6.

Hair Restraints. □ Serious accidents can occur if long hair becomes tangled in drill presses or other turning equipment. When hair is long or loose, it should be contained by one of the following methods:

- wear a woolen hat
- use a head band
- wear a hardhat
- wear a hairnet

Figure 4–6. Clear face shields provide protection to the entire face. They also provide extra protection to the eyes when worn over glasses.

Protective Clothing

No part of the body is safe from injury in shop accidents. Suitable protective clothing that fits properly helps to prevent or reduce injuries. There should be no cuffs, strings, or ties for turning machinery to catch. Clothing should be fire resistant and provide protection from scrapes and abrasions. Protective clothing must be easily cleaned. Keeping clothing clean keeps clothing more fire resistant.

Coveralls. □ Coveralls are the most versatile and all-around item of clothing for agricultural mechanics. Coveralls protect the arms, body, and legs. They can be buttoned or zipped to the neck for maximum protection. Coveralls should fit well and be easily removable. Coveralls are safer if they do not have sleeve or pant cuffs. Cuffs are hazardous because flying sparks can be caught and trapped, resulting in the garment catching on fire. Bulky cuffs can catch in machinery and other objects.

Coveralls have the advantage of many pockets for pencil, pad, small tools, and objects which the worker uses. The pockets may be covered with flaps to provide a smooth body cover that will not catch sparks and other objects. Coveralls should have an elastic waist, sleeves that button, and flaps that cover all buttons and zippers. When jackets, pants, or shirts are worn instead of coveralls, it is important to select clothing that fits snugly and has the same safety qualities as coveralls.

Aprons. □ Some shop teachers permit students to wear aprons in the shop or laboratory. Heavy cloth or leather aprons provide good protection for the front of the body and upper legs. Aprons are recommended only for limited shop work at benches or as additional covering over coveralls.

Vinyl or rubber aprons should be worn when liquids are used. Aprons are economical to buy and easy to store.

Shop Coats. □ Shop coats have the benefits of aprons plus additional body protection. The arms, body, and upper legs are protected for most work where the operator is standing. The shop coat is easy to put on. Therefore, it is used frequently by instructors and others who need to put on and remove protective clothing frequently throughout the day.

Footwear. □ Leather shoes with steel toes are recommended when working in the shop and when using machinery, figure 4–7. Many workers prefer shoes with six-inch tops; others prefer the higher boot-type shoes. Leather is the preferred material for footwear

Figure 4–7. Leather shoes with steel reinforced toes provide excellent protection to the feet.

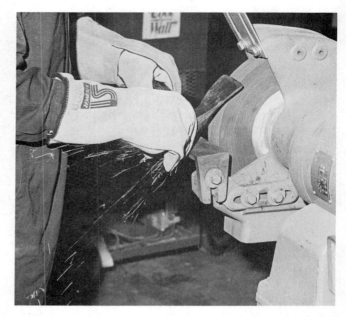

Figure 4–8. Leather gloves provide good protection for the hands.

because of its strength, durability, and comfort. The ability of leather to breathe explains why it is more healthful, cool, and comfortable than other materials. Farmers, ranchers, foresters, carpenters, plumbers, and all who handle heavy or hot objects need sturdy shoes. The popular soft, lightweight vinyl or canvas shoes that many wear for casual activities are not safe for agricultural work.

Rubber boots are needed when working in water or using pesticides. Rubber is more durable than leather in extremely wet conditions and offers better protection against wet feet. Rubber will not absorb pesticides and can be washed free of such materials.

Gloves. □ Gloves are used to keep the hands warm as well as to protect them from excessive abrasion, heat, liquids, or chemicals. Cloth gloves are suitable for warmth, but leather is needed where protection from heat, abrasion, or impact is needed, figure 4–8. Only rubber or vinyl gloves are suitable where liquids or chemicals are involved.

Hardhats. □ Hardhats are needed when working where objects are above head level or flying objects could be encountered, figure 4–9. Hardhats are made from special lightweight and impact-resistant materials. It is important that hardhats be approved by the Occupational Safety and Health Administration. Standards set by this agency help protect workers.

Masks and Respirators. □ Masks that cover the nose and mouth are needed to filter out particles of dust or spray paint, figure 4–10. Such materials irritate the nostrils and sinuses. Continuous inhaling of dust leads to lung diseases such as black lung and cancer. Effective dust masks are not expensive and should be worn when sanding, painting, welding, mixing soil, shoveling grain, or whenever dust is encountered.

Figure 4–9. Hard hats are worn where there is danger from falling objects or other sources of potential head injury.

Figure 4–10. Face masks should be worn when dust is encountered in the work area.

Respirators that cover the nose and mouth and contain special filters are needed for certain jobs. When using pesticides, it is important to use the specific type respirator recommended by the pesticide manufacturer.

Ear Muffs and Ear Plugs. □ Ear protection is recommended when working in certain types and levels of noise. Equipment such as the radial arm saw, planer, router, chain saw, tractor, and lawnmower can produce noise that may damage hearing.

Ear muffs or plugs are recommended when the intensity, frequency, or duration of noise reaches certain levels. *Noise intensity* refers to the energy in the sound waves. *Noise duration* refers to the length of time a person is exposed to a sound.

Distance has a great effect on sound pressure or intensity. A person standing five feet from a machine can reduce the sound pressure to 25 percent by moving away another five feet. The *decibel* (dB) is the standard unit of sound. The Occupational Safety and Health Act (OSHA) established a 90 dB noise level for an eight hour period as the maximum safe limit. An 85 dB limit is safer. The sound level in decibels and duration of time are illustrated in figure 4–11. A sound level meter is used to determine noise levels.

Time is an important factor on the effect of noise on hearing. The ears can stand loud noises for a few

DURATION OF TIME PERMITTED AT VARIOUS SOUND LEVELS	
DURATION PER DAY IN HOURS	**SOUND LEVEL IN dB**
8	90
6	95
3	97
2	100
1½	102
1	105
½	110
¼ or less	115
none	over 115

DECIBEL (dB) LEVELS OF COMMON SOUNDS AT TYPICAL DISTANCE FROM SOURCE	
0	Acute threshold of hearing
15	Average threshold of hearing
20	Whisper
30	Leaves rustling, very soft music
40	Average residence
60	Normal speech, background music
70	Noisy office, inside auto 60 mph
80	Heavy traffic, window air conditioner
85	Inside acoustically insulated protective tractor cab in field
90	OSHA limit—hearing damage on excess exposure to noise above 90 dB
100	Noisy tractor, power mower, all-terrain vehicle, snowmobile, motorcycle, in subway car, chain saw
120	Thunderclap, jackhammer, basketball crowd, amplified rock music
140	Threshold of pain—shot gun, near jet taking off, 50 hp siren (100')

Figure 4–11. The Occupational Safety and Health Act (OSHA) establishes maximum safe levels of noise, in decibels, for specified periods of time. Workers should not be exposed to a higher level of noise or noise for more than the recommended amount of time without wearing ear muffs or plugs *(courtesy of OSHA).*

minutes. That same noise may damage the ears if exposed for longer periods of time.

□ STUDENT ACTIVITIES

1. Interview the school nurse or medical assistant to determine what kind of accidents have occurred in the school.

2. Search recent newspapers for articles on home, shop, farm, or work accidents and discuss them in class. Classify each accident or incident described in the newspaper articles as preventable or not preventable.

3. Make a bulletin board using newspaper clippings about home, shop, farm, or work accidents.

4. Make safety posters depicting the causes of accidental injury.

5. Contact your Cooperative Extension Service or other agencies and request current accident information regarding your community.

6. Study every machine and work area in your home shop or school shop and list all hazards that should be corrected. If at home, correct them; if at school, assist your teacher in correcting them.

7. Examine your school agricultural mechanics shop for proper use of safety and focal colors. Ask your teacher what plans there are to correct any incorrectly painted areas.

8. Survey your agricultural mechanics shop for the following items: safety glasses for all students, goggles, face shields, welding helmets, protective clothing, respirators, hearing protectors, and fire extinguishers. Ask your teacher to acquire any of the items that are not on hand.

□ **SELF-EVALUATION**

A. Multiple Choice. Select the best answer.

1. Accidents among farm workers most often involve
   a. burns
   b. drowning
   c. falls
   d. machinery

2. For safety purposes, moving parts on machines should be
   a. labeled
   b. oiled
   c. painted
   d. shielded

3. Color coding is used in the shop to
   a. alert people to dangers and hazards
   b. make the shop a pleasant place to work
   c. help people react quickly to emergencies
   d. all of these

4. Which of the following is *not* regarded as a major type of accident that causes injury?
   a. assault and battery
   b. electrical contact
   c. falling
   d. inhaling

5. The national organization(s) that helped to develop safety color coding is/are the
   a. American Society of Agricultural Engineers
   b. American Vocational Association
   c. National Safety Council
   d. all of these

6. The safety color used to identify wheels, levers, or knobs that control or adjust machines is
   a. red
   b. yellow
   c. orange
   d. none of these

7. Fire equipment and safety switches are indicated by the color
   a. orange
   b. purple
   c. red
   d. bright green

8. The number of safety colors in the shop color coding system is
   a. nine
   b. eight
   c. seven
   d. four

9. The number of focal colors in the shop color coding system is
   a. one
   b. two
   c. three
   d. four

10. Suitable eye protection must be worn when working with
    a. chemicals
    b. grinding machinery
    c. welding equipment
    d. all of these

11. Protective clothing used in the shop must
    a. be fire resistant
    b. fit properly
    c. be clean
    d. all of these

12. The best item of protective clothing for agricultural workers is
    a. an apron
    b. a shop coat
    c. jeans
    d. coveralls

13. The length of time a person is exposed to sound is called
    a. noise intensity
    b. noise duration
    c. decibels
    d. sound pressure

14. Hearing damage may occur if excessively exposed to noise above
    a. 30 decibels

b. 60 decibels

c. 75 decibels

d. 90 decibels

B. Matching. Match the meaning in column I to the correct color code in column II.

Column I	Column II
1. indicates warning	a. yellow
2. indicates danger	b. red
3. indicates informa-tion	c. green
	d. blue
4. indicates radioactiv-ity	e. black and yellow stripes
5. indicates caution	f. orange
6. indicates safety	g. black and white stripes
7. indicates traffic ar-eas	h. white
8. indicates traffic markings	

C. Completion. Fill in the blanks with the word or words that will make the following statements correct.

1. Many accidents leave the victim partially or to-tally _____ .

2. All electrical devices should be installed accord-ing to the _____ _____ Code.

3. All machinery should be installed according to the _____ specifications.

4. The standard unit of sound is the _____ .

5. Minimum eye protection is provided by wearing _____ _____ or _____ .

6. The hazards of breathing dust and paint spray can be reduced by wearing a _____ .

7. The body parts most often injured in shop acci-dents are the _____ and _____ .

8. When working with objects above head level a _____ should be worn.

Unit 5
Reducing Hazards in Agricultural Mechanics

Objective

To recognize and reduce hazards in agricultural mechanics settings, and react effectively in case of fire or other emergencies.

Competencies to Be Developed

After studying this unit, you will be able to:
- Reduce hazards in agricultural mechanics.
- State the three conditions necessary for combustion.
- Match appropriate types of fire extinguishers to each class of fire.
- Use a fire extinguisher.
- Interpret labels on hazardous materials.
- Describe appropriate action in case of fire, accident, or other emergency

Terms to Know

fire triangle	extinguish
fuel	slow moving vehicle
combustion	SMV
oxygen	cardiopulmonary
heat	resuscitation

Materials List

Different types of fire extinguishers
Labels from agricultural chemicals

Fire, slow moving vehicles, highway crossings, and chemicals create unique hazards in agricultural mechanics. Each may hold dangers to the worker and to others in the area. Fortunately, there are ways to reduce hazards and take action quickly if accidents do occur.

☐ REDUCING FIRE HAZARDS

The discovery of fire and how to create it was one of man's most important achievements. Fire is used to heat homes, cook food, generate electricity, melt ore to refine metals, heat metals to bend and form them, and to cut metals. Yet, fire has not really been tamed. Fire breaks out of control at unexpected times causing injury and loss of property and lives. Burns are probably the most painful of all injuries.

It is known how to prevent uncontrolled fire, except in the case of volcano and lightning. Even lightning can be directed to a safe ground and thus reduce its destruction. The important thing is that most losses from fire can be prevented; however, it requires attention and knowledge of how fire works.

The Fire Triangle

To produce fire, three components must be present at the same time. These three components are fuel, heat, and oxygen; they are known as the *fire triangle*, figure 5–1.

Fuel is any combustible material that will burn. Combustible comes from the word *combustion*, which means to burn. Common fuels are gasoline, kerosene, diesel fuel, wood, paper, acetylene, and propane. Most materials will burn if they are made hot enough in the presence of oxygen.

Heat simply refers to a type of energy which causes the temperature to rise. If the temperature of a room is changed from 50 degrees to 70 degrees, it is done by using heat. Remember, most materials will burn if they are made hot enough in the presence of oxygen.

Oxygen is a gas in the atmosphere. It is not a fuel, but must be present for fuels to burn. Oxygen is nearly always present except in airtight conditions. This fact is important to remember in fire safety and control.

Preventing Fires in Agricultural Mechanics

If any one of the three components of the fire triangle (fuel, heat, oxygen) is eliminated fire will be prevented from starting, or it will be stopped if it has started.

TO PRODUCE FIRE, FUEL, HEAT, AND OXYGEN MUST BE PRESENT AT THE SAME TIME.

IF ANY ONE OF THE THREE COMPONENTS IS MISSING, A FIRE CANNOT BE STARTED.
WITH THE REMOVAL OF ANY ONE COMPONENT, THE FIRE WILL BE EXTINGUISHED.

Figure 5–1. The fire triangle

Therefore, in order to prevent, control, or stop fires

- store fuels in approved containers
- store fuels away from other materials that burn easily
- store materials in areas that are cooler than their combustion temperature
- use fire only in safe surroundings
- put out fires by removing one or more elements in the fire triangle

The prevention of fire goes hand-in-hand with safe use of equipment and efficient management of work areas. For instance, the proper use of a gas cutting torch decreases the likelihood of fire resulting from its use. Proper storage of materials decreases the chance of fire and keeps materials readily available when needed. Clean work areas prevent people from slipping or tripping, and damage to parts or projects. A clean work area also decreases the chance of a fire. Special paint booths provide a clean area for paint jobs and also decrease the likelihood of fire.

□ EXTINGUISHING FIRES

Fires are *extinguished* or put out by adding water to cool them, covering them to cut off the oxygen, or by removing the fuel. This could mean wrapping a person whose clothing is on fire with a blanket. It could mean stopping fire in a field by raking grass and leaves out of the path of the fire or by throwing soil on the fire to smother it. Fire at a gas torch or hoses may be stopped by shutting off the gas at the cylinder. A burning container of paper may be extinguished by cooling it with water from a hose or bucket.

Classes of Fires

To effectively and safely put out a fire with a fire extinguisher, the class of fire must be known. This is determined by the material and the surroundings as follows:

- Class A–Ordinary Combustibles. Ordinary combustibles include wood, paper and trash. Class A combustibles do not include any item in the presence of electricity or any type of liquid.
- Class B–Flammable Liquids. Flammable liquids include fuels, greases, paints, and other liquids, as long as they are not in the presence of electricity.
- Class C–Electrical Equipment. Class C fires involve the presence of electricity.
- Class D–Combustible Metals. Combustible metals are metals that burn. Burning metals are very difficult to extinguish. Only Class D extinguishers will work on burning metals.

Fire classification is based on how to safely and cheaply extinguish each type of material. Water is generally the cheapest material to use in fire control, but it may not be safe or effective. A firefighter can be electrocuted if the stream of water hits exposed electrical wires, plugs, appliances, or controls. Water is not suitable on fires involving petroleum products since the fuel floats to the top of the water and continues to burn.

Types of Fire Extinguishers

The proper fire extinguisher can put out a fire within seconds. However, such results occur only if the fire is extinguished when it first bursts into flames. The key is the proper extinguisher, used immediately, and in the proper way. This combination may make the difference between a mere frightful moment or a multimillion dollar fire loss with serious injuries and death.

Students should learn to recognize extinguishers by their type, figure 5–2. Common types of extinguishers are

- water with pump or gas pressure (used for Class A fires)
- carbon dioxide gas (CO_2) (used for Class B and C fires)
- dry chemical (used for Class A, B and C fires)
- foam (used for Class A and B fires)
- blanket (used for smothering fires on humans or animals)

Extinguishers are marked according to the class or classes of fires on which they will safely work. Fire extinguisher labels contain standardized symbols to help the reader act quickly in an emergency, figure 5–3. The symbols are as follows:

- Green triangle—for Class A, ordinary combustibles
- Red square—for Class B, flammable liquids
- Blue circle—for Class C, electrical equipment
- Yellow star—for Class D, combustible metals

Location of Fire Extinguishers

The location of fire extinguishers is very important. A few seconds lost in looking for the right type of extinguisher could allow a fire to rage out of control. Class A extinguishers should be placed in areas where Class A fires are likely to occur. Class B extinguishers should be placed in areas where Class B fires are likely to occur, and so on. Placing a water-type extinguisher in an area where an electrical fire is likely to occur is of little value. The extinguisher should not be used on an electrical fire.

Extinguishers should be placed in clean, dry locations near exits within easy reach. The extinguisher should be hung on the wall so that the top of the extinguisher is not more than 3½ to 5 feet above the floor, figure 5–4. The bottom of the extinguisher should be at least 4 inches above the floor. The extinguisher should be positioned so it can be removed quickly.

Everyone should be familiar with the locations and use of all types of extinguishers.

Using Fire Extinguishers

Generally, fire extinguishers are held upright and operated by a lever. However, some types are activated by inverting the tank, causing chemicals to mix inside the container. Sometimes the lever is blocked by a pin to prevent accidental discharge. Before operating an extinguisher, the instructions on the container should be read. For most extinguishers, a pin is pulled and a lever pressed.

Before discharging an extinguisher, the firefighter should move to within 6 to 10 feet of the fire and

STORED PRESSURE FIRE
EXTINGUISHER USING WATER

CARBON DIOXIDE
FIRE EXTINGUISHER

DRY CHEMICAL
FIRE EXTINGUISHER

Figure 5–2. Types of fire extinguishers (*courtesy of Amerex Corporation*)

ORDINARY

FLAMMABLE

ELECTRICAL

COMBUSTIBLE

COMBUSTIBLES · LIQUIDS · EQUIPMENT · METALS

Figure 5–3. Standardized symbols on fire extinguishers indicate the type of fire for which they are used.

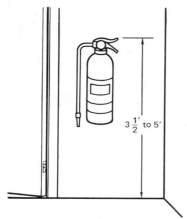

$3\frac{1}{2}'$ to 5'

Figure 5–4. Fire extinguishers should be hung on the wall so the top of the extinguisher is not more than 3½ to 5 feet above the floor.

direct the extinguisher nozzle toward the base of the fire, figure 5–5. The extinguisher will be empty in a matter of seconds, so any material that misses the fuel at the base of the fire will be wasted. A monthly inspection of all fire extinguishers should be made to assure that the extinguishers are useable in case of an emergency, figure 5–6.

Always think before acting. Call for help immediately. Be sure that any fire is completely out before leaving the area. The local fire department is the best source of help on fire safety and prevention.

☐ SIGNS OF DANGER

There are many signs to warn of possible hazards. Signs which say stop, yield, caution, and crossing, alert drivers to dangers on the highway. Signs such as danger, no trespassing, condemned, and keep out warn of dangers around old buildings.

On the farm and in other agricultural mechanics settings warning signs are found on machinery. Such signs may say not to remove a shield, not to overspeed, to keep bolts tight, or to position a part in a certain way, figure 5–7.

The Slow Moving Vehicle Emblem

A very important sign for safety on the highway is the slow moving vehicle (*SMV*) emblem, figure 5–8. This is a reflective-type emblem consisting of an orange triangle with a red strip on each of the three sides. It glows brightly when only a small amount of light hits it. Therefore, it is generally the first item to be seen on a vehicle. When operators of fast moving vehicles know the meaning of the emblem, they have time to slow down before running up on slow vehicles. SMV emblems are required on all vehicles that travel at 25 miles per hour or less on public roads.

(1) HOLD UPRIGHT. PULL RING PIN.

(2) START BACK 10 FEET. AIM AT BASE OF FIRE.
←— 10 Ft. —→

(3) SQUEEZE LEVER. SWEEP SIDE TO SIDE.

Figure 5–5. Instruction for operating a fire extinguisher (*courtesy of Amerex Corporation*)

MONTHLY FIRE EXTINGUISHER CHECK

- See that the proper class extinguisher is in the area of fire class risk.

- See that the extinguisher is in its place.

- See that there is no obvious mechanical damage or corrosive condition to prevent safe reliable operation.

- Examine or read visual indicators (safety seals, pressure indicators, gauges) to make certain the extinguisher has not been used or tampered with.

- Check nameplate for readability and lift or weigh extinguisher to provide reasonable assurance extinguisher is fully charged.

- Examine nozzle opening for obstruction. If equipped with shut-off type nozzle at the end of the hose, check the handle for free movement.

CHECK LIST

☐ Located in a proper place	☐ Gauge or Indicator in Operable Range
☐ Safety Seals	☐ Proper Weight
☐ Free of Dirt and Rust	☐ No Nozzle Obstruction

Figure 5–6. Items to include in the monthly inspection of fire extinguishers

Figure 5–7. Warning signs on machinery are placed there by the manufacturer to alert operators of danger. It is important to follow the instructions in order to prevent personal injury or property damage.

Figure 5–8. The slow moving vehicle (SMV) emblem is commonly used on farm and industrial equipment, road trucks, and other vehicles that travel 25 miles per hour or less on public roads (*courtesy of the National Safety Council*)

It is important to place such an emblem on every piece of machinery that is transported on the highway. Drivers of automobiles and trucks must always be prepared to slow down when the SMV emblem comes into view.

Package Labels

All commercial products have a label to provide the user with certain information. Food product labels tell about nutrient content. Repair parts come with instructions for installation. Paint labels name the components, and clothing labels tell how they are to be cleaned.

Labels on hazardous products may be a matter of life or death in the event of accident. Common products such as kerosene and turpentine are poisonous if taken internally. Agricultural pesticides are products designed to be poisonous to pests; they may cause illness or death to humans if misused. If used according to the instructions on the label, pesticides are generally safe. It is very important to keep materials in their original containers. Original containers are the correct type of container. They carry the label that describes the product, its hazards, and procedures to use in case of an emergency with the product.

Pesticide labels are legal documents. Label directions are required by law to be accurate and complete. Labels should be read carefully before buying a pesticide. Labels should be read again before handling, opening, mixing, using, or disposing of pesticides.

Pesticide labels have at least sixteen different items of information. Study the pesticide label shown in

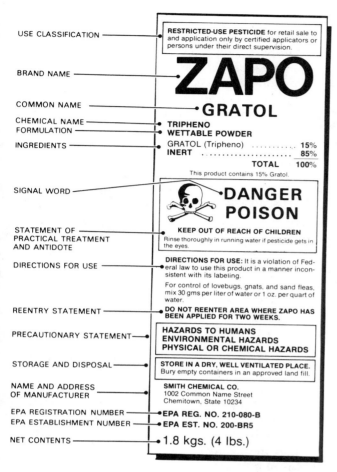

USE CLASSIFICATION

BRAND NAME

COMMON NAME

CHEMICAL NAME
FORMULATION

INGREDIENTS

SIGNAL WORD

STATEMENT OF
PRACTICAL TREATMENT
AND ANTIDOTE

DIRECTIONS FOR USE

REENTRY STATEMENT

PRECAUTIONARY STATEMENT

STORAGE AND DISPOSAL

NAME AND ADDRESS
OF MANUFACTURER

EPA REGISTRATION NUMBER
EPA ESTABLISHMENT NUMBER

NET CONTENTS

Figure 5–9. Agricultural pesticides and other hazardous chemicals have labels that warn the user of any hazards associated with the material (*courtesy of American Association for Vocational Instructional Materials*).

figure 5–9 to see how the sixteen items of information are displayed on the label. Pesticides should always be stored in a locked cabinet or area, figure 5–10.

□ EMERGENCIES OR ACCIDENTS

Quick action of the correct type can change the outcome of an emergency from a tragedy to simply a frightful moment. If fire should break out

- notify the teacher
- keep everyone calm
- set off the fire alarm
- call the fire department
- clear the area
- use fire extinguishers if this seems logical under the circumstances

If an injury occurs, quick action is in order. Such action must be correct and based on good thinking;

Figure 5–10. Pesticides should always be stored in a locked cabinet or other locked area.

otherwise, additional injury may result. For many emergencies the following procedures should be helpful.

Procedure

1. Call or send for help.
2. Call for emergency police or ambulance service. **Note:** Some communities now have ambulance service and medical helicopters standing by that can reach an accident scene with paramedics and emergency equipment in minutes. These services often make the difference between life and death.
3. Do not move the victim unnecessarily.
4. Try to arouse the victim by talking.
5. Treat for shock. Keep the victim lying down. Elevate the victim's feet 8 to 10 inches if there are no signs of bone fractures or head or back injuries. Place a blanket under and over the victim to maintain body heat. If the victim complains of being thirsty, moisten a clean cloth and wet the victim's lips, tongue and inside of the mouth.
6. If the victim is bleeding, stop the bleeding by wrapping or pressing clean cloth or gauze directly on the wound.
7. If the victim is not breathing, clear the air passage

and find someone to administer rescue breathing or cardiopulmonary resuscitation. *Cardiopulmonary resuscitation* (CPR) is a first aid technique to provide oxygen to the body and circulate blood when breathing and heart beat stops. CPR training is given to many teachers and students. The technique requires a special course. To save a victim's life, CPR must be started within 4 to 6 minutes of drowning, electrocution, suffocation, smoke inhalation, gas poisoning, or heart attack.

8. Do not move the victim if broken bones are suspected, unless problems with breathing, bleeding, or other life-threatening factors exist.

Qualified people should be called to give medical aid to accident victims. In schools, there are specific procedures for teachers and students to follow in case of injury or other emergency. Each student should learn these specific policies and procedures.

□ STUDENT ACTIVITIES

1. List the types of fire extinguishers which are needed for classroom, shop, and laboratory. Check each fire extinguisher that is present for state of charge, type, proper hanger, and appropriate location. Ask your teacher to correct all problems regarding fire extinguishers.

2. Ask your teacher to arrange for a representative of the local fire department to visit your class and demonstrate appropriate fire safety practices.

3. Install a slow moving vehicle (SMV) emblem on all school equipment that needs one.

4. Install an SMV emblem on all equipment at home that needs one.

5. As an FFA chapter project, conduct an SMV emblem sale and campaign to increase the use of the emblem in your community.

6. List on your class chalkboard all the signs of danger which you and your classmates can think of.

□ SELF-EVALUATION

A. Multiple Choice. Select the best answer.
  1. Which is *not* part of the fire triangle?
     a. fuel
     b. combustion
     c. oxygen
     d. heat
  2. A commonly used fuel is
     a. acetylene
     b. acetone
     c. oxygen
     d. magnesium
  3. Fire can always be prevented or stopped by eliminating
     a. combustible gases in the area
     b. congestion in the shop
     c. improper storage of fuels
     d. any item in the fire triangle
  4. Fire hazards associated with painting can be reduced by
     a. using a spray gun instead of a brush
     b. using newspaper to protect bench surfaces
     c. using a special paint booth
     d. painting with several people in the area
  5. Effective fire control techniques include
     a. cooling a fire with water
     b. wrapping a blanket around a person whose clothes are on fire
     c. raking dead leaves and grass away from an advancing fire
     d. all of these
  6. Fires are classified according to
     a. materials involved and techniques that safely extinguish them
     b. size and duration of the fire
     c. season of the year when the fire occurs
     d. the amount of material being burned
  7. A green triangle on a fire extinguisher means that the extinguisher can be used to put out burning
     a. metals
     b. liquids
     c. wood
     d. electrical wires
  8. Most fire extinguishers will discharge when the
     a. pin is pulled and the lever pressed
     b. extinguisher is inverted
     c. either a or b, depending on the extinguisher
     d. none of these
  9. SMV means
     a. small mechanical vehicle
     b. stop! moving vehicle
     c. slow moving vehicle
     d. none of these
  10. SMV emblems are required when
     a. vehicles are standing
     b. vehicles travel 25 miles per hour or slower
     c. vehicles travel 30 miles per hour or slower
     d. vehicles travel over 30 miles per hour
  11. Pesticide labels are
     a. legal documents
     b. used only on insecticides
     c. used primarily on powdered chemicals
     d. generally written in two or more languages

B. Matching. Match the class of fire in column I with the burning material in column II.

Column I

1. Class A fire
2. Class B fire
3. Class C fire
4. Class D fire

Column II

a. paper and wood
b. flammable liquids
c. combustible metals
d. electrical equipment

C. Matching. Match the type of extinguisher in column I with the type of fire for which it is used in column II.

Column I

1. dry chemical extinguisher
2. water extinguisher
3. carbon dioxide gas extinguisher
4. foam extinguisher

Column II

a. Class B and C fires
b. Class A fires
c. Class A and B fires
d. Class A, B and C fires

D. Brief Answers. Briefly answer the following questions.

1. Name four items of information provided on pesticide labels.
2. Name four signs that may be seen along highways that alert people of possible hazards.
3. Name three warning signs often found on machinery.
4. Describe the slow moving vehicle emblem.
5. Name five things to do if a fire breaks out.
6. Name six emergency procedures used for accident victims.
7. Explain how to properly extinguish a fire using a fire extinguisher.

Unit 6
Shop Cleanup and Organization

Objective

To work cooperatively with classmates to clean the shop efficiently and store all tools and materials correctly.

Competencies to Be Developed

After studying this unit, you will be able to:
• Use shop cleaning equipment properly.
• Clean benches, machines, and floors.
• Store materials properly.
• Store tools properly.
• Do assigned tasks.
• Work cooperatively with others.

Terms to Know

silhouette
flammable materials
  cabinet
vertical rack
floor broom
dust mop
bench brush

scoop shovel
dust pan
cleanup wheel
cleanup skills checklist
cleanup assignment
  sheet

Materials List

Bench brush
Floor broom
Dust mop
Vacuum cleaner
Rag can
Scrap wood box
Scrap metal can
Shop cleanup wheel chart
Shop cleanup assignment sheet

□ A CLEAN AND ORDERLY SHOP

Each student should have a clear vision of what is meant by a clean and orderly shop. All students working under the direction of the teacher should help produce and maintain a clean and orderly shop. The signs of a properly cleaned shop are as follows:

• A signal is given to stop work and start cleanup at a specified time.
• Every student helps with cleanup.
• Benches are cleared and clean.
• Machines are clean.
• Paint brushes and spray equipment are properly cleaned and stored.
• Solvents, paints, and greases are properly stored.
• Tools are in their places.
• Lumber, metal, and other construction materials are stored.
• Projects and related materials are in approved places.
• Floor is clean and trash is in containers.
• Cabinets and storage areas are locked.
• Every job is checked for completeness.
• Every student is graded according to the quality of his or her cleanup contribution.
• Sinks and restrooms are clean and orderly.
• Students are waiting in an orderly manner for dismissal by the teacher.

□ REASONS FOR KEEPING THE SHOP CLEAN

A quick and efficient cleanup procedure is important to the safety of students. It adds greatly to the success of an agricultural mechanics shop program. There are good reasons for cleaning the shop after each class, every day. Some of these are related to personal safety, some to learning efficiency, and some to student comfort and convenience. A properly organized shop cleanup procedure is important for the following reasons:

• Each student's projects and possessions are stored properly. When projects are stored properly, they do not interfere with the work of other students and are not damaged by other students using the shop. Projects or project parts may be stored in drawers, lockers, storage cabinets, storage rooms, or fenced areas. In well-managed shops, project parts may be stored in containers by, on, or under large projects such as tractors or machinery.

- All project parts are stored together. This enables students to see if any items need to be brought the next day in order to continue the project.
- Shop spaces are cleared so that other classes can safely use the areas.
- Tools are returned to their proper places. Tools should be mounted on panels over colored outlines of each tool, figure 6–1. These outlines are called *silhouettes;* they make it easy to check for missing tools at the end of each class. Tools can also be easily checked to see if they are in need of repair.
- Each student learns to put tools and materials in their proper places and can expect to find them quickly when needed. This eliminates lost time looking for tools and materials.
- The hazards of fire and explosion are reduced by proper storage of materials.
- Students learn cooperation and teamwork.

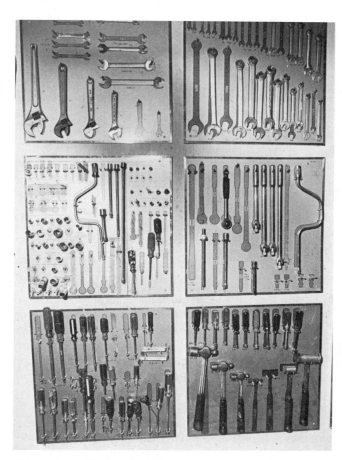

Figure 6–1. Tools should be mounted on panels over colored outlines of each tool.

□ EQUIPMENT AND CONTAINERS USEFUL FOR SHOP CLEANUP

It is important to have enough cleanup equipment and materials on hand so that all students can participate in cleaning and storage activities. Each student must do his or her part to make the cleanup easier for everyone. It is also important to have storage containers for every type of material used in the shop, figure 6–2. Many shops have excellent commercial flammable materials cabinets. Flammable liquids, such as grease, oil, and solvents are stored in these cabinets which are made of steel and close automatically in the presence of fire, figure 6–3. Such cabinets must meet all safety requirements.

Racks for lumber and metal provide safe and convenient storage for these materials, figure 6–4. Vertical racks permit the storage of both long and short items. On such racks, material can be reached with little moving of other items.

There are many items of equipment that are necessary to clean a shop quickly and efficiently and to store materials safely. These items include the following:

- floor brooms
- floor dust mops
- bench brushes
- shop vacuum cleaner(s)
- dust collection and chip removal system
- metal cans for rag storage
- large metal trash cans
- storage cabinets for combustible materials
- shovels and dust pans to pick up dirt and trash
- varsol for cleaning up grease and oil spills
- sawdust to absorb liquids
- commercial material to sprinkle on the floor to control dust
- clean rags

Figure 6–2. Storage containers are needed for different types of materials.

Figure 6–3. Flammable liquids such as grease, oil, and solvents must be stored in special cabinets that close automatically in the presence of fire.

- storage cabinets for tools and hardware
- storage racks for lumber and metal
- steel cans for metal used for practice welding
- suitable containers for scrap wood
- cabinets, lockers, fenced areas, etc., for project storage

The soft-bristled brush and shop vacuum cleaner are the standard tools for removing dirt, saw dust, and trash from benches and machines, figure 6–5. The floor broom and dust mop are important floor cleaning equipment. The dust pan and standard scoop shovel are commonly used to move the trash from the floor to the trash can.

□ TECHNIQUES FOR EFFECTIVE CLEANING

Effective cleaning skills cannot be taken for granted; they must be developed. Many students go through the motions of cleaning, but many do not do a good job. Shop cleanup tasks include

- removing tools and materials from benches and floor before cleaning
- cleaning all paint brushes
- first cleaning high areas such as racks, machines, and bench tops

Figure 6–4. Vertical racks for lumber, steel rods, bands, angle iron, and pipe make good use of wall space. These racks permit the storage of both long and short items. Material can be reached without moving other materials.

- cleaning the shop starting at far sides and ends and working toward the trash collecting area(s)
- using brushes and brooms in short strokes and lifting intermittently (brushes and brooms should be tapped against the surface or floor frequently to shake out dirt particles)
- using a commercial dust absorbing material, if available
- using a vacuum cleaner to clean machines whenever possible (the vacuum cleaner is very desirable because it does not create dust)
- putting oily rags in closed metal containers
- using sawdust or commercial materials to absorb liquids such as oil spills
- sweeping fine waste particles into floor drops of dust collection systems, if provided
- putting all trash in suitable metal containers
- placing trash containers in the proper place
- storing all cleanup equipment properly
- cleaning all sink areas and picking up paper towels
- helping others to finish the cleanup

Figure 6–5. The shop vacuum cleaner makes cleanup easier and does not create dust.

□ SHOP CLEANUP SYSTEMS

Organization is the key to a clean shop. Without good organization, students who see the need for a clean and orderly shop soon lose their willingness to put the shop in good order. This is natural since the value of teamwork and fair play is learned at a very young age. Therefore, a system that involves every student on an equal basis is needed. Several systems have been developed to get the job done.

All-Pitch-In Method

For lack of a better term, the most simple system of shop cleanup is called the "all-pitch-in" method. With this system, the teacher announces cleaning time verbally, or by whistle, bell or other device. Students start by putting their own materials away. They then do cleanup, arrangement, or storage tasks according to their knowledge and maturity. This system generally fails because students lack the knowledge and skills necessary to do a good job at all cleanup tasks. Students and teacher soon discover that certain important cleanup tasks are left undone. This is due to lack of organization.

Cleanup-Wheel Method

Many shops use a cleanup wheel, figure 6–6. This system uses a chart shaped like a wheel. The teacher specifies all cleanup tasks in equal sections on the outer section of the wheel chart. Students are placed in groups or given a group number. Either names or group numbers are placed in the inner section of the wheel chart. Each group of students is assigned the task or tasks listed in line with their name or group number on the wheel chart. The tasks remain stationary at the edge of the chart. The wheel itself can be periodically rotated so each group has a chance to do all cleanup tasks.

For the cleanup-wheel method to work, some kind of check sheet is needed. It is important that a student who is respected by his or her classmates and is a good judge of achievement be given this job. A check sheet with a format that works well is shown in figure 6–7. The checklist should be placed in a prominent location so students can see how they are being evaluated. Students with zeroes (0) and ones (1) should be encouraged to improve their cleanup skills.

When using a checklist, the foreman's name is listed

NOTE: 1. ALL SECTIONS IN THE WHEEL ARE OF EQUAL SIZE (ALL SPOKES ARE EQUAL DISTANCE AROUND THE WHEEL)
2. THE HUB AND INNER RING WITH STUDENTS' NAMES WILL ROTATE TO CHANGE ASSISMENTS
3. THE OUTER SECTION OF THE WHEEL DOES NOT ROTATE

Figure 6–6. An example of a shop cleanup wheel chart. All sections in the wheel are of equal size. The hub and inner ring with students' names will rotate to change assignments. The outer section of the wheel does not rotate.

SHOP CLEANUP SKILLS CHECKLIST

KEY:
  3 = Done well
  2 = Done satisfactorily
  1 = Done poorly
  0 = Job not done
  ab = Student is absent

DATE _4/4/86_

SHOP FOREMAN _Walker_

NAME **DATE AND RATING**

NAME	3/5	3/6	3/7	3/8	3/9	3/12	3/13	3/14	3/15	3/16	AVERAGE SCORE	COMMENTS
Avery	3	3	2	3	3	3	3	3	3	3	2.9	
Brown	3	2	1	1	1	ab	0	0	1	1	1.0	
Cochran	1	3	0	3	3	3	3	3	3	3	2.5	
Daniels	ab	3	3	2	2	2	2	2	3	2	2.1	
Walker (Foreman)	Jw	Jw	Jw	Jw	Jw	Jw	Jw	Jw	Jw	Jw		

Figure 6–7. An example of a shop cleanup skills checklist. Notice in this example that the teacher initialed the box by the foreman's name. This indicates that the teacher observed the general conditions of the shop and approved the job done by the foreman.

along with all others. During the shop cleanup time, the foreman carries the checklist on a clipboard and evaluates all students except himself or herself. The teacher then evaluates how well the foreman has done the evaluations.

This method helps students see how well they are developing their cleanup skills. It also helps to develop management and supervisory skills. The checklist has only enough columns for a limited number of shop periods. New cleanup tasks should be assigned to students every two weeks. This assures that everyone will do all jobs sometime during the year and learn how to do each one in a skillful manner.

Both the teacher and the students must realize that the shop cleanup skills are valuable. Future employment demands that such skills be developed in vocational education. Therefore, every effort should be made to teach and learn procedures and tasks used in effective cleanup operations.

Assignment-Sheet Method

The assignment sheet method may be used in place of the cleanup wheel and the shop cleanup skills checklist, figure 6–8. One important advantage of the assignment sheet is that as many students as necessary can be assigned to any given task in order to get it done. The equal-sized groups necessary for the cleanup wheel are not needed for the assignment sheet. A disadvantage of assignment sheets is the time required to periodically reassign students.

With the shop cleanup assignment sheet, the teacher develops a list of cleanup tasks. Class members

are then assigned to these tasks in numbers needed to get the job done. For instance, if it takes two persons to sweep off benches, four to clean the welding areas, and five to sweep the floor, the appropriate number of students can be assigned to each task. The teacher decides who will do the tasks although students can volunteer for various tasks. Assignments may be changed as often as desired. A new shop cleanup assignment sheet must be prepared each time a rotation is made.

A student foreman and the check-off system are used to evaluate cleanup skills. The check-off, or evaluation, process is the same as described for the shop cleanup-wheel method.

Choosing a System

The shop cleanup system used is largely up to the teacher and students. Each method requires a cooperative attitude on the part of all. Each student must do the assigned task every day or the system will not work. This is because others must do the job for the negligent student. Jobs of absent students must also be covered by others assigned to the same task. To avoid this problem, several students may be designated as substitutes to do the tasks assigned to others who are absent on any given day.

The following must be provided for if a shop cleanup system is to work well.

• assignment of every task to some person
• fair assignments based on desirability of the task and effort required to do it

SHOP CLEANUP ASSIGNMENT SHEET

KEY: 3 = Done well
       2 = Done satisfactorily
       1 = Done poorly
       0 = Job not done
       ab = Student is absent

DATE _____

SHOP FOREMAN _____

TASK	PERSON(S) RESPONSIBLE	DATE AND RATING									AVERAGE SCORE	COMMENTS
	Date:											
Sweep Benches	Avery											
	Brown											
Clean Welding Areas	Daniels											
	Elson											
	Graham											
Sweep Floor	Holt											
	Iddings											
	Jones											
	Kirk											
Foreman	Mozier											

Figure 6–8. An example of a shop cleanup assignment sheet. This is similar to the shop cleanup skills checklist except that a task column is added.

- rotation of assignments so all students learn all tasks
- cooperation by all parties
- genuine honesty in evaluating performance
- a record of individual performance
- a clean and safe shop by the end of each period

By working together, agricultural mechanics students can enjoy the benefits of a clean and safe shop. This provides satisfaction, good workmanship, cooperative effort, and personal gain.

□ **STUDENT ACTIVITIES**

1. List the items in the school agricultural mechanics shop that are used for the storage of

   - lumber and metal
   - fasteners such as nails and screws
   - flammable liquids
   - waste materials
   - tools

2. Examine the tools and equipment in your school agricultural mechanics shop that are used to clean the shop. Learn to use each item to clean thoroughly.

3. Check all containers used in your agricultural mechanics shop to see if they are of the proper type and properly labeled. Report your findings to the teacher.
4. Help improve the cleanup equipment and facilities in your shop.
5. Volunteer for the shop cleaning job of your choice.

□ **SELF-EVALUATION**

A. Multiple Choice. Select the best answer.

1. Oily rags should be stored in a
   a. card board box
   b. plastic bag
   c. wooden box
   d. closed metal can
2. A clean, organized shop reduces the chance of
   a. fire
   b. lost tools
   c. damage to projects
   d. all of these
3. Brushes and brooms work better if
   a. pushed in a continuous path
   b. pushed and lifted intermittently
   c. pushed back and forth
   d. pushed in long strokes

4. Sawdust is useful in shop cleanup to
   a. absorb liquids on the floor
   b. reduce dust in the trash container
   c. condition bristles on floor brooms
   d. none of these
5. A recommended material for cleaning grease from the floor is
   a. water
   b. gasoline
   c. varsol
   d. sawdust
6. The foreman's job in the cleanup process is
   a. supervision
   b. reward
   c. evaluation
   d. assigning jobs
7. The best item for cleaning nongreasy machines is
   a. rags
   b. brush
   c. air gun
   d. vacuum cleaner
8. The shop cleaning method which gives the best control over the cleanup process is the
   a. all-pitch-in method
   b. cleanup wheel method
   c. assignment sheet method
   d. honor system method
9. The main advantage of the shop cleanup assignment sheet over the shop cleanup wheel is the
   a. flexibility in assigning students to tasks
   b. ease in reassigning tasks
   c. use of a checklist for evaluation
   d. use of a foreman for evaluations
10. Rotating shop cleanup duties
    a. enables everyone to learn the various cleaning tasks
    b. promotes fairness in assigning undesirable tasks
    c. involves every student on an equal basis
    d. all of these

B. Brief answers. Briefly answer the following questions.
1. List ten signs of a properly cleaned shop.
2. List seven reasons for using a properly organized cleanup procedure.
3. List fifteen tasks included in shop cleanup.
4. Name two cleaning tools that are used to move trash from the floor to the trash cans.
5. What is the best cleaning tool to use when cleaning machines?

Section 3
HAND WOODWORKING AND METALWORKING

Unit 7
Hand Tools, Fasteners, and Hardware

The use of hand tools remains the foundation of agricultural mechanics. Large and efficient power tools are used to do most of the work in society. However, hand tools are used to do the small jobs and to do the work where large machines cannot function.

The use of hand tools is basic and essential for work in design, construction, maintenance, and repair. Hand tools are used by all who construct buildings, install landscape structures, wire computers, or repair tractors. In fact, all people in society use hand tools

Figure 7–1. Hand tools are instruments powered by human hands to do work.

or pay others to do jobs that require their use. A *tool* is any instrument used in doing work. A *hand tool* is any tool operated by hand to do work, figure 7–1. This is contrasted with a *power tool* which is operated by some source of power other than human power.

☐ TOOL CLASSIFICATIONS

There are various methods of classifying tools. Therefore, the student needs to be prepared to recognize tools by their various classifications. Tools may be classified according to who uses them, e.g., carpenters' tools, masons' tools, mechanics' tools, or machinists' tools. Tools from all such trades are used in agriculture. Therefore, classification of tools by use or function is more meaningful to workers in agricultural mechanics.

The major classifications of tools according to use or function are layout tools, cutting tools, boring tools, driving tools, holding tools, turning tools, digging tools, and other tools. Students should study tools and learn how to classify and correctly spell them. This is helpful when buying or using tools in the shop, home, or on the job.

Layout Tools (L)

A *layout tool* is a tool used to measure or mark wood, metal, and other materials, figure 7–2. Layout tools may be designated by using a capital "L." Rules, squares, scratch awls, calipers, measuring tapes, and dividers are classified as layout tools. These tools are used to measure and mark materials before cutting or shaping is done. Measuring and marking are done when laying out work so other functions can follow according to plan.

Cutting Tools (C)

A *cutting tool* is a tool used to cut, chop, saw, or otherwise remove material. This permits the user to shape the material. A capital "C" may be used to designate cutting tools. Tools such as saws, chisels, hatchets, and planes are classified as cutting tools, figure 7–3. Since there are so many cutting tools in common use, those that are used to make holes are classified in a separate group and are known as boring tools. Boring tools are designated by the capital letter "B."

Boring Tools (B)

A *boring tool* is a tool used to make holes or to change the size or shape of holes. Boring tools include bits, drills, reams, and the devices used to turn them, figure 7–4. In addition to many common hand tools, this group includes a large variety of specialized tool bits for power machines.

Driving Tools (Dr)

Another group of tools classified by their use is driving tools. These tools may be designated by the letters "Dr." A *driving tool* is a tool used to move another tool or object, figure 7–5. Driving tools rely on their weight and speed to provide force to move an object when it is hit. Using a hammer to move a sliding belt tightener or striking a cold chisel to cut a rivet are uses for driving tools. Driving tools are used to force nails into wood, or stakes into soil. Some examples of driving tools are hammers, sledges, punches, and various auto body tools.

Holding Tools (H)

A *holding tool* is a tool used to grip wood, metal, plastic, and other materials, figure 7–6. They may be designated by the capital letter "H." Holding tools are used to hold material while other tools are used to cut, shape, modify, or turn threaded items like screws. Holding tools are also used to grip objects such as bolts or pieces of wire. They may be used to bend or shape such objects as needed. Holding tools include vises, pliers, and clamps.

Turning Tools (T)

Engine and machinery mechanics make frequent use of turning tools. Turning tools are designated by the capital letter "T." A *turning tool* is a tool used to turn

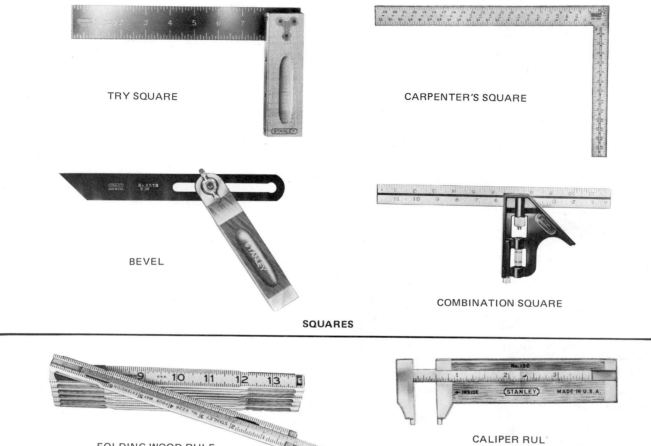

TRY SQUARE

CARPENTER'S SQUARE

BEVEL

COMBINATION SQUARE

SQUARES

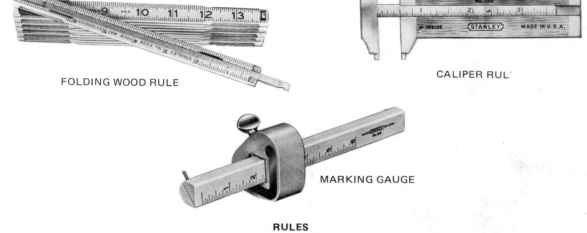

FOLDING WOOD RULE

CALIPER RUL'

MARKING GAUGE

RULES

LEVEL

LINE LEVEL

LEVELS

Figure 7–2. These common layout tools are used to measure and mark materials (*courtesy of Stanley Tools;* inside caliper *courtesy of Mac Tools, Inc.*). Figure continues on page 56.

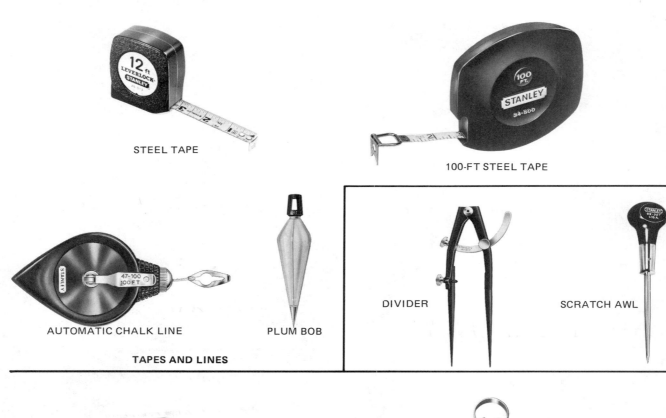

STEEL TAPE

100-FT STEEL TAPE

AUTOMATIC CHALK LINE

PLUM BOB

DIVIDER

SCRATCH AWL

TAPES AND LINES

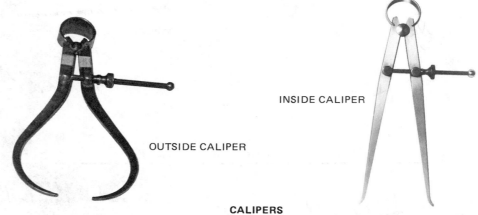

INSIDE CALIPER

OUTSIDE CALIPER

CALIPERS

Figure 7–2. *Continued*

1. PRUNING SAW

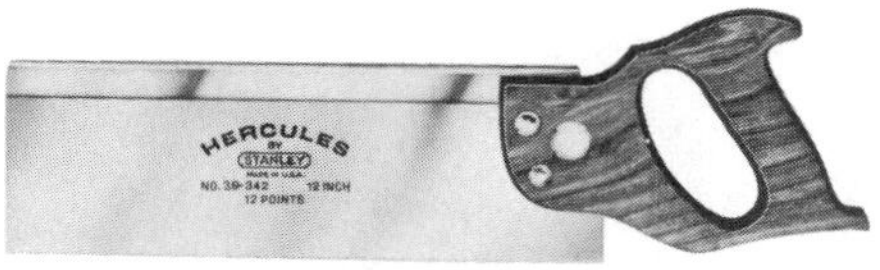

2. BACK SAW

3. BOW SAW

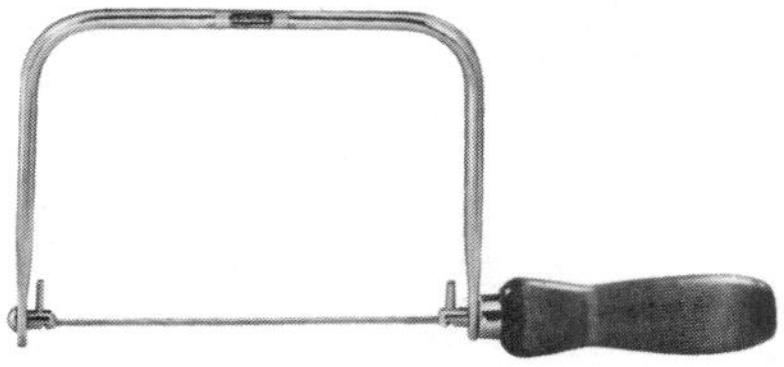

4. COPING SAW

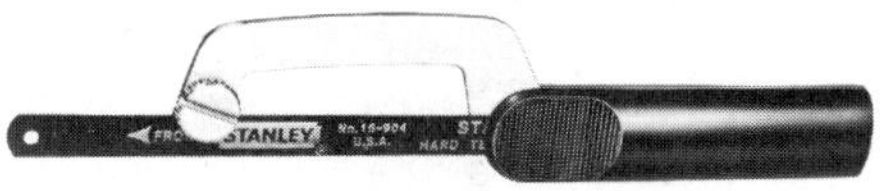

5. MINI HACKSAW

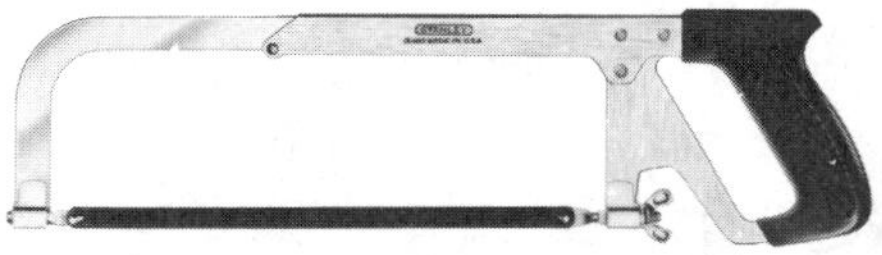

6. HACKSAW

7. HANDSAW

8. HOLE SAW

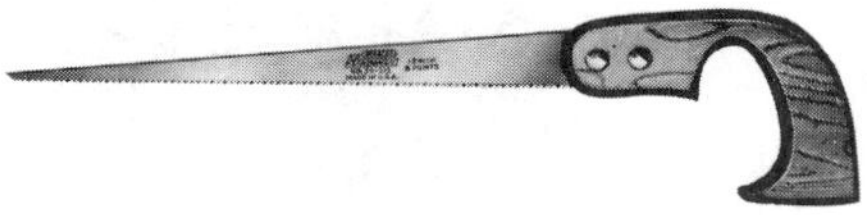

9. COMPASS SAW

SAWS

10. WOOD CHISEL

11. COLD CHISEL

13. CAPE CHISEL

12. FLOORING CHISEL

14. DIAMOND POINT CHISEL

CHISELS

Figure 7–3. Cutting tools (*courtesy of Stanley Tools*, 1–15, 16–27, 31–32; *Klein Tools, Inc.*, 29, 33; *Mac Tools, Inc.*, 34–40, 42) Figure continues on pages 58–60.

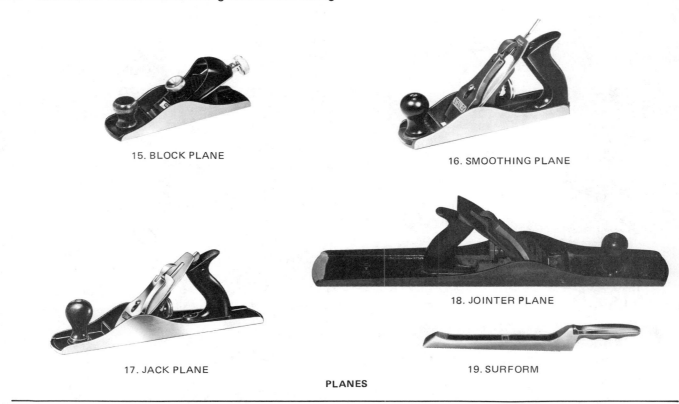

15. BLOCK PLANE

16. SMOOTHING PLANE

17. JACK PLANE

18. JOINTER PLANE

19. SURFORM

PLANES

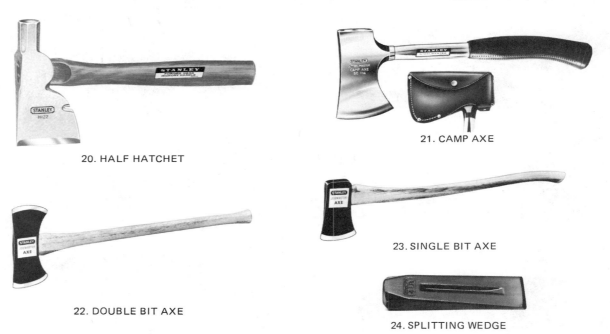

20. HALF HATCHET

21. CAMP AXE

22. DOUBLE BIT AXE

23. SINGLE BIT AXE

24. SPLITTING WEDGE

AXES AND WEDGES

Figure 7–3. *Continued*

25. COMPOUND SNIPS

26. TIN SNIPS

27. DIAGONAL CUTTERS

28. PRUNING SHEARS

29. BOLT CUTTERS

31. GLASS CUTTER

30. PIPE CUTTERS

32. UTILITY KNIFE

33. NIPPERS

CUTTERS

Figure 7–3. *Continued*

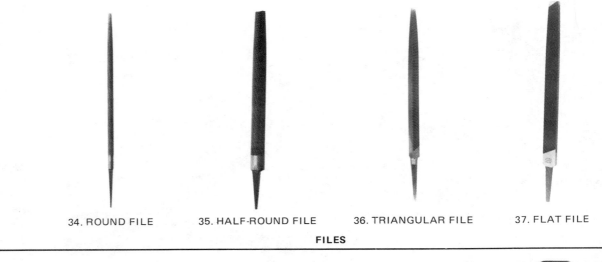

34. ROUND FILE   35. HALF-ROUND FILE   36. TRIANGULAR FILE   37. FLAT FILE

FILES

40. TAP WRENCH

41. DIE STOCK

38. TAP   39. T-HANDLE TAP WRENCH

42. DIE

TAPS AND DIES

Figure 7–3. *Continued*

nuts, bolts, or screws. Turning tools include wrenches, sockets, and screwdrivers, figure 7–7. The handles and extensions used with sockets are also classified as turning tools.

Digging Tools (D)

Many people working in agriculture use a group of tools known as digging tools (designated by the capital letter "D"). A *digging tool* is any device used to turn up, loosen, or remove earth, figure 7–8. Digging tools include shovels, mattocks, hoes, rakes, post hole diggers, and garden trowels.

Other Tools (O)

All tools do not fit into neat categories. Therefore, the term *other* is used to classify tools used in agricultural mechanics that do not fit into standard catego-

ries. These tools are designated by the capital letter "O." Pry bars, scrapers, and masonry tools may be classified as other tools, figure 7–9.

☐ FASTENERS

A *fastener* is any device used to hold two or more pieces of material together or in place. The carpenter creates a building out of individual pieces by using fasteners such as nails. By using fasteners such as bolts and nuts, the manufacturer assembles a machine from many parts. Fasteners are used to hang pictures and laminate panels. *Laminate* means to fasten two or more flat pieces together with an adhesive. An *adhesive* is a sticky substance such as glue. The most common fasteners used in agricultural mechanics are nails, screws, bolts, nuts, washers, and machine screws.

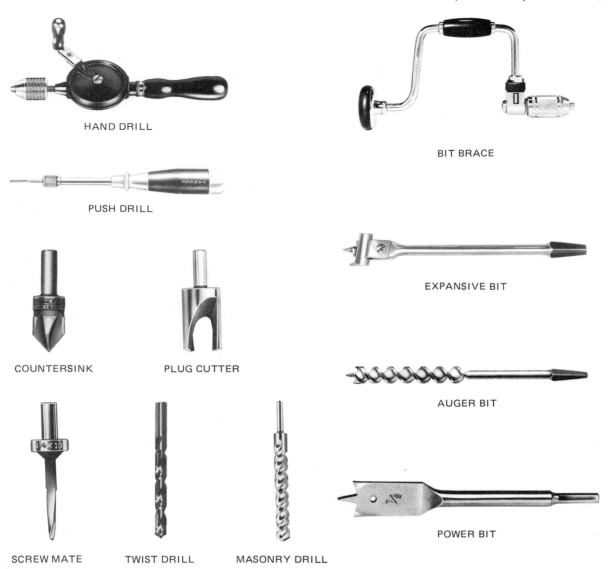

HAND DRILL

PUSH DRILL

COUNTERSINK

PLUG CUTTER

SCREW MATE

TWIST DRILL

MASONRY DRILL

BIT BRACE

EXPANSIVE BIT

AUGER BIT

POWER BIT

Figure 7–4. Boring tools (*courtesy of Stanley Tools*)

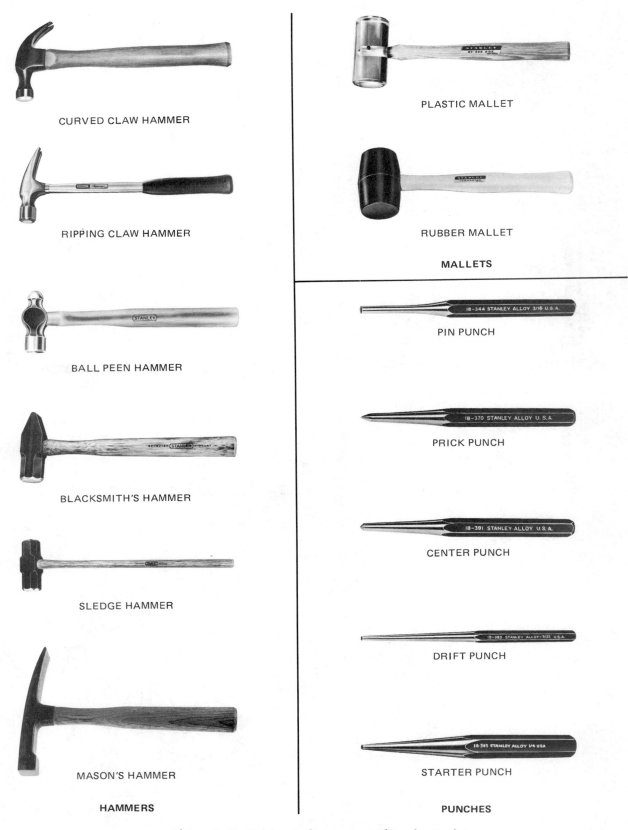

CURVED CLAW HAMMER

RIPPING CLAW HAMMER

BALL PEEN HAMMER

BLACKSMITH'S HAMMER

SLEDGE HAMMER

MASON'S HAMMER

HAMMERS

PLASTIC MALLET

RUBBER MALLET

MALLETS

PIN PUNCH

PRICK PUNCH

CENTER PUNCH

DRIFT PUNCH

STARTER PUNCH

PUNCHES

Figure 7–5.   Driving tools (*courtesy of Stanley Tools*)

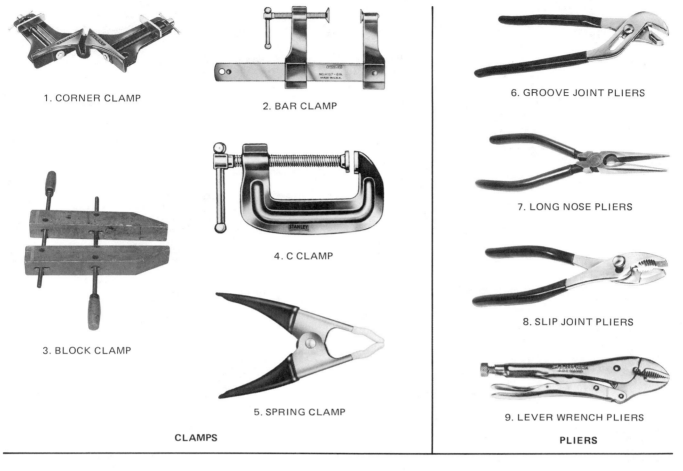

1. CORNER CLAMP

2. BAR CLAMP

6. GROOVE JOINT PLIERS

3. BLOCK CLAMP

4. C CLAMP

7. LONG NOSE PLIERS

5. SPRING CLAMP

8. SLIP JOINT PLIERS

9. LEVER WRENCH PLIERS

CLAMPS

PLIERS

10. MACHINIST'S VISE

11. BENCH VISE

12. PIPE VISE

VISES

Figure 7–6. Holding tools (*courtesy of Stanley Tools,* 1–2, 4–8, 11; *Klein Tools, Inc.,* 9; *The L. S. Starrett Company,* 10)

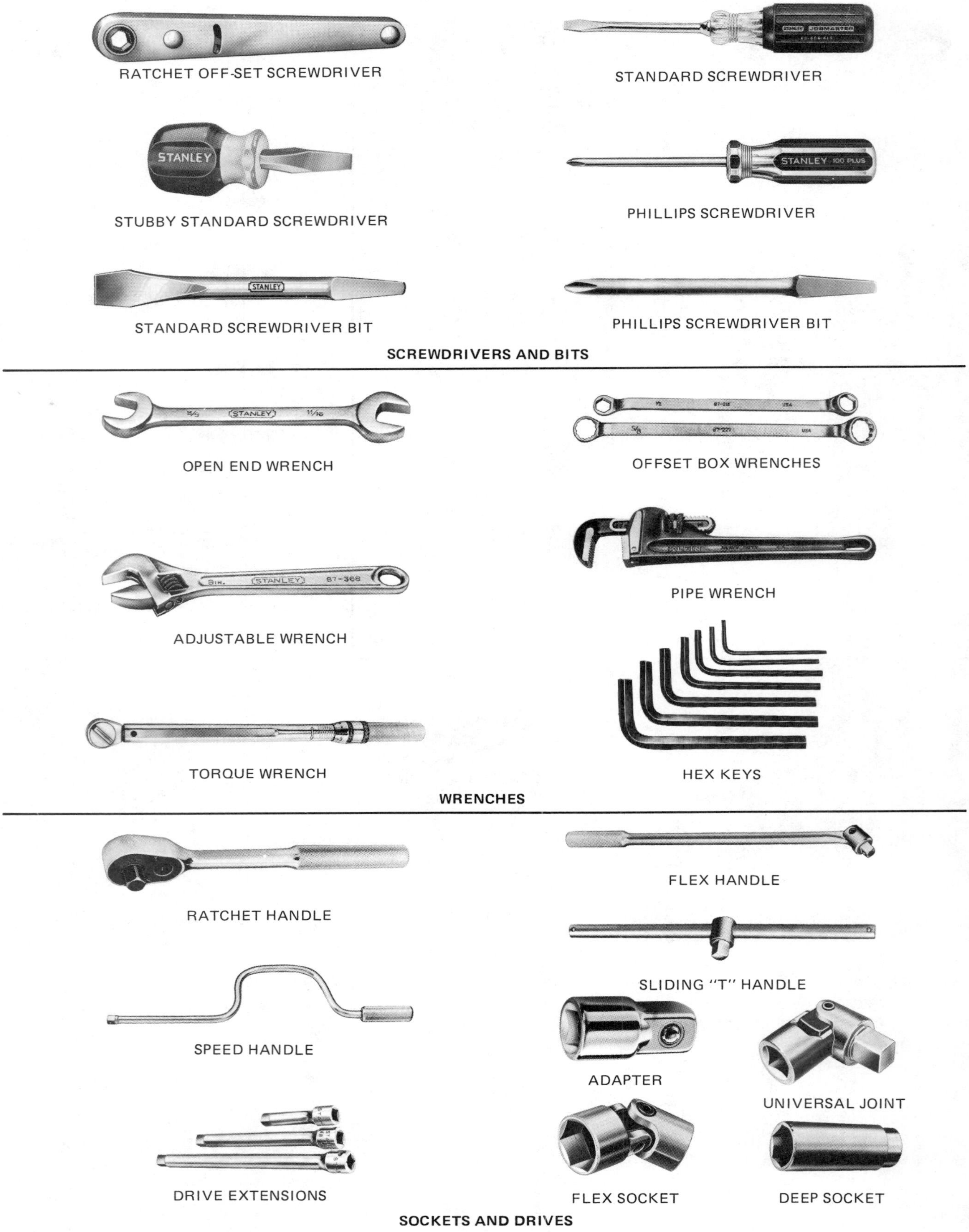

RATCHET OFF-SET SCREWDRIVER

STANDARD SCREWDRIVER

STUBBY STANDARD SCREWDRIVER

PHILLIPS SCREWDRIVER

STANDARD SCREWDRIVER BIT

PHILLIPS SCREWDRIVER BIT

SCREWDRIVERS AND BITS

OPEN END WRENCH

OFFSET BOX WRENCHES

ADJUSTABLE WRENCH

PIPE WRENCH

TORQUE WRENCH

HEX KEYS

WRENCHES

RATCHET HANDLE

FLEX HANDLE

SLIDING "T" HANDLE

SPEED HANDLE

ADAPTER

UNIVERSAL JOINT

DRIVE EXTENSIONS

FLEX SOCKET

DEEP SOCKET

SOCKETS AND DRIVES

Figure 7–7. Turning tools (*courtesy of Stanley Tools;* pipe wrench *courtesy of Klein Tools, Inc.*)

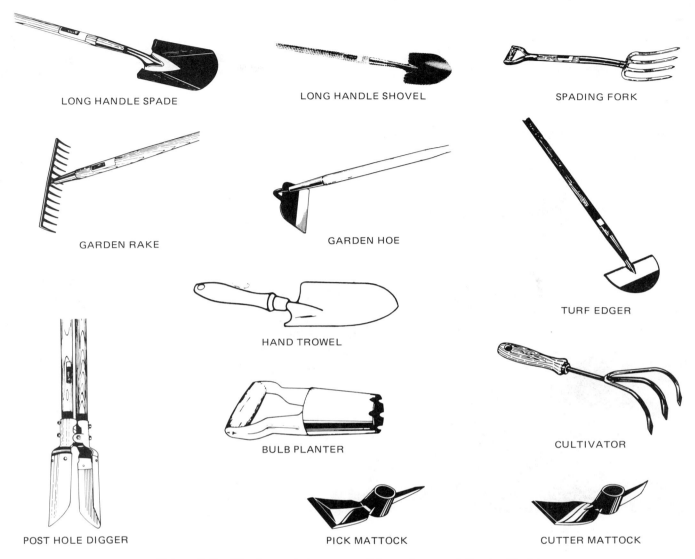

LONG HANDLE SPADE

LONG HANDLE SHOVEL

SPADING FORK

GARDEN RAKE

GARDEN HOE

TURF EDGER

HAND TROWEL

POST HOLE DIGGER

BULB PLANTER

CULTIVATOR

PICK MATTOCK

CUTTER MATTOCK

Figure 7–8. Digging tools (*courtesy of Ames Lawn and Garden Tools*)

Nails

A *nail* is a fastener which is driven into the material it holds. There are many types of nails classified generally by their use or form, figure 7–10. When buying nails, the purchaser must know the use for the nail, the desired length, and the desired thickness or diameter. *Diameter* refers to the distance across the center of a round circle or object. The *shank* is the long stem part of a nail or screw; the *head* is the enlarged part on top. Probably the most familiar nail is the common nail. It has a fairly thick shank and a medium-sized head. Nails that are flat and tapered are called cut nails. These are cut from steel and used for wood flooring, or they may be hardened and used as fasteners in masonry materials.

Nail Lengths. □ The unit of measure used to designate the length of most nails is the *penny*, figure 7–11. The symbol for penny is the lowercase letter "d." The term penny was originally used to indicate the number of English pennies needed to purchase 100 nails of a given size. The lengths of common nails, box nails, finishing nails, cut nails, and spikes are designated by penny. The size range for these nails are

- common nail—2d to 60d
- box nail—2d to 40d
- finishing nail—2d to 20d
- cut nail—2d to 20d
- spike—16d to 12 inches

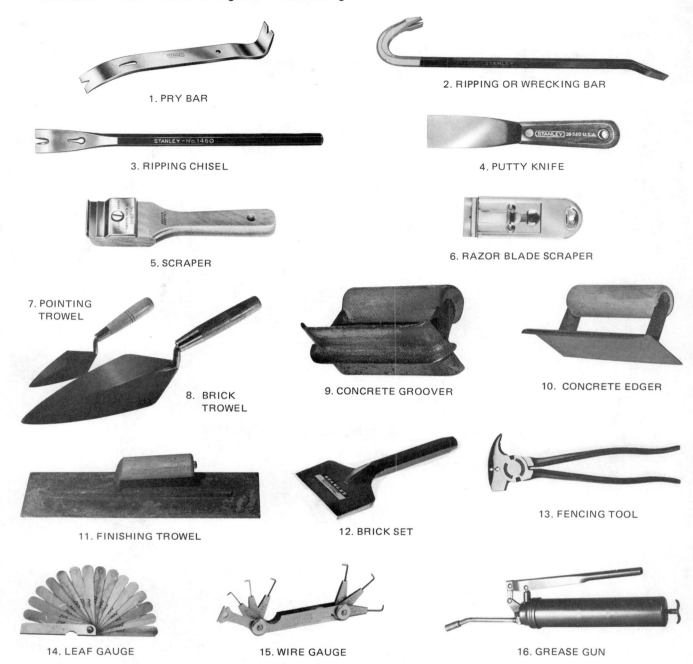

1. PRY BAR

2. RIPPING OR WRECKING BAR

3. RIPPING CHISEL

4. PUTTY KNIFE

5. SCRAPER

6. RAZOR BLADE SCRAPER

7. POINTING TROWEL

8. BRICK TROWEL

9. CONCRETE GROOVER

10. CONCRETE EDGER

11. FINISHING TROWEL

12. BRICK SET

13. FENCING TOOL

14. LEAF GAUGE

15. WIRE GAUGE

16. GREASE GUN

Figure 7–9. Many tools do not fit into categories and are referred to as other tools (*courtesy of Stanley Tools,* 1–6; *Diamond Tool and Horseshoe Company,* 13; *Sears, Roebuck and Company,* 14–15; *Mac Tools, Inc.,* 16).

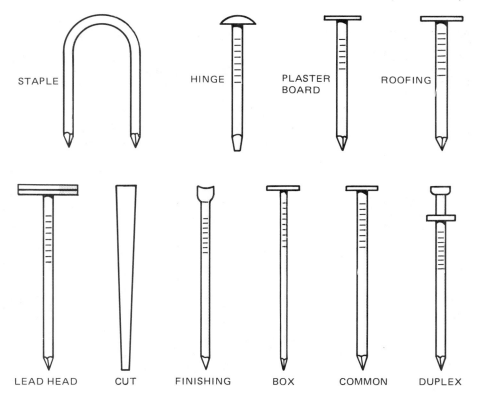

Figure 7–10. Types of nails

Uses for Nails. □ Nails vary in thickness of shank and diameter of head according to their use. If the material being held is soft, then a large headed nail is needed. Otherwise, the material will pull over the head of the nail. If the material being held is heavy, then a thick, strong shank is required. Following are some important types of nails and their uses.

- common nail—used for general construction; nailing sheeting, shiplap, and board fencing
- cut nail—used for nailing tongue and groove flooring; if hardened, used for nailing in masonry materials
- box nail—used for light household construction; nailing siding on buildings, nailing into the end grain of boards
- finishing nail—used for interior finishing of buildings; trim, cabinet, and furniture work, when countersinking is needed
- shingle nail—used for nailing shingles
- roofing nail—used for nailing rolled roofing and composition shingles
- plaster board nail—used to attach plaster board to studs in buildings
- hinge nail—used to fasten hinges on doors and cabinets

- duplex nail—used for construction of forms for concrete work and for nailing insulators on wooden posts for electric fencing
- wire staple—used in wire fence construction
- lead head nail—used for nailing galvanized steel roofing and siding

Improved Nails. □ Changes in nail forms have come about with the development of new types of building materials. Some new materials require special fastening devices. For instance, soft insulating boards should be nailed with special nails having large square heads.

Improved nails, sometimes referred to as thread nails, are basically the same as regular nails. They differ in that a portion of the nail shank is threaded with annular or helical threads, figure 7–12. The indentations in the threads along the nail shank provide grooves into which the wood fibers can expand. Thus, both friction and sheer resistance make it very difficult to remove the nail. Nails that must be driven into very hard substances are heat treated to make them hard.

Nails are generally made of steel or aluminum. Steel nails will rust and must be galvanized if exposed to moisture. *Galvanized* means coated with zinc. Steel

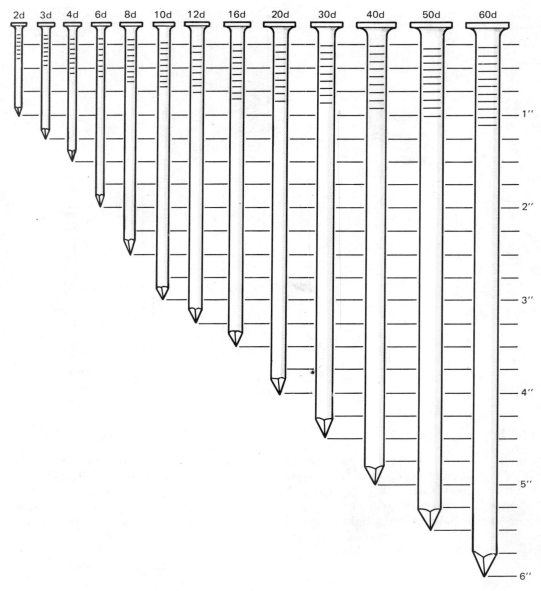

Figure 7–11. Length of nails by penny size

nails can be coated with a varnishlike cement to make them hard to remove from wood. Nails used with aluminum should be made of aluminum. This is necessary since steel nails will rust and galvanized nails will react with and destroy the aluminum that they touch.

Screws

A *screw* is a fastener which has threads that bite into the material it fastens. The term *threads* refers to grooves of even shape and taper that wrap continuously around a shank or hole. Screws are made for use in wood, sheet metal, plastic, and any material solid enough to hold them. Screws generally cut

threads into the material to which they hold, figure 7–13.

Classifications of Screws. ☐ Screws can be classified according to the material they hold. Wood screws have threads designed to bite into wood fibers which draw the screw into the wood when turned. Sheet metal screws have threads that are wide enough to permit the thin metal to fit between the ridges of the threads. Cap screws are designed to thread into thick metal that has matching threads cut into the metal. Lag screws, also called lag bolts, have very coarse threads designed for use in structural timber or lead wall anchors.

Screws can be classified according to the metal they

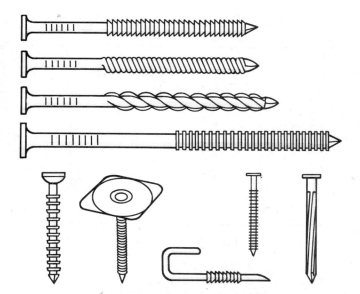

Figure 7–12. Improved nails have threaded shanks and offer more holding power than regular nails.

are made from or the finish used. Steel screws may be coated with a blued, galvanized, cadmium, nickel, chromium, or brass finish. Solid brass screws are rust proof and are used where severe moisture problems destroy coated or plated screws.

Screws can be classified by the shape of their heads, figure 7–14. Flat head screws have tapered heads designed to fit into a hole countersunk in the material. The result is a screw head that is flush with the surface. Round head screws extend above the material. They do not need the special countersunk hole required by flat head screws. Oval head screws extend both above and below the surface of the material. Pan head screws have heads which look like an upside-down frying pan. They are used on sheet metal. A phillips head screw has a flat head with slots in the head that look like a plus sign (+). Lag screws have unslotted heads that are either square or hex shaped. (A *square head* has four equal sides; a *hex*

head has six equal sides.) Lag screws are turned with a wrench instead of a screwdriver.

Screws can be classified by the type of tool needed to turn them. A *slotted head screw* requires a standard screwdriver. A *phillips head screw* requires a phillips screwdriver. An *allen screw* is one with a six-sided hole in the head. A six-sided hex or allen wrench is used to turn allen screws.

Size of Screws. □ The size of the screw is specified by the diameter and length of shank. Diameter is expressed by gauge numbers that run from 2 through 24. Numbers 6, 8, and 10 are common sizes. Common lengths range from ¼ inch through 4 inches, figure 7–15.

Bolts

A bolt is a fastener with a threaded nut, figure 7–16. Types of bolts that are often used in agricultural mechanics include machine bolts, carriage bolts, stove bolts, plow bolts, and special bolts. Bolts and nuts have national course, national fine (SAE), or metric threads.

Bolts are generally straight with round, square, or hex heads. However, some are bent, modified, or shaped for special uses. These include the eye bolt, hook or J bolt, U bolt, turn buckle, and plow bolt. A lag screw, when used with an expansion plug in a masonry wall, is known as an expansion bolt. A toggle is a collapsible winged nut which opens when inserted in a hollow wall. A toggle with bolt is called a toggle bolt.

The *machine bolt* is a fastener with a square head or hex head with threads on just the last one inch or more of the shank. Machine bolts are used extensively in engines and other machines. Special hardened machine bolts with extra strength and toughness have heads marked with short lines that point from the shoulders toward the center of the head. Three lines designate a very tough bolt and six lines

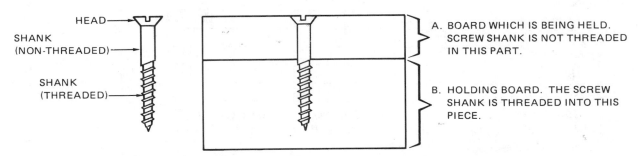

Figure 7–13. Screws hold and resist pulling by threading into the holding material as shown in B. The screw must not be threaded in the material being held. This permits the material in A to draw tightly against material B.

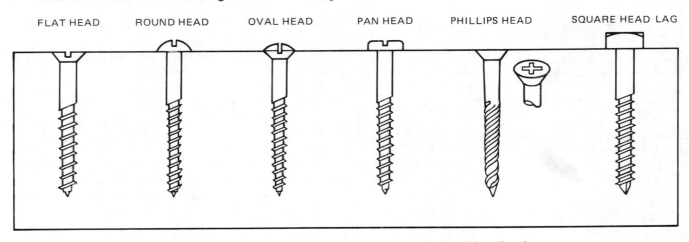

Figure 7–14. Screws can be classified by the shape of their heads.

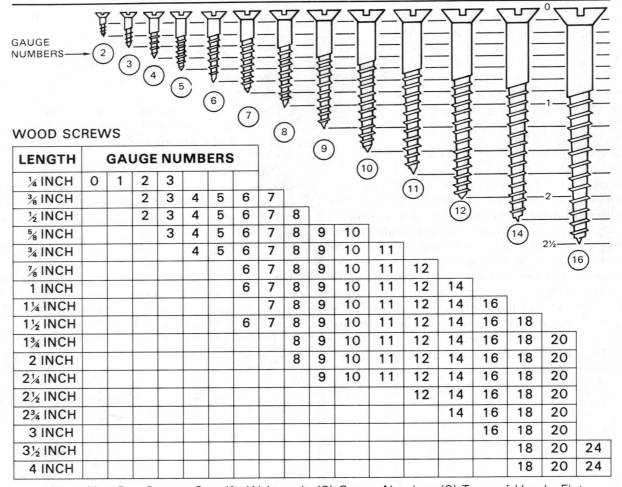

WOOD SCREWS

LENGTH	GAUGE NUMBERS													
¼ INCH	0	1	2	3										
⅜ INCH			2	3	4	5	6	7						
½ INCH			2	3	4	5	6	7	8					
⅝ INCH				3	4	5	6	7	8	9	10			
¾ INCH					4	5	6	7	8	9	10	11		
⅞ INCH							6	7	8	9	10	11	12	
1 INCH							6	7	8	9	10	11	12	14
1¼ INCH								7	8	9	10	11	12	14	16
1½ INCH							6	7	8	9	10	11	12	14	16	18
1¾ INCH									8	9	10	11	12	14	16	18	20
2 INCH									8	9	10	11	12	14	16	18	20
2¼ INCH										9	10	11	12	14	16	18	20
2½ INCH													12	14	16	18	20
2¾ INCH														14	16	18	20
3 INCH															16	18	20
3½ INCH																18	20	24
4 INCH																18	20	24

When You Buy Screws Specify (1) Length, (2) Gauge Number, (3) Type of Head—Flat, Round, or Oval, (4) Material—Steel, Brass, Bronze, etc., (5) Finish—Bright, Steel Blued, Cadmium, Nickel, or Chromium Plated.

Figure 7–15. Screws are classified by a gauge number according to the diameter of the screw. Length of screws is expressed in inches.

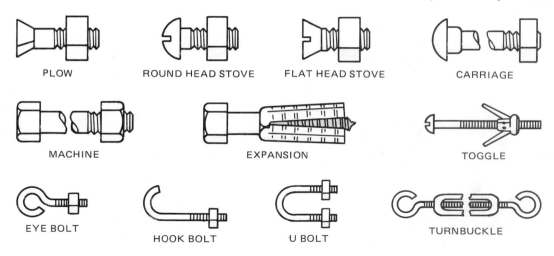

PLOW ROUND HEAD STOVE FLAT HEAD STOVE CARRIAGE

MACHINE EXPANSION TOGGLE

EYE BOLT HOOK BOLT U BOLT TURNBUCKLE

Figure 7–16. Common types of bolts

designate a very, very tough bolt. The tougher the bolt, the more expensive it is to manufacture.

The *cap screw* looks like a machine bolt with the following exceptions. A cap screw

- usually is threaded over its entire length
- generally is two inches or shorter in length
- may have a special head such as one requiring an allen wrench
- threads into an object rather than a nut

The *carriage bolt* is a fastener with a round head over square shoulders and is used with wood. The shoulders are drawn down into the wood to prevent the bolt from turning. The round, low profile of the head makes it almost flush with the surface of the wood. Carriage bolts are used for construction of wagon bodies, feeders, doors, and other wooden projects.

The *stove bolt* is a round headed bolt with a straight screwdriver slot. It is threaded the entire length. When purchased, stove bolts generally come with square nuts. They are generally available in sizes up to ⅜ inch diameter and up to 6 inches in length.

The *plow bolt* has a square tapered head. When placed in a hole designed for a plow bolt, the head is flush with the surface. The plow bolt is used to hold shares and other parts in place on tillage implements.

Nuts

A *nut* is a device with a threaded hole, figure 7–17. Nuts are the moveable part of bolts which are used to fasten two items together. Nuts may be square with four sides, or hexagonal with six sides. Hexagonal is also referred to as simply, hex. Wing nuts are nuts

with winged extensions for tightening with the fingers.

Special nuts are available with slots on one side to permit the use of cotter pins or keys to lock the nut in place. Another technique used to lock a nut in place is to tighten a second nut against the first. The second nut is then referred to as a lock nut. Some nuts are said to be self-locking. These have special design features that make them difficult to turn on or off a bolt.

Nuts are available with several types of threads. The most popular threads used on bolts and nuts for agricultural applications are national coarse (NC), national fine (NF) or (SAE), and metric.

Washers

A *washer* is a flat device with a hole in the center and is used as part of a fastener, figure 7–18. Flat washers are used to prevent bolt heads or nuts from penetrating material. Special lock washers are used to prevent nuts or bolts from loosening due to vibration and use. Lock washers may be split and made

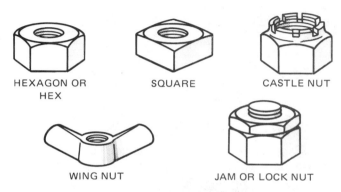

HEXAGON OR HEX SQUARE CASTLE NUT

WING NUT JAM OR LOCK NUT

Figure 7–17. Types of nuts

FLAT WASHER

EXTERNAL LOCK WASHER

SPLIT LOCK WASHER

COUNTERSUNK LOCK WASHER

Figure 7-18. Types of washers

with spring steel to exert back pressure when pressed. They may also have spring steel fingers that bite the surfaces which press against them. The spring pressure or biting capability of lock washers prevents nuts or bolts from loosening.

Washers are designated by the size of the bolt for which they are designed. For instance a washer with a hole that is correct for fitting on a ⅜-inch bolt is called a ⅜-inch washer. Washers are sold by the piece or by the pound.

Machine Screws

A machine screw is a small bolt with a hex nut. Machine screws range in size from very tiny up to less than ¼ inch in diameter, and up to several inches in length. The diameter of machine screws is expressed by a number, figure 7-19. The most common machine screws range from number 4 (very small) to number 12 (nearly ¼ inch in diameter).

Threads on machine screws are designated by the number per inch. Finer threads have more threads per inch than coarse threads and are less likely to

DIAMETER NUMBER	THREADS PER INCH
4	40
6	32
6	40
8	32
8	36
10	24
10	32
12	24
12	28
¼*	20
¼	28

* Same as ¼-inch stove bolt or cap screw with standard threads

Figure 7-19. Machine screw sizes and threads

vibrate loose. On the other hand, coarse threads permit greater ease of starting and greater speed in applying a nut.

When buying machine screws, the diameter by number, the threads per inch, and the length must be specified. For example, a 4–40 × ½ machine screw is number 4 diameter, 40 threads per inch, and ½ inch long.

☐ HARDWARE

The term *hardware* is used in agricultural mechanics for special fasteners. These include hinges, brackets, plates, and miscellaneous metal objects. Hardware is used to support doors, lock windows, secure sideboards on trucks, and anchor roofs. Whenever two objects need to be connected, hardware is generally used. Hardware, as the term suggests, is hard because it is usually made from steel or brass.

Hinges

A *hinge* is an object that pivots and permits a door or other object to swing back and forth or up and down. Small hinges used on cabinets and furniture are referred to as cabinet hinges. Hinges are classified according to their type and size. Some common types of hinges are butt hinges, strap hinges, T hinges, screw hook and strap hinges, and hasps, figure 7–20. All hinges consist of two parts plus a pin. The parts are fastened to the stationary object and the moving object. The pin is then inserted to permit movement between the parts, figure 7–21.

Butt Hinges. ☐ *Butt hinges* are used to mount doors in a butted position. This means the door and the strip it is mounted on are flush and form a smooth surface. Butt hinges are fairly short in length as compared to their height or the length of their pins. Butt hinges may be mounted on the surface or may be mounted in a concealed position between the door and its mounting strip. If the hinges are concealed or out of sight, they must be set in the material and have countersunk screw holes with flat head screws.

Some surface mounted butt hinges are designed to be attractive. An example is the butterfly hinge. It is so named because it is shaped like a butterfly in flight.

Strap Hinges. ☐ *Strap hinges* are much longer than they are high. Their long, thin, straplike appearance gives them their name. The strap hinge is used where the hinge must reach across a long surface; it provides additional support for the door. This feature is useful where the door or its mounting strip

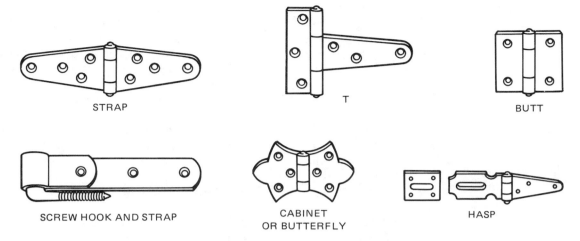

STRAP

T

BUTT

SCREW HOOK AND STRAP

CABINET
OR BUTTERFLY

HASP

Figure 7–20.   Types of hinges

is not very strong. Strap hinges are also useful where excessive weight or pressure is used on a door.

T Hinges. □ The *T hinge* is made by combining the features of the butt hinge and the strap hinge. One leg of the hinge is short and high like a butt hinge. The other leg is narrow and long like a strap hinge. The T hinge is useful where the mounting strip

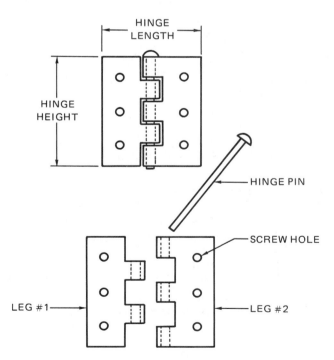

Figure 7–21.   A hinge has three parts: 2 legs and a pin. Leg 1 is interlocked with leg 2 using the hinge pin. Hinges may be purchased with removable pins for use on doors where the door is intended for occasional removal. However, many hinge pins are not removable.

for a door is thin but strong, and the door is flimsy and needs extra hinge support.

Screw Hook and Strap Hinges. □ The *screw hook and strap hinge* is also called a gate hinge. Two special screw hooks are screwed into a post or door jamb. The two straps are then mounted in the correct positions on the door. The door with hinge straps mounted is set or hung on the screw hooks. This completes the installation.

Hasps. □ The *hasp* is not really a hinge but is constructed like one and is sometimes classified with hinges. It is used to hold doors closed. One part of the hasp is attached to the door. The hinged part fits over a metal loop attached to the mating surface. A padlock or other device is placed in the loop to hold the door closed.

Brackets and Flush Plates

Hardware suppliers can provide many types of brackets, plates, door closers, door catches, and other hardware items for construction and maintenance. Special devices called *gussets* are available to reinforce joints in wood and metal. Wall brackets and special mounting channels with prefinished shelving are some hardware items which make construction fast and easy.

Some common devices used for strengthening corners of frames and doors are flush plates, figure 7–22. A *flush plate* is made of metal about $\frac{1}{16}$ inch thick and $\frac{1}{2}$ inch wide. They are drilled and have countersunk holes to receive flat head screws. Flush plates are available as straight, tee, inside, or flush corners.

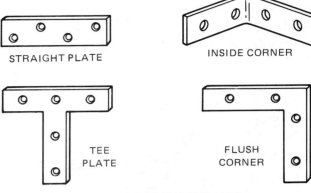

STRAIGHT PLATE

INSIDE CORNER

TEE PLATE

FLUSH CORNER

EXAMPLES OF FLUSH PLATES

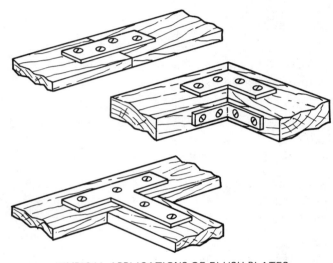

TYPICAL APPLICATIONS OF FLUSH PLATES

Figure 7–22. Flush plates are used to strengthen corners of frames and doors.

□ STUDENT ACTIVITIES

1. Study individual tools provided by the instructor and classify them according to their use.
2. Prepare a bulletin board with captions showing the major classes of tools: layout, cutting, boring, driving, holding, turning, digging and other. Cut pictures of tools from old issues of magazines and catalogs and place them under their caption on the bulletin board.
3. Examine the tool storage area in your agriculture department. On a sheet of paper, list all tools that you see. Place a letter after each tool to indicate its classification as follows: L = layout; C = cutting; B = Boring; Dr = Driving; H = Holding; T = Turning; D = Digging; and O = Other.
4. Make up a board with samples of various types of nails attached; properly label the nails.

5. Make up a board with samples of screws and bolts attached; properly label them.
6. Visit a hardware store and examine various nails, screws, bolts, and other fasteners. Ask the store manager for manufacturer charts or tables that describe fasteners.
7. Ask your instructor if there are some activities which you can do to improve the arrangement and storage of nails, screws, bolts, and other fasteners in the agriculture department.

□ SELF-EVALUATION

A. Multiple Choice. Select the best answer.
1. The use of hand tools is
   a. for those who cannot afford power tools
   b. for a limited number of highly specialized jobs
   c. primarily for engine and machinery mechanics
   d. the foundation of agricultural mechanics
2. Tools are generally classified according to
   a. use
   b. color
   c. construction
   d. origin
3. An example of a layout tool is
   a. claw hammer
   b. outside calipers
   c. hand saw
   d. plug cutter
4. Saws are classified as
   a. kerf tools
   b. push tools
   c. flexing tools
   d. cutting tools
5. Taps and dies are classified as
   a. holding tools
   b. digging tools
   c. cutting tools
   d. turning tools
6. Wrenches are classified as
   a. turning tools
   b. digging tools
   c. other tools
   d. cutting tools
7. The lowercase letter d is used to designate sizes of
   a. lumber
   b. screws
   c. nails
   d. bolts
8. Which is *not* the name of a type of nail?
   a. lumber

b. plaster board
c. roofing
d. duplex

9. The term *improved* means a nail
   a. is made of copper
   b. is easy to remove
   c. has a thick shank
   d. holds better

10. Screws are classified
    a. according to the material they hold
    b. by the metal from which they are made or the finish used
    c. by the shape of their heads
    d. all of these

11. The number of a screw refers to its
    a. diameter
    b. length
    c. head type
    d. use

12. The difference between a bolt and a screw is
    a. a bolt has threads
    b. a bolt has a nut
    c. a screw has a slotted head
    d. a screw is suitable for use in wood

13. A bolt used in wood that has a round head over square shoulders is a
    a. stove bolt
    b. machine bolt
    c. carriage bolt
    d. none of these

14. How many sides are there on a hexagon nut?
    a. four
    b. six
    c. eight
    d. twelve

15. A 4–40 × ½ machine screw
    a. has four threads per inch
    b. has forty threads per inch
    c. is four inches long
    d. comes four to the package

16. Which is *not* a type of hinge?
    a. butt
    b. strap
    c. T
    d. N

17. The hinge which contains a feature from each of the two other hinges is the
    a. butt hinge
    b. strap hinge
    c. T hinge
    d. N hinge

18. The best hinge to use if it is not to be seen is the
    a. butt hinge
    b. strap hinge

c. T hinge
d. N hinge

19. The best hinge for a very large and extra heavy door or gate is the
    a. butt hinge
    b. T hinge
    c. N hinge
    d. Screw hook and strap hinge

20. Corners of frames and doors may be strengthened by using a
    a. hasp
    b. butt hinge
    c. flush plate
    d. hook and eye bolt

B. Brief Answers. Briefly answer the following questions.

1. What is a hand tool?
2. Name and correctly spell 6 layout tools.
3. Name and correctly spell 6 cutting tools.
4. Name and correctly spell 6 boring tools.
5. Name and correctly spell 6 driving tools.
6. Name and correctly spell 6 holding tools.
7. Name and correctly spell 6 turning tools.
8. Name and correctly spell 6 digging tools.

C. Identification. Name, correctly spell and classify each of the following tools.

1.

2.

3.

4.

5.

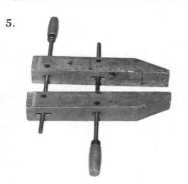

6.

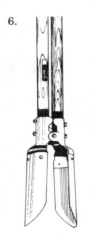

7.

8.

9.

10.

D. Identification. Name, correctly spell and classify as to type of fastener each of the following items.

1.

2.

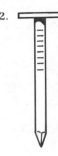

3.

4.

5.

6.

7.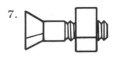

8.

9.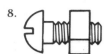

10.

2.

3.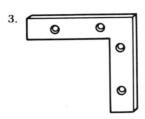

E. Identification. Name, correctly spell and classify each of the following items of hardware.

1.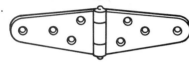

4.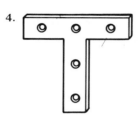

Unit 8
Layout Tools and Procedures

Objective

To select and use appropriate layout tools and procedures for wood and metal working.

Competencies to Be Developed

After studying this unit, you will be able to:
- Select appropriate tools for layout procedures in wood and metal working.
- Use layout tools correctly and accurately.
- Make and use a pattern.
- Lay out wood and metal for cutting and shaping.

Terms to Know

layout	gauge
gradations	framing square
scale	heel
aluminum	try square
plastic	combination square
craftsman	sliding T bevel
English system	bevel
U.S. Customary System	right triangle
linear	dividers
metric system	scribers
meter	level
centimeters	spirit level
millimeters	line
kilometer	chalk line
tape	plumb line
folding rule	pattern
caliper	

Materials List

Examples of tools or tool parts made from steel, aluminum, wood, plastic, and cloth
Examples of tools of poor quality that have been bent or broken with normal use
A meter stick and a 12-inch ruler
12-foot tape (with English and metric numbers preferred)
Steel or wooden scale
Folding rule
Inside calipers
Outside calipers
A variety of gauges (sheet metal, wire, drill, feeler, and spark plug)
Framing square
Combination square
Try square
Sliding T bevel
Dividers and scribers
Carpenter's level and line level
Plumb line
Chalk line
Tag board and scissors
String for 12-foot lines
Hammer and three slender 12-inch stakes

Layout tools are used to guide the worker when cutting, sawing, drilling, shaping, or fastening. The term *layout* means to prepare a pattern for future operations. A truck driver uses a map to determine the correct route between two points. The map requires the use of many roads, numerous turns, and changes in direction to get to the destination. The situation is similar when workers design and construct projects, machinery, equipment, or buildings. They place marks on their work and then cut, shape, dig, or fasten according to the marks.

Many workers must use layout tools in performing their work. These workers include plumbers, electricians, designers, engineers, surveyors, farmers, landscapers, builders, and many others. The student will do well to learn how to use layout tools carefully. Layout tools have many uses.

When laying out work it is important to choose the right tool. A second important factor is to use the tool correctly. For many jobs, there are several tools that will work, but usually one works better than others. For instance, a one-foot ruler or a yardstick may be used to measure a house that is 60 feet long. However, a 100-foot tape is quicker and more accurate.

☐ MATERIALS USED FOR LAYOUT TOOLS

Measuring devices are generally made of steel, aluminum, wood, plastic, or cloth. Steel tools are generally more durable than those made of other materials.

When numbers and lines, called *gradations*, are stamped into steel tools, they stay readable for a long time. Numbers and gradations make up a *scale* on measuring tools. Another advantage of steel is that it bends without breaking. Steel can stand the rough use found at construction sites. Steel is the first choice of material for calipers, dividers, and other slender tools that must measure very accurately and not bend easily.

Aluminum is the second choice of materials for use in many layout tools. Aluminum is a very tough metal that is light and durable. It is used extensively in levels and chalkers.

Wood is a cheap, relatively soft and light material. It is good for handles of layout tools. Wood is not used extensively in tools because it breaks easily and does not wear well. An exception is the wooden folding rule which is compact, useful, and popular. However, it is very easily broken if not handled correctly.

Plastic is a term used for a group of materials made from chemicals and molded into objects. Some plastics are tough and light. They can be molded into any shape and are cheap to produce. A major disadvantage of plastics is their tendency to melt if touched by a hot object or flame. They may also be damaged by solvents found in shops. Some layout tools such as rulers are made of plastic, but the use of plastic is limited.

Cloth is a material used in some 50-foot tapes. It is cheap and lightweight. Cloth tapes are not very accurate as they stretch under stress and moisture. Any stretching of a measuring device reduces its accuracy.

LIQUID MEASURE

2 cups	1 pint
2 pints	1 quart
4 quarts	1 gallon
31½ gallons	1 barrel
2 barrels	1 hogshead
7½ gallons of water	1 cu ft
U.S. Gallon	231 cu in

AVOIRDUPOIS WEIGHT

16 ounces	1 pound
100 pounds	1 hundredweight
20 hundredweight	1 ton

DRY MEASURE

2 pints	1 quart
8 quarts	1 peck
4 pecks	1 bushel

1 bushel contains 2150.42 cubic inches or approximately 1¼ cubic feet.

SQUARE MEASURE

144 square inches	1 square foot
9 square feet	1 square yard
30¼ square yards	1 square rod
160 square rods	1 acre (43,560 sq ft)
640 acres	1 square mile
1 square mile	1 section

WEIGHT

Gram	15.432 grains
Gram	.0353 ounce
Kilogram	2.2046 lb
Kilogram	.0011 ton (short)
Metric ton	1.1025 ton (short)
Grain	.064 gram
Ounce	28.35 grams
Pound	453.5 grams
Ton (Short)	907.18 kilograms
Ton (Short)	.907 metric tons
Ton (Short)	2,000 lb

LINEAR MEASURE

12 inches	1 foot
3 feet	1 yard
5½ yards (16½ feet)	1 rod
320 rods	1 mile
1760 yards (5,280 feet)	1 mile

CUBIC MEASURE

1728 cubic inches	1 cu ft
27 cubic feet	1 cu yd
128 cu ft (8′ × 4′ × 4′)	1 cord
1′ × 1′ × 1″	1 bd ft

LIQUID DATA

1 gallon of water weighs 8.34 pounds
1 gallon of milk weighs 8.6 pounds
1 cubic foot of water is equal to 7.48 gallons
1 inch of rainfall amounts to 27,154 gallons per acre

Figure 8–1. The English system, now known as the U.S. Customary System, has long been the standard of weights and measures in the United States.

A COMPARISON OF THE INTERNATIONAL METRIC SYSTEM AND THE ENGLISH SYSTEM OF MEASUREMENT

1 Centimeter = .3937 inch	1 Kilometer = 1000 meters	1 Gallon = 3.785 liters
1 Inch = 2.54 centimeters	1 Kilometer = .62137 mile	1 Gram = 15.43 grains
1 Foot = 30.48 centimeters	1 Sq. Centimeter = .155 Sq. inch	1 Ounce = 28.35 grams
1 Meter = 39.37 inches	1 Sq. Decimeter = 100 cu. centimeters	1 Kilogram = 1000 grams
1 Meter = 100 centimeters	1 Cu. Centimeter = .061 cu. inch	1 Kilogram = 2.205 pounds
1 Meter = 1.094 yards	1 Cu. Decimeter = 1000 cu. centimeters	1 Pound = 7000 grains
1 Meter = 1000 millimeters	1 Cu. Meter = 100 liters	1 Pound = .4536 kilogram
1 Millimeter = .001 meter	1 Fluid Ounce = 29.54 milliliters	1 Kilogram = 1000 milliliters
1 Yard = .9144 meter	1 Liter = 1000 cu. centimeters	1 Kilogram = 1 liter
1 Mile = 1609.344 meters	1 Liter = 1.057 quarts	

METRIC PREFIXES

Names of multiples of Metric units are formed by adding a prefix to "meter," "gram," or "liter." These prefixes are used also with units other than these three metric ones.

tera	- T	10^{12}	1,000,000,000,000.	ONE TRILLION
giga	- G	10^{9}	1,000,000,000.	ONE BILLION
mega	- M	10^{6}	1,000,000.	ONE MILLION
kilo	- k	10^{3}	1,000.	ONE THOUSAND
hecto	- h	10^{2}	100.	ONE HUNDRED
deka	- dk	10^{1}	10.	TEN
—	-	10^{0}	1.	ONE
—	-	10^{0}	1.	ONE
deci	- d	10^{-1}	0.1	ONE TENTH
centi	- c	10^{-2}	0.01	ONE HUNDREDTH
milli	- m	10^{-3}	0.001	ONE THOUSANDTH
*micro	- μ	10^{-6}	0.000001	ONE MILLIONTH
nano	- n	10^{-9}	0.00000001	ONE BILLIONTH
pico	- p	10^{-12}	0.00000000001	ONE TRILLIONTH

* The special case of one millionth of a meter is called a micron.

CONVERSION TABLE

MULTIPLY	TO OBTAIN	MULTIPLY	TO OBTAIN	MULTIPLY	TO OBTAIN
Bushels by .8	Cu. Feet	Cu. Yards by 764600	Cu. Centimeters	Meters by 100	Centimeters
Bushels by 4	Pecks	Cu. Yards by 22	Bushels	Meters by 3.281	Feet
Bushels by .04545	Cu. Yards	Cu. Yards by 27	Cu. Feet	Meters by 39.37	Inches
Centimeters by 0.3937	Inches	Cu. Yards by 46.656	Cu. Inches	Meters by .01	Kilometers
Centimeters by 0.01	Meters	Cu. Yards by 0.7646	Cu. Meters	Meters by 1000	Millimeters
Centimeters by 10	Millimeters	Cu. Yards by 202.0	Gallons	Meters by 1.094	Yards
Cu. Centimeters by .00003531	Cu. Feet	Cu. Yards by 764.6	Liters	Microns by .00001	Meters
Cu. Centimeters by .06102	Cu. Inches	Cu. Yards by 1616	Pints (Liq.)	Miles by 160900	Centimeters

From × Factor = To	From × Factor = To	From × Factor = To
Cu. Centimeters by .00001 . Cu. Meters	Cu. Yards by 807.9 . Quarts (Liq.)	Miles by 5280 .... Feet
Cu. Centimeters by .0000001308 ... Cu. Yards	Fathoms by 6 .... Feet	Miles by 1.609 .... Kilometers
Cu. Centimeters by .0002642 .. Gallons	Feet by 30.48 ..... Centimeters	Miles by 1760 .... Yards
Cu. Centimeters by .001 .... Liters	Feet by 12 .... Inches	Miles per Hr. by 44.70 .... Centimeters per Sec
Cu. Centimeters by .002113 Pints (Liq.)	Feet by 0.3048 .... Meters	Miles per Hr. by 88 ...... Feet per Min.
Cu. Centimeters by .001057 ... Quarts (Liq.)	Feet by ⅓ .... Yards	Miles per Hr. by 1.467 ... Feet per Sec.
Cu. Feet by .0002832 .. Cu. Centimeters	Gallons by 3785 .... Cu. Centimeters	Miles per Hr. by 1.609 ..... Kilometers per Hr.
Cu. Feet by 1728 .... Cu. Inches	Gallons by 0.1337 .... Cu. Feet	Miles per Hr. by 0.8684 .... Knots
Cu. Feet by 0.02832 ..... Cu. Meters	Gallons by 231 .... Cu. Inches	Miles per Hr. by 26.82 . Meters per Min.
Cu. Feet by 0.03704 ..... Cu. Yards	Gallons by .003785 .... Cu. Meters	Millimeters by 0.03937 .... Inches
Cu. Feet by 7.48052 .... Gallons	Gallons by .004951 .... Cu. Yards	Ounces by 2 .... Tablespoons (Liq.)
Cu. Feet by 28.32 .... Liters	Gallons by 3.785 .... Liters	Ounces by 6 ...... Teaspoons (Liq.)
Cu. Feet by 59.84 .... Pints (Liq.)	Gallons by 8 .... Pints (Liq.)	Ounces by 3 .... Tablespoons (Dry)
Cu. Feet by 29.92 ..... Quarts (Liq.)	Gallons by 4 .... Quarts (Liq.)	Ounces by 9 ...... Teaspoons (Dry)
Cu. Feet by 1.25 .... Bushels	Gallons Water by 8.3453 Lbs. of Water	Ounces by 28.349527 .... Grams
Cu. Inches by 16.39 ... Cu. Centimeters	Grams by 0.03527 .... Ounces	Ounces by 0.9115 ...... Ounces (Troy)
Cu. Inches by .0005787 ...... Cu. Feet	Grams by 0.03215 .... Ounces (Troy)	Ounces (Fluid) by 1.805 ..... Cu. Inches
Cu. Inches by .00001639 .. Cu. Meters	Grams by .002205 .... Pounds	Pounds by 16 .... Ounces
Cu. Inches by .00002143 ... Cu. Yards	Inches by 2.540 .... Centimeters	Pounds of Water by 0.01602 .. Cu. Feet
Cu. Inches by .004329 .... Gallons	Kilometers by 100000 .... Centimeters	Pounds of Water by 27.68 .. Cu. Inches
Cu. Inches by .01639 .... Liters	Kilometers by 3281 .... Feet	Pounds of Water by 0.1198 ... Gallons
Cu. Inches by .03463 ...... Pints (Liq.)	Kilometers by 1000 .... Meters	Tablespoons (Liq.) by 0.5 ..... Ounces
Cu. Inches by 0.01732 .... Quarts (Liq.)	Kilometers by 0.6214 .... Miles	Tablespoons (Dry) by 0.3333 .. Ounces
Cu. Meters by 10000 .. Cu. Centimeters	Kilometers by 1094 .... Yards	Tablespoons by 3 .... Teaspoons
Cu. Meters by 35.31 .... Cu. Feet	Liters by 1000 .... Cu. Centimeters	Teaspoons (Liq.) by 0.1666 ..... Ounces
Cu. Meters by 61.023 .... Cu. Inches	Liters by 0.03531 .... Cu. Feet	Teaspoons (Dry) by 0.1111 ..... Ounces
Cu. Meters by 1.308 .... Cu. Yards	Liters by 61.02 .... Cu. Inches	Teaspoons by 0.3333 ..... Tablespoons
Cu. Meters by 264.2 .... Gallons	Liters by .01 .... Cu. Meters	Temp (C) + 17.78 by 1.8 ..... Temp (F)
Cu. Meters by 1000 .... Liters	Liters by .001308 ..... Cu. Yards	Temp (F) − 32 by 5/9 ...... Temp (C)
Cu. Meters by 2113 .... Pints (Liq.)	Liters by 0.2642 .... Gallons	Tons (Long) by 2240 .... Pounds
Cu. Meters by 1057 .... Quarts (Liq.)	Liters by 2.113 .... Pints (Liq.)	
	Liters by 1.057 ...... Quarts (Liq.)	
	T (in) by W (ft) by L (ft) by Board Feet	

Figure 8–2. The metric system of measurement and factors for converting units in the English and metric systems.

Cost vs. Quality

Cheap tools are seldom a bargain. A cheap tool is likely to be inaccurate and unsatisfactory after little use. Good tools may seem expensive; however, they last a lifetime if used properly. High-quality tools are an excellent investment. Craftsmen use only good tools. A *craftsman* is a skilled worker. Good tools permit the worker to become a good craftsman if proper techniques are used. On the other hand, even skilled workers cannot do good work with poor tools.

□ SYSTEMS OF MEASUREMENT

The *English system* has always been the standard system of measurement in the United States. This system is now called the *U.S. Customary System*, figure 8–1. It uses the inch, foot, yard, rod, and mile as units of linear measure. *Linear* means in a line. The English system uses units of various size, based on similarity to a certain familiar object or part of the human body such as the foot or hand. Even today, horses are said to be so many "hands" high.

There are twelve inches in a foot, three feet in a yard, sixteen and one-half feet in a rod and 5,280 feet in a mile. The English system does not seem very logical.

In the early times, people could function satisfactorily with approximate units of measure such as a hand to measure the height of a horse. However, as the technological age evolved, measurements had to be standardized. Can you image the results if the human foot was used as the method for measuring when constructing buildings?

Today, the foot and most other linear units of measure are very precise lengths. If layout tools are constructed properly, a foot is the same length on all tools. Variations may be observed on poor-quality tools and tools that measure long distances such as tapes. The longer the instrument the greater the problem of inaccuracy due to stretching, expansion, and contraction. The foot is represented by one mark (') and the inch by two marks ("). Hence, 2'6" means two feet and six inches.

The *metric system* of measurement has always been used for scientific work in the United States, figure 8–2. Only recently has the metric system been used for nonscientific work. Linear measurements in the metric system are based on the meter. The *meter* (m) equals 39.37 inches. The meter contains 100 *centimeters* (cm) and 1000 millimeters (mm). One thousand meters equal a *kilometer* (km). Notice how the units of measure in the metric system relate to one another by multiples of ten. The metric system is mathematically logical and easy to use.

Figure 8–3. This tape has both U.S. Customary and metric units of measurement. The 1, 2, 3 values refer to inches. The 20, 30, 40 values refer to millimeters.

The millimeter, centimeter, meter, and kilometer are the most used metric units of linear measurement in agricultural mechanics. The inch, foot, yard, rod, and mile are the most useful units of linear measurement in the U.S. Customary system. The U.S. Customary system is still the most used system in the United States. However, the metric system is increasing in use and many layout tools use both systems, figure 8–3.

The Inch as a Unit of Measurement

The inch is the traditional unit of measurement for wood and metal working in the United States. It must be divided into smaller units to be useful for most applications. Some fine rules or scales may have as many as 32 marks per inch. Each mark is $\frac{1}{32}$ of an inch apart. One-sixteenth inch is more commonly used as the smallest unit on a rule. Lines of different lengths are used to show ½, ¼ and ⅛ of an inch on many measuring devices such as rules and squares, figure 8–4.

The Millimeter as a Unit of Measurement

The *millimeter* (mm) is slightly smaller than $\frac{1}{16}$ of an inch. It is a very convenient unit for linear measurement without using fractions. It has the advantage of being $\frac{1}{1000}$ of a meter and $\frac{1}{10}$ of a centimeter. This means one meter plus 250 millimeters equal 1¼ meters. It is more convenient to write 1¼ meters using a decimal rather than a fraction. Using the decimal, 1¼ meters reads 1.250 meters. Similarly, 1½ meters reads 1.500, or simply 1.5 meters. On many metric rules, each centimeter division contains 10 marks to represent millimeters, figure 8–5. Centimeters can be changed to millimeters by multiplying by 10. Simply adding a 0 to the centimeter's value obtains the millimeter value, e.g., 4 centimeters is 40 millimeters.

When working on small projects, measurements may be made entirely in millimeters to avoid the use of fractions or decimals. If a board is 1.5 meters in length, this value can be written as 1500 millimeters. To change meters to millimeters multiply by 1000.

□ MEASURING TOOLS

Measuring tools include tapes, rules, calipers, and other devices used to determine specific distances. They are used to measure length, width, height, depth, thickness, spacing, and clearances.

When measuring, the last number on the scale is read and then the fraction of inches or number of millimeters are added, depending on the scale. If very exact measurements are necessary, a thin tape is used or the measuring device is turned up on its edge to get an exact reading.

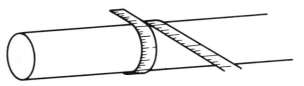

MEASURING THE CIRCUMFERENCE OF A ROUND OBJECT.

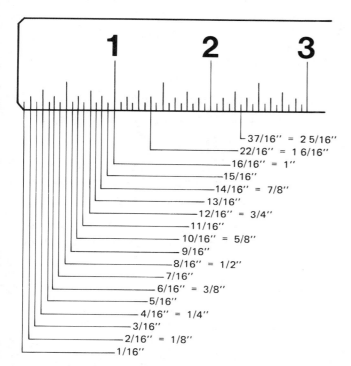

Figure 8–4. Rules, squares and other measuring devices are divided into halves, quarters, eighths, and sixteenths of an inch. These divisions are frequently illustrated by lines of different lengths.

37/16″ = 2 5/16″
22/16″ = 1 6/16″
16/16″ = 1″
15/16″
14/16″ = 7/8″
13/16″
12/16″ = 3/4″
11/16″
10/16″ = 5/8″
9/16″
8/16″ = 1/2″
7/16″
6/16″ = 3/8″
5/16″
4/16″ = 1/4″
3/16″
2/16″ = 1/8″
1/16″

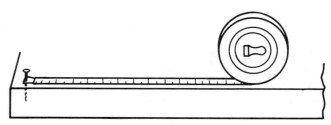

TAKING A MEASUREMENT STARTING AT A NAIL.

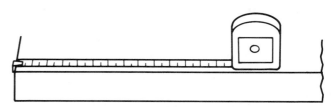

TAKING A MEASUREMENT STARTING FROM THE END OF AN OBJECT.

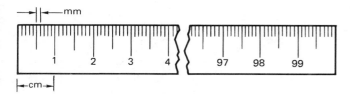

Figure 8–5. The marks that appear between each centimeter division are millimeters. A meter has 100 centimeters and 1000 millimeters.

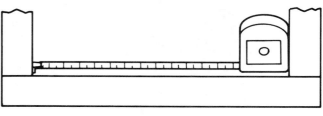

MEASURING THE DISTANCE BETWEEN TWO INSIDE SURFACES.

Figure 8–6. Tapes can be used to measure in various ways. The hook provided on one end of most tapes provides for accurate measurement whether it is hooked onto an object or pushed against a surface.

Tapes

The term *tape* refers to flexible measuring devices that roll into a case. They range in length from several feet to hundreds of feet and are made of steel, cloth, or fiber materials. Some tapes are self-retracting; they have a spring inside to wind up the tape when released. Some have locks to hold the tape out while in use. Some have a button on the side to control the movement of the tape.

Most tapes have a hook on the end which slides a distance equal to the thickness of the hook. This permits the tape to provide accurate measurements whether hooked to an object or pushed against a surface. Figure 8–6 shows various ways of measuring using a tape measure. Students must handle tapes carefully as they are easily broken. Tapes break if pulled out too far, if forced back into the case, or if bent excessively.

Folding Rules

A *folding rule* is a rigid rule of two to eight feet in length. It can be folded for handling and storage. At one time the folding rule was the standard measuring device used by carpenters. However, the compactness and length of modern tapes make them preferable to the folding rule for many jobs. Folding rules are made of wood, plastic, or metal.

Good quality wooden folding rules have spring loaded buttons to keep them open; these rules are

Figure 8–8. Wooden bench rules and steel scales are useful in the shop. They are not often used on job sites since they cannot be folded.

called spring-joint rules. A good wooden folding rule is made of tough wood and is easy to read. Some folding rules have a sliding metal insert in one end, figure 8–7. The insert permits accurate inside measurements and depth measurements.

Special care must be used in opening and closing folding rules. A careless twist can cause the rule to break. However, careful workers, who oil the joints periodically, may use the same folding rule day after day for many years without breakage.

Scales and Bench Rules

The term *scale* as used here means a rigid steel or wooden measuring device, figure 8–8. Scales are generally one to three feet in length and ¾ inch to 1 inch in width. Wooden scales are sometimes called bench rules. They are about ¼-inch thick. Metal scales are relatively thin and, therefore, better than wooden scales where very accurate work is required.

Scales and bench rules are handy in the shop. They are not used much on other jobs since they cannot be folded.

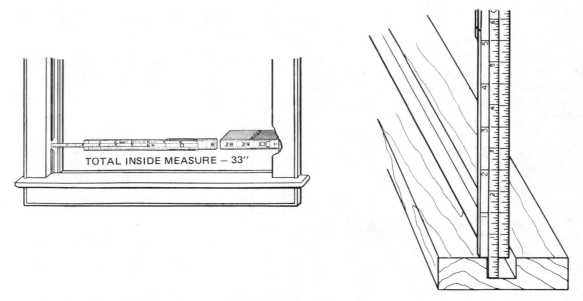

TOTAL INSIDE MEASURE — 33"

Figure 8–7. The insert in the end of a wooden folding ruler makes the ruler useful for accurate inside measurements (*courtesy of McDonnell & Kaumeheiwa, Use of Hand Woodworking Tools, Delmar Publishers Inc., 1978*).

Calipers

A *caliper* is an instrument used to measure the diameter or thickness of an object. There are inside calipers and outside calipers, figure 8–9. Inside calipers are so named because they measure inside distances such as the diameter of a cylinder. Outside calipers are used to measure the outside of round objects such as pipes.

With the most economical caliper a rule or scale may be needed to measure its settings. Other calipers have their own dimension scales. Another type of caliper has a dial to indicate very small differences in the size of objects measured.

Gauges

A *gauge* is a device used to determine the thickness, gap in space, or diameter of materials. Examples are sheet metal gauges, wire gauges, drill gauges, spark plug gauges, and feeler gauges.

☐ SQUARES AND THEIR USES

Squares are used to guide the builder. A square is a device used to draw angles for cutting and to check the cuts for accuracy. The squares most often used are the framing square, try square, combination square, and the sliding T bevel.

Framing Square. ☐ The *framing square* is a flat square with a body and tongue. It is usually made of steel. The body is 24 inches long and the tongue is 16 inches long. The *heel* is the place where the tongue and body meet to form a 90° angle.

The framing square is also called a carpenter's square and a steel square. It may contain several tables. One table found on many framing squares helps the user to calculate the board feet in a piece of lumber. Another table may help the user to lay out rafters. Special booklets and chapters in advanced texts provide more information on the use of these tables.

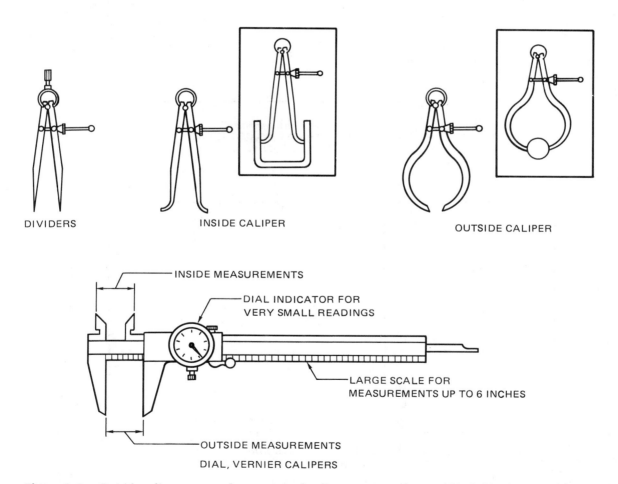

DIVIDERS   INSIDE CALIPER   OUTSIDE CALIPER

INSIDE MEASUREMENTS

DIAL INDICATOR FOR VERY SMALL READINGS

LARGE SCALE FOR MEASUREMENTS UP TO 6 INCHES

OUTSIDE MEASUREMENTS

DIAL, VERNIER CALIPERS

Figure 8–9. Outside calipers are used to measure the distance across the outside of objects. Inside calipers are used to measure the inside of objects.

Framing squares have measurements that start at the inside of the heel and run out the body and tongue. Inside measurements of the tongue are 1 to 14 inches. Inside measurements of the body are 1 to 22 inches. Another set of measurements starts at the outside of the heel and runs out the body and tongue. Outside measurements of the body are 1 to 24 inches. Outside measurements of the tongue are 1 to 16 inches. It is important to study the scales carefully before using a framing square. Figure 8–10 shows how the tongue and the body are used to measure different materials.

Try Square. □ The *try square* is so named because it is used to try or test the accuracy of cuts that have been made. It is also used to mark lines on boards in preparation for cutting. Try squares have wood, steel, or plastic handles and steel blades.

The try square is a good tool for marking lines on boards up to 12 inches wide. Small try squares are handy to mark lines across the edge or end of boards. They may be used to draw both 90° or 45° lines on boards, figure 8–11.

Combination Square. □ The *combination square* is so named because it combines many tools into one, figure 8–12. It contains a bubble that permits the tool to be used as a level. Another bubble permits it to be used for plumbing objects. An object is *plumb* when it is vertical to the axis of the earth or in line with the pull of gravity. In other words, when something is straight up and down it is said to be plumb.

The combination square contains a removable blade which can be used as a steel scale. When the blade is in place, it may be used as a depth gauge. The

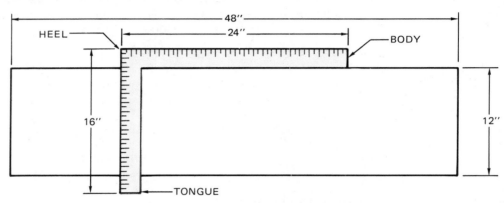

A. THE TONGUE OF THE FRAMING SQUARE IS USED TO SQUARE BOARDS UP TO 14 INCHES WIDE.

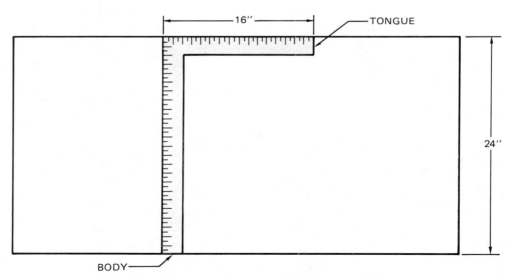

B. THE BODY OF THE FRAMING SQUARE IS USED TO SQUARE MATERIALS UP TO 24 INCHES WIDE.

Figure 8–10. The tongue and the body of the framing square are used to measure materials of different size.

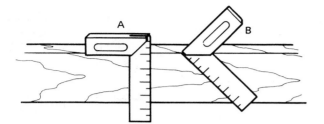

Figure 8–11. The try square can be used to draw 90° or 45° lines across the face, edge, or ends of boards. To mark a 90° angle, the try square is positioned as shown in A. To mark a 45° angle, the 45° shoulder of the try square is placed as shown in B. After cutting the board, the try square is used to check the exactness of the cut (*courtesy of McDonnell & Kaumeheiwa, Use of Hand Woodworking Tools, Delmar Publishers Inc., 1978*).

tool is very handy as an ordinary square and as a 45° miter square. A *miter* is an angle. Special heads make the tool useful for locating the center of round objects. The combination square is probably the most used square because of its many functions.

Sliding T Bevel. □ The *sliding T bevel* is a device used to lay out angles. It is also called a bevel square. A *bevel* is a sloping edge. If the corner is cut from a board, a bevel is created, figure 8–13. To use a sliding T bevel, first loosen the wing nut or thumb screw. Then adjust the blade to the desired angles and length, and retighten.

Using a Square to Mark a Board

A square is used when marking boards to be sawed exactly square or at 90° to the edge. When using a square to mark a board the following procedure is used:

Procedure

1. Measure and mark the length of board desired. The mark should be thin and at a 90° angle to the board's edge.
2. Place the handle of the square firmly against the edge of the board.
3. Move the square until the blade is against the mark.
4. Hold a sharp pencil or knife against the blade and draw a line through the mark and across the board, figure 8–14. The board is now ready to be sawed.

□ SQUARING LARGE AREAS

When working large areas such as laying out a building, care must be taken when squaring corners. An error as small as one degree will cause serious construction problems. One way to establish 90° corners is to create a right triangle with sides that are 3 feet, 4 feet and 5 feet in length. Any right triangle with sides that are multiples of 3 feet, 4 feet and 5 feet will work. A *right triangle* is a three-sided figure with one angle of 90°.

To lay out a 90° corner for a large area, use 6, 8, and 10 feet, respectively, and the following steps.

Procedure

1. Drive a thin stake at the point of the right angle or corner.
2. Lay the heel of a framing square against the stake.
3. Attach a piece of string and stretch it along the line of one leg of the square. Stake the other end of the string and call this line A.
4. Attach a second string at the point of the right angle and stretch it along the other leg of the square. Call this line B.
5. Measure along line A a distance of six feet. Tie a piece of string at this 6-foot mark.
6. Measure along line B and tie a piece of string at the 8-foot mark.
7. Measure the distance from the string tied on line A and the string tied on line B. The two points should be exactly 10 feet apart. If they are not, move either line A or B in or out until the distance is exactly 10 feet (within $\frac{1}{16}$ of an inch). The angle will then be square or 90°, figure 8–15.

A useful procedure for determining if four corners of a rectangle are square is to measure the distance diagonally from corner to corner. If all four corners are square, the diagonals will be equal in length, figure 8–16. To be considered equal, two measurements must be exactly the same.

□ DIVIDERS AND SCRIBERS

Dividers and scribers are instruments used to make circles and other curved lines. They are also used to transfer equal measurements from a scale or piece of material to another item. Dividers and scribers are very similar. One difference is that the scriber has one steel leg and a pencil for the other leg. The divider has two sharp steel legs, figure 8–17.

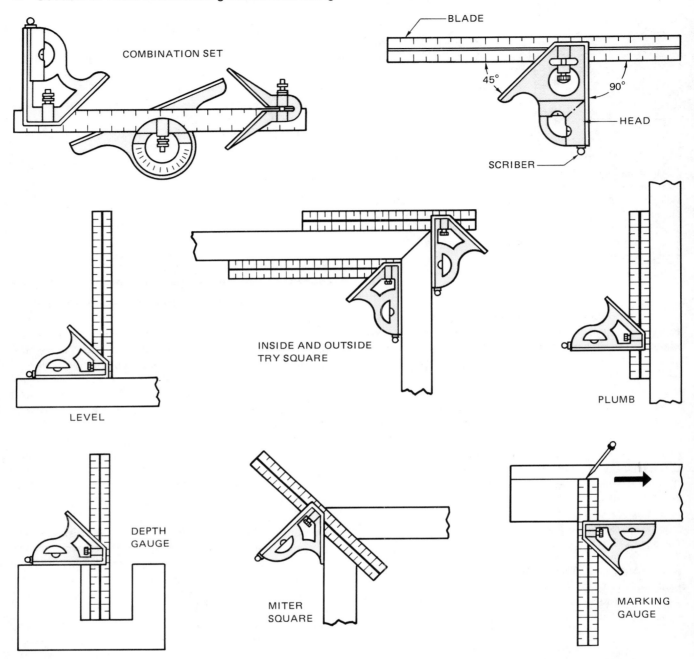

Figure 8–12. The combination square is many tools in one. It is a plumb, level, scriber, scale, depth gauge, marking gauge, miter square and regular square.

□ LEVELS

A *level* is a device used to determine if an object has the same height at two or more points. To determine this, a spirit level is generally used. A *spirit level* consists of alcohol in a sealed, curved tube with a small air space or bubble mounted in a bar of wood or aluminum. When both ends of the level are exactly the same height, the bubble flows to the high point of the tube. This point is indicated by the bubble fitting exactly between two lines marked on the glass. If the tube is mounted cross ways in the bar, the level may be used to plumb an object.

Levels may be made to attach to a string or line stretched between two distant points. This is called a line level. When the string is pulled tight, the level

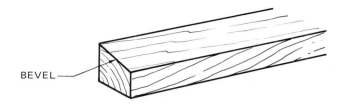

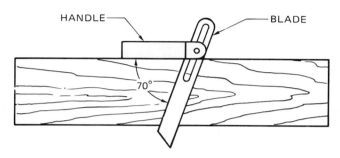

Figure 8–13. A bevel is a sloping edge made on boards by cutting down a corner. The sliding T bevel is used to lay out bevels and other angles.

indicates when both ends are the same height, figure 8–18. Line levels are useful for leveling the corners when building with block and setting stakes to gauge the depth of concrete.

□ LINES

Strong cotton or nylon string is used as a guide when laying block or flooring, cutting rafters, applying ceiling tiles and other jobs. A *line* is any thin material stretched tightly between two or more points.

A *chalk line* is a cord with chalk applied by rubbing or by drawing the line through powdered chalk. When stretched and snapped, the line leaves a thin trail of chalk to mark straight lines up to 50 feet or more.

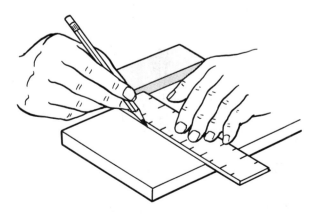

Figure 8–14. Holding the square against the mark, a line is drawn through the mark across the board.

A *plumb line* is a string with a round and pointed piece of metal called a *plumb bob* attached. The plumb line hangs perfectly vertical. It is useful when building greenhouses, sheds, barns, and other buildings to check for plumbness of walls and structures.

□ USING PATTERNS

Layout tools and instruments can be used to create almost any object. However, much time may be saved by copying the shapes of objects or designs rather than creating them. A pattern is useful for copying objects. A *pattern* is a model or guide from which to make an object.

In order to build a wall a carpenter may cut one 2″ by 4″ piece of lumber exactly 7 feet 6 inches long with both ends square. This piece will then be used as a pattern to mark and saw several dozen pieces exactly like the pattern.

In the agricultural mechanics shop, students may wish to develop simple woodworking skills by making a decorative cutting board as a project. The teacher may have a pattern made of a thin piece of masonite cut the shape of a pig or other object. The masonite is laid on a piece of wood and the outline of the pattern is drawn. Within minutes, the project is ready to be cut.

Individual designs for a project or project part can be created. When creating designs, it is always wise to begin by making a pattern out of paper. Several techniques may be used to make a pattern.

• *Tracing a pattern.* If a picture of the desired design is available, the pattern can be made by tracing. This is done by placing thin tracing paper over the picture and drawing an outline of the object. The tracing paper is then placed over a piece of heavy paper, such as tag board. Both pieces are then cut out at one time.
• *Using carbon paper.* Carbon paper may be placed between a picture and a piece of heavy paper. The picture is then outlined with a ballpoint pen and the image is transferred to the heavy paper. The paper is then cut into a pattern.
• *Drawing a pattern.* A pattern can be made by using a pencil, ruler, compass or other drawing instruments. A pattern can also be drawn free hand without instruments. If the object is to be the same on both the left and the right, the paper should be folded in half to make the pattern symmetrical. One-half of the object is then drawn. By cutting the doubled paper, a pattern is obtained with both halves being exactly alike. This procedure is especially helpful if the pattern has curved lines.

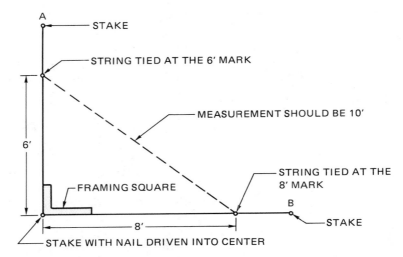

Figure 8–15. The 6, 8, 10 method is useful in establishing a square corner in a large area. The framing square provides a means to place the strings at approximate right angles to each other. Measurements of 6 feet, 8 feet, and 10 feet provide more accuracy in placing stakes A and B to create an angle that is exactly 90°.

Enlarging a Pattern

If a pattern is smaller than desired, an enlarged version can be made. First, a grid of vertical and horizontal lines is drawn over the original pattern or a copy of the pattern. The lines must all be spaced the same distance apart. One-quarter inch spacings work well for objects smaller than a standard sheet of paper.

A similar grid is then drawn on a piece of heavy drawing paper, but the space between lines in the

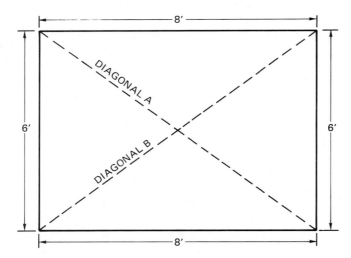

Figure 8–16. When the length of diagonal A is the same as the length of diagonal B, all corners are square if the figure is a rectangle. To be a rectangle, the two sides must be of equal length and the two ends must be of equal length.

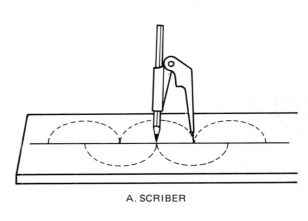

A. SCRIBER

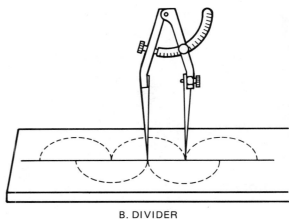

B. DIVIDER

Figure 8–17. Both scribers and dividers are used to mark circles. Scribers have one steel leg and a pencil. Dividers have two steel legs which are useful for marking dark metal.

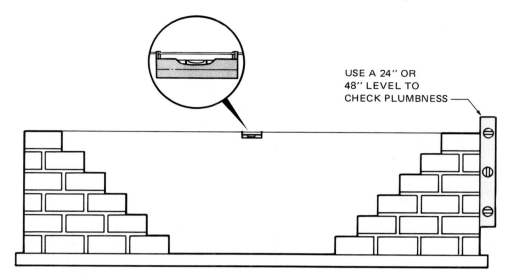

Figure 8–18. A level attached to a string is called a line level. It is used to check the levelness of two distant points. The corners of the walls shown are 16 feet apart. A carpenter's level may be used to check the wall for plumbness.

grid is increased. This increase is in proportion to the desired increase in size of the pattern. For example, suppose the original is a picture in a book that is 6 inches high and 4 inches wide. If the pattern is to be twice as large, the desired height would be 12 inches and the width 8 inches. In this case, the distance between the lines of the grid in the pattern is doubled. Suppose the picture has identical left and right sides, such as a Christmas tree. After the grid on the paper is prepared, the paper is folded in half. Observe how the lines of the original picture cut across the grid. An outline of half of the tree is drawn onto the new grid using the corresponding lines and squares of the original to guide the drawing on the pattern. The pattern is then cut out and unfolded. It should be a perfect enlargement of the original, figure 8–19.

An alternate method is to use an overhead or opaque projector to project an enlarged pattern onto paper. The image is then marked with a pencil or pen.

□ **STUDENT ACTIVITIES**

1. Obtain a steel tape, steel or wooden scale, combination square, try square, and framing square. Examine the tools and record the information needed to complete the following table.

Tool	Basic Units of Measure	Marks Per Unit	Length
a. Steel tape	Inch/Foot	16 per inch	12 inch
b. Steel or wooden scale	_____	_____	_____
c. Combination square	_____	_____	_____
d. Try square	_____	_____	_____
e. Framing square	_____	_____	_____

2. Using a ruler, draw 10 straight lines 5 inches long on a piece of paper. Place short marks across each line at distances exactly as follows:
   a. Line 1: Distance between marks = 3″
   b. Line 2: = 4½ inches
   c. Line 3: = 2⅛ inches
   d. Line 4: 2¾ inches
   e. Line 5: 3⅞ inches
   f. Line 6: ¹⁵⁄₁₆ inches
   g. Line 7: 7 centimeters
   h. Line 8: 84 millimeters
   i. Line 9: 7.3 centimeters
   j. Line 10: 7 centimeters and 3 millimeters

3. Use a tape and measure the length and width of your classroom, your desk top, and the classroom door.

4. Examine a folding rule. Carefully unfold and refold the rule being careful to swing each section open and closed, using the spring joint as a hinge.

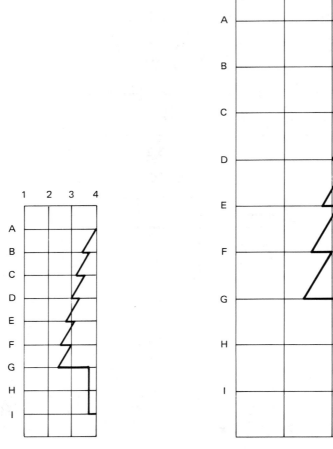

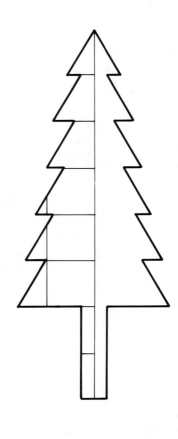

STEP 1: DRAW A GRID OVER THE ORIGINAL PICTURE. USE 1/4" SPACES BETWEEN LINES. NUMBER AND LETTER THE LINES IN THE GRID; THEN FOLD THE ORIGINAL IN HALF.

STEP 2: FOLD A PIECE OF TAG BOARD IN HALF. ADD A GRID USING 1/2" SPACINGS BETWEEN LINES. DRAW THE TREE ONTO THE ENLARGED GRID AND CUT OUT THE TREE AS OUTLINED.

STEP 3: THE PATTERN IS NOW READY FOR DRAWING THE OUT-LINE ONTO WOOD OR METAL. THE GRID LINES ON HALF OF THE PATTERN MAY BE IGNORED.

Figure 8–19. A pattern of any size may be made from a drawing of an object such as a Christmas tree. In this example, a pattern is made which doubles the size of the original picture.

5. Examine inside calipers and outside calipers. Measure the inside and outside diameter of a round object such as a can, pipe fitting, or piece of pipe.

6. Obtain a square of your choice and a board that is 6 inches to 10 inches wide. Draw a line across the board that is straight and square to the edge of the board.

7. Obtain a 12 inch, 24 inch, or 48 inch level. a) Place the level on the window sill in the classroom, shop or other room. Is the window sill level? b) With the level resting on the window sill, lift the right end of the level. Which way did the bubble move? c) Lift the left end of the level. Which way did the bubble move? d) Place the level in a vertical position against the classroom or shop door. Is the door plumb? Do all bubbles in the level read alike? e) With the level still in a vertical position against the door, hold the bottom of the level against the door but move the top away from the door. Which way did the bubble move? f) Hold the top of the

level against the door and move the bottom away. Which way did the bubble move?

8. Use a combination square to:
   a) draw a 90° line on a board
   b) draw a 45° line on a board
   c) measure the width of a board
   d) measure the depth of a groove
   e) test the levelness of a bench

9. Create a 90° corner in the lawn using the 6', 8', 10' method. You will need three stakes, a hammer, two lengths of string each 12 feet long, two string ties, a framing square and a 12-foot tape.

10. Use a scriber to create a 4-inch circle on a piece of paper. Use a divider to create a 3-inch circle on a scrap piece of wood or metal.

11. Prepare a pattern for some small object which you would like to make out of wood or metal.

□ SELF-EVALUATION

A. Multiple Choice. Select the best answer.

1. On most rules and tapes used in the shop, the shortest line represents
   a. ⅛ inch
   b. 1⁄16 inch
   c. 1⁄64 inch
   d. 1 inch

2. Of the following, the smallest unit of metric measurement is
   a. meter
   b. millimeter
   c. kilometer
   d. centimeter

3. Tapes generally break due to
   a. frequent use
   b. use of spring retractors
   c. use outdoors
   d. forcing the tape back into the case

4. Wooden scales are sometimes called
   a. depth gauges
   b. bench rules
   c. marking gauges
   d. scribers

5. The best tool for measuring the outside diameter of a pipe in its middle section is the
   a. try square
   b. tape
   c. calipers
   d. dividers

B. Matching. Match the word or phrase in column I with the correct word or phrase in column II.

Column I	Column II
1. high quality tools	a. m
2. English system	b. cm
3. metric system	c. mm
4. millimeter	d. board foot table
5. centimeter	e. km
6. meter	f. good investment
7. kilometer	g. yards, feet, inches
8. framing square	h. check high structures
9. combination square	i. meters, centimeters, millimeters
10. plumb line	j. includes a sliding scale and level
11. line level	k. requires tight string
12. 90°	l. used to lay out angles
13. body, tongue and heel	m. right triangle
14. sliding T bevel	n. dividers
15. 6', 8', 10'	o. bubble
16. scriber	p. grid
17. glass tube and alcohol	q. framing square
18. tracing paper	r. right angle
19. pattern enlargement	s. contains a pencil
20. makes a circle	t. pattern making

C. Completion. Fill in the blanks with the word or words that will make the following statements correct.

1. When laying out work it is important to choose the right tool and to use the tool _____.

2. The materials tools are made of include _____, _____, _____, _____, and _____.

3. Tools used to guide workers when cutting, sawing, drilling, shaping, or fastening are called _____ tools.

4. The official system of measurement for non-scientific purposes in the United States is the _____ system.

5. The metric system of measurement is regarded as a logical system because its units of measure relate to each other in multiples of _____.

6. Use the following drawing and read the rule to the nearest one-eighth inch.

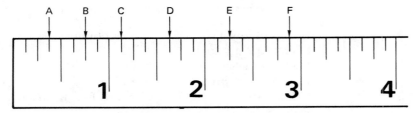

A _____

B _____

C _____

D _____

E _____

F _____

7. Use the following drawing and read the rule to the nearest one-sixteenth inch.

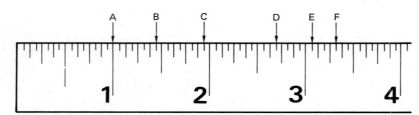

A _____

B _____

C _____

D _____

E _____

F _____

Unit 9
Selecting, Cutting, and Shaping Wood

Objective

To select wood and use tools and procedures in cutting and shaping wood.

Competencies to Be Developed

After studying this unit, you will be able to:
- Name and correctly spell 12 species of lumber that may be used in agricultural mechanics shops.
- Select lumber for agricultural projects.
- Determine the most useful hand tools for cutting and shaping wood.
- Cut and shape wood.

Terms to Know

wood	bore
lumber	drill
annual rings	bit brace
species	tang
kiln dried	chuck
warping	hand drill
planer	push drill
nominal	automatic drill
handsaw	clockwise
cross cut	counterclockwise
rip	plane
grain	plane iron
crosscut saw	chamfer
rip saw	file
kerf	rasp
back saw	file card
miter box	dado
coping saw	rabbet
keyhole saw	perpendicular

Materials List

Twelve specified species of lumber
Planed and rough lumber
Dried and green lumber
Samples of S2S, S4S, and sanded lumber
Samples of standard cuts of lumber
Combination square
Handsaws of various kinds
Boring tools
Planes
Files
Miter box
¾-inch wood chisel
Sandpaper

The agricultural mechanics student should know some basic information about wood and its source, figure 9–1. *Wood* is the hard, compact fibrous material that comes from the stems and branches of trees. Nutrients flow up and down a tree trunk in bundles of tubelike tissue that are formed just inside the bark of trees each year. It is this yearly addition of new bundles that causes a tree to have annual rings. *Annual rings* are patterns caused by the hardening and disuse of the bundles; they form the grain in wood. Annual rings can be seen by looking at the end of a tree trunk or tree stump.

□ CHARACTERISTICS OF WOOD AND LUMBER

It is important to know the characteristics of wood from various species of trees, figure 9–2. The word *species* means plants or animals with the same permanent characteristics. Some woods are hard and some are relatively soft. Some woods resist rotting better than others. Some are very attractive while other woods are plain. Some woods are stronger and tougher than others.

When building with wood, it is important to select the correct type. Common species of trees used for fine furniture are oak, walnut, maple, cherry, and mahogany. Other woods commonly used for furniture are white pine, red cedar and poplar. Wood used for building construction includes white pine, yellow pine, hemlock, fir, redwood, cypress, and oak. Each species has its advantages and shortcomings.

HOW A TREE GROWS

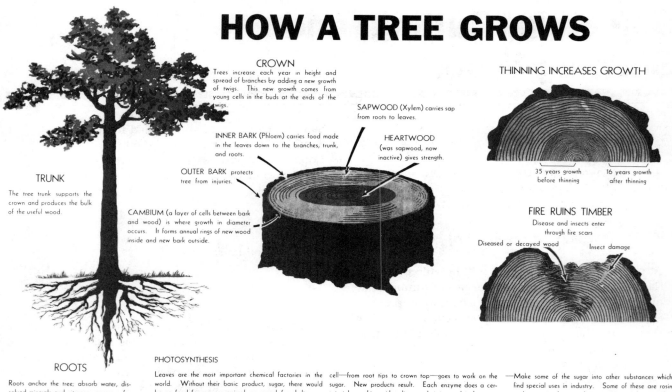

CROWN
Trees increase each year in height and spread of branches by adding a new growth of twigs. This new growth comes from young cells in the buds at the ends of the twigs.

SAPWOOD (Xylem) carries sap from roots to leaves.

INNER BARK (Phloem) carries food made in the leaves down to the branches, trunk, and roots.

HEARTWOOD (was sapwood, now inactive) gives strength.

OUTER BARK protects tree from injuries.

TRUNK
The tree trunk supports the crown and produces the bulk of the useful wood.

CAMBIUM (a layer of cells between bark and wood) is where growth in diameter occurs. It forms annual rings of new wood inside and new bark outside.

THINNING INCREASES GROWTH

35 years growth before thinning

16 years growth after thinning

FIRE RUINS TIMBER
Disease and insects enter through fire scars

Diseased or decayed wood

Insect damage

ROOTS
Roots anchor the tree; absorb water, dissolved minerals and nitrogen necessary for the living cells which make the food; and help hold the soil against erosion. A layer of growth cells at the root tips makes new root tissue throughout the growing season.

PHOTOSYNTHESIS
Leaves are the most important chemical factories in the world. Without their basic product, sugar, there would be no food for man or animal, no wood for, shelter, no humus for the soil, no coal for fuel.

Inside each leaf, millions of green-colored, microscopic "synthetic chemists" (chloroplasts) manufacture sugar. They trap radiant energy from sunlight for power. Their raw materials are carbon dioxide from the air and water from the soil. Oxygen, a byproduct, is released. This fundamental energy-storing, sugar-making process is called photosynthesis.

What happens to this leaf-made sugar in a tree? With the aid of "chemical specialists" (enzymes), every living cell—from root tips to crown top—goes to work on the sugar. New products result. Each enzyme does a certain job, working with split-second timing and in harmony with the others. In general, they break down sugar and recombine it with nitrogen and minerals to form other substances.

ENZYMES

—Change some sugar to other foods such as starches, fats, oils, and proteins, which help form fruits, nuts, and seeds.

—Convert some sugar to cell-wall substances such as cellulose, wood, and bark.

—Make some of the sugar into other substances which find special uses in industry. Some of these are rosin and turpentine from southern pines; syrup from maples; chewing gum from chicle trees and spruces; tannin from hemlocks, oaks, and chestnuts.

—Use some of the sugar directly for energy in the growing parts of the tree—its buds, cambium layer, and root tips.

TRANSPIRATION
Transpiration is the release of water-vapor from living plants. Most of it occurs through the pores (stomates) on the underside of the leaves. Air also passes in and out.

Figure 9–1. How a tree grows (*courtesy of the U.S. Department of Agriculture*)

Lumber Grade and Wetness

Lumber is wood that has been cut into boards. Lumber is graded according to its appearance and soundness. The fewer and less obvious the scars, knots, or blemishes, the more the lumber costs. For instance, clear pine shelving costs more than knotty pine shelving. Lumber with knots that are loose or missing costs less than lumber with tight knots. The lowest grade of lumber must simply hold together under ordinary handling.

When lumber is first sawed it is said to be green or wet. This is due to the high moisture content. Wood with too much moisture will split, bend, cup, and twist if it is not dried evenly. This is called *warping*. Excessive moisture will evaporate from lumber if narrow strips of wood called stickers are placed between all boards in a stack. The process is called air drying and takes about a year. Moisture can be driven from lumber within several weeks if placed in special ovens called kilns. The lumber is then said to be *kiln dried*. Kiln dried lumber must be used when constructing furniture or the wood pieces will warp or split. Warping can also be controlled by constructing an object quickly and then sealing the wood to prevent the escape of moisture.

Lumber Finish

Lumber is generally purchased with all sides and edges smooth. The *planer* is a machine that cuts lumber down to an exact size and leaves it smooth. Lumber may be purchased in the following four categories.

- **Rough**–not planed; delivered as it comes from the sawmill; width and thickness varies from piece to piece.

[handwritten: White Ash / HARD / Looks Similar to OAK / Furniture Oares & Baseball bats]

SPECIES	HARDNESS	KNOWN FOR	SOME MAJOR USES
Birch	Hard	Surface veneer for panels	Cabinets and doors
Cedar, red	Medium	Pleasant odor	Furniture, chests and closet linings
Cherry	Hard	Red grain	Fine furniture
Cypress	Medium	Rot resistance	Structural material in wet places
Fir and Hemlock	Soft	Light, straight, strong	Construction framing, siding, sheathing
Locust, black	Hard	Rot resistance	Fence posts
Maple	Hard	Light grain	Floors, bowling alleys, durable furniture
Mahogany	Medium	Reddish color	Fine furniture
Oak	Hard	Toughness, strength	Floors, barrels, wagon bodies, feeders, farm buildings
Pine, yellow	Medium	Wear resistance, tough	Floors, stairs, trim
Pine, white	Soft	Easy to work, straight	Shelving, siding, trim
Redwood	Soft	Excellent rot resistance	Yard posts, fences, patios, siding
Walnut, black	Hard	Brown grain	Fine furniture
Willow, black	Soft	Brown grain, easy to work, walnut look	Furniture

Figure 9–2. Characteristics of common woods

[handwritten notes: Lumber - wood dressed or processed for use in building / - graded / - ungraded - local Sawmills / Commercial Lumber - wood that has been dressed & graded / - Boards - < 2" thick, > 2" wide - Timbers ≥5" THICK & wide / - Dimension lumber 2-5" thick, > 2" wide]

- **S2S**–surfaced two sides; all pieces are the same thickness; edges are not planed; widths vary.
- **S4S**–surfaced four sides; the sides and edges are planed to exact dimensions.
- **Sanded**–width and thickness dimensions are exact on all pieces; all surfaces are sanded.

Many agricultural students buy rough, green lumber to make livestock feeders, wagon bodies and other projects for the farm. The rough lumber is satisfactory for such uses and much cheaper than finished lumber.

Standard Lumber Sizes

Rough pieces of lumber, as they come from a sawmill, are not all the same size. What is called a 2 x 4 may be 2¼" x 4 or 4¼". When a 2 x 4 is planed, the actual size is only 1½" x 3½". The name given to cuts of lumber reflects the *nominal* or approximate size. The real or actual sizes are ½ inch less than the nominal sizes, figure 9–3.

☐ CUTTING AND SHAPING WOOD WITH SAWS

Wood has become quite expensive, therefore, much care should be taken when measuring and cutting. A good rule is to measure twice and cut once. If a piece of wood must be recut because of an error, it generally means a new piece of wood must be used or all pieces must be recut and the project made smaller.

Saws are used to cut boards to length and width. Saws are also used to cut curves, make holes, and cut panels to size. The more common saws are the handsaw, back saw, coping saw, and compass saw. The compass saw is very similar to the keyhole saw.

Handsaws

The *handsaw* is used to cut across boards or to rip boards and panels. The words *cross cut* mean to cut across the grain of a board. The word *rip* means to

STANDARD LUMBER SIZES

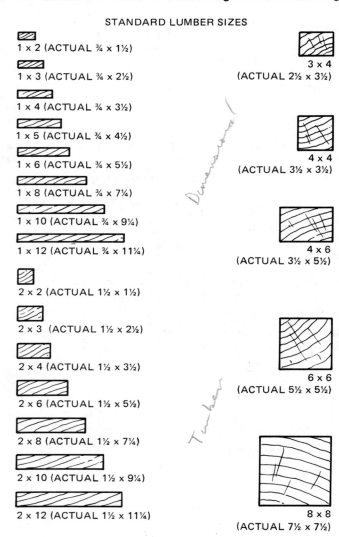

1 x 2 (ACTUAL ¾ x 1½)

1 x 3 (ACTUAL ¾ x 2½)

1 x 4 (ACTUAL ¾ x 3½)

1 x 5 (ACTUAL ¾ x 4½)

1 x 6 (ACTUAL ¾ x 5½)

1 x 8 (ACTUAL ¾ x 7¼)

1 x 10 (ACTUAL ¾ x 9¼)

1 x 12 (ACTUAL ¾ x 11¼)

2 x 2 (ACTUAL 1½ x 1½)

2 x 3 (ACTUAL 1½ x 2½)

2 x 4 (ACTUAL 1½ x 3½)

2 x 6 (ACTUAL 1½ x 5½)

2 x 8 (ACTUAL 1½ x 7¼)

2 x 10 (ACTUAL 1½ x 9¼)

2 x 12 (ACTUAL 1½ x 11¼)

3 x 4 (ACTUAL 2½ x 3½)

4 x 4 (ACTUAL 3½ x 3½)

4 x 6 (ACTUAL 3½ x 5½)

6 x 6 (ACTUAL 5½ x 5½)

8 x 8 (ACTUAL 7½ x 7½)

Figure 9–3. The actual thickness and width of planed lumber is less than its nominal size.

cut along the length of the board or with the grain. *Grain* in a board refers to lines caused by the annual rings in the tree.

The teeth of a handsaw determine how the saw should be used, figure 9–4. Teeth cut and filed to a point are designed to cut across the grain of boards. Such saws are called *crosscut saws.* Teeth filed to a knifelike edge are designed to cut with the grain. These saws are called *rip saws.*

Both crosscut and rip saws may be purchased in sizes 20 to 28 inches in length. The shorter saws are easier to use by smaller people. Shorter saws are frequently the choice for saws with small teeth designed for finer cuts.

Saws may be purchased with various tooth sizes. These range from 6 teeth per inch to 14 teeth per inch on handsaws. The tooth size is designated by the number of tooth points per inch. For example, an 8-point saw has 8 large teeth per inch of blade and is said to be a coarse saw. A 12-point saw has 12 small teeth per inch and is considered a fine saw. When sawing a board the saw removes wood and leaves an opening called a *kerf.*

When sawing wood, the material should rest on a solid work area or be held in a vise. Most carpenters use saw horses to support the material being sawed. The following procedure is recommended when using a crosscut saw or a rip saw.

Procedure

1. Select a piece of 1″ x 6″ lumber, several feet long.
2. Use a square and draw a thin pencil mark across the board one inch from the end.
3. Place the stock (board) on a saw horse with the marks over the end to the right. This assumes that the operator is right-handed. (Reverse if left-handed.) Hold the stock down on the saw horse with the left knee.
4. Grasp the saw handle in the right hand. The saw can be better controlled by placing the right forefinger along the outside of the handle.
5. Place the left hand on the board and use the thumb nail against the saw as a guide to start the cut.
6. Place the heel of the saw on the mark at the edge of the board and pull it toward your body to start the cut. Stand so your eyes are in line with the cut.
7. After the cut or kerf is started, push the saw forward, applying a light, downward pressure. Since the saw is designed to cut on the downward stroke only, apply no pressure on the return stroke. Operate the saw at a 45° angle (60° angle when ripping with a rip saw) to the surface of the board, figure 9–5.
8. Use long, slow strokes to complete the cut. The last few strokes should be very slow with no pressure on the saw.

The Back Saw

A *backsaw* is similar to a crosscut handsaw. It has very, very fine teeth and a stiff metal back from which it gets its name. These features make it an excellent tool for making very accurate cuts in smooth lumber. The back saw may be placed in a tool called a miter box to guide the saw, figure 9–6. A *miter box* is a device used to cut molding and other narrow boards at any desired angle. An adjustable miter box is operated by simply squeezing a lever and swinging the

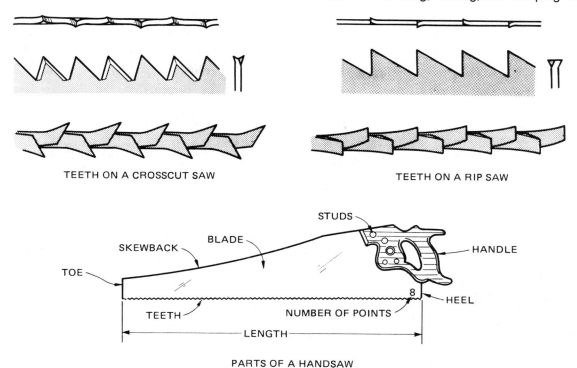

TEETH ON A CROSSCUT SAW

TEETH ON A RIP SAW

PARTS OF A HANDSAW

Figure 9–4. Handsaws are used for crosscutting or ripping, depending upon the shape of the teeth (*courtesy of McDonnell & Kaumeheiwa,* Use of Hand Woodworking Tools, *Delmar Publishers Inc., 1978*).

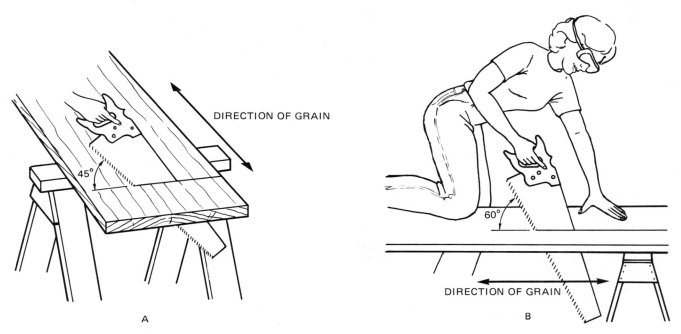

Figure 9–5. When using a crosscut saw, hold the saw at a 45° angle. When using a rip saw, hold the saw at a 60° angle. Start all cuts with short strokes, then use full, even strokes for good cuts. The saw kerf should be on the waste side of the line.

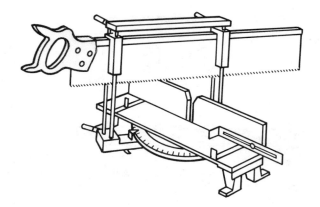

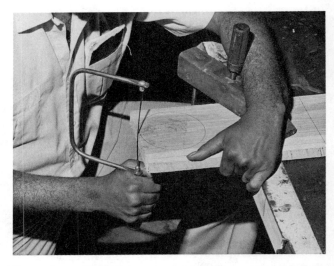

Figure 9–7. The coping saw is useful for cutting irregular shapes in wood. The blade can be removed from the frame, inserted into a hole, and reinstalled in the frame for the inside cutting.

Figure 9–6. The backsaw can be used alone or in a miter box. It is excellent for making very fine and accurate cuts.

saw to the desired angle. The lever is released and the saw locks at that angle. The wood is inserted, held firmly and then sawed.

The Coping Saw

The *coping saw* has a very thin and narrow blade supported by a spring steel frame. This design permits the blade to be removed from the frame, inserted into a hole, and placed back into its frame, figure 9–7. This makes it possible to use the saw to cut large holes or other shapes that are totally surrounded by wood. The coping saw is useful for cutting any kind of irregular, curved cuts in wood or other soft materials.

Compass and Keyhole Saws

The *compass saw* and *keyhole saw* are designed for making cuts starting from a hole, figure 9–8. A hole about one inch in diameter is bored and the slim

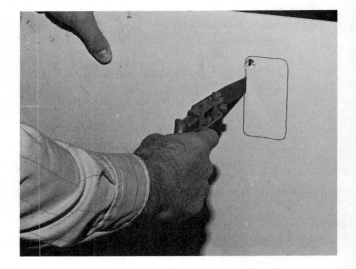

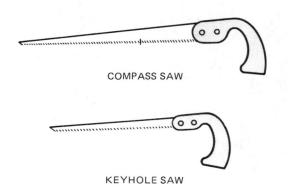

COMPASS SAW

KEYHOLE SAW

Figure 9–8. The compass and keyhole saws are used to make cuts starting from holes.

blade of the saw is inserted in the hole. The operator saws outward and follows the lines of the desired cut. The compass saw is similar to the keyhole saw, but its blade is wider at the base.

□ CUTTING AND SHAPING WOOD WITH BORING AND DRILLING TOOLS

The words *bore* and *drill* are often used to mean the same thing—to make a hole. However, the word boring generally refers to low speed, while drilling is a high speed operation. The most popular hand tools for boring and drilling are the bit brace, hand drill, and push drill.

The Bit Brace

The *bit brace* has a large crank-type handle. This permits the worker to turn bits which make holes up to 2 inches or more in diameter. Bit braces are also available with ratchet drives. A ratchet drive permits use of the tool in tight places such as against a wall. The handle drives the bit forward and ratchets backward for another stroke.

The jaws of the bit brace receive auger bits, expansive bits, forstner bits, screwdriver bits, and countersink bits, figure 9–9. These bits all have a tang. A *tang* is a tapered shank with four corners. A bit with a tang will not slip in the jaws of a bit brace.

The best hand tool for making holes larger than ¼ inch is the bit brace and auger bit, figure 9–10. Auger bits range in size from ¼ inch to one or more inches. The size is designated by a number equal to the diameter of the bit, in sixteenths of an inch. For example, a number 6 auger bit is ⁶⁄₁₆ inch or ⅜ inch in diameter.

The auger bit has a screw on its end to pull the bit into the wood, figure 9–11. Care must be taken not to damage this screw point. The bit also has four knifelike edges to cut into the wood. Moderate pressure is needed by the operator to keep the bit cutting. The cutting edges of the auger bit must be sharp. If the screw point is damaged or the cutting edges are not sharp, excessive pressure must be applied by the operator to make a hole. Both the screw point and the cutting edges will be ruined if the bit hits bolts, screws or nails in the wood. They will also be damaged if they strike hard surfaces such as a vise, metal bench, or concrete floor.

When boring a hole and the tip of the bit pierces the back side of the board, the operator should stop boring, figure 9–12. The hole should be completed by boring from the other side of the board. Completing a hole by boring from the back side of the board is a very important step when using an auger bit. This procedure prevents the bit from splitting the face of the board when it breaks through.

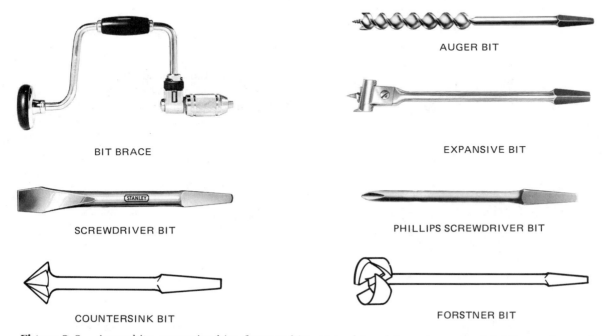

BIT BRACE

SCREWDRIVER BIT

COUNTERSINK BIT

AUGER BIT

EXPANSIVE BIT

PHILLIPS SCREWDRIVER BIT

FORSTNER BIT

Figure 9–9. Auger bits, expansive bits, forstner bits, screwdriver bits and countersink bits can be used with the bit brace (*photos courtesy of Stanley Tools*).

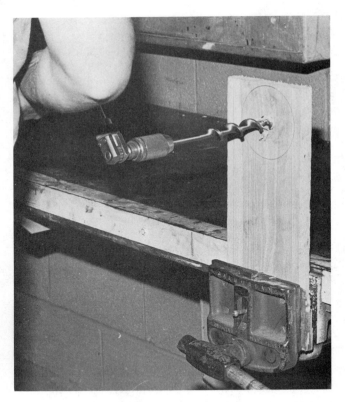

Figure 9–10. The bit brace and auger bit is the most popular hand tool for boring holes larger than ¼ inches in diameter.

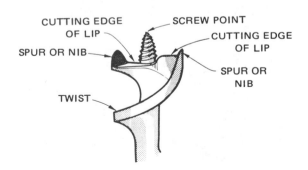

Figure 9–11. Parts of an auger bit (*courtesy of McDonnell & Kaumeheiwa, Use of Hand Woodworking Tools, Delmar Publishers Inc., 1978*).

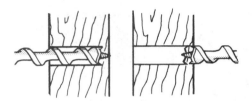

Figure 9–12. When the tip of the bit pierces the back side of the board, boring should be stopped. The hole should be completed by boring from the back side of the board. (*courtesy of McDonnell & Kaumeheiwa, Use of Hand Woodworking Tools, Delmar Publishers Inc., 1978*).

The Hand Drill

The *hand drill* is a device with gears that drive its bit much faster than the handle turns. This increased speed, however, means the turning power delivered to the bit is reduced. Therefore, the hand drill is useful only for making holes up to about ¼ inch in diameter in soft materials.

The hand drill must be held carefully to prevent injury to the hands and to assure that the hole is straight. Care must be taken to keep hands and clothing away from the turning cogwheels of the tool. When cogwheels come together, they can catch cloth or skin and severely pinch them between the teeth of the cogwheels.

The bit of the hand drill is held in place by a chuck. A *chuck* is a device with jaws that open and close to receive and hold bits with smooth shanks, figure 9–13. Bits with smooth shanks include the twist drill, countersink bit, and combination wood drill and countersink. In order to drill holes of uniform depth, a wooden dowel can be placed on the bit to act as a depth gauge, figure 9–14. If a hole is to be drilled the whole way through a board, pressure should be eased on the drill when the bit starts to show on the back side of the board. Easing pressure prevents the wood from splitting.

In order to remove the bit from a hand drill, the chuck is held with one hand. The crank is then turned backwards with the other hand, figure 9–15.

The Push Drill

The *push drill* is also called an *automatic drill*. It has a spiral-shaped shaft that turns clockwise when pushed against an object. *Clockwise* means turning in the direction that the hands of a clock turn. A spring causes the shaft to turn *counterclockwise* or backwards when the pressure is released. Therefore, the pumping of the handle causes the bit of the drill to turn.

The push drill is very fast and handy for drilling holes up to ³⁄₁₆ of an inch in diameter. It is generally used to make holes for wood screws. The bit for a push drill has a round shank with flat places on its end, figure 9–16. This enables the bit to lock quickly and securely in the chuck of the push drill. Bits are stored in the handle of most push drills, figure 9–17.

Downward pressure drives the bit of a push drill. Therefore, if the operator drills all the way through a piece of wood using normal pressure, the bit will split the wood when it breaks through. Such damage

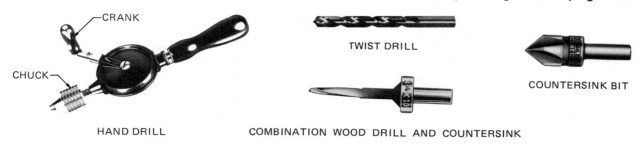

TWIST DRILL

COUNTERSINK BIT

CRANK

CHUCK

HAND DRILL

COMBINATION WOOD DRILL AND COUNTERSINK

Figure 9–13. The chuck of a hand drill receives bits with smooth shanks (*courtesy of Stanley Tools*).

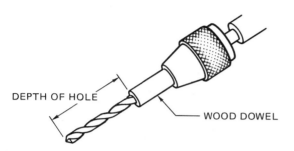

DEPTH OF HOLE

WOOD DOWEL

Figure 9–14. A wood dowel can be placed on the bit to act as a depth gauge in order to drill holes of uniform depth.

Figure 9–15. To remove a bit from the hand drill, hold the chuck with one hand and turn the crank backward with the other hand (*courtesy of McDonnell & Kaumeheiwa, Use of Hand Woodworking Tools, Delmar Publishers Inc., 1978*).

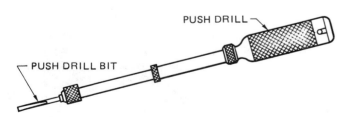

PUSH DRILL

PUSH DRILL BIT

Figure 9–16. Push drills have spiral shanks that turn the bit when pressure is applied to the handle. The operator's up and down motion keeps the bit turning at a high rate of speed.

is avoided if the push drill is used only for making holes to install screws.

The procedure for using a push drill is as follows.

Procedure

1. Mark the spot where the center of the hole will be.
2. Install the correct bit.
3. Place the point of the bit on the wood.
4. Hold the push drill at a 90° angle to the work.
5. Push the handle up and down until the hole is the desired depth.
6. Pull the drill bit out of the hole.

□ **SHAPING EDGES**    *SQUARING A BOARD*

Some projects have all straight edges and square ends. In this case a crosscut handsaw or backsaw is used to cut the board to length. The task of cutting and shaping wood is greatly simplified if S4S lumber is used. When surfaced on four sides it has smooth, straight edges from which to work. Planes are used to work the edges of boards.

A *plane* is a tool that shaves off small amounts of wood and leaves the surface smooth. The most common planes are the jointer, jack, smoothing, and block planes, figure 9–18. The jointer, jack, and smoothing planes require the operator to use both hands, figure 9–19. The block plane is small enough to be used with one hand.

The cutting part of a plane is called the *plane iron*. Plane irons must be very sharp and properly adjusted, figure 9–20. The procedure for adjusting a plane iron follows.

Procedure

1. Hold the plane upside down and look down the bottom of the plane from front to rear.
2. Move the adjusting lever or adjusting screws to set the plane iron. The plane iron should project

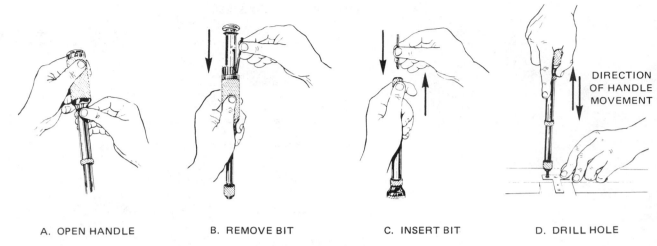

A. OPEN HANDLE   B. REMOVE BIT   C. INSERT BIT   D. DRILL HOLE

DIRECTION OF HANDLE MOVEMENT

Figure 9–17. Bits are stored in the handles of most push drills. This feature, plus the easy push action of the tool, makes it a favorite for making holes for screws (*courtesy of McDonnell & Kaumeheiwa, Use of Hand Woodworking Tools, Delmar Publishers Inc., 1978*).

less than ⅟₁₆ of a inch through the bottom of the plane. Both sides should be projecting the same distance.

3. Make a trial cut by pushing the plane along the edge to be planed. Plane with the grain, i.e., push the plane in the direction that makes the smoothest cut.

4. Readjust the plane until it cuts smoothly and a thin curl of wood is produced by each stroke.

□ CUTTING A BOARD TO WIDTH

The procedure for cutting a board to width involves several tools and steps. The following procedure is used when cutting boards to width. The board used as an example in the procedure is to be cut to a width of 4 inches.

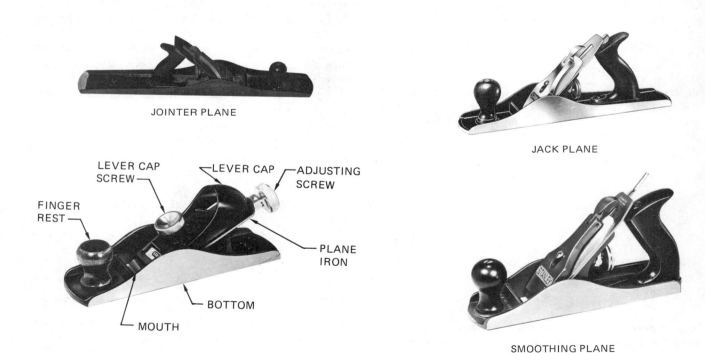

JOINTER PLANE

JACK PLANE

LEVER CAP SCREW   LEVER CAP   ADJUSTING SCREW

FINGER REST   PLANE IRON

BOTTOM

MOUTH

SMOOTHING PLANE

Figure 9–18. Common planes used for making straight, smooth edges (*courtesy of Stanley Tools*).

Figure 9–19. The jointer plane, jack plane, and smoothing plane require the operator to use both hands.

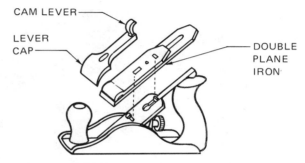

CLAMP THE LEVER CAP WITH THE CAM.

VARY THE CLAMPING PRESSURE BY TURNING THE LEVER CAP SCREW.

Procedure

1. Set a combination square so its blade extends out from the handle 4 inches.
2. Place the handle of the square against a smooth edge of the board.
3. Place a sharp pencil at the end of the blade.
4. Move the pencil and the square along the entire length of the board making a line, figure 9–21. The pencil should be held straight up and down and the angle not changed at any point along the way.
5. Reset the square at $4\frac{1}{16}$ inches and draw a second line.
6. Use a hand rip saw to saw the board. Make the saw kerf on the outside or waste side of the second line. The kerf should just touch the line.
7. Use a plane to smooth the saw cut. Lift the plane on each back stroke to avoid dulling the blade.
8. Check the edge for squareness.
9. Plane off any remaining wood until the edge is straight, smooth, square, and exactly down to the first line drawn.

□ MAKING FINAL CUTS IN CURVED EDGES

When cutting curved lines in wood and other soft materials a coping saw is used. A cut is made about $\frac{1}{16}$ inch outside the line. The saw must be kept moving exactly parallel to the line. After sawing out the project, a block plane, file, and sandpaper are used to work the wood down to the line. After using sandpa-

Figure 9–20. Adjusting the plane iron

per no cutting tool should be used. This is because sand particles in the wood will dull cutting tools.

A block plane is useful for gently curving outside lines. It is also good for cutting chamfers, figure 9–22. A *chamfer* results when a corner between the edge and face of a board is cut. A chamfer is similar to a bevel except that a bevel extends all the way across the edge of the board.

Using Surforms

A relatively new tool, called a Surform®, is available. Surform® is a trademark of Stanley Tool Works. The Surform® may be used instead of a plane or file, figure 9–23. It consists of several styles of handles and blades. The blades have sharp ridges that cut wood, plastic,

Figure 9–21. When marking the width of a board, the pencil and combination square are moved along the board together.

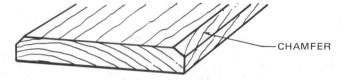

CHAMFER

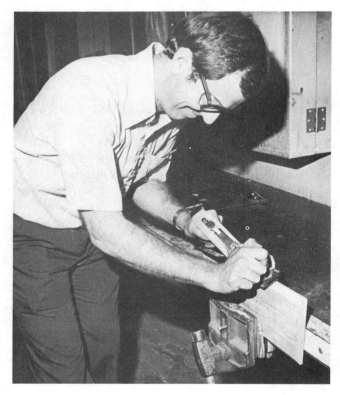

Figure 9–22. A plane is used to create a chamfer.

Figure 9–23. A Surform® can be used instead of a plane or file to shape edges of boards.

or body fillers. Small curls of material are produced much like a plane.

Surforms® are handy since they do not need to be adjusted or sharpened. The blade is simply replaced. Surforms® come in a variety of shapes. Some are useful for cutting material to form straight and flat surfaces. Others are useful for rounding edges and curves. A variety of surfaces can be produced by simply changing the direction of stroke, figure 9–24. Surforms® will remove large amounts of material if pushed at a 45° angle to the material. Pushing the tool in a straight line in the direction of the work produces a fine, smooth cut.

Using Files

Various files are used to dress down and finish off curved cuts on boards. Files are handy for smoothing edges and shaping materials to odd shapes, figure 9–25. A *file* may be flat, round, half-round, square, or three-sided. Files have teeth that cut wood or metal. Files designed for cutting wood and soft metals have coarse teeth. Those designed for cutting steel have smaller teeth. Most files are hard enough to cut either wood or metal. The *rasp* is a type of file with very, very coarse teeth. It is designed to be used only on wood and other soft materials. Rasps are good for making rough cuts where a lot of material must be removed.

When filing wood, the file must not be pushed across the edge of the board as this would tear down the corner and leave it ragged. Best results are obtained when the file is pushed lengthwise along the board. File teeth cut only when the file is pushed forward. Therefore, it is advisable to apply pressure on the forward stroke only. The file should be lifted on the return stroke. The teeth of the file may become filled with the material being filed. Material may be

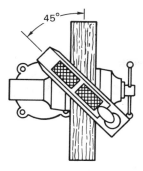

THE MAXIMUM AMOUNT OF MATERIAL
IS REMOVED BY PUSHING THE TOOL
AT A 45° ANGLE TO THE BOARD.

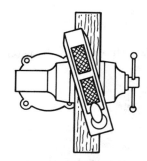

LESS MATERIAL IS REMOVED AND
A SMOOTHER SURFACE OBTAINED
BY REDUCING THE ANGLE.

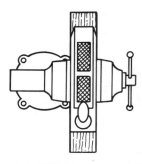

A SMOOTH SURFACE IS OBTAINED
BY PUSHING THE TOOL IN A
STRAIGHT LINE.

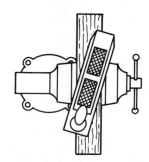

AN ALMOST POLISHED EFFECT IS
OBTAINED BY PUSHING THE TOOL
AT A SLIGHTLY REVERSED ANGLE.

Figure 9–24. By changing the direction of the stroke, Surforms® can produce a variety of surfaces. The maximum amount of material is removed by pushing the tool at a 45° angle to the board. Less material is removed and a smoother surface obtained by reducing the angle. A smooth surface is obtained by pushing the tool in a straight line. An almost polished effect is obtained by pushing the tool at a slightly reversed angle.

Figure 9–25. Files are handy for smoothing edges and shaping materials to odd shapes.

removed by tapping the handle up and down on a bench top. A special wire brush called a *file card* may also be used, figure 9–26.

A file has a sharp end called the tang. The tang is designed to go into a wooden or plastic handle. Files should not be used without handles. Serious injury to the hand can result from using a file without a

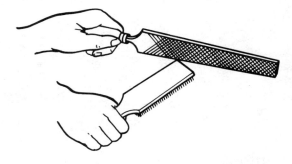

Figure 9–26. A wire brush called a file card is used to clean material from the teeth of a file.

handle. A file handle can be installed by driving the handle on the tang using a soft mallet. Another method is to place the tang in the handle and tap the handle up and down on the bench top until the tang is totally hidden in the handle.

Using Sandpaper

A small piece of sandpaper can be wrapped around a file for the finishing touches, figure 9–27. The file provides a thin, rigid core of the desired shape to support the sandpaper. The sandpaper is wrapped with the rough surface out.

To use the sandpaper and file, pinch the two lightly at both ends. Then, simply move the two as a unit. Sandpaper can be pushed across edges without splitting the corner of boards. Medium sandpaper should be used first and then very fine sandpaper can be used to finish.

□ CUTTING DADOS AND RABBETS

A *dado* is a square or rectangular groove in a board. The purpose of the dado is to receive the end or edge of another board to make a dado joint. A *rabbet* is a cut or groove made at the end or edge of a board to receive another board. The result is a rabbet joint when the two are fastened together. Dado and rabbet joints are used wherever strong 90° joints are needed, figure 9–28. To make a dado, the following procedure is used. (A rabbet is made using a similar procedure.)

Figure 9–27. A file wrapped with sandpaper can be used for finishing edges.

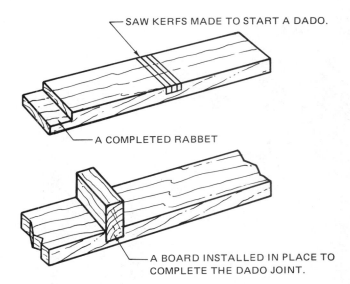

Figure 9–28. Dados and rabbets are made by making multiple saw cuts and finishing with a wood chisel and file. A groove made in the end and/or edge of a board to create a joint is called a rabbet. A groove made elsewhere in the board to create a joint is called a dado.

Procedure

1. Using a combination square, mark the face of the board with two lines ¾ inch apart to indicate the position and width of the dado.

2. Use a square to mark down across the edge of the board with lines perpendicular to the lines on the face of the board. *Perpendicular* means at a 90 degree angle.

3. Set the blade on the combination square to extend a distance equal to ½ the thickness of the board. This distance will be ⅜ inch for a board that is ¾ inch thick.

4. Lay the blade of the square against the edge of the board and mark the bottom of the dado by drawing a line at the end of the blade.

5. Mark the depth and sides of the dado on the other edge of the board.

6. Carefully saw four kerfs into the dado stopping with the kerf exactly at the bottom of the dado. This can be seen on both edges of the board. A backsaw or miter box saw should be used for this operation.

7. Use a wood chisel to remove the wood between the kerfs, figure 9–29. Be careful to chisel the wood to a uniform depth at the bottom of the kerf.

8. Smooth the bottom of the dado with a flat file. Be careful not to rock the file. Rocking will round the edges at the bottom of the dado and make a poor joint.

Figure 9–29. A wood chisel is used to remove the wood between the kerfs when making a dado.

□ STUDENT ACTIVITIES

1. Examine the stump of a recently cut treee. Count the number of rings in the stump. How old was the tree? Examine the distance between the rings. Wide distances indicate a year of good growth, perhaps plenty of rain.

2. Examine samples of oak, walnut, maple, cherry, mahogany, white pine, red cedar, poplar, fir, redwood, and cypress. Record the color and degree of hardness of each.

3. Learn to identify and correctly spell the various species of wood.

4. Examine lumber that is warped and has split ends. Try to explain the reason for the warping.

5. Describe the steps in preparing lumber from log to sanded board.

6. Measure the width and thickness of five different types of finished lumber. Record the actual size and nominal size of each.

7. Obtain a board 18 inches long and 5½ inches wide. Soft wood is recommended. Place five lines across the board ½ inch apart, near the end and square to the edge. Use a crosscut saw to saw off the piece nearest the end. Check the cut with a try square. Is it exactly square when checked in both directions? Saw and check each of the other four cuts, trying to make perfect cuts. Report the results to your instructor.

8. Use a combination square and mark a line ½ inch from the edge of a board one to two feet long. Rip the board on the side of the line closest to an edge. Check the cut for straightness and squareness. What are the results?

9. Use a plane to improve the edge of the board sawed in number 8. Now is the edge straight? Is it square?

10. Set a compass at one-half the width of the board used in number 9. Place the steel point on the board so the pencil can be swung around and just touch both edges and an end of the board. Swing the compass in an arc and mark a rounded end on the board. Saw the end round with a coping saw and finish it with a file or Surform®.

11. Using the board from number 10, bore a ½ inch hole with a brace and bit at the mark left by the steel compass point.

12. Using the board from number 11, remove ½ inch from the nonrounded end by using a miter box. Cut a rabbet in the end which is ⅜ inch deep and ¾ inch wide.

13. Measure and mark the board (used in number 11) at 3 inches and 3¾ inches from one end. Cut a dado ⅜ inch deep and ¾ inch wide.

14. Set a combination square so the blade extends ¼ inch. Choose one of the four corners of the board which you used in number 13. Place the square and draw a line on the edge of the board ¼ inch down from the corner you chose. The line should run the entire length of the board. Reposition the square and draw a line from one end to the other on the face of the board. Use a block plane or a smoothing plane to make a chamfer.

15. Discuss your workmanship in the preceding procedures with your instructor. Save the board for an exercise in Unit 10.

□ SELF-EVALUATION

A. Multiple Choice. Select the best answer.
1. Grain in lumber is caused by the
   a. age of the board
   b. annual rings
   c. special drying techniques
   d. stain
2. Lumber is graded according to its
   a. appearance and soundness
   b. color and species
   c. strength and durability
   d. cost and length

3. A crosscut handsaw with very coarse teeth would have how many teeth per inch?
   a. 6
   b. 10
   c. 12
   d. 14

4. A crosscut handsaw with very fine teeth would have how many teeth per inch?
   a. 6
   b. 8
   c. 14
   d. 20

5. The wood removed by a saw blade leaves an opening called a
   a. bevel
   b. channel
   c. chamfer
   d. kerf

6. The back saw gets its name from
   a. its use as a back-up tool
   b. its fine teeth
   c. its stiff back
   d. its original use in making chair backs

7. A suitable tool for cutting curves is the
   a. compass saw
   b. coping saw
   c. keyhole saw
   d. all of these

8. A bit brace uses
   a. an auger bit
   b. an expansive bit 2 twist bit
   c. a screwdriver bit
   d. all of these

9. The plane small enough to be used with one hand is the
   a. jointer plane
   b. jack plane
   c. smoothing plane
   d. block plane

10. The teeth on a file cut
    a. best when oiled
    b. only in soft materials
    c. only on the backward stroke
    d. only on the forward stroke

11. The groove cut across the end of a board to receive another board is called a
    a. dado
    b. rabbet
    c. miter
    d. ratchet

12. A file handle is necessary for
    a. operator protection
    b. file functioning
    c. compliance with the law
    d. none of these

B. Matching. Match the type of wood in column II to its major use in column I.

Column I	Column II
1. fence posts	a. red cedar
2. reddish, fine furniture	b. cypress
3. barrels, wagon bodies, farm buildings	c. fir and hemlock
4. shelving, siding, trim	d. black locust
5. floors, stairs, trim	e. mahogany
6. excellent rot resistance; patios	f. maple
7. structural material in wet places	g. oak
8. brown, fine furniture	h. white pine
9. floors and bowling alleys	i. yellow pine
10. chests and closet linings	j. redwood
11. construction framing, siding, sheathing	k. black walnut
12. surface veneer for cabinets and doors	l. birch

C. Completion. Fill in the blanks with the word or words that will correctly complete the statement.

1. A piece of lumber that is planed and measures 1½″ x 3½″ is called a _____ by _____.

2. A rip saw is designed to cut _____ the grain and the crosscut saw is designed to cut _____ the grain.

3. When boring a hole with a twist drill, it is very important to _____ the pressure when the bit is coming through the reverse side.

4. Two advantages of Surform® tools over planes are they do not need to be _____ or _____ .

D. Brief Answers. Briefly answer the following questions.

1. Explain why "measure twice, cut once" is a good rule when cutting wood.

2. What does the designation S2S for lumber mean?

3. Explain the difference between boring and drilling.

Bevel

Unit 10
Fastening Wood

Objective

To use nails, screws, bolts and glue in assembling wood.

Competencies to Be Developed

After studying this unit, you will be able to:
- Drive nails.
- Set screws.
- Use bolts.
- Use glue.
- Assemble wood parts.

Terms to Know

joint	end nailing
butt joint	flat nailing
lap joint	clinch
dado joint	staple
rabbet joint	countersink
miter joint	shank hole
dovetail joint	pilot hole
setting	dowel
nail set	member
toe nail	assembly

Materials List

Samples of common wood joints (butt, lap, dado, rabbet, and dovetail)
Common and finishing nails: 4d, 8d, 10d
Scraps of 2 x 4s in hard and soft woods
Claw hammers, nail set
Stapling gun and staples
Flat head wood screws (#8 x 1¼")
Hand drill and drill bit set; countersink
Push drill and drill bits
Bit brace and auger bit (⅜ inch recommended)
Standard screwdriver to fit #8 screws
Hinge set with screws
Liquid glue
Assortment of dowel pins

Glue
Clamps
Clean, soft rags

Wood and wood products are attractive, strong, and long lasting. They are used for framing, siding, walls, doors, cabinets, counters, and furniture. Wood is used extensively on farms for construction of buildings, fences and machinery such as wagons and trailers. Wood is easy to fasten. Common materials for fastening wood include nails, screws, bolts, and glue.

When fastening wood, a joint that is strong enough to do the job should be choosen. A *joint* is the union of two materials. Joints may be secured with nails, screws, bolts or glue. The most popular joints are butt, lap, dado, rabbet, miter, and dovetail, figure 10–1.

Butt Joints. ☐ The *butt joint* is formed by placing two pieces end to end or edge to edge in line or at a 90° angle. Butt joints may be strengthened by applying thin wood or metal plates at corners or across flat surfaces where parts meet.

Lap Joints. ☐ The *lap joint* is formed by fastening one member face-to-face on another member of an assembly. *Member* means piece; *assembly* means pieces fastened together. The members may be offset or may be cut so the two form a flat surface. In other words, the members are set into each other. Lap joints are stronger than butt joints.

Dado and Rabbet Joints. ☐ The *dado joint* is a rectangular groove cut in a board. The end or edge of another board is then inserted in this groove. The *rabbet joint* is similar to a dado except that it occurs at the end or edge of a board.

Miter Joints. ☐ The *miter joint* is formed by cutting the ends of two pieces of lumber at a 45° angle. The two pieces are then joined to form a 90° angle.

Dovetail Joints. ☐ The *dovetail joint* is formed by interlocking parts of two pieces. This is the strongest joint. However, it is also the most difficult to make. Dovetail joints are used extensively in high-quality and expensive furniture.

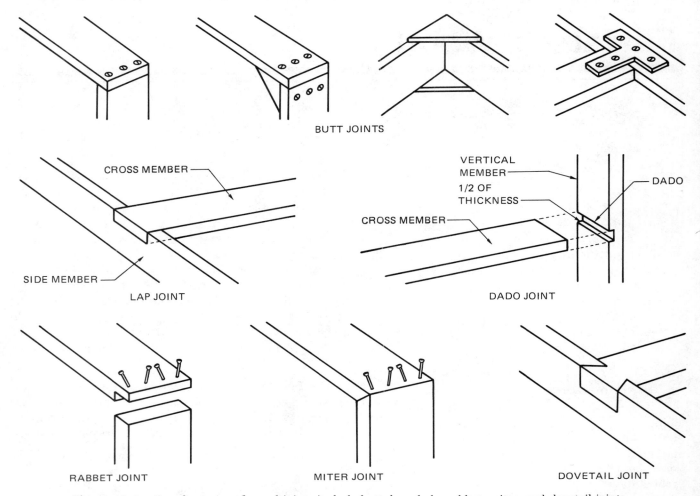

BUTT JOINTS

CROSS MEMBER

SIDE MEMBER

LAP JOINT

VERTICAL MEMBER

1/2 OF THICKNESS

CROSS MEMBER

DADO

DADO JOINT

RABBET JOINT

MITER JOINT

DOVETAIL JOINT

Figure 10–1. Popular types of wood joints include butt, lap, dado, rabbet, miter, and dovetail joints.

□ FASTENING WITH NAILS

Nailing is the fastest way to fasten wood. However, it is the least rigid and has the least strength compared to the other methods of fastening wood. Carpenters use nails to build frames of houses, attach siding and trim out house interiors. Farmers and others in agricultural mechanics find that proper nailing techniques are a very useful skill. A few nails, carefully placed and carefully driven, make quick and effective repairs.

Driving Nails

Nails can be driven into flat surfaces easily if the material is soft. To drive a nail the following procedure is used.

Procedure

1. Hold the nail between the thumb and index finger and place the point of the nail on the material, figure 10–2A. The fingers should be placed high on the nail to permit them to be knocked free rather than smashed if the hammer accidentally hits them.

2. Hold a claw hammer in the other hand with the hand near the end of the handle, figure 10–2B.

3. Keep your eyes focused where you want the hammer to hit.

4. Tap the nail with the hammer until it stands up itself.

5. Use the wrist and arm to deliver firm blows using the weight of the hammer to drive the nail.

6. Drive the nail until the head is flush with the wood and the two pieces of wood are tightly fastened.

Match the hammer to the size of the nail. Common weights of claw hammer heads are 7, 13, 16, and 20 ounces. The 7-ounce hammer is excellent for brads, tacks, and small finishing nails. The 13-ounce hammer is good for light general nailing. For most farm applications, the 16- or 20-ounce hammer is preferred because it is heavy enough to drive large nails and spikes.

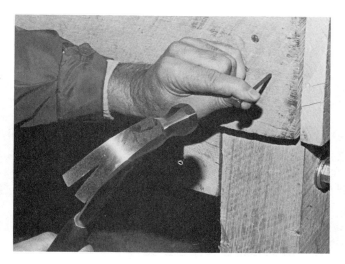

A. Start a nail by holding it with one hand and tapping it with the hammer held in the other hand.

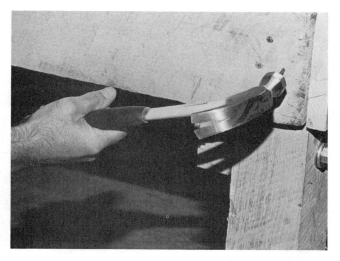

B. Drive the nail with long, even strokes. The hammer handle should be parallel with the work when the head strikes the nail.

Figure 10–2. Proper technique for driving a nail.

Pulling Nails. □ Claw hammers and ripping bars are used to pull nails and to rip boards from surfaces. When pulling a nail with a claw hammer, place a block under the hammer to prevent breaking the hammer handle, figure 10–3.

Setting Nails

When using finishing nails, the head of the nail is hidden in the wood. This is done by setting the nail. *Setting* a nail means driving the head below the surface, figure 10–4. To set a nail, the nail is driven with a hammer until the head touches the wood. A nail set is then used to drive the head below the surface about 1/16 inch. A *nail set* is a punchlike tool that has

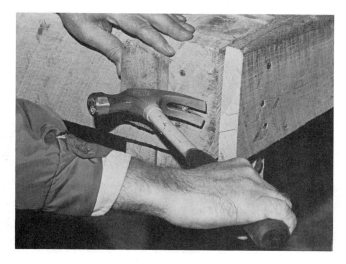

Figure 10–3. When a claw hammer is used to pull a nail, a block is placed under the hammer head to prevent breaking the hammer handle.

a cupped end rather than the flat or pointed end of a punch.

Toe Nailing

When two large pieces of wood must be fastened at right angles, they may be toe nailed. To *toe nail* means to drive a nail at an angle near the end of one piece and into the face of another piece, figure 10–5. Toe nailing is done extensively in framing using 2 x 4 or 2 x 6 lumber and 8d common nails.

End Nailing

End nailing is done by nailing through the thickness of one piece of lumber and into the end of another piece, figure 10–5. The nail is driven with the grain into the end of the receiving piece. Common nails or spikes are used for end nailing; box nails can be used for end nailing thin materials. End nailing has poor holding power. The technique is most useful when weight is permanently resting on the assembly. When end nailing small projects, it is advisable to use glue as well.

Flat Nailing

Flat nailing means to fasten two flat pieces to each other. One flat piece may also be nailed to a thicker piece. When two thin pieces are nailed together, the nails must be clinched. To *clinch* means bending the nail over and driving the flattened end down into the wood, figure 10–5. If splitting is likely, the nail should be clinched across the grain. Clinching results in a very strong nailed joint. To clinch nails, use the following procedure.

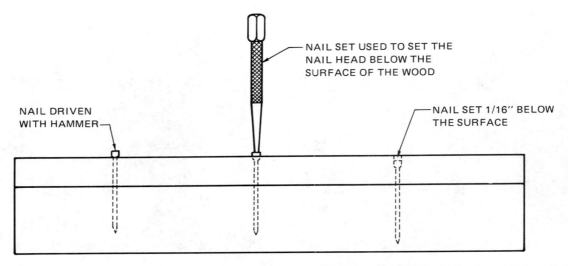

Figure 10–4. When appearance is important, finishing nails are used and the heads set below the surface. The holes are then filled with putty before the wood is finished.

Procedure

1. Choose a common nail that is about 1 inch longer than the thickness of the two boards.
2. Drive the nail through the boards until the head is flush with the surface of the wood. Be careful to support the wood so there is a space under the area where the nail comes through. If a space is not provided, the nail will damage the surface or the support, or the pieces will be nailed to the bench or saw horse.
3. Turn the two boards over and place the head of the driven nail on a hard surface. A steel vise works well. When nailing large assemblies, a

heavy sledge hammer can be held against the nail head.
4. Hit the end of the nail sideways until it is bent flat to the surface.
5. Drive the nail until it is embedded in the wood.

Stapling

A *staple* is a piece of wire with both ends sharpened and bent to form two legs. Staples are generally used to attach fencing or electrical wiring to wood. Staples are driven until they just touch the wire. If they are driven too tight, they will cause a short in electrical wire. It is also possible to weaken or break fence wire if staples are driven tightly against the wire.

Staples are also used to fasten ceiling tile and back panels on furniture. This type of staple is driven with a staple gun.

□ FASTENING WITH SCREWS

Screws hold better than nails. They are installed with a screwdriver which is a turning tool with a straight tip, phillips tip, or special tip. The phillips screwdriver end is shaped like a plus sign (+). Screws may be driven quickly and easily with power screwdrivers or variable speed drills. Screws are used extensively in doors, windows, wall systems, and furniture. The flat head screw is the one used more extensively in woodworking. Its head is countersunk and flush with the surface after it is installed. The phillips head is preferred because it is easier to keep the screwdriver bit in the head of the screw.

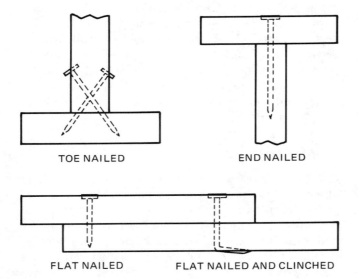

TOE NAILED   END NAILED

FLAT NAILED   FLAT NAILED AND CLINCHED

Figure 10–5. Four methods of nailing lumber

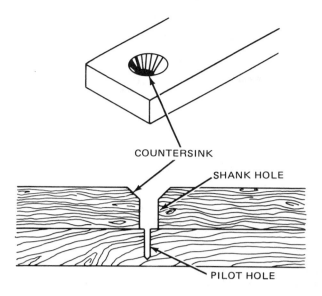

COUNTERSINK

SHANK HOLE

PILOT HOLE

A. WOOD PIECES PROPERLY DRILLED TO RECEIVE A FLAT HEAD WOOD SCREW.

B. A COMBINATION DRILL THAT MAKES THE PILOT HOLE, SHANK HOLE, AND COUNTERSINK IN ONE OPERATION MAY BE PURCHASED FOR THE MORE POPULAR SIZES OF WOOD SCREWS..

Figure 10–6. Holes must be predrilled for screws.

Preparing Wood for Flat Head Screws

A hole must be drilled in the wood before a screw is inserted. Holes made in wood to receive screws must be the correct diameter and depth; otherwise, the screw will not hold properly. Screws have a head, shank, and core with threads—all require exact holes, figure 10–6. The flat head must have a hole with tapered sides that fit. The hole for the head is called the *countersink*. The shank requires a hole which will permit it to drop through. The hole for the shank is called a *shank hole*. The core requires a hole small enough to permit the threads to screw into the wood. The hole for the core is called a *pilot hole*. When drilling holes for screws, it is recommended that the drill sizes be determined using a drill size chart, figure 10–7.

Attaching Wood with Screws

When using screws to fasten two pieces of wood together, the following procedure is recommended.

Procedure

1. Measure the thickness of the material to be attached.
2. If possible, use a screw three times as long as the thickness of the board being attached. If the screw will reach all the way through the second board, use a shorter screw.
3. Determine the appropriate screw spacing and mark the spots for all screws.
4. The diameter of the screw should look in balance with the spacing. Close screws should be smaller

Screw Gauge	0	1	2	3	4	5	6	8	10	12	14	16	18
Shank Hole	$\frac{1}{16}$	$\frac{5}{64}$	$\frac{3}{32}$	$\frac{7}{64}$	$\frac{7}{64}$	$\frac{1}{8}$	$\frac{9}{64}$	$\frac{11}{64}$	$\frac{3}{16}$	$\frac{7}{32}$	$\frac{1}{4}$	$\frac{17}{64}$	$\frac{19}{64}$
Pilot Hole	$\frac{1}{64}$	$\frac{1}{32}$	$\frac{1}{32}$	$\frac{3}{64}$	$\frac{3}{64}$	$\frac{1}{16}$	$\frac{1}{16}$	$\frac{5}{64}$	$\frac{3}{32}$	$\frac{7}{64}$	$\frac{7}{64}$	$\frac{9}{64}$	$\frac{9}{64}$

Figure 10–7. Chart of drill sizes for wood screws *(courtesy of M. Huth, Introduction to Construction, Delmar Publishers Inc., 1980)*.

in diameter. The most frequently used screw sizes are numbers 6, 8, 9, and 10.

5. Use a chart to determine the drill size for the shank and pilot holes. An alternative method is to hold different drills under the screw shank until you have one which appears to be the same diameter. For the pilot hole, the drill must be the size of the screw core, not the diameter of the threads.

6. Insert the pilot hole drill into a hand drill chuck. Adjust the length of the exposed drill until it equals the length of the screw. The bit may be too long. If so, mark the drill with a sliver of masking tape or use a piece of wooden dowel to limit the exposed length of the drill.

7. Place two pieces of wood together and drill the pilot hole through both pieces of wood.

8. Install the shank hole drill in a hand drill. Use the shank hole drill to enlarge the hole in the first piece. **Note:** The practice of enlarging a hole with a drill is acceptable in wood only. If attempted in metal, it may damage the bit.

9. Use a countersink tool to create a countersink which exactly fits the screw head. The screw should drop through the piece to be attached. The head of the screw should be perfectly level with the surface of the board.
   Note: Steps 7, 8, and 9 can be done in one operation using a combination drill designed for this purpose.

10. Screw the two pieces together. Use a screwdriver that fits the screw slot. Turn the screw until it is snug. Do not over tighten.

11. Drill pilot holes for all additional screws.

12. Drill the shank holes for all additional screws. The depth of the shank hole is very critical. If it is drilled deeper than the thickness of the board, the screw will not hold properly.

13. Countersink all holes.

14. Set all screws.
    Note: A project looks better if the screws are evenly tightened and all of the screw slots are aligned.

Installing Round Head Screws

The procedure for using round head screws is similar to the procedure for using flat head screws. Countersinking is simply eliminated. When tightening screws, screws should not be forced. If all holes are the proper depth and diameter, the screw should be easy to turn. In hard wood, it may be necessary to drill the pilot hole $\frac{1}{64}$ inch larger than the chart size. Screws are easier to turn if the threads are lubricated with soap, wax, oil, or water.

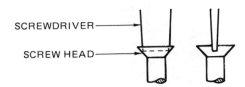

Figure 10–8. The screwdriver should be the same width as the screw head. It is very important that the end of the screwdriver be squared off so that it fits the screw slot exactly.

If the screwdriver slips out of the screw slot, the following should be checked:

- be sure the screwdriver fits the screw, figure 10–8
- be sure the holes are the proper depth and diameter
- lubricate the screw threads
- check the condition of the end of the screwdriver—reshape if necessary

A standard screwdriver should have a flat end with square edges. If the end is rounded, it should be reshaped with a file or another screwdriver should be used.

If a phillips screwdriver end is damaged or the end does not fit the screw properly, another screwdriver should be used. Damaged or worn phillips screwdrivers should be discarded.

Attaching Metal with Screws

Metal objects, such as hinges, require shorter screws than wood. This is due to the thinness of the material being attached. When setting screws to attach hinges to wood, the following procedure is recommended.

Procedure

1. Hold the door in place.
2. Hold the hinge in place.
3. Draw a circle in one hole of the hinge.
4. Make a pilot hole with a push drill, hand drill, or pointed awl pushed by hand.
   Note: The shank hole is already drilled in the hinge.
5. Install one screw.
6. Repeat the process on the other end of the hinge.
7. Install the second hinge using two screws.
8. Check the door for proper closing and clearance.
9. Reset one or more screws if necessary.
10. Install all remaining screws.

☐ FASTENING WITH BOLTS

Bolts can be the strongest method of fastening wood—with the possible exception of glue. The strength of bolts and the addition of large washers make them especially useful at high stress points. To install bolts, use the following procedure.

Procedure

1. Select the type of head needed. The carriage bolt is designed specifically for wood. It leaves a fairly smooth head.
2. Drill a hole in both members the size of the bolt.
3. Place a washer on the bolt if a machine bolt is being used.
4. Insert the bolt.
5. Add a flat washer and nut.
6. Tighten the nut until the members are tight.

Lock washers are not needed when using bolts in wood. The wood creates back pressure on the nut. When tightening bolts, crushing the wood fibers or drawing the head below the surface of the wood should be avoided.

☐ FASTENING WITH GLUE

Gluing is used extensively as a fastener of wood. Glue is used in the manufacture of plywood and particle boards. It is used extensively in manufacturing furniture. Even the rafters of wooden barns and large gymnasiums require glue to laminate the pieces into arched shapes. Glue is often used in combination with nails, screws, or bolts to provide extra strong joints.

There are many kinds of glue. Some examples are resorcinol, urea, polyvinyl, epoxy, contact cement, casein, and animal glues. The white liquid polyvinyl glue combines the advantages of several other glues. Polyvinyl fills cracks, dries colorless, sets quickly, and gives good results if protected from moisture.

Another important adhesive for agricultural uses is epoxy. Epoxy produces a strong, waterproof bond on wood, plastics, ceramics, and other materials. Epoxy materials are also available for difficult masonry repairs. The disadvantage of expoxy is that it is supplied as two materials. These materials must be mixed in the correct proportions and used promptly once mixed.

Using White Polyvinyl Glue

A properly glued wood joint will be as strong as the wood itself. When gluing two pieces of wood together with polyvinyl glue, the following procedure is used.

Procedure

1. Cut the two pieces of wood so all surfaces match perfectly.
2. Put the two dry pieces of wood together to check that the fit is good. Modify the pieces to make them fit, if needed.
3. Drill all screw or bolt holes.
4. Obtain all clamps that may be needed.
5. Apply a small bead of glue on the mating surfaces of both pieces of wood.
6. Spread the glue evenly with a flat object.
7. Put the two pieces together and nail, screw, bolt, or clamp the material, figure 10–9.
8. Check the joint and project to be certain every part is correctly aligned.
9. Retighten all clamps.
10. Remove all glue runs when slightly dry with a putty knife or wood chisel, figure 10–10.
11. Wipe all glue marks a second time using a clean wet rag. Then wipe the wood with a dry one. **Note:** If all glue is not removed, the wood will not stain evenly. If too much water is used in the cleanup process, the grain of the wood will get rough and will need sanding before finishing.
12. Clamps may be removed after one-half hour. For better results, leave the clamps on for about 12 hours.

Figure 10–9. Bar clamps can be used to draw the pieces of wood together until the glue sets. The clamps should be only moderately tightened to avoid glue-starved joints.

Figure 10–10. When the correct amount of glue is used, only small droplets of excess glue will be squeezed from the joint. These droplets are removed with a putty knife after the glue dries.

or std glue

Glue will not adhere to paint, grease or wax. Therefore, it is best to cut, file, or sand the areas to be glued.

Using Dowel Pins

A *dowel* is a round piece of wood. Dowels are generally sold in 36 inch lengths. They range from $\frac{1}{16}$ inch to one inch or larger in diameter. Popular sizes are $\frac{1}{8}$, $\frac{3}{16}$, $\frac{1}{4}$, $\frac{5}{16}$, $\frac{3}{8}$, $\frac{7}{16}$, $\frac{1}{2}$, $\frac{5}{8}$, $\frac{3}{4}$, $\frac{7}{8}$ and 1 inch. An important use of dowels is to strengthen glue joints by extending from one part to the other, figure 10–11.

Dowels are also sold in lengths of about 2 inches. These lengths have tapered ends that make them easy to use when gluing parts together. They have spiral grooves that permit glue to make an excellent bond between the dowel and the board.

□ STUDENT ACTIVITIES

1. Obtain two pieces of scrap wood 2 x 4 or 2 x 6, each about 12 inches long, cut from pine, spruce, hemlock, redwood, poplar or other soft wood. Place one piece on top of the other to double the thickness. Practice driving 4d, 8d, and 10d common nails and finishing nails. Try different sizes of hammers.
2. Drive several nails part way into a piece of wood. Place a block of wood (about $\frac{2}{3}$ the height of the nail) beside a nail. Place the hammer claws under the head of the nail with the hammer head on the block. Pull the nail. Practice this skill being

Figure 10–11. Dowel pins that extend into the two pieces of wood greatly strengthen glue joints. The dowel pins should be covered with a film of glue before installation.

careful to use wooden blocks to avoid breaking the hammer handle.

3. Repeat the exercises for #2 using oak, locust, maple, or other hard wood. What differences do you observe? Nails are easier to drive into hard wood if a hole smaller than the nail is drilled throught the first piece.
4. Use a nail set to drive the heads of several finishing nails below the surface.

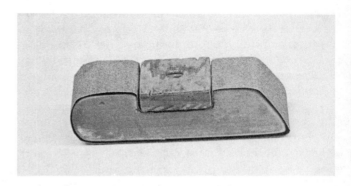

Figure 10–12. This sanding block works well because it fits comfortably in one hand. It uses a strip of sandpaper 2 inches wide and the length of a standard sheet of sandpaper. The cleat is held down with a wood screw to keep the paper in place. This sanding block is made from a piece of soft wood $\frac{3}{4}''$ x $2''$ x $4''$. The dado (groove) is $\frac{3}{8}$ inch deep and $1\frac{1}{2}$ inches wide. The cleat is $\frac{3}{8}''$ x $1\frac{3}{8}''$ x $2''$. The screw is No. 8 x $\frac{3}{4}$ inch long.

5. Practice toe nailing a 2 x 4 to a 2 x 4.
6. Study some pieces of furniture and identify butt, lap, dado, rabbet, and dovetail joints.
7. Attach two small boards together using glue and two flat head screws. Be careful to provide the correct countersink, shank hole, and pilot hole.
8. Take out the board you used to cut a dado in Unit 9. Cut a ¾-inch piece of wood several inches wide to fit into the dado groove. Glue and clamp the board to form a dado joint. Save this board for an exercise in Unit 11.
9. Make a sanding block with a hold down cleat like the one shown in figure 10–12.

□ SELF-EVALUATION

A. Multiple Choice. Select the best answer.
1. Two pieces of wood joined together is called
   a. an angle
   b. a joint
   c. a lap
   d. a splice
2. The tool used to push nail heads below the surface of wood is a nail
   a. driver
   b. press
   c. punch
   d. set
3. The strongest nailing method is
   a. clinching
   b. end nailing
   c. flat nailing
   d. toe nailing
4. An advantage of epoxy glue is
   a. ease of use
   b. no mixing
   c. waterproof bond
   d. none of these

B. Matching. Match the word or phrase in column I that best matches the word or phrase in column II.

Column I	Column II
1. groove in end of board	a. butt joint
2. interlocking extensions	b. dado joint
3. groove in middle of board	c. dovetail joint
4. face-to-face	d. lap joint
5. edge-to-edge	e. rabbet joint
6. flat head of screw	f. three holes at once
7. smooth stem of screw	g. countersink
8. threaded core of screw	h. shank hole
9. combination drill	i. pilot hole
10. shank hole for No. 8 screw	j. $^{11}/_{64}$ inch
11. shank hole for No. 10 screw	k. $^{3}/_{16}$ inch
12. pilot hole for No. 8 screw	l. $^{5}/_{64}$ inch
13. pilot hole for No. 10 screw	m. $^{3}/_{32}$ inch

Unit 11
Finishing Wood

Objective

To prepare wood for finishing and to apply attractive and durable finishes.

Competencies to Be Developed

After studying this unit you will be able to:
- Sand wood.
- Remove dents.
- Fill holes.
- Select finishing materials (except paint).
- Select solvents for thinning.
- Apply stain and clear finishes.
- Clean brushes.

Terms to Know

finish	tack rag
shellac	polyurethane
varnish	glossy
spar varnish	satin
marine varnish	wood filler

Materials List

Sandpaper
Steel wool
Heating iron
Clean, soft rags
Putty, glazing compound or plastic wood
Clean paint brush (1 inch)
Sealer (shellac suggested)
Finishing material (varnish suggested)
Stain
Paste furniture wax
Paint thinner
Shellac thinner

Wood objects and buildings that were properly finished have remained durable and beautiful for centuries. There are many, many ways to finish wood to protect the surface.

Wood can be truly beautiful if finished so the grain shows. If painted, wood offers only its shape and strength. The secret to an attractive wood project is to prepare the wood for finishing. This means removing dents, leveling high spots, removing glue stains, and sanding to a smooth surface.

This unit covers the finishing of wood with stain and clear finishes. Painting will be covered in another unit.

☐ PREPARING WOOD FOR FINISHES

Soft wood is easily damaged until a finish is applied. A *finish* is a chemical layer that protects the surface of a material. Lumber should always be handled carefully. Walking on clean boards or laying them on dirty benches or floors should be avoided. Sanded boards should be protected with cloth or paper. Pencil marks should be made lightly and then erased. If pencil marks are sanded, they become embedded in the wood.

Removing Dents

If wood is accidentally dented, it may be repaired unless the wood fibers are cut. The procedure is as follows.

Procedure

1. Heat a soldering copper to about 400° Fahrenheit or a clothes iron to its hottest setting.
2. Place a damp, soft, clean cloth over the dent.
3. Apply the hot iron to the cloth until steam rises from the cloth, figure 11–1.
4. Let it steam for about 5 seconds, but do not burn the wood.
5. Remove the iron.
6. Remove the cloth.
7. Examine the dent. The steam should have caused the compressed wood fibers to expand. The previous dent should be level.
8. If there is still a small dent, redampen the cloth and repeat steps 2–7.
9. Sand the area when the surface dries.

Figure 11–1. To remove a dent in wood, lay a wet cloth over the dent. Press the flat area of a hot iron on the cloth. The steam from the water in the cloth should cause the compressed fibers of the wood to expand, thus eliminating the dent.

Filling Holes

A word of caution is in order about filling holes. Some students expect to correct mistakes with filling materials. It is very difficult to improve the looks of a project after bad cuts are made or it has been inadequately sanded. In this regard, wood filling materials work well only in holes where dowels or nails are set below the surface, or where a puncture or blemish is small in diameter.

Holes made by installing finishing nails may be filled with putty, glazing compound, or plastic wood. Putty and glazing compound are used to hold window panes and fill cracks and crevices before painting.

Putty and Glazing Compound. □ To use either putty or glazing compound to fill holes, the procedure is as follows.

Procedure

1. Seal the wood with a sealer or primer.
2. Take a small amount of putty or compound from the can. Work it in the palm of your hand until it is soft.
3. If you wish to color the material, work in some oil base tinting pigment to obtain the desired color.
4. Place a bit of the material on a screwdriver tip and press it into the hole, figure 11–2A.

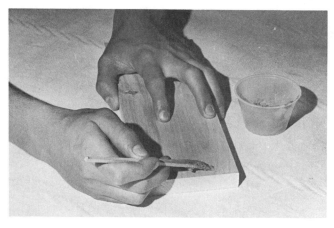

A. Place a bit of the filling material on a screwdriver tip or other implement and press it into the hole.

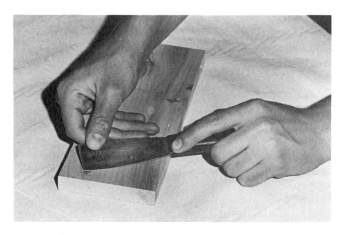

B. Smooth the filler with a putty knife.
Figure 11–2. Holes in wood may be filled with putty, glazing compound, or plastic wood.

5. Smooth and level the material by wiping the surface firmly with a flexible putty knife or your finger, figure 11–2B.
6. Permit the material to firm up overnight.
7. Apply the desired finishing materials to the wood.

Plastic Wood. □ Plastic wood is a product which dries very quickly. It has the advantage of becoming very hard in a matter of minutes. Plastic wood cannot be colored. However, it can be purchased in colors to match various woods such as pine, oak, walnut and mahogany. To apply plastic wood, the following procedure is recommended.

Procedure

1. Select a can or tube of plastic wood of the color to match the finished color of the wood.

2. Put a small amount of plastic wood on the tip of a screwdriver and close the can or tube.
3. Quickly press the plastic wood into the hole.
4. Smooth the surface by drawing the flat part of the screwdriver blade back over the plastic wood.
5. Add more plastic wood and smooth out if necessary.
6. Recheck to be sure the can or tube is tightly closed.
7. Scrape off any plastic wood on the surface around the hole.
8. After 3 to 5 minutes, sand the area thoroughly with fine sandpaper. Otherwise, every spot touched by plastic wood will leave an ugly stain on the wood.
9. Apply the desired finishing materials to the wood.

Putty Stick. □ A more recent product for filling holes is the putty stick. Putty sticks may be purchased in dozens of shades to match any color of wood finish. The procedure for using a putty stick follows.

Procedure

1. Apply a stain (if desired) and a final finish on the wood.
2. Select a putty stick to match the finish of the wood.
3. Rub the stick back and forth over the hole until the hole is filled with putty.

4. Smooth the putty so it is level with the surface of the wood.
5. Rub the area with a cloth or paper towel to remove all excess putty. The job is then done since all finishes are applied before using the putty stick.

Selecting the Finish

Wood products must be carefully prepared before applying stains and other finishing materials. Paint and opaque stains require less sanding than penetrating stains and clear finishes. The type of finish must be selected before the project is completed.

If expensive wood with an attractive grain is used, a penetrating stain with a clear finish is desirable. If the grain is not attractive, an opaque stain or paint can be used. This requires less time than preparing the wood for a finish which lets the grain show.

Sanding

All edges and ends of boards should be worked down with a file or other cutting tool to the desired shape. The wood is then sanded with medium sandpaper and finally with fine or very fine sandpaper. Figure 11–3 gives guidelines for selecting sanding materials.

For best results, always do the final sanding with the grain; otherwise, ugly sanding marks will be left in the wood, figure 11–4.

Faces of boards should be sanded with a power

MATERIAL	RECOMMENDED USE	SPECIAL COMMENTS
Flint	Painted or pitchy surfaces; jobs where clogging is a problem	Low cost; paperback abrasive wears off quickly; use where paper clogs easily
Garnet	Hand sanding of wood	Durable; low cost; paper backed
Emery	Hand sanding of metal to remove rust or smooth rough areas	Expensive; cloth backed; disintegrates when wet
Aluminum Oxide	Machine sanding of wood or metal; finishing of bronze and steel	Fast cutting; long lasting
Silicone Carbide (wet and dry)	Wet: remove automotive–type finishes; feather edges of chipped paint. Dry: finish brass, copper, aluminum; sand plastics, glass and ceramics	Has a waterproof backing to permit rinsing of paint from the abrasive; very effective when used with water; durable; expensive
Steel Wool	Remove old finishes; smooth and polish fine wood finishes; polish brass	Coarse and medium grades work well with paint removers. Very fine is used to cut the gloss and provide a dead smooth surface after applying wood finishes

Figure 11–3. Guidelines for selecting sanding materials and steel wool

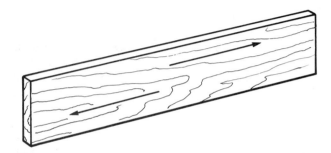

Figure 11–4. Final sanding should always be done with the grain.

sander or sandpaper wrapped around a block of wood. This helps keep the surface of the wood level. It also spreads the use across a large area of the sandpaper. A sanding block with a beveled edge permits sanding in corners. Again, start with medium sandpaper and end with very fine. After final sanding, the wood should be wrapped in cloth or paper to protect the surface from dirt and damage until the finish is applied.

□ APPLYING CLEAR FINISHES

Clear finishes protect the wood and let the beauty and color of the wood show through. An excellent finish is obtained by simply rubbing boiled linseed oil into wood. The same is true for paste-type furniture wax. However, since four to ten coats are required, most people use finishes that are easier to apply.

The procedures given for applying finishes are general instructions. The student is reminded to always read the label and follow the specific directions given with the product.

Shellac and Varnish

Shellac is a product that comes from an insect and is used to seal wood. After sealing with shellac, varnish is used to provide a protective finish. *Varnish* is a clear, tough, water-resistant finish. Special varnishes such as *spar varnish* and *marine varnish* are designed for use on objects exposed to a great deal of moisture. The procedure for a durable shellac-varnish finish is as follows.

Procedure

1. Sand the wood until all marks are absent, the wood is extremely smooth, and the grain shows in all areas.

2. Wipe the wood with a rag dampened with shellac thinner (denatured alcohol). A rag dampened with a solvent is called a *tack rag.*

3. Choose white shellac for a clear finish. Choose orange shellac for an amber finish.

4. Thin the shellac according to the directions on the container.

5. Apply the shellac quickly with a clean, pure bristle brush. Do not brush excessively and do not go over the surface a second time.

6. Hang up the project or support it with small nails until dry.

7. Clean the brush with alcohol and then wash it with warm water and soap. Wrap the bristle end in a paper towel to permit the bristles to dry and maintain the brush shape.

8. Allow 12 hours or more for the project and brush to dry.

9. Rub the dried shellac lightly with 000 grade steel wool or very fine sandpaper. The wood surface should become very smooth.

10. Choose a work area that is clean and especially free of dust in the air or on work areas.

11. Wipe the project with a rag dampened with turpentine, varsol, or paint thinner.

12. Drive four small nails in the bottom of the project to serve as temporary legs. An alternate method is to drive a nail with a head in the bottom of the project and then use the nail to suspend the project on a wire for drying.

13. Select a good quality, perfectly clean, pure bristle brush. Brushes ranging from 1 inch to 2½ inches are recommended for shop projects and furniture.

14. Open the can of varnish and stir it slowly with a clean object. Avoid causing bubbles due to vigorous stirring.

15. Rewipe the project to remove dust.

16. Apply the varnish by starting in hard-to-reach areas and moving toward the easier-to-reach areas.

17. Frequently check previously varnished areas. Smooth out any areas where the varnish runs.

18. Do not skip any areas; cover the end grain and undersides as well as visible areas.

19. Set the project in a dust-free area to dry at least 12 hours (24 or more in damp weather).

20. When thoroughly dry, rub with 000 steel wool or sand lightly with very fine sandpaper.

21. Apply a second coat of varnish. Runs may be harder to control on the second coat, so extra care is needed when applying the varnish.

22. Rub again with 000 steel wool.

23. Restore the shine with a paste-type furniture wax. Wipe on a thin coat of wax and allow it

to dry to a haze—about 10 minutes. Then polish the surface with a clean, soft, dry rag. The wax provides a hard, waterproof and durable finish.

Polyurethane

Polyurethane is a clear, durable, water-resisant finish. It requires no separate sealer. It seals the pores of the wood and provides a very durable finish. Polyurethane can be purchased in a high gloss form or *satin* form. The satin form looks like a hand rubbed varnish finish without the need for steel woolling or sanding between coats. A satin finish has a low sheen and is regarded by many as more desirable than a very shiny or glossy finish.

Polyurethane dries quickly and is applied like varnish. Two or three coats are recommended. Turpentine, varsol, or paint thinners are satisfactory thinners and brush cleaners. Once dried, polyurethane is not easily removed even with chemical paint and varnish removers.

Lacquers

Lacquer is a clear finish that requires its own special thinner. It seals wood and provides a good protective finish. Lacquer and lacquer thinners are identified by their bananalike odor. Lacquer thinner softens paint so it is important not to apply lacquer over other finishes.

Lacquer is popular because it is a clear finish which dries almost instantly if sprayed. A special slower drying lacquer is available for brushing. The procedure for using the brush-type lacquer follows.

Procedure

1. Sand and clean the wood properly.
2. Use a clean, pure bristle brush.
3. Wipe the wood with a rag dampened with lacquer thinner.
4. Quickly apply a coat of lacquer. Move forward into new wood and do not brush over previously coated areas.
5. Replace the lid on the can promptly.
6. Clean the brush immediately in lacquer thinner followed with soap and water.
7. Allow the finish to dry thoroughly.
8. Apply a second coat.
9. After drying, rub paste wax into a pad of 000 steel wool. Rub the finish briskly with the wax and steel wool pad. Before the wax dries, polish the surface vigorously with terrycloth or a coarse towel.

□ APPLYING WOOD STAINS

Wood may be stained to many shades of black, brown, red, blonde, and white. Stains may be purchased in forms that penetrate the wood and enhance the grain. Another type of stain colors the surface of the wood and hides the grain. A third type is thick and fills the pores of open grained woods. This is called a *wood filler*.

Penetrating oil stains are popular and easy to use. To apply penetrating oil stains, use the following procedure.

Procedure

1. Sand and prepare the wood for finishing.
2. Stir the stain thoroughly.
3. Use a brush or soft, clean cloth to cover all parts of the wood with stain.
4. Allow about 5 minutes for the stain to penetrate the wood.
5. Wipe off the excess stain with clean, soft, absorbent rags.
6. Wipe and lightly polish the wood until no stain comes off when touched.
7. Set the project aside to dry for at least 24 hours.
8. Clean the brush in turpentine, varsol, or paint thinner followed by soap and warm water.
9. Clean stain off your hands with the brush cleaner.
10. Apply a clear finish as described previously.

□ SOLVENTS AND THINNERS

Many problems can be traced to a lack of knowledge about solvents and thinners. The five solvents that are used extensively in the shop and home are denatured alcohol, turpentine, paint thinners, varsol, and lacquer thinner, figure 11–5.

Denatured Alcohol

Denatured alcohol or wood alcohol is sold as shellac thinner. It will thin shellac and finishes that contain shellac. For the most part, alcohol is a single product solvent in the finishing room. However, it is used elsewhere in the shop as a cleanup solvent when overhauling brake cylinders.

Turpentine, Paint Thinners and Varsol

These products are all suitable solvents or thinners for oil-based stains, paints, and varnishes. While some

THINNER/ SOLVENT*	USED WITH	CLEANS BRUSHES USED IN
Denatured Alcohol	Shellac (both white and orange) Sealers that contain shellac	Shellac Sealers that contain shellac
Turpentine and Paint Thinners	Oil base stains and wood fillers Oil base (alkyd) paints Varnish Polyurethane	Oil base stains, fillers, and paints Varnish Polyurethane
Varsol	Oil base stains, wood fillers, and alkyd paints Solvent for grease and oil	Oil base stains, wood fillers, and alkyd paints
Lacquer Thinner**	Brush lacquer Spray lacquer Finishes containing lacquer	Lacquer and finishes containing lacquer Oil base stains, fillers, and alkyd paints (Also softens hard lacquer, oil base stains and fillers, and alkyd paints in brushes)
Water	Latex paints and other latex products	Latex paint and products (when water is warm and used with a detergent) Other thinners (when water is warm and used with soap)

* **Caution:** Protect eyes with safety devices and use all solvents in well-ventilated area.
** **Caution:** Lacquer thinner is especially hazardous because it evaporates quickly and dissolves the natural oils from skin. Avoid excessive contact with skin.

Figure 11–5. Thinners and solvents commonly used with finishing materials

cost more than others, most product labels indicate equally good results from all three. However, varsol is especially desirable as a solvent in the shop. Varsol is safer than most solvents from the standpoint of fire hazard. Secondly, it is a good grease solvent and can be rinsed away with water.

Lacquer Thinner

Lacquer thinner is a fairly universal solvent. Not only does it thin lacquer, but it softens hard paint as well. This makes it useful for cleaning paint brushes used for lacquer as well as other materials.

Lacquer thinner evaporates quickly. This creates problems with fumes in poorly ventilated areas. Extra care is needed to avoid breathing lacquer thinner fumes. Similarly, special care must be taken to avoid sparks or fire when using lacquer or any other finishing material.

□ USE AND CARE OF BRUSHES

High-quality brushes make it possible to apply high-quality finishes. Without good brushes, high-quality finishes are impossible. Unfortunately, a good brush can be ruined the first time it is used. Following is the procedure for correctly using a brush.

Procedure

1. Choose the right brush for the material you are using. Nylon brushes work best with latex paints. Pure bristle brushes are best for most other stain, varnish, and painting products.
2. Dip only about ½ inch of the brush into the material.
3. Brush back and forth to keep the bristles together. Do not bend or spread the bristles excessively.
4. Once a job is started, alternate dipping the brush into the material and applying it so the material does not dry in the brush.
5. When the job is finished, clean the brush promptly with the proper solvent and then wash it with soap and warm water.

After use, brushes must be properly cleaned. The cleaning procedure should leave the brush perfectly clean. When dry, the bristles should be soft, compact

A. Remove excess material from the brush.

B. Use thinner to dissolve material in the brush.

C. Use soap and warm water to remove the thinner and dissolved material from the brush.

Figure 11–6. Brushes must be cleaned properly after use.

D. A brush that is clean and ready to dry

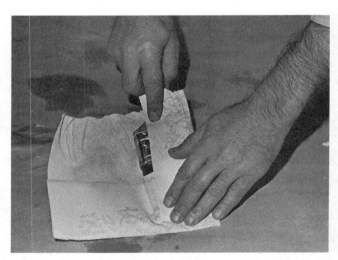

E. Wrap the brush to keep the bristles compact and in shape while drying.

F. A clean brush, properly wrapped and ready to be left to dry and ready for reuse. Like a good tool, a good brush, properly used and properly cleaned, can provide many years of excellent service. The procedure for cleaning a brush follows and is shown in figure 11–6.

Procedure

1. Remove as much material from the brush as possible. Rags, newspapers, or paper towels may be used to wipe the brush.
2. Pour about ½ inch of solvent for the product being used into a container just a little bigger than the width of the brush. Using a small container reduces the amount of solvent needed.
3. Insert the brush and push it up and down bending the bristles slightly, but not flattening them.
4. Pour the used solvent into an empty solvent can marked "used."
5. Pour some more clean solvent into the container and again flush the brush.
6. Pour the solvent into the "used" solvent can. Replace the caps on all solvent cans. After several days the used solvent will become clear and usable again for cleaning brushes.
7. Wipe all excess solvent from the brush with a rag or paper towel.
8. Place a bar of soap in the palm of your hand and run a small stream of warm water on the soap.
9. Rub the brush on the soap and in the water until all solvent and material are flushed away and the soap creates a good lather.
10. Rinse all soap from the brush. The bristles should feel like clean hair.
11. Lay a paper towel on a bench.
12. Position the brush on the left lower quarter of the towel.
13. Wrap the brush one or two turns, fold the paper down over the top of the brush, continue the wrap until all of the towel is used. Then twist the towel around the handle of the brush.
14. Lay the brush in a warm place to dry. It will take about 24 hours.

□ STUDENT ACTIVITIES

1. Take out the board that you used for Student Activity # 9 in unit 10 and perform the following procedures.

   • Drill two ³⁄₁₆ inch holes about ¼ inch deep.
   • Fill the holes with natural-colored plastic wood.
   • Use a file and sandpaper to smooth all edges of the board.
   • Sand all surfaces, remove all marks and create clean, smooth surfaces on the board.
   • Apply stain to the back side (side without the dado).
   • Apply a sealer, such as shellac.
   • Apply one or two coats of finish, such as varnish.
   • Apply wax.

2. Examine the finished board. What procedures did you do well? What errors did you make? How can the errors be avoided in the future?

□ SELF-EVALUATION

A. Multiple Choice. Select the best answer.
   1. A dent may be removed from wood with
      a. a special tool
      b. a grinder
      c. sandpaper
      d. steam from a wet rag
   2. Holes in wood may be filled with
      a. glazing compound
      b. plastic wood
      c. putty
      d. all of these
   3. Oil base stains, varnish, and alkyd paints should be thinned with
      a. alcohol
      b. lacquer thinner
      c. turpentine
      d. water
   4. Wood may be colored by applying
      a. polyurethane
      b. stain
      c. varnish
      d. wax
   5. The material used to fill holes after wood has a final finish is
      a. glazing compound
      b. plastic wood
      c. putty
      d. putty stick
   6. Plastic wood is recommended for
      a. filling nail holes
      b. hiding bad cuts
      c. removing sanding marks
      d. all of these
   7. Final sanding should be with
      a. coarse sandpaper
      b. the grain
      c. medium sandpaper
      d. a file

8. The correct sequence when applying finishes is
   a. wax, shellac, varnish, stain
   b. stain, shellac, varnish, wax
   c. stain, varnish, wax, shellac
   d. shellac, varnish, wax, stain
9. A clear, durable, water-resistant finish that is brushed on and requires no separate sealer is
   a. polyurethane
   b. shellac
   c. stain
   d. wax
10. The solvent that dissolves the greatest number of finishing products is
    a. alcohol
    b. lacquer thinner
    c. turpentine
    d. varsol

B. Matching. Match the finishing materials in column I with the thinner/solvent they are used with in column II.

Column I	Column II
1. oil base stain and varnish	a. denatured alcohol
2. latex paint	b. lacquer thinner
3. shellac	c. turpentine and varsol
4. lacquer	d. water

C. Brief Answers. Briefly answer the following questions.

1. What is a rag dampened with solvent called?
2. List the steps to follow when cleaning a varnish brush.

Unit 12
Identifying, Marking, Cutting, and Bending Metal

Objective

To identify, mark, cut, and bend cold metal.

Competencies to Be Developed

After studying this unit, you will be able to:
* Identify metals.
* Mark metal.
* Cut and file metal.
* Bend square, round, and flat steel.
* Form sheet metal.

Terms to Know

ferrous	scriber
nonferrous	dividers
alloy	center punch
malleable	soapstone
casting	hacksaw
ductile	single cut file
malleable cast iron	double cut file
annealed	stock
tempering	drawfiling
galvanized steel	snips
corrosion	shears
solder	cold chisel
scratch awl	

Materials List

Samples of
    cast iron
    wrought iron
    mild steel (various shapes)
    tool steel
    stainless steel
    galvanized steel
    aluminum
    copper
    lead

Scratch awl
Scriber
Assorted files
Soapstone
Chalk
Dividers
Center punch
Hack saw
Metal snips
Cold chisel
10 inch or 12 inch adjustable wrench
Machinist's vise
Anvil
Magnet
Combination square with scriber
Stove bolt with nut (¼" x 2")
⅛" x 1" x 6" flat steel
Blacksmith's or heavy ballpein hammer

Metals form much of the structures in the world around us. Skyscraper buildings became a reality only after it became possible to make structural steel. The transportation industry depends on metal for vehicles, rails, and reinforcement for concrete roads. Even the high technology of electronics and computers relies extensively on copper wire, silver connectors, and other metal components.

Most agricultural machines and equipment are made of steel and other metals. A successful career in many areas of agricultural mechanics depends upon a knowledge of metals and their uses in agriculture.

☐ IDENTIFYING METALS

All metals can be classifed as either ferrous or nonferrous. *Ferrous* metals come from iron ore. *Nonferrous* metals do not contain iron. This distinction is necessary because ferrous metals are used differently than nonferrous metals.

Most metals are not used in their pure metallic state. They are usually combined with one or more other metals. This combination of two or more metals is an *alloy.* Alloys have characteristics that make them different from the original metals that were used to

METAL	ORIGIN	CHARACTERISTICS	MAJOR USES
Cast iron	Iron ore	Forms into any shape; brittle	Machinery parts; engine blocks
Wrought iron	Iron ore	Malleable; tough; rust-resistant	Decorative fences; railings
Mild steel	Iron ore	Malleable; ductile; tough	Structural steel
Tool steel	Iron ore	High carbon; heat treatable; expensive	Tools; tool bits
Stainless steel	Iron ore, nickle, and chromium	Very corrosion resistant; bright appearance; hard; tough	Food handling equipment; milktanks; restaurant equipment
Galvanized steel	Steel; zinc	Zinc coated steel	Water tanks; towers; fencing; roofing; siding
Aluminum	Ore	Light; tough; relatively soft; good electrical conductor; silver-white color	Roofing; siding; truck bodies; automobiles; electric wires and cables
Copper	Ore	Tough; malleable; corrosion resistant; excellent heat and electrical conductor; reddish brown color	Pipe; electrical wire and cables; rain spouts and gutters; electrical equipment; bronze; brass
Brass	Copper and zinc	Soft; malleable; corrosion resistant	Water valves; boat accessories; ornaments
Bronze	Copper, zinc, and tin	Soft; malleable; corrosion resistant	Ornaments
Lead	Ore	Soft, very heavy; bluish gray	Batteries; cable coverings; shot; solder
Tin	Ore	Very malleable; corrosion resistant; silver color	Plating; bronze; solder

Figure 12–1. Common metals and alloys

form the alloy. The alloy is made to improve the strength or some other quality of the original metals. For example, carbon is added to iron to make steel. Figure 12–1 gives the characteristics and major uses of some common metals and alloys.

Ferrous Metals

Ferrous metals are mostly iron. Iron is refined from iron ore. Iron ore is found in abundance throughout the world. Because of the abundant supply, iron is relatively inexpensive and, therefore, used extensively. However, ferrous metals rust very easily and must be protected by coatings that resist corrosion. Ferrous metals are easy to identify because they are magnetic and give off sparks when ground on an emery wheel.

Iron is the most useful metal for making tools and machinery. Its usefulness is increased with the addition of carbon to make steel. Further, when combined with certain other metals, iron becomes a very corrosion-resistant material called stainless steel.

Cast Iron. □ Cast iron is a type of iron that is grainy and not easily shaped or bent. It is not malleable. *Malleable* means workable. Parts with odd shapes may be made by pouring molten iron into a mold called a *casting*. When the metal cools and the form is removed, the item is said to be made of cast iron. Pure cast iron will break before it will bend.

Cast iron can be treated so it is somewhat workable. This is done by special heat treatments. The result is iron with a ductile outer layer. *Ductile* means that the metal can be bent slightly without breaking. The combination of the cast iron core and ductile metal outer layer is called *malleable cast iron*. Malleable cast iron is used extensively in farm and factory machinery.

Wrought Iron. □ Wrought iron is almost pure iron. It is very malleable. It can be bent, shaped, welded, drilled, sawed and filed. Wrought iron also resists rust. These qualities make it a popular material for fences and other ornamental uses.

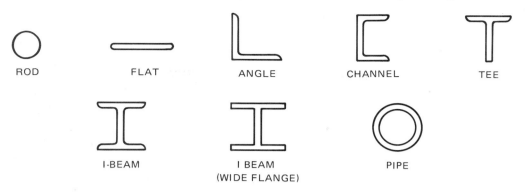

Figure 12–2. Mild steel is formed into many shapes for use in agriculture and industry.

Mild Steel. □ Mild steel is the workhorse of metals. It is made by adding small amounts of carbon with iron. Steel is tough, strong, ductile, and malleable. It is rolled into many shapes such as flat bands, angles, channels, tees, I-beams, rods, and pipe, figure 12–2. Mild steel is used extensively in the agricultural mechanics shop.

Tool Steel. □ Tool steel contains a specific amount of carbon which permits it to be hardened. Tool steel can be *annealed. Anneal* means to heat a metal to the proper temperature and then slowly cool it. The process of annealing softens and toughens steel.

Tool steel can be hardened by heating to the proper temperature and then rapidly cooling the steel. The degree of hardness is determined by controlling the temperature of the metal and the speed of cooling after heating. Only tool steel can be tempered. *Tempering* is carefully controlled reheating and cooling of steel after it has been shaped. Tempering results in a specified degree of hardness, relieves stress, and prevents cracking in steel. A tempered product is hard, tough, and not easily bent.

Tool steel can generally be identified by the exploding, sparkling nature of the sparks given off when the metal is ground.

Stainless Steel. □ Stainless steel is made by adding nickel and chromium to steel. It is very tough, won't rust and resists corrosion of all types. Stainless steel is used extensively for milk tanks, kitchen equipment, and factory equipment where food is being processed.

Plated Steel. □ Steel is made rustproof by applying a coating of metal such as tin or zinc. At one time tin was used on rolled steel roofing material. This gave rise to the term "tin roof." When a relatively thick coating of tin was used, the roof stayed bright for many years without rusting. When rust started to appear, the roof was painted. Today, most steel for roofs and siding is coated with zinc. Zinc-coated steel is called *galvanized steel.* The rust resistance of galvanized steel depends upon the thickness of the zinc coating. The thickness is rated in ounces of zinc per square unit of metal surface.

Galvanized steel sheets, pipe, buckets, and tanks must be handled with care in the shop. When heated or welded, the zinc gives off a poisonous gas. Therefore, such operations must be done with great care in a well-ventilated area.

Nonferrous Metals

Nonferrous metals and their alloys do not contain iron. They are more expensive than ferrous metals on a per weight basis and their supply is rather limited. However, nonferrous metals are needed and used extensively in industry and agricultural mechanics. Nonferrous metals include aluminum, copper, lead, tin, zinc, and other metals. Some nonferrous metals are light in weight.

Aluminum. □ Aluminum is identified by its toughness, light weight, and silver-white color. The combination of toughness and light weight makes aluminum the first choice for truck bodies, aircraft, ladders, certain equipment, containers, roofing, and siding.

Aluminum is a very good electrical conductor. Therefore, aluminum wire is often used when copper wire is too expensive. Aluminum resists corrosion. *Corrosion* is the reaction of metals to liquids and gases that cause them to deteriorate or break down. Aluminum is often painted to provide choices in color when used in doors, siding, gutters, awnings, and toys.

Pure aluminum is very soft and malleable. Other materials such as silicon, manganese, and magnesium are added to give aluminum more strength and other qualities. Special techniques must be used to weld aluminum.

Copper. □ Copper is an excellent electrical conductor and does not corrode much in air or water. Therefore, copper is used extensively in electrical wiring, electric motors, and plumbing systems.

Copper makes up the bulk of the metal in the alloys brass and bronze. Brass is made of 60 to 90 percent copper and the rest is zinc. Bronze is copper and zinc plus about 10 percent tin. These alloys have been used for centuries. They are used extensively today for their decorative and corrosion-resistant qualities.

Lead. □ Lead is a very soft and heavy metal. It is used in the manufacture of batteries, lead pipe, wire coverings, and in combination with tin to make solder. *Solder* is a soft alloy used to join many kinds of metal. Lead is also used in washers on roofing nails to provide a watertight seal between the nail and roofing.

Tin. □ Tin is used to coat steel for temporary rust protection. It is also a component of solder and bronze.

Tin has long been used to coat steel used in cans, but the coating generally provides protection for only a short period of time. Steel, coated with tin, does not rust until it is discarded outdoors. The tin coat then wears away quickly and the steel rusts.

□ MARKING METAL

Steel is the most commonly used metal in agricultural mechanics. Therefore, most metalworking skills will focus on steel. The student should realize, however, that most procedures used in working with steel apply to other metals as well.

Tools for marking metal and wood are discussed in the unit on layout tools and procedures. The main difference between marking wood and marking metal is the instrument used for marking. Metals are either very shiny or very dark in color. Therefore, a lead pencil mark does not show up well. The scratch awl, scriber, file, soapstone, dividers, and center punch are tools used to mark metal.

Scratch Awl. □ The *scratch awl,* or simply awl, is a tool with a wood or plastic handle and a pointed steel shank. It is held in the hand and used like a pencil. The awl is used to scratch the surface of metal to make a mark.

Figure 12–3. Soapstone can be used to mark metal.

Scriber. □ The name *scriber* is generally given to very small, metal, sharp-tipped markers. Many combination squares have a scriber in the handle that can be removed and used for marking metal.

Files. □ Any file will work for marking metal. The sharper the edge or corner, the finer and more accurate the mark can be. To mark with a file, the sharp corner of the file is pushed over the flat side or the corner of the metal. The file works better on structural steel than it does on plate or sheet steel.

Soapstone. □ Effective markers are made by cutting soapstone into thin pieces resembling pencils. *Soapstone* is a soft, gray rock that shows up well on most metals, figure 12–3. It can be wiped off, but not easily. Soapstone can be sawed to a knifelike edge, if desired. This sharp edge makes a narrow line on metal.

Common chalk can also be used to mark metal. However, chalk will not leave a narrow mark and it brushes off easily.

Dividers. □ *Dividers* have two steel legs with sharp points that are used to make arcs and circles on metal. The tool is also useful for transferring measurements from one item to another. A third important use is to divide any measurement into two equal parts, hence the name dividers.

Center Punch. □ The *center punch* is a steel punch with a sharp point. The center punch is used to make a small indentation or hole in metal. This hole is used to start a twist drill. The procedure for using the center punch is as follows.

Procedure

1. Place the point of the center punch exactly where the center of the hole should be.
2. Hold the punch straight.
3. Give the punch a light blow with a steel hammer.
4. Check to see if the dent is exactly where the center of the hole should be.
5. Place the punch in the dent and give the punch one solid blow with the hammer.

□ CUTTING METAL

Metal can be cut or formed by sawing, shearing, filing, or grinding. Since metals are harder to work than wood the choice of tools is somewhat more limited for working metal. The hacksaw, file, snips, and cold chisel are tools used to cut metal.

The Hacksaw

The tool used most often for cutting metal is the hacksaw, figure 12–4. The term *hacksaw* means a device that holds a blade designed for cutting metal. Such tools may be as simple as a handle that grips a com-

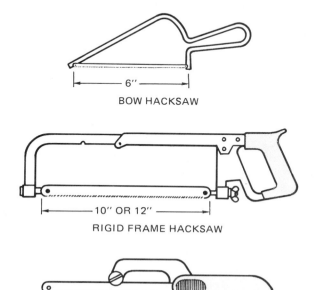

Figure 12–4. Commonly used hacksaws (*courtesy of Stanley Tools*)

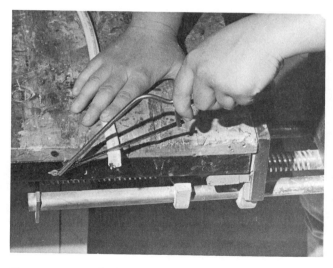

Figure 12–5. The bow-shaped hacksaw is often used by electricians to cut heavy wire.

mon hacksaw blade. Another type of hacksaw is shaped like a minature bow saw and holds a 6 inch blade with very fine teeth. Electricians favor such saws for cutting heavy copper and aluminum wire, figure 12–5.

When using a hacksaw, a blade should be selected that is fine enough to allow at least three teeth at a time on the metal, figure 12–6. Standard hacksaw blades come in 10 inch or 12 inch lengths. They may be bought with 14, 18, 24 or 32 teeth per inch. The blade with 32 teeth per inch has very small teeth and should be used to cut very thin metal. The thicker the metal, the coarser the teeth should be so the blade will cut faster.

Hacksaws are designed to cut on the forward stroke. Therefore, blades must be installed with the teeth pointing away from the handle. When using hacksaws, a small notch should be filed to help start the cut. The blade is pulled backwards and then pushed forward with slight pressure. After starting the cut, long full strokes with pressure on the forward stroke should be used. No pressure is used on the backward stroke.

Metal must be held tightly when sawing to prevent it from flexing or twisting and to prevent damage to the hacksaw blade. The best way to secure metal for sawing is to use a metal vise. The metal should be placed in the vise so the cut will be only ½ inch from the vise jaws, figure 12–7. The metal should be positioned so that three or four teeth are cutting at one time.

A used hacksaw blade should not be replaced after the cut has been started. A new blade will have a wider set in its teeth. When a new blade is forced down into the kerf left by an old blade, the teeth

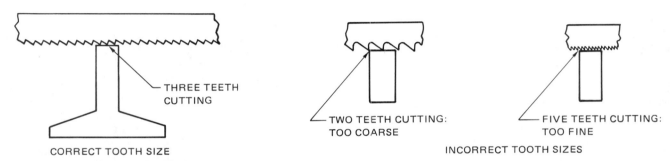

THREE TEETH CUTTING

CORRECT TOOTH SIZE

TWO TEETH CUTTING: TOO COARSE

FIVE TEETH CUTTING: TOO FINE

INCORRECT TOOTH SIZES

Figure 12–6. When sawing with a hacksaw, three teeth at a time should be on the metal. If less than three teeth are in use, the teeth will be stripped off the blade. If more than three teeth are being used, a lot of pressure will be needed to push the saw forward.

are damaged. To avoid such damage, a cut with the new blade should be started on the opposite side of the metal.

Band saws, power hacksaws, and saber saws all may be used to saw metal. However, the correct blades must be used. The use of power saws with hacksaw blades is discussed in another unit.

Files

Files are classified by length, shape, design of teeth, and coarseness of teeth, figure 12–8. Files used in the agricultural mechanics shop are generally 8 inches to 12 inches long. The shop usually has one or more of each of the various file shapes (triangular, half-round, round and flat) so that any kind of filing job can be done.

If a file has straight teeth going in one direction, it is said to be a *single cut file.* If a file has teeth going in two directions, it is a *double cut file.* A double

cut file cuts faster than a single cut file. Single cut and double cut files can be purchased in bastard (coarse), second cut (medium), and smooth (fine) grades.

The rasp cut file has raised, sharp, individual teeth. Rasps are used for shaping wood, horses' hooves for shoeing, and soft metal. They are not suitable for steel.

The curved tooth file has teeth that follow a half-round pattern. The teeth are very coarse. The curved tooth file is designed to cut soft metals such as aluminum and copper.

Filing Metal. □ When filing metal, every effort should be made to secure the metal firmly, figure 12–9. Filing will cause metal to vibrate if it is not held tightly. A file cuts on the forward stroke only. Filing may be done across metal or along its length.

To file, the point of the file is grasped between the thumb and index finger of one hand. The handle is gripped with the other hand. The file is pushed forward while light pressure is applied. The file should be slightly lifted and brought back to the starting position.

Drawfiling □ *Drawfiling* is done by placing the file at a 90° angle to the metal and pushing or pulling the file in this 90° position. Pressure is applied with a push or pull movement of the file over the length of the stock. *Stock* refers to a piece of material such as wood or metal.

Cleaning Files. □ Files are cleaned by tapping the handle sharply on the bench top. Another cleaning method is to use a file card or wire brush when particles are difficult to remove.

Snips and Shears

The terms snips and shears are terms used for the same tool. *Snips* or *shears* are large scissorlike tools

Figure 12–7. Place the metal to be sawed one-half inch from the vise jaws (*courtesy The L. S. Starrett Company*).

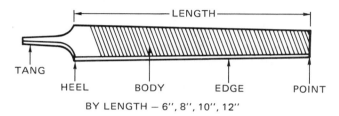

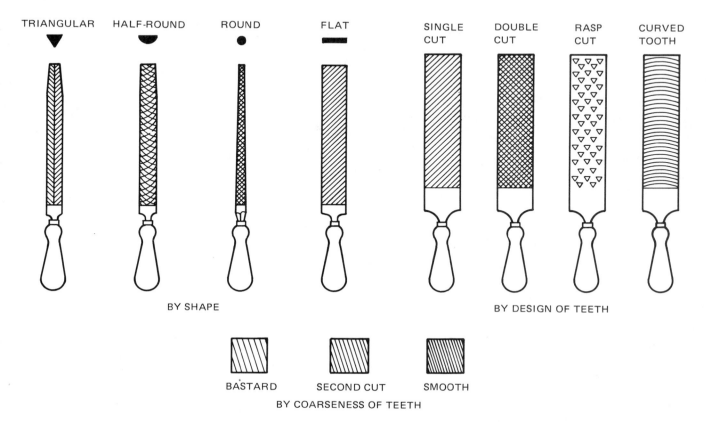

Figure 12–8. Files are classified according to the length and shape of the file, the design of the teeth, and the coarseness of the teeth.

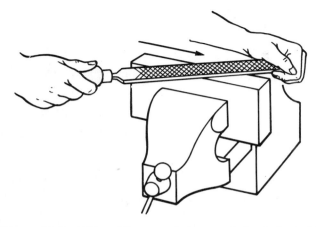

Figure 12–9. When filing metal, the metal should be held tightly in a vise.

for cutting sheet metal and fabrics. They may be purchased for cutting straight, left-hand, or right-hand curves. Regular snips are designed like scissors. Snips for heavy cutting have compound handles and are also called aviation snips, figure 12–10.

To cut sheet metal, a pair of snips that are sharp and free of nicks is selected. A piece of scrap should be cut first to be sure the snips are heavy enough. The snips are used like scissors. For snips to work well, the metal must curl or lift up and out of the snips as the cut progresses. Since curled metal is sharp and will cut, it is advisable to wear gloves when handling sheet metal. If cutting is difficult when using regular snips, let the lower handle rest on the bench top. Exert extra pressure by pushing down on the top handle.

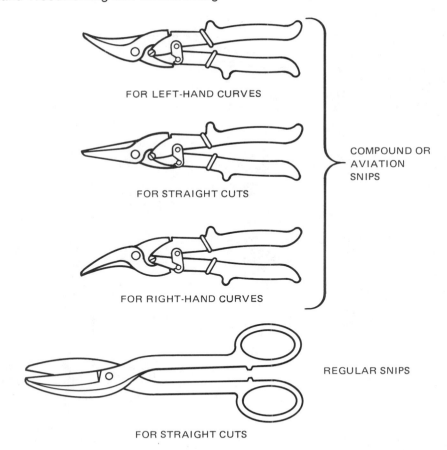

FOR LEFT-HAND CURVES

FOR STRAIGHT CUTS

FOR RIGHT-HAND CURVES

COMPOUND OR AVIATION SNIPS

REGULAR SNIPS

FOR STRAIGHT CUTS

Figure 12–10. Several types of snips are available. Compound or aviation snips provide extra leverage for heavy cutting.

Cold Chisels

A *cold chisel* is a piece of tool steel shaped, tempered, and sharpened to cut mild steel when driven with a hammer. A regular cold chisel has a wide cutting edge. Special chisels such as the cape, round nose, and diamond point have cutting edges designed to cut grooves, figure 12–11.

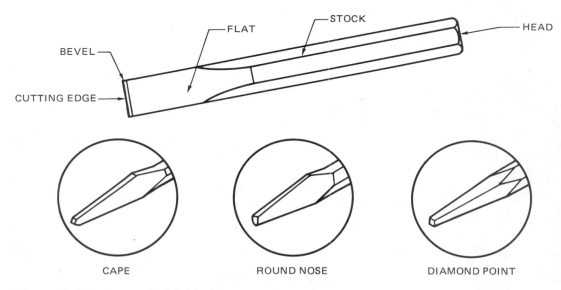

FLAT — STOCK — HEAD

BEVEL

CUTTING EDGE

CAPE   ROUND NOSE   DIAMOND POINT

Figure 12–11. Types of cold chisels

Cold chisels are used to cut mild steel rods and bands, and to cut rusted nuts and bolts. A cold chisel may also be used to loosen nuts that have rounded shoulders. The procedure for loosening such nuts is as follows.

Procedure

1. Saturate the nut with a rust solvent.
2. Position the cutting edge of the chisel on a shoulder of the nut.
3. Angle the chisel so it will turn the nut counterclockwise (backwards) when struck.
4. Drive the chisel with a ball pein or blacksmith's hammer.

When using the cold chisel to cut nuts, the following procedure is recommended.

Procedure

1. Select a sharp chisel.
2. Examine the way in which the bolt and nut are mounted.
3. Place the chisel on the nut so it will be driven against a solid structure. The nut can be split by cutting downward parallel to the bolt. The nut may also be cut into at a right angle to the bolt.
4. Cut through the nut as far as possible.
5. Change the position of the chisel and drive it in such a way as to turn the nut off the bolt. The nut should either turn off or split off the bolt.

Sometimes it is easier to cut a bolt than to cut a nut. This may be true when the bolt is ⅜ inch or smaller. For best results, the bolt should be tight in its hole. The procedure for cutting a bolt is as follows.

Procedure

1. Place a sharp chisel at the base of the nut and at a right angle to the bolt.

2. Drive the chisel so it is forced under the nut. The lifting action of the chisel should cause the bolt to be exposed under the nut.
3. Continue driving until the bolt is cut.

The hacksaw is the best tool for cutting mild steel. However, there are times when the cold chisel may be preferred. When cutting rod or flat iron, the following procedure is recommended.

Procedure

1. Mark the metal where it is to be cut.
2. Place the steel in a heavy vise. The cutting mark should be even with the jaws of the vise.
3. Lay a heavy cold chisel on top of the moveable jaw of the vise. The cutting edge of the chisel is to rest against the steel to be cut. If cutting flat iron, the cutting edge of the chisel should be placed on the corner of the iron.
4. Hold the chisel so it rests at about a 30° angle to the bench top. The flat of the chisel should be parallel to the bench top.
5. Drive the chisel until the metal is cut. The metal is cut by a shearing action. The vise acts as one-half of the shear; the cold chisel is the other half, figure 12–12.

Always use special face and hand protection when using driving tools. The occasional piece of metal that may fly from a chisel or hammer may become embedded in an eye or other parts of the face. Hands may be bruised or seriously injured by a chisel that slips or a hammer that misses its mark. Therefore, every precaution should be taken to protect the body.

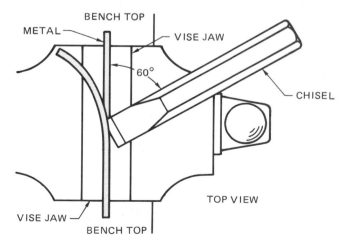

Figure 12–12. When cutting with a cold chisel, drive the chisel so that it acts as the second half of a pair of shears.

□ BENDING METAL

Mild steel can be bent cold in sizes up to ½ inch square, ½ inch round and ³⁄₁₆″ x 1″ flat. A vise anchored to a heavy bench makes the bending of cold metal possible. Metal can be twisted or bent for ornamental purposes such as for porch railings. Metal may also be bent at various angles to fit some object or to finish an assembly. Round metal may be bent to form an eye. An *eye* is a piece of metal bent into a small circle.

Bending Flat Metal

To bend a flat piece of metal up to one inch wide, or a rod up to ½ inch in diameter, the following procedure is recommended.

Procedure

1. Mark the metal where the bend is to occur.
2. Place the metal in a vise with the mark at the top of the vise jaws. It is best to have the longest part of the metal extending above the vise.
3. Tighten the vise securely.
4. Hold a heavy ball pein or blacksmith's hammer in one hand.
5. Grasp the metal near its end with the other hand.

Caution:

Wear a glove or hold the chisel with a cloth to protect the hand.

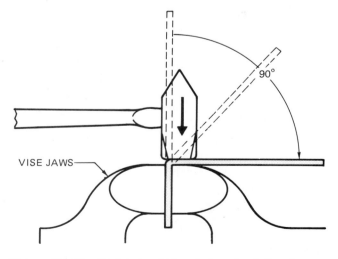

Figure 12–13. Metal should bend sharply at the vise.

6. Push the metal with one hand while hitting it with the hammer near the vise. Coordinate the movements so the metal bends sharply at the vise and not above it, figure 12–13.

Rounding Metal

Metal can be rounded by using a vise and a piece of round stock or pipe. The procedure for rounding metal is as follows.

Procedure

1. Select a piece of round stock or pipe with a diameter equal to the desired bend.
2. Place the end of the material to be bent tightly between the round stock and a vise jaw, figure 12–14A.
3. Using a hand and a hammer, bend the metal around the round stock.
4. Open the vise and move the material around the round stock, figure 12–14B.
5. Tighten the vise.
6. Bend and hammer the material to fit the round stock.
7. Repeat the above steps until the part is bent to the desired shape.
8. Reposition the metal as needed and use the closing power of the vise to tighten and shape the bend.

Using an Anvil □ An *anvil* is a heavy steel object that is used to help cut and shape metal, figure 12–15. Metal may be rounded by holding the long part of the metal in one hand and bending the short end of the metal over the horn of an anvil, figure 12–16.

Twisting Metal

Another useful procedure in agricultural mechanics is the twist. To make a 90° twist in flat iron, the following procedure is used.

Procedure

1. Mark the metal where the twist should start.
2. Measure up the metal a distance equal to 1½ times the width of the metal and make a second mark.
3. Place the metal in a vise with the twist mark at the top of the vise jaws.
4. Take a 10-inch or 12-inch adjustable wrench and adjust it to fit the thickness of the metal.
5. Place the wrench on the metal so the jaws extend

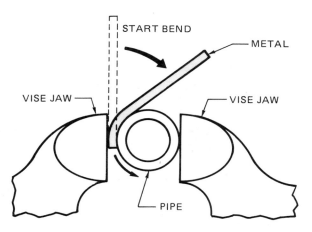

A. Place metal tightly between the pipe and the jaws of the vise.

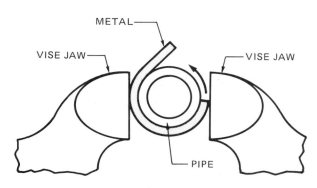

B. Move the material around the round stock and continue the bending process.

Figure 12–14. Rounding metal using a pipe and a vise

the width of the metal. Move the wrench up the metal until the bottom of the wrench is on the upper line.

6. Hold the metal straight up in the air with one hand while turning the wrench with the other, figure 12–17.

7. Stop when the upper section of the metal is at 90° to the lower section.

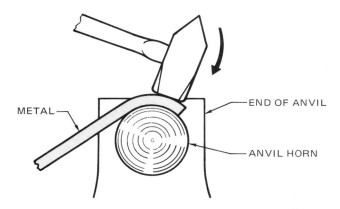

Figure 12–16. Using the horn of the anvil and a blacksmith hammer to round metal.

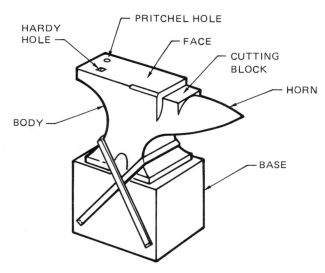

Figure 12–15. The parts of an anvil. The tapered square shank of an anvil chisel, or hardy, fits into the hardy hole. Metal is placed on the cutting edge of the hardy and is struck with a hammer to cut the metal. The pritchel hole is used to help punch holes in metal. Red hot metal is placed over the hole and the round, tapered part of the punch is driven through the metal and into the hole.

Figure 12–17. Twisting flat iron using an adjustable wrench

It may be necessary to move the wrench up or down the metal. Move it up if the metal starts to shear at the vise. Move it down if the bend is too long or gentle.

Bending Sheet Metal

Sheet metal is relatively easy to bend. Commercial benders are available. However, most agricultural shops use hand tools for this task. To bend sheet metal without special tools, the following procedure is used.

Procedure

1. Mark the metal where the bend should occur.
2. Find a bench with a sharp 90° angle at the edge. If the edge is not sharp, place a 90° piece of angle iron over the edge and clamp the angle to the bench top.
3. Place the sheet metal so the metal to be bent extends out over the bench. Be sure the bend line is directly over the edge of the bench.
4. Push the metal down along the edge to start the bend.
5. Complete the bend by tapping the metal with a hammer along the length of the bend line.

For small projects, place an angle iron or heavy piece of flat steel of the suitable size in a vise. Form the metal using the steel to support the metal as it is hammered and shaped.

□ STUDENT ACTIVITIES

1. With the use of a magnet, classify the following items as ferrous or nonferrous. Determine what metal or metals are in the items.

   - Penny
   - Quarter
   - Door knob
   - Door hinge
   - Framing square
   - Automatic chalker
   - Piece of electric wire
   - Wrench
   - Water pipe

Caution:

Use goggles for all metalworking procedures.

2. Break a piece of cast iron that has been discarded. Describe the appearance of the metal exposed by the break.
3. Examine a combination square. Does it have a scriber? If so, remove, examine, and replace it.
4. Examine five or six different files. Indicate the shape and type of teeth for each file.
5. Cut and bend flat steel.
   a. Obtain a piece of ⅛″ x 1″ flat mild steel, 6 inches long.
   b. Measure 1 inch in from one end.
   c. Use a square and scratch awl to make a line across the metal.
   d. Saw off 1 inch using a hacksaw. Check the work for squareness.
   e. Again, measure 1 inch from the end.
   f. Square a line across the metal with soapstone. (If soapstone is not available, use chalk).
   g. Shear off 1 inch using a vise and cold chisel.
   h. Again, measure 1 inch in from one end. Square across using a scriber or scratch awl.
   i. Measure in 1½ inches from the line, and square a line across.
   j. Place a 90° twist between the two lines using a vise and adjustable wrench.
   k. Secure 1 inch of the piece of metal in the vise and make a 90° bend.
6. Cut bolts.
   a. Obtain a ¼-inch stove bolt, 2 inches or more in length, and a ¼-inch nut.
   b. Screw the nut on the bolt about ½ inch.
   c. Grip the nut tightly in a machinist's vise with the bolt in a horizontal position.
   d. Use a hacksaw and saw off the part of the bolt that extends beyond the nut.
   e. Smooth up the end of the bolt with a fine file.
   f. Unscrew the nut to straighten the threads.
   g. Screw the nut back on the bolt about ½ inch.
   h. Place the bolt vertically in the vise with the nut resting against the top of the vise. Tighten the vise.
   i. Use a cold chisel to shear off the nut. Drive the chisel slowly.

Caution:

Warn others to stay away from the area, since the end of the bolt and the nut can become a projectile if driven hard.

j. Replace the bolt in the vise leaving about ½ inch extended above the vise.

k. Shear off the end with a cold chisel.

7. Shear and bend sheet metal.

a. Lay out a piece of sheet metal 4 inches square.

b. Cut out the 4-inch square sheet metal with snips.

c. Use a combination square and scratch awl to lay out a line 1 inch from one side of the sheet metal. Extend the line from edge to edge.

d. Set a dividers with the legs 1 inch apart. Place one leg of the dividers on the line 1 inch from the end. Make an arc across the corner of the metal.

e. Repeat step d on the other end of the line.

f. Round the two corners with snips.

g. Using a hammer and the edge of the bench, make a 90° bend on the line laid out in step c.

□ SELF-EVALUATION

A. Multiple Choice. Select the best answer.

1. When using a hacksaw, the number of teeth cutting at one time should be
   a. one
   b. two
   c. three
   d. five

2. Another name for snips is
   a. aviation
   b. combination
   c. scissors
   d. shears

3. Standard hacksaw blades come in lengths of
   a. 6 and 10 inches
   b. 7 and 10 inches
   c. 9 and 11 inches
   d. 10 and 12 inches

4. Choice of teeth per inch for hacksaw blades are
   a. 14, 18, 24 and 32
   b. 14, 20, 26 and 32
   c. 14, 24 and 36
   d. none of these

5. A hacksaw cuts on
   a. the backward stroke only
   b. the forward stroke only
   c. both the forward and backward strokes
   d. the stroke recommended by the manufacturer

6. Single cut and double cut refer to
   a. teeth design on files
   b. teeth on a hacksaw blade
   c. speed designations for blades
   d. width of saw kerfs

7. Placing a file at a 90° angle to the metal and pushing or pulling is
   a. burnishing
   b. drawfiling
   c. push filing
   d. none of these

8. It is very important to wear face protection when using a chisel because
   a. a chip of metal may hit you
   b. the hammer head is likely to come off
   c. chisels frequently break in half
   d. sight is improved by good face shields

B. Matching. Match the shape of the steel in column I with the correct name in column II.

Column I	Column II
1. ●	a. channel
2. —	b. tee
3. ∟	c. pipe
4. ⊏	d. rod
5. T	e. I-beam
6. I	f. flat
7. ◎	g. angle

C. Matching. Match the description in column I with the type of metal in column II.

Column I	Column II
1. malleable iron	a. cast iron
2. flats, I-beams, pipe	b. wrought iron
3. zinc on steel	c. mild steel
4. any shape; brittle	d. tool steel
5. electrical wire; pipe	e. stainless steel
6. batteries	f. galvanized steel
7. alloy of copper, zinc and tin	g. aluminum
8. plating for steel cans	h. copper
9. alloy of copper and zinc	i. brass
10. light, tough, roofing and siding	j. bronze
11. food equipment, shiny	k. lead
12. chisels, punches	l. tin

D. Completion. Fill in the blanks with the word or words that will make the following statements correct.

1. A soft, gray rock that is used to mark metal is _____.

2. A tool with a handle and a pointed steel shank used to mark metal is the _____ _____.

3. Ferrous metals are easily identified with a _____.

4. Thin metal may be bent into curves and circles by using a piece of _____ in a vise.

5. A 90° twist may be made in flat iron by using a vise and an _____ _____.

Unit 13
Fastening Metal

Objective

To fasten metals using procedures frequently used in agricultural mechanics.

Competencies to Be Developed

After studying this unit, you will be able to:
- Drill holes in metal.
- Tap threads in holes.
- Cut threads on bolts and pipe.
- Fasten metal with bolts and screws.
- Fasten metal with rivets.
- Solder sheet metal.
- Sweat copper pipe.

Terms to Know

soldering	pipe joint compound
brazing	teflon tape
high-speed drill	rivet
jobbers-length drill	pop rivet
tap	solder
tap wrench	soldering
die	50-50 solder
die stock	hollow-core solder
tap and die set	flux core solder
screw plate	soldering copper
plug tap	sweating
taper tap	flux
bottoming tap	tinning
bolt threads	sal ammoniac
pipe fittings	

Materials List

Center punch
Set of twist drills
Drill press or power hand drill
Tap and die set or selected parts
Oil for cutting
Rags
Pipe cutter

Tubing cutter
Pipe wrenches, adjustable wrench, and screwdriver
Pipe reamer
Pipe dies and stock
A length of steel pipe—½ inch, ¾ inch or 1 inch diameter and 1 foot or more in length
A pipe fitting for the length of steel pipe
A length of copper tubing or pipe—⅜ inch or ½ inch in diameter and 6 inches or more in length
Copper fitting for the length of copper tubing or pipe
Metal countersink
Sheet metal screws
Stove bolt, ⁵⁄₁₆″ x ½″ long
Steel rivets, round head and countersink head
50-50 solder and flux
Metal fruit or vegetable can
One foot of #12, #14, or #16 stranded, covered copper wire
Hot rolled band iron—¼″ x 1″ x 18″ long
Cold rolled round stock, ⅜ inch diameter x 3 inch long
Soldering copper and/or soldering gun
Teflon tape or pipe compound

Metals are fastened with various devices such as screws, bolts, and rivets. Fasteners are needed to assemble machines, equipment, tractors, tools, and buildings. Screws are used for such things as holding the metal panels on the backs of appliances. Bolts are used to hold metal parts together on an automobile engine and transmission. Rivets may be used to hold the rain gutters and downspouts together on a garage or barn. Rivets are also used in the large bridges that cross rivers and highways.

Metal pieces are also fastened together by melting a second metal between two pieces. This procedure is called *soldering* or *brazing*, depending upon the material that is used. The copper pipes in houses, schools, and other buildings are held together with solder.

A third important way to hold metals together is to simply melt them together. This is called welding. A filler rod may be used to add metal to strengthen the bond.

☐ MAKING HOLES IN METAL

Holes are drilled in metal to permit screws to pass through a shank hole in the first piece. A pilot hole

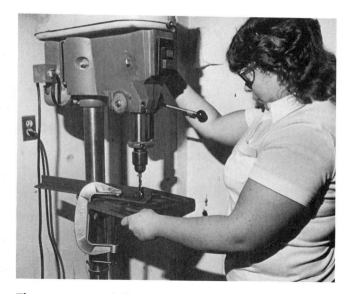

Figure 13–1. A drill press is commonly used to drill holes in metal.

is needed in the second piece to accommodate the threaded part of the screw. In sheet metal, the pilot hole in the second piece is drilled so that the solid core of the screw will pass through. The threads of the screw hold by straddling the metal.

Holes are also drilled in metal to receive bolts and rivets. When threads are needed in a hole, the hole must be of a very specific size; otherwise, it cannot be threaded.

Tools for Drilling Holes

Holes are usually drilled in metal with a portable electric drill or a power drill press, figure 13–1. Some farm shops still have a blacksmith's post drill. This is a hand-powered machine that is very effective for drilling metal.

Only high-speed twist drills are recommended for drilling metal. A *high-speed drill* is a drill that is

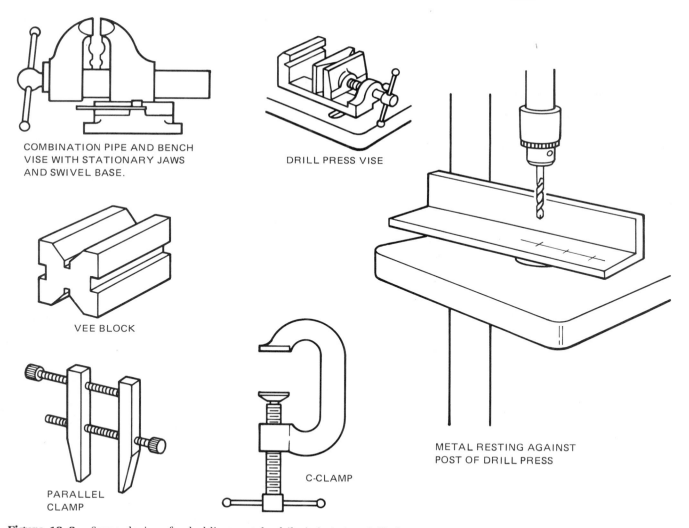

COMBINATION PIPE AND BENCH VISE WITH STATIONARY JAWS AND SWIVEL BASE.

DRILL PRESS VISE

VEE BLOCK

PARALLEL CLAMP

C-CLAMP

METAL RESTING AGAINST POST OF DRILL PRESS

Figure 13–2. Some devices for holding metal while it is being drilled

hardened for use in metal. Such drills may be purchased in regular length, or longer length known as *jobbers-length drills.* It is especially important to use eye protection when drilling wood or metal.

Drilling Holes in Metal

To drill a hole in metal, the following procedure is used.

> **Caution:**
>
> **It is especially important to wear eye protection for all metalworking procedures.**

Procedure

1. Use a center punch to make a small dent to guide the center of the drill.
2. Secure the metal by clamping it to the drill press table, holding it in a metal vise, or resting it against the post of the drill press, figure 13–2.
3. Use a power drill that turns at the correct speed for the size of drill being used (fast for small drills and slow for large drills) figure 13–3.
4. Select a high-speed twist drill of the desired size. **Note:** For holes larger than ⅛ inch in diameter, first drill a small hole (pilot hole) about the size

Figure 13–4. Tighten the chuck using the chuck key in at least two holes. This will prevent the chuck from slipping and damaging the drill shank.

of the dead center of the twist drill. Then, proceed to drill with the full-sized bit.

5. Place the drill bit in the drill chuck. Tighten the chuck by using the key in at least two holes in the chuck (three holes are preferred), figure 13–4.

> **Caution:**
>
> **Be sure to remove the key from the chuck after tightening.**

6. Apply cutting oil to carry heat away from the drill point. Cutting oil is a special oil which does not have additives used in oils designed for engines.
7. Apply firm, even pressure to the drill when starting and drilling.
8. Keep the drill perpendicular to the metal.
9. A uniform ribbon of metal is cut when the drill cuts steadily.
10. Use very little pressure as the drill breaks through the metal.
11. Be alert to the possibility of the bit seizing the metal as the drill breaks through. Hold portable drills very firmly at this point.
12. Avoid any side pressure on the drill bit. Small drill bits are brittle and break very easily.
13. After drilling the hole, remove any metal burrs with a file.

Punching Holes in Metal

Holes may be made in metal with punches. Soft, thin metal, like aluminum in storm windows, is punched rather than drilled. Punching is faster than drilling.

DRILL SPEEDS IN RPM FOR STEEL

DRILL SIZE	RECOMMENDED DRILL SPEED
⅛	2100–2450
3/16	1400–1600
¼	1000–1200
5/16	850–950
⅜	700–800
7/16	600–700
½	450–600
9/16	425–550
⅝	400–500
11/16	375–450
¾	350–400
13/16	325–375
⅞	300–350
15/16	275–325
1	250–300

Figure 13–3. Selecting the correct speed for the size of drill being used ensures long drill life and good results.

Steel may be punched if it is very thin. Steel up to ½ inch thick can be punched with hand tools if the metal is red hot. Blacksmith's tools are needed to heat and punch steel. Heavy stationary punches driven by hand, hydraulics, or electricity may be used to punch cold steel. These machines are used when many holes of the same size are needed.

□ THREADING METAL

Threads may be cut into a hole in metal with a tap, figure 13–5. A *tap* is a hardened, brittle, fluted tool that cuts threads into the hole. A tap is turned with a *tap wrench*. Threads are cut onto a rod or bolt with a *die*, figure 13–6. A die is turned with a *die stock*. A set of tools for making threads is called a *tap and die set* or a *screw plate*, figure 13–7. Common types of bolt threads are National Coarse (NC), National Fine (NF/SAE) and metric.

Tapping Threads

When tapping threads in a hole, the following procedure is used.

Procedure

1. Select the bolt that is to be threaded into the hole.
2. Determine the outside diameter of the bolt. Machine screw diameters are stated as Numbers 2,

Figure 13–6. Threads are cut onto a rod with a die.

4, 6, 8, 10 or 12. Other bolts are ¼ inch, ⁵⁄₁₆ inch, ⅜ inch, ⁷⁄₁₆ inch, ½ inch, etc. Metric bolts have diameters ranging from about 3 to 20 millimeters.

3. Determine the type of threads on the bolt by using a screw pitch gauge. Most bolt threads are National Coarse (standard), National Fine (SAE) or standard metric. Metric bolts are also available with metric fine threads.

Figure 13–7. A set of taps and matching dies and the handles to turn them is called a tap and die set.

Figure 13–5. Threads are cut into a hole with a tap.

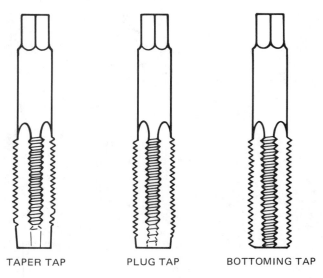

TAPER TAP   PLUG TAP   BOTTOMING TAP

Figure 13–8. When tapping threads to the bottom of a blind hole, first use a taper tap, then a plug tap, and finally a bottoming tap.

4. Select a taper tap of the correct size and thread type. The *taper tap* will start threads and complete them if the tap can extend through the material. If the hole does not go through the metal, a *plug tap* is used, followed by a *bottoming tap*. These taps will cut threads to the bottom of a blind hole, figure 13–8.

5. Center punch the location of the hole to be drilled.

6. Obtain a drill of the exact size specified for the tap selected, or use a tap and drill size chart, figure 13–9. When selecting a drill size, the exact number or fractional size is needed when the hole is to be threaded for machine screws or bolts.

7. Carefully drill the hole. The hole must be drilled at a 90° angle to the surface of the metal. Do not permit the drill bit to wobble.

8. Place the tap in a tap wrench.

9. Place the end of the tap in the hole. Be careful to keep the tap in line with the hole during the entire tapping process, figure 13–10.

10. Apply enough cutting oil (if available) or motor oil to cover the threads of the tap and the hole in the metal.

11. Apply downward pressure on the tap and turn

NATIONAL COARSE THREADS (NC)			NATIONAL FINE THREADS (NF)		
SIZE OF BOLT OR SCREW AND TAP	SIZE OF DRILL TO USE	THREADS PER INCH	SIZE OF BOLT OR SCREW AND TAP	SIZE OF DRILL TO USE	THREADS PER INCH
#1	#53 or 1/16"	64	#1	1/16"	72
#2	#50	56	#2	#50	64
#3	#47 or 5/64"	48	#3	#45	56
#4	#43	40	#4	#42 or 3/32"	48
#5	#38	40	#5	#37 or 7/64"	44
#6	#36 or 7/64"	32	#6	#33	40
#8	#29	32	#8	#28 or 9/64"	36
#10	#25	24	#10	#21 or 5/32"	32
#12	#16	24	#12	#14	28
1/4"	#6 or 13/64"	20	1/4"	#3 or 7/32"	28
5/16"	1/4"	18	5/16"	17/64"	24
3/8"	5/16"	16	3/8"	21/64"	24
7/16"	23/64"	14	7/16"	25/64"	20
1/2"	27/64"	13	1/2"	29/64"	20
9/16"	31/64"	12	9/16"	33/64"	18
5/8"	17/32"	11	5/8"	37/64"	18
3/4"	21/32"	10	3/4"	11/16"	16
7/8"	49/64"	9	7/8"	13/16"	14
1"	7/8"	8	1"	15/16"	14

Figure 13–9. Taps and appropriate drill sizes

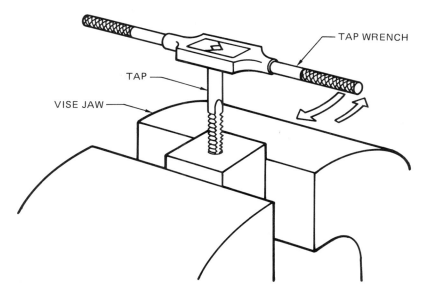

Figure 13–10. The tap must be kept in line with the hole and turned carefully. Keep the tap and the area being threaded covered with oil so tapping proceeds smoothly. Turn the tap forward about one-half turn, then turn it backward one-quarter turn to break the chip. Proceed until the tap turns freely, indicating the entire hole is threaded.

the tap forward (clockwise) about one-half turn. Do not exert any sideways pressure or cause the tap to wobble. It will break very easily.

12. Turn the tap backward (counterclockwise) one-quarter turn to break the chip of metal that forms.

13. Turn the tap forward one or two turns.

14. Turn the tap backward to break the chip.

15. Repeat steps 13 and 14 until the tap is through the hole or to the bottom of a blind hole. Keep the tap and metal covered with oil.

16. Remove the tap by turning it backward.

17. Wipe oil and metal chips from the parts and work area using a cloth or paper towel.

18. Dispose of oily materials in the proper metal container.

Caution:

Do not brush metal chips away with bare hands as the chips are sharp and will cause injury.

Threading a Rod or Bolt

The principles involved in threading a rod or bolt are similar to tapping threads. There is one major difference, however. Some dies are adjustable, but taps are not adjustable. With an adjustable die, the threads on a rod can be cut to a greater or lesser depth as desired. This adjustment permits a rod to be threaded to fit a nut perfectly. The procedure for cutting threads on a rod is as follows.

Procedure

1. Clamp a rod of cold rolled steel in a vise (hot rolled steel will not thread satisfactorily).

2. Bevel the end of the rod with a file, figure 13–11.

3. Select a die of the correct size and type of thread.

4. Mount the die in a stock (handle).

5. Place the die over the rod with its tapered side toward the rod.

6. Oil the die and rod.

7. Apply pressure on the die and turn forward about one-half turn.

8. Back up one-quarter turn to break the chip.

9. Repeat steps 7 and 8 until a short length of rod has been threaded. Add oil as needed to keep the die lubricated and cutting smoothly. If the die is adjustable, check to see if the nut threads on properly. If not, adjust the die to obtain a good fit before cutting threads over the entire area.

10. Continue cutting threads until the desired threaded length is obtained.

11. Remove the die and wipe all chips and oil from the rod, tools, and work area.

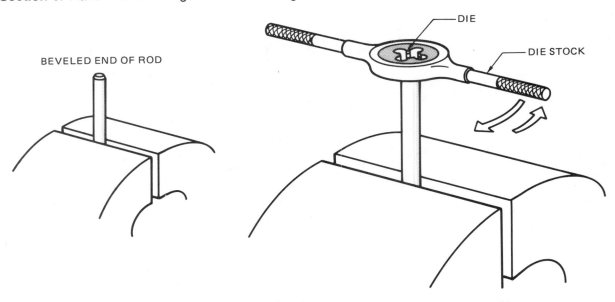

BEVELED END OF ROD

DIE

DIE STOCK

Figure 13–11. Bevel the end of the rod with a file. The die must be the correct size.

Cutting and Threading Pipe

Cutting Pipe. ☐ Pipe is cut with pipe cutters that consist of one steel cutting wheel plus two rollers. Some pipe cutters have three cutting wheels and no rollers. A small version of the pipe cutter is used for cutting copper and steel tubing. When using pipe or tubing cutters, the following procedure is used.

Procedure

1. Unscrew the cutter handle until the pipe fits between the rollers and wheels.
2. Slide the cutter onto the pipe until the cutter wheel(s) is on the mark.
3. Tighten the handle until the cutter wheel starts biting into the pipe.
4. Roll the pipe cutter around the pipe one or two times. The addition of oil will help the cutting process.

5. Repeat steps 3 and 4 until the pipe is cut. Do not tighten the handle more than one-half turn before rotating the cutter each time.
6. Insert a pipe reamer in the end of the pipe. Turn the reamer until the inside ridge is removed.

Threading Pipe. ☐ Pipe threads are not the same as bolt threads. *Bolt threads* are straight and can be cut the entire length of a rod. Pipe threads are tapered and can only be cut a short distance on the end of a piece of pipe, figure 13–12. The die gets tight and binds if too many threads are attempted. The tapered threads permit pipe connections to be gas and liquid tight.

Pipe fittings screwed onto pipe create joints that hold air, steam, and liquids under pressure without leaking. *Pipe fittings* are hollow connectors designed to attach to pipe. Threaded fittings for steel pipe require some material to seal the areas between threads if they are to be used for gases or liquids. Materials for this purpose are *teflon tape* and *pipe joint com-*

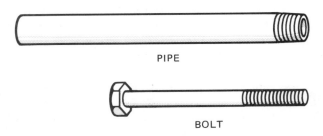

PIPE

DIAMETER OF THREADED AREA IS LESS THAN THE DIAMETER OF THE PIPE. THE THREADS TAPER TO THE END OF THE PIPE.

BOLT

DIAMETER OF THREADED AREA IS UNIFORM.

Figure 13–12. Pipe threads are tapered. Approximately one inch only can be threaded at the end of the pipe. In contrast, bolt threads are not tapered and can be cut along the entire length of the bolt.

pound also referred to as pipe dope. The procedure for threading pipe is as follows.

Procedure

1. Select the correct die size for the pipe to be threaded.
2. Place the die in the stock.
3. Tighten the pipe in a pipe vise with 8 to 10 inches extending beyond the pipe vise. A regular vise may crush the pipe.
4. Slide the tapered side of the die onto the end of the pipe.
5. Apply cutting oil to the die and pipe as needed.
6. Turn the die until the end of the pipe starts to extend through the die.
7. Remove the die. Check to see if a pipe fitting threads on and becomes tight after 4 to 6 turns. If not, cut more threads on the pipe.
8. Use a cloth to wipe all oil and chips from the pipe, tools, and work area. Remove all chips and oil from the inside of the pipe. Such materials interfere with the use of the pipe if they are not properly removed.

□ FASTENING METAL USING BOLTS AND SCREWS

Flat pieces of metal are fastened quickly and easily with bolts. One of two methods is generally used. A nut is used with the bolt, or the bolt is screwed into the second piece of metal after threads are tapped.

Using a Bolt and Nut

When using a bolt and nut, the following procedure is recommended.

Procedure

1. Drill or punch a hole through both pieces of metal. The holes should be the same diameter as the bolt. Oversized holes may be used if there is a poor fit.
2. Select the desired type of bolt.
3. Countersink to accommodate the bolt head, if using a flat head stove bolt.
4. Insert the bolt.
5. Add a lock washer.
6. Thread on the nut.
7. Line up all parts.
8. Tighten the nut until the lock washer is flattened.

Using Bolts without Nuts

Cap screws, machine bolts, and machine screws may be used without nuts. The procedure is as follows.

Procedure

1. Select the desired size and type of bolt.
2. Drill a hole in the first piece of metal the diameter of the bolt.
3. Drill the second piece of metal with the exact drill specified for the bolt size and thread type.
4. Tap threads into the second piece.
5. Place a lock washer on the bolt.
6. Push the bolt through the first piece.
7. Screw the bolt into the second piece.
8. Tighten the bolt until the lock washer is flattened.

Using Sheet Metal Screws

Sheet metal screws are used to attach thin pieces quickly. The procedure is as follows.

Procedure

1. Select the desired type and size of sheet metal screw.
2. Drill or punch a shank hole in the first piece of metal.
3. Drill or punch a pilot hole in the second piece.
4. Screw the two pieces together, but do not tighten the screw.
5. Align the parts.
6. Tighten the screw snugly. Be careful not to over-tighten. If the screw is overtightened, the pilot hole may be stripped and the screw will not hold.

□ FASTENING WITH RIVETS

A *rivet* is a fastening device which is held in place by spreading one or both ends. A special tool called a rivet set makes it easy to install rivets. Many shops do not have a rivet set and the peining method is used.

Most rivets are purchased with flat, round, or countersunk heads; they may be solid, tubular, or split, figure 13–13. Tubular and split rivets have the advantage of being easy to *rivet*. This means they can be spread or shaped by hammering. They will not hold as well as solid rivets. Steel rivets, copper rivets, and pop rivets are commonly used.

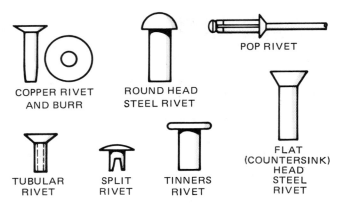

Figure 13–13. Rivets may be solid, tubular, or split and are available in different head styles. They are generally made of steel, aluminum, or copper.

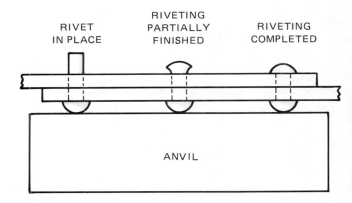

Figure 13–14. Riveting procedure using a ball peen hammer

The Steel Rivet

The round head steel rivet is used to fasten structural and sheet steel in bridges, buildings, and truck bodies. An important use of steel rivets on the farm is to attach mower knives. For this application the flat head rivet is used. To install a steel rivet, the following procedure is used.

Procedure

1. Mark the spot and center punch the metal.
2. Drill a hole in both pieces of metal the exact size of the rivet.
3. Determine the length of the rivet. The length should be the thickness of the two pieces of metal plus the diameter of the rivet. For example, when fastening two pieces of metal each ⅛ inch thick with a rivet of ⅛ inch diameter, the rivet length should be ⅜ inch (⅛" + ⅛" + ⅛" = ⅜").
4. Cut the rivet to length with a hacksaw.
5. Insert the rivet through the two pieces of metal.
6. Place the head of the rivet on a solid steel object such as an anvil or metal vise.
7. Place a rivet set, or other object with a hole the size of the rivet, over the protruding rivet end.
8. Drive the metal down tight against the rivet head.
9. Use a ball pein hammer to create a rivet head on the extended end. This is done by striking or peining the rivet stem on the edges until they are rounded down and mushroomed against the metal, figure 13–14. Or, a rivet set may be used to form the riveted end.

The Copper Rivet

Copper rivets with a tight fitting washer called a burr are used to fasten harness and other leather products.

A burr is as large as the head so soft materials are held without pulling the rivet through the material.

The copper rivet with burr is installed much like a steel rivet. The rivet is inserted in the material; the washerlike burr is then added to the rivet. The burr and material are hammered together. Finally, the end is riveted to keep the burr in place.

The Pop Rivet

The *pop rivet* is so named because it pops when installed. It is especially useful in situations where it is difficult to hold one end of the rivet and hammer the other end. An example of this is repairing sheet metal on an automobile.

When purchased, the pop rivet appears to be mostly stem. However, during installation, the special pop rivet tool pulls the stem until its ball-like head forces the end of the rivet to expand, figure 13–15. This enlargement holds the rivet in place. As the stem is pulled further by the tool, it breaks with a popping sound. The stem pulls out of the hollow rivet and is thrown away. The riveting process is then complete.

□ FASTENING METALS WITH SOLDER

Solder is a mixture of tin and lead. The joining of two materials with solder is known as *soldering*. Solder is frequently used to join thin metal, to make electrical connections, and to join copper tubing and pipe. The following general procedure is used when soldering.

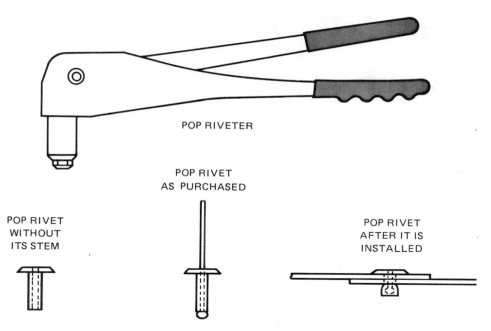

POP RIVETER

POP RIVET
AS PURCHASED

POP RIVET
WITHOUT
ITS STEM

POP RIVET
AFTER IT IS
INSTALLED

Figure 13–15. Pop rivets are installed from one side of the metal only. A built-in stem with a round ball is pulled into a hollow rivet. The ball causes the rivet to spread on the opposite side of the metal.

Procedure

1. Remove all dirt and corrosion from the metal with sandpaper or emery cloth. The metal should be shiny and bright.
2. Coat the tip of the soldering copper with solder. This process is called *tinning*. Tinning is done by cleaning the soldering copper with steel wool, sandpaper, or a file. The copper is heated and flux is applied. After the flux melts, solder is applied to the copper until it covers the tip. *Flux* is a material which removes tarnish or corrosion, prevents corrosion from developing, and acts as an agent to help the solder spread over the metal. Three fluxes that are commonly used are acid, rosin, and sal ammoniac.
3. Apply heat to the metal and apply flux.
4. Apply additional heat to the metal being soldered until that metal will melt solder.
5. Tin the metal by applying solder from a wire spool or a solid bar.
6. Apply heat to both pieces of metal until the solder melts and flows between them.
7. Wipe off excess solder with a cloth or sponge dampened with water while the solder is still molten.
8. Do not permit movement in the joint until the solder solidifies or hardens.

When soldering, several points should be kept in mind. It is important to select the appropriate flux and solder for the metal being soldered. Directions on the labels of commercial flux products as to the metals they will clean should be followed. Most soldering jobs in agricultural mechanics can be done with solder composed of 50 percent tin and 50 percent lead. This solder is called *50-50 solder.* Other solders are 30-70, 40-60, 60-40, and 80-20. Notice that each combination of numbers equals 100. Different solders are recommended for various special jobs.

Solder may be purchased with flux inside. It is referred to as *hollow core* solder or *flux core* solder. Many people find the flux in the core of the solder makes soldering more convenient and easy.

Tinning a Soldering Copper

A *soldering copper* is a tool consisting of a wooden or plastic handle, steel shank, and copper tip, figure 13–16. It may also have an electric heating element. The procedure for tinning a soldering copper is as follows.

Procedure

1. Remove burned residues with sandpaper, emery cloth, steel wool or a file.
2. Heat the soldering copper until it will melt the flux or make smoke when rubbed in sal ammo-

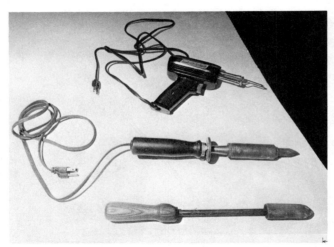

Figure 13–16. A soldering copper consists of a wooden or plastic handle, steel shank, and copper tip.

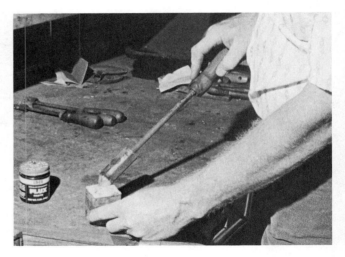

Figure 13–17. A block of sal ammoniac is used to clean the soldering copper.

niac. *Sal ammoniac* is a cube or block of special flux for cleaning soldering coppers, figure 13–17.

3. Rub the heated copper in sal ammoniac, or apply paste flux with a small wooden stick. The copper should get shiny.

4. Unroll about 6 inches of solder from the spool and touch the end to the copper. The solder should melt and flow over all flat surfaces of the copper if the copper is the correct temperature.

5. If the solder does not melt, heat the copper to a higher temperature. Then repeat steps 3 and 4.

6. Brush the copper lightly with a cloth dampened with water. The soldering copper should be a shiny silver color. The silver color indicates the copper is tinned.

Soldering Sheet Metal

To solder sheet metal, a large soldering copper is used. A soldering copper may be heated using a gas torch, a gas furnace, or an electric element. Electric units are called soldering irons; however, the tip of the unit is made of copper, not iron. When soldering sheet metal, the following procedure is used.

Procedure

1. Clean the area to be soldered using an abrasive or a file.

2. Heat each piece of metal with the flat part of a tinned soldering copper.

3. Add flux to the metal as it heats up.

4. As the metal gets hotter, keep the flat part of

the soldering copper flat on the metal near one edge.

5. Touch the solder wire to the metal, not the soldering copper.

6. Watch the point when the solder melts and begins to flow.

7. Then move the soldering copper slowly across the metal, figure 13–18.

8. Follow the soldering copper with the solder as needed to tin the area to be soldered.

9. Repeat steps 2 through 8 on the second piece of metal.

10. Lay the two tinned surfaces face to face on a piece of scrap wood if support is needed.

11. Heat both sheets of metal with the soldering copper until the solder starts to flow.

Figure 13–18. When soldering, the soldering copper is moved slowly across the metal.

12. Feed in additional solder to fill the joint as needed.
13. Wipe off excessive molten solder with a damp cloth.
14. Hold the two pieces together with a piece of scrap wood until the solder cools. The solder is cool when it changes from a shiny to a dull finish.

Some of the steps in this procedure must be done at the same time, or they must be done in close sequence. Therefore, exact instructions from an experienced teacher may be needed to learn the technique.

Wood is recommmended to support or hold the metal because wood does not carry heat away from the metal. Metal table tops, vises, and metal tools interfere with heat control. The control of heat is the heart of the soldering process.

Soldering Electric Wires

Copper wires are soldered by first removing about 1 inch of insulation. The soldering procedure is as follows.

Procedure

1. Twist the two bare ends of the wire together.
2. Insert one of the wires in a vise and gently tighten the vise jaws against the insulation.
3. Hold a hot, tinned soldering copper under the twisted wires, figure 13–19.
4. Hold rosin core solder on top of the wire.
   Note: Acid flux is not recommended for electrical connections.

Figure 13–19. When soldering an electrical wire, a hot, tinned soldering copper is held under the twisted wires.

Caution:

Do not use an acid flux as this interferes with the function of the splice.

---

5. Wait for the heated wire to cause the rosin to flow and clean the wire.
6. Watch the solder melt and run into and around the strands of wire.
7. Remove the heat.
8. Apply several thicknesses of vinyl electrical tape to the joint.

An electric soldering pencil or gun works well for soldering copper wires. These special soldering tools provide the correct amount of heat for electrical applications.

Soldering Copper Pipes and Fittings

The process of soldering a piece of copper pipe into a fitting is called *sweating*, figure 13–20. The ability to sweat copper fittings is useful for repairing and installing water lines at home, on the job, or on the farm.

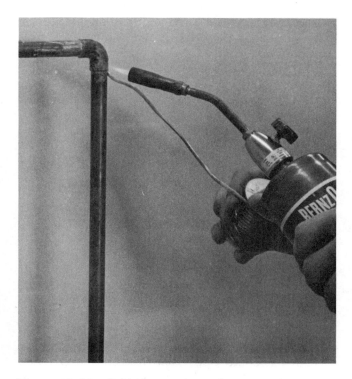

Figure 13–20. Soldering a piece of copper pipe into a fitting is called sweating.

Common pipe fittings include elbows, tees, couplings, unions, adaptors, and others. Fittings are used to change the direction of a pipe line. They are also used to adapt one type of pipe to another such as copper to plastic.

Fittings are used to connect pipe to faucets, valves and other fixtures. To sweat a fitting onto a piece of copper pipe, the following procedure may be used.

Procedure

1. Use very fine sandpaper or steel wool to polish about ¾ inch at the end of the pipe.
2. Place the sandpaper or steel wool across the end of your index finger and polish the inside of the fitting.
3. Apply a paste flux to the polished surfaces of the pipe and fitting.
4. Press the fitting onto the pipe.
5. Use a propane or other gas torch to heat the outside of the fitting near the opening where the pipe is inserted.
6. Heat the fitting ½ inch in from the opening. Use the tip of the inner flame of the torch to heat the underside of the fitting. Heat will rise to the upper areas.
   Note: It is important to heat the fitting rather than the pipe to draw the solder into the joint.
7. Hold a length of 50-50 solid core solder at the joint between the fitting and pipe.
8. Continue to heat the pipe until the solder starts flowing into the joint.
9. Move the solder around the joint to permit solder to fill all areas between the pipe and fitting.
10. Remove the heat and solder when solder is seen all the way around the joint and the joint is full of solder.
11. Before the solder hardens, wipe around the joint with a damp cloth.
12. Cool the joint.
13. Turn on the water to pressurize the system.
14. Check for leaks.

When using a propane torch several precautions are important. First, wear goggles and insulated gloves. Second, work only in areas that are free of paper, grease, fuel, and other materials that burn easily. Third, point the torch away from you when you turn it on and light it. Fourth, be fire safe.

Whenever any flame is used, be sure there is a fire extinguisher nearby. To reduce fire hazards, place sheet metal behind pipes where a torch is used. When finished with the torch, it is best to extinguish the flame. A new type of propane torch has a high flame trigger. When the propane unit is set down, the flame become low.

□ STUDENT ACTIVITIES

1. Make a bench stop using the following procedure.
   a. Cut a piece of band iron ¼ inch thick x 1 inch wide and 5 inches long.
   b. Make the ends square; slightly round all corners to remove sharp edges.
   c. Center punch the metal ½ inch from each end and each side.
   d. Drill a $\frac{5}{16}$ inch hole through the metal at the two punch marks.
   e. Tap National Coarse (NC) threads in each hole to receive a ⅜ inch threaded rod.
   f. Cut two pieces of round cold rolled steel ⅜ inch in diameter and 1½ inches long. Remove all sharp corners with a file.
   g. Thread each rod down ¼ inch.
   h. Screw the two threaded rods into the ¼" x 1" x 5" piece. The rods should thread in until tight. The end of the rods should be flush with the bottom of the metal strip.
   Note: The bench stop is useful to stop lumber when sanding, planing, or nailing on a bench. To use the stop, drill two holes in the bench top ⅜ inch or larger in diameter. The holes are drilled so the two rods drop into them. The bench stop is then ready for use.
2. Cut and thread a piece of steel pipe.
   a. Obtain a piece of steel pipe 1 foot or more in length. Pipe with an inside diameter of ½ inch, ¾ inch or 1 inch is recommended.
   b. Cut off any existing threads with a pipe cutter.
   c. Use a pipe reamer to remove any inside ridge.
   d. Cut threads on the pipe.
   e. Apply teflon tape or pipe joint compound.
   f. Screw on a pipe fitting such as a tee or elbow.
   g. Have the teacher check your work.
   h. Remove the fitting, wipe off any pipe joint compound and store the pipe.
3. Bolt and rivet two pieces of band iron.
   a. Cut two pieces of band iron ¼" x 1" x 6".
   b. Mark one piece A and the other B.
   c. Center punch piece A at 1½ inches, 3 inches and 4½ inches from one end and along the center line of the metal.
   d. Lay A over B.
   e. Drill a hole through both pieces near one end with a ⅛-inch drill.
   f. Redrill A with a $\frac{5}{16}$-inch drill.
   g. Redrill B with a ¼-inch drill.

h. Tap ⁵⁄₁₆ inch NC threads into the hole in B.

i. Screw the two pieces together with a ⁵⁄₁₆″ x ½″ stove bolt, machine bolt, or cap screw.

j. Drill a ¼-inch hole through both pieces at the center and end marks.

k. Install a round head rivet in the center hole.

l. Countersink the third hole and install a flat head steel rivet.

4. Solder sheet metal.

a. Obtain a clean steel fruit can and its lid.

b. Cut the outer corrugations off the lid so only the flat center remains.

c. Turn the empty can upside down.

d. Solder the round piece from the lid onto the flat round area on the bottom of the can.

e. Punch several holes in the corrugations outside the patched areas with a 4d nail.

f. Close the holes by soldering.

5. Solder electric wires.

a. Obtain two pieces of number 12, 14, or 16 stranded and insulated copper wire about 4 inches long.

b. Remove 1 inch of insulation from one end of each wire.

c. Cross the two bare wires and twist them around each other to form a smooth splice.

d. Solder the joint.

e. Tape the joint.

6. Sweat copper pipe.

a. Obtain a short length of ⅜ or ½ inch copper pipe.

b. Sweat a copper fitting onto the pipe.

c. Have the instructor inspect the job.

d. Remelt the solder and remove the fitting with pliers.

e. Wipe off the hot solder with a damp cloth.

☐ SELF-EVALUATION

A. Multiple Choice. Select the best answer.

1. Metal may be fastened by
   a. bolts
   b. rivets
   c. screws
   d. all of these

2. The process of joining metal by melting a different metal between two pieces is known as
   a. gluing
   b. soldering
   c. washing
   d. welding

3. Holes are usually made in heavy metal by using
   a. an auger bit
   b. a forge
   c. a punch
   d. a high-speed twist drill

4. When drilling, the lightest pressure should be placed on the drill when
   a. breaking through
   b. midway drilling
   c. starting the hole
   d. none of these

5. Threads are cut onto a rod with a
   a. die
   b. ream
   c. stock
   d. tap

6. When tapping threads, always start with a
   a. bottoming tap
   b. plug tap
   c. taper tap
   d. die

7. The first tool to use in drilling a hole in metal is the
   a. center punch
   b. drill
   c. file
   d. ream

8. When cutting threads, oil is used to
   a. clean the tool
   b. harden the threads
   c. lubricate the tool
   d. soften the metal

9. Pipe threads differ from bolt threads in that
   a. pipe threads are tapered, bolt threads are not
   b. oil is needed to cut pipe threads but not bolt threads
   c. pipe is threaded with a die, a bolt is not
   d. all of these

10. If you cannot get to both sides of sheet metal, the rivet to use is the
    a. copper rivet and burr
    b. pop rivet
    c. split rivet
    d. solid steel rivet

B. Matching. Match the word or phrase in column I with the correct word or phrase in column II.

Column I	Column II
1. drill	a. rivet
2. pop	b. solder
3. solder	c. jobbers-length
4. countersink rivet	d. National Fine
5. tin and lead	e. seal pipe threads
6. N.C.	f. flux
7. N.F. or S.A.E.	g. 50-50
8. sal ammoniac	h. mower knives
9. teflon tape	i. National Coarse
10. sweating	j. soldering copper pipe

C. Completion. Fill in the blanks with the word or words that will make the following statements correct.

1. The screw used to fasten thin metal is called a _____ _____ screw.

2. The metal used in the ends or tips of all soldering irons, guns or pencils is _____ .

3. The process of covering the tip of the soldering copper with solder is called _____ .

4. When soldering electrical wire or connections, _____ core solder must be used.

5. After cleaning with an abrasive, metal must be treated with a _____ before soldering.

Section 4
POWER TOOLS IN THE AGRICULTURAL MECHANICS SHOP

Unit 14
Portable Power Tools

Objective

To select and safely use major portable power tools in agricultural mechanics.

Competencies to Be Developed

After studying this unit, you should be able to:
- State recommended procedures for using portable power tools.
- Write a description of the uses of portable power tools.
- Name and properly spell the names of common portable power tools used in agricultural mechanics.
- Identify and spell the names of the major parts of portable power tools.
- Safely operate a portable power drill, belt sander, disc sander/grinder, finishing sander, sabre saw, reciprocating saw, and circular saw.

Terms to Know

saw horse	variable speed
trestle	reversible
double insulated	hammer
ground-fault interrupter	cordless
air tool	pilot hole
compressed air	orbital
duty cycle	belt sander
continuous duty	disc sander

grinders	blind cut
finishing sander	reciprocating saw
orbital sander	tiger saw
power handsaw	portable circular saw
sabre saw	bushing
bayonet saw	

Materials List

Examples of portable electric and air-driven tools
Unlabeled diagrams of portable power tools:
> Drill
> Belt sander
> Disc sander
> Grinder
> Finishing sander
> Sabre saw
> Reciprocating saw
> Circular saw

Note: Before proceeding with this unit, review the material in Unit 4 on personal safety.

Portable power tools save labor and are relatively inexpensive to buy. Most people prefer to invest in portable power tools rather than stationary ones for agricultural use. Since work on the farm and in other agricultural settings requires that tools be taken to the job, portable tools are especially useful and efficient.

When using portable power tools, it is important that the work be well secured. There are several ways to do this. If possible, place the work in a vise or some other kind of holding device. If large pieces are being used, saw horses, also called trestles, are recommended. A *saw horse* or *trestle* is a wood or metal beam or bar with legs. It is used for the temporary support of materials. Large flexible panels require special handling to prevent the binding of saws and equipment. If two or three 2 x 4s are placed on saw horses, flexible panels can then be placed across them for support. Large panels can be cut safely when supported in this manner.

□ SAFETY PRECAUTIONS

Several safety problems are inherent with portable electric tools. When working with power tools, sharp blades, bits, and abrasives should be used to reduce the pressure needed to make the tool function. Reduced pressure decreases the likelihood of the tool binding or slipping.

Wet areas are dangerous areas for using electrical power tools. Work areas should always be set up in dry locations. Wooden floors are the best, but concrete is safe if it is dry. Shoes with rubber soles or rubber boots reduce electrical hazards when power tools are being used. A *ground-fault interrupter* (*GFI*) should be used in circuits where the tool operator is working with wet hands, feet, or body. A GFI breaks the electrical circuit when the operator is threatened by electrical shock.

Another important safety consideration in using portable power tools is proper balance and footing.

It is important to stay balanced at all times. Excessive reaching which may cause lose of balance or control should be avoided.

Eye and face protection are very important when using power tools. Under no circumstances should power tools be used without wearing safety glasses or goggles. Face shields are also recommended.

In addition to eye protection, coveralls and other protective clothing are recommended when using power tools. It is important to avoid all loose clothing. Leather shoes are recommended as a minimum. Steel-toed shoes are needed when heavy materials are being handled.

Figure 14-1 lists some basic safety rules to follow when using electrical power tools.

Power Cords

The electrical cord of a power tool must be in good condition. The cord should be checked for broken insulation, broken plugs, bare wires, or other evidence of cord damage. Power cords often break, short out, or become electrical hazards due to breaks in the insulation where the cord leaves the power tool. It is advisable not to turn the first wind of the power cord too sharply if cords are wrapped around tools for storage.

All motors must have some method for protecting the operator against electrical shock in case of an internal electrical problem. Motors with metal housings should have a special ground wire and a plug with a ground prong, figure 14-2. The cords of such tools contain three wires: two wires carry current for the

---

1. If the tool housing is metal, be sure to use a three-wire grounded power source.
2. Be certain the power cord, switch, and all electrical parts are in good condition.
3. Be certain that blades, bits, and other cutting units are clean and sharp.
4. Support work carefully to avoid any tendency for it to bend, buckle, or bind when power tools are used.
5. Keep power tools clean and free of dirt.
6. Keep vent holes free of dirt to permit ventilation and cooling of the motor.
7. Wear complete body protection.
8. Exercise care to avoid accidental electrocution caused by working in wet areas or by cutting power cords.
9. Do not force power tools to cut, drill, or otherwise work faster than designed to work.
10. Hold every power tool firmly and in control at all times.
11. In school, always obtain the instructor's permission before using a power tool.
12. If there is any question about the condition of a power tool, check with the instructor before proceeding.
13. Announce to others around you when you are ready to start a power tool.
14. Lay the power tool down safely so it will not damage cutting parts or injure others.
15. Report any faulty condition of the power tool and its cutting parts to the instructor as soon as it is noticed.

Figure 14-1. Basic safety rules for using electrical power tools

Figure 14–2.   A three-prong, grounded plug

Figure 14–3.   Couple and uncouple air hoses with care. The short burst of air during coupling or uncoupling can cause injury to the operator or bystanders.

motor and the third wire is a safety ground wire. The ground wire connects to the longer prong on a 120-volt plug. This ground prong should never be broken off or damaged. If a tool in the agriculture shop is not grounded properly, the instructor should be notified immediately.

Motors with plastic housings generally are double insulated and do not need the special ground wire and ground prong. *Double-insulated* tools use two-wire, nongrounded cords. The electrical parts are insulated or separated from the user by special insulation inside the motor and by the plastic motor housing. Some electric motors have internal parts insulated in two different internal locations.

Extension Cords. □ In most cases an extension cord is needed with a portable power tool. Before a power tool is used, it should be determined if the tool is meant to be used with a three-wire extension cord. It is important that a three-wire grounded extension cord be used with tools that have metal bodies.

Extension cords should always be checked before use. Cords with frayed insulation or damaged ends should not be used. Extension cords longer than needed to reach the job should be used. If short cords are used, it is possible for tools to be pulled from benches if the cord is moved by passersby. Workers must always be aware of the location of the power cord. Otherwise, electrical shock may result due to damage to the cord by sawing, drilling, sanding, binding, or crushing.

Air-Driven Tools

Some agricultural mechanics shops have air tools. An *air tool* is a tool that is powered by compressed air. *Compressed air* is air pumped under high pressure and carried by special hoses. Compressed air provides pressure for spray guns; it also drives portable tools such as drills.

Compressed air can be dangerous if not handled properly. It is particularly hazardous when air hoses are being coupled and uncoupled, figure 14–3. High pressure air lines must not be uncoupled when people are nearby. Streams of air must never be directed toward a person's face or body. Dust and dirt may be driven into the eyes, or the compressed air may damage ear drums and other organs of the body. Serious accidents of many types have been reported from the misuse of compressed air. Compressed air is not recommended for cleanup purposes unless the pressure is less than 30 pounds per square inch.

□ PORTABLE DRILLS

A portable drill is a small tool that can be easily moved to the work. The major parts of a portable drill are the

• power cord
• handle
• motor housing
• gear chuck
• vents
• trigger switch
• trigger switch lock
• reversing switch
• chuck wrench or key

These parts are shown in figure 14–4.

Portable drills may be classified by the chuck size. A chuck is the device used to hold a drill or tool bit in the machine. Common portable electric drill sizes

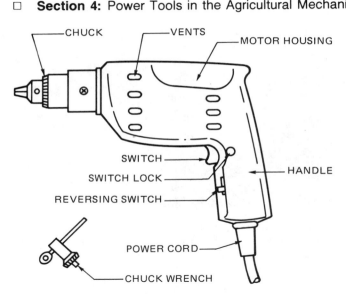

Figure 14–4. Major parts of a portable power drill

used in agricultural mechanics are ¼ inch, ⅜ inch, and ½ inch. The ¼-inch and ⅜-inch drills usually turn quite fast. The ½-inch drills run more slowly. This type of drill has reduction gears that generate more torque than is provided by the drill motor itself.

Another way of classifying drills is by their power rating. Power drills typically draw from two to five amperes of electricity. A 115-volt motor using five amperes of electricity develops approximatey ½ horsepower. Such a drill is considered a powerful portable drill.

Portable drills may also be rated by duty cycle. *Duty cycle* refers to the amount of time a motor can run versus the time it needs to cool off. *Continuous duty* means a tool can be used all the time for a 6- or 8-hour day. Most drills, however, are not continuous duty cycle drills. Therefore, it is possible to overwork and overheat a power drill. When a drill gets too warm to hold comfortably, it is time to stop work and let it cool off. If overused, it is possible to burn out the motor. This same principle applies to all portable electric power tools.

Types of Drills

Drills may be single speed or variable speed. *Variable speed* means the speed of the motor can be controlled by the operator. Variable speed drills are useful for special purposes such as running slow enough to drive screwdriver bits. These bits are used to install and remove screws. A variable-speed drill will turn fast for small drill bits and more slowly as needed for larger ones.

Some drills are reversible. *Reversible* means they will run backward as well as forward. A variable speed, reversible drill is useful to back screws out as well as drive them in.

Another feature of some drills is their ability to hammer. The capacity to *hammer* means a drill will turn a bit and also provide a rapid striking action on the bit to speed up drilling in masonry materials. This feature is especially useful for workers who must make holes in brick, concrete, or stone walls, or who must install plugs and anchors in such walls.

Some drills are cordless. A *cordless* drill means it contains a rechargeable battery pack to drive the unit when it is not plugged into an electrical outlet. Cordless drills operate for only a few hours. The tool must then be plugged into an electrical outlet or rechargeable unit to recharge the battery. The cordless drill is handy because it can be used in the field or any other location even though no electrical wiring exists.

Uses for Drills

Portable power drills have many uses, figure 14–5. The most basic is the drilling of holes. Power drills can turn many different types of bits. Some power drills can turn screws in and out. Masonry drill bits permit the drilling of holes in brick, block, or stone walls. Hole saws are driven by portable electric power drills. Some people use sanding discs and polishing heads on portable electric drills. When using a portable power drill, the following procedure is recommended.

Figure 14–5. The portable power drill may be used to drive drills, wood bits, hole saws, sanding discs, polishing bonnets, screwdriver bits, and other accessories.

Procedure

1. Use only straight-shank bits.
2. When tightening a drill chuck, place the key into one hole and tighten the chuck securely. Then place the key into a second hole and, again, tighten securely.

Caution:

Always remove the chuck key from the chuck after tightening a drill bit. Otherwise, the chuck key will be thrown when the drill is started.

---

3. Always center punch metal to help start a bit.
4. Hold materials in a vise or other secure device.
5. Use slow-turning drills for large bits.
6. Use even pressure on the drill.
7. Ease off the pressure when the drill is breaking through the material.
8. Always hold the drill so as to avoid binding the drill bit.
9. Keep operator positioned so that balance is always maintained.
10. Always remove the drill bit from the chuck when finished.
11. Store the portable power drill in its own case or in a special storage rack.

When drilling large holes, a pilot hole is used. A *pilot hole* is a small hole drilled in material to guide the center point of larger drills. By drilling a pilot hole, the bit stays exactly where planned and cuts with less power and pressure.

If a drill bit is not cutting, check the reversing switch to see if it is turned on. If the drill is turning counterclockwise, the bit will not cut. If the drill is turning clockwise and is not cutting the drill bit is dull and must be sharpened.

□ PORTABLE SANDERS

Three types of portable sanders are used in agricultural mechanics: They are the portable belt sander, portable disc sander, and portable finishing sander. Belt sanders and disc sanders are used for coarse sanding; finishing sanders are used for the last operation before applying finishes.

Power sanders do the same work as hand sanding. However, power sanders remove wood or other materials faster and easier than hand sanding. It is the speed that makes power tools desirable for sanding or grinding. However, the best work often can be done by hand sanding. Hand sanding permits the operator to control the sanding process better.

Effective Sanding

When sanding, it is important to sand with the grain of the wood for fine work. However, crossgrain sanding may be useful

- if boards are uneven.
- if extremely rough boards are encountered.
- if very difficult finishes must be removed.

Generally, coarse sandpaper is used first, then medium sandpaper, and finally, fine or very fine sandpaper is used to complete the job.

Sometimes, workers choose to use a belt or disc sander for the rapid removal of material. A finishing sander or hand sanding is then used to complete the job. The finishing sander moves forwards and backwards. Therefore, the sanding can always be with the grain of the wood.

A slight variation of the forward and backward movement is provided by an orbital sander. The *orbital* sander moves in a circular pattern. This results in a faster cut, but still leaves a fine finish. After using a finishing sander, it is desirable to hand finish using fine or very fine sandpaper. This process finishes wood to the point of maximum smoothness and beauty.

Safe Use of Portable Belt Sanders

The *belt sander* is a tool with a moving sanding belt, figure 14–6. Major parts include the

- motor housing
- handles
- belt
- belt rollers
- belt adjustment
- trigger switch
- switch lock
- power cord
- dust bag

Each part should be learned by the student.

The belt sander is often used in agricultural mechanics shops. It is a relatively safe tool to use, but care must be taken as with all power tools. When using the belt sander, the work should be secured well and both hands used on the machine to manage it very carefully, figure 14–7.

When using belt sanders, the following general pro-

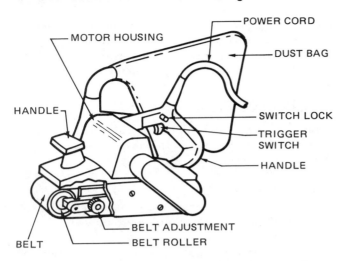

MOTOR HOUSING

POWER CORD

DUST BAG

HANDLE

SWITCH LOCK

TRIGGER SWITCH

HANDLE

BELT

BELT ADJUSTMENT

BELT ROLLER

Figure 14–6. Major parts of a portable belt sander

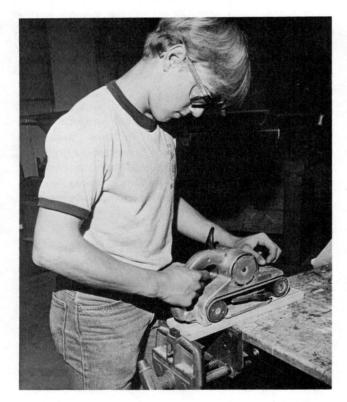

Figure 14–7. When using a belt sander, both hands are used to manage the machine so sanding is controlled.

cedures are recommended for safe use and effective operation.

Procedure

1. Wear suitable face protection and protective clothing.

2. Check the power cord and extension cords for safety.

3. Install a sanding belt of suitable coarseness.

4. Lay the sander on its side when not in use. This prevents the sander from running off the bench if accidently turned on.

5. Be sure the dust bag is empty or nearly so before starting to sand.

6. Always start the machine while holding it slightly above the material.

7. Keep the power cord out of the way of the belt.

8. After turning on the sander, touch the work with the front part of the belt first, then slowly settle the rest of the belt down onto the work.

9. Operate the machine with two hands at all times.

10. Sand with the grain. Move the machine from one end of the board to the other in a straight path; then move it slightly sideways and draw the machine back over new area. Gradually work across the board by slightly overlapping the forward and backward passes.

11. Keep the machine in motion. If permitted to sand in one spot, it will cut a depression in the wood.

12. The final movement is to lift the machine off the work while it is still running.

13. Examine the work carefully. If necessary, resand in order to create a perfectly level surface that is smooth.

14. Install a fine sanding belt and resand. This resanding leaves the work in its smoothest possible form using the belt sander.

15. Use a finishing sander or hand sand to obtain the degree of fineness desired.

Safe Use of Portable Disc Sanders and Grinders

Some portable tools may be used only as sanders or only as grinders. Others are designed to be used as

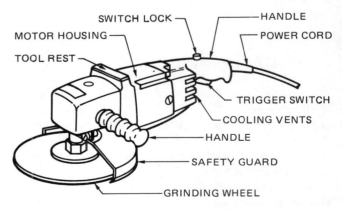

SWITCH LOCK

HANDLE

MOTOR HOUSING

POWER CORD

TOOL REST

TRIGGER SWITCH

COOLING VENTS

HANDLE

SAFETY GUARD

GRINDING WHEEL

Figure 14–8. Major parts of a portable sander/grinder

either, figure 14–8. By simply changing the sanding disc to a grinding wheel, many disc sanders become grinders. The manufacturer's instructions should be followed regarding any single tool. It is important to use the proper guards and wear protection with either the sanding disc, grinding wheel, or wire brush.

Some important parts of portable sanders and/or grinders are the

- motor housing
- handles
- power cord
- switch
- switch lock
- cooling vents
- wheel
- spindle
- safety guard
- tool rest

A *disc sander* is a tool with sanding materials, called grit, on a revolving plate. Disc sanders can be used for sanding wood or metal. However, only certain types of sanding discs are suitable for metal. Discs made from aluminum oxide may be used for sanding wood or metal. Discs made from flint paper are useable only for wood sanding.

Grinders have rigid grinding wheels instead of flexible discs. They cut metal only. Grinders may be used to shape metal, grind down welds, and remove metal as needed. Portable grinders may also be used to turn wire brushes used for cleaning metal. In agricultural mechanics, wire brushes are often used to remove rust and scaling paint.

Some special safety precautions are in order when using a grinding wheel. Some of these are as follows.

- Wear a face shield.
- Always check the grinding wheel for cracks or damage before use. Do not use a wheel that shows any sign of damage.
- Be sure to use wheels that are designed for the machine.
- Tighten the wheel securely and carefully.
- Never use a grinding wheel that is less than one-half of its original diameter.
- When preparing to grind small pieces, secure them in a vise, if possible.
- Do not grind metal in areas of combustible gases or materials.
- Hold the machine with both hands at all times.
- Do not discharge sparks against persons, clothing, or other combustible materials.

The procedure for using a sander or grinder is as follows.

Procedure

1. Select the correct sanding disc, grinding wheel, or wire brush for the job.

> **Caution:**
>
> **Be sure the wheel or disc is rated to turn at speeds higher than that of the machine.**

2. Install the appropriate guard for the job being done.
3. Wear appropriate face protection and protective clothing.
4. Be sure the work is properly secured.
5. Keep the power cord out of the way of the machine.
6. Grip the machine firmly with both hands and turn on the switch.
7. Settle the turning sanding disc, grinding wheel, or wire brush, onto the work slowly.
8. Touch the work gently with the wheel. This is to avoid the wheel catching the work and throwing metal particles toward the operator, bystanders, or flammable materials.
9. Do not apply pressure to the machine. The weight of the machine is generally sufficient to sand, grind, or brush properly. Keep the wheel clean and sharp. Keep a fresh abrasive disc on the machine to grind or sand quickly and efficiently.
10. After turning the switch off, do not lay the machine down until it has completely stopped.
11. Most machines have a rest or a flat spot to rest upon. Do not lay the machine down on its disc or wheel.
12. Remove the grinding wheel or sanding disc and store the machine properly after use.

Safe Use of Finishing Sanders

The *finishing sander* is a tool with a small sanding pad driven in a forward-backward or circular pattern. A straight-line sander moves backward and forward in a straight line, figure 14–9. An *orbital sander* is a finishing sander that moves in a circular pattern. The finishing sander is generally the last power tool used

Figure 14–9. The pad on a straight-line finishing sander moves backward and forward in a straight line.

on a project. It cuts slowly but gives the work a very smooth finish.

Important parts of the finishing sander are the

- motor housing
- handles
- switch
- switch lock
- power cord
- paper clamps
- pad
- sandpaper

These parts are shown in figure 14–10.

The recommended procedure for using a finishing sander is as follows:

Procedure

1. Use appropriate face and body protection.
2. Check the sander to determine if it is a straight-line or orbital type. Some sanders can be set to operate in either mode. Remember, straight sanding is the smoothest sanding.

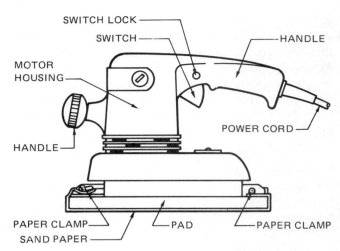

Figure 14–10. Major parts of a finishing sander

3. When starting with rough work, use coarse sandpaper for the first sanding. Switch to medium sandpaper, and then to fine sandpaper for the final sanding.
4. To prepare sandpaper, cut the paper to the appropriate size so it fits the machine. Precut sandpaper can also be purchased.
5. To install sandpaper, loosen the clamps on the pad, insert the paper, and close the clamps.
6. Proceed to sand applying only slight pressure on the sander. The sander should be in constant movement over the work.
7. Remove the dust from the work frequently to keep the material being removed from clogging the sandpaper.
8. Store the machine properly when finished.

□ PORTABLE SAWS

Portable saws are useful for carpentry projects and repair activities on the farm and in other agricultural settings. Their use will speed jobs and permit the operator to move tools to the job quickly and easily.

Portable saws include the sabre saw, reciprocating saw, and circular saw. The action of sabre and reciprocating saws is up and down or back and forth. They are compact, portable, and useful for cutting curves in plywood, paneling, dry wall, and other sheet materials. The circular saw is generally known simply as a power hand saw. The *power hand saw* has a circular blade and is used extensively for cutoff work.

Safe Use of Sabre Saws

The *sabre saw* is also referred to as a *bayonet saw*. It is used primarily to cut curves or holes in wood,

Figure 14-11. The sabre saw is a convenient tool for making curved and irregular cuts.

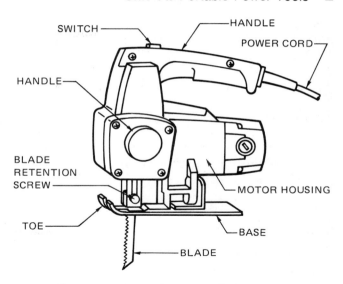

Figure 14-12. Major parts of a sabre saw

metal, cardboard, and similar materials, figure 14–11. Sabre saws are compact. The saw blade motion is up and down rather than circular and continuous. This feature makes sabre saws less hazardous than circular saws; they are dangerous none the less if not handled properly. The major parts of the sabre saw are the

- motor housing
- base
- blade
- toe
- blade retention screw
- handles
- switch
- power cord

These parts are shown in figure 14–12. A safe procedure for using the portable sabre saw is as follows.

Procedure

1. Wear proper eye protection and protective clothing.
2. Select the correct blade for the job. Narrow blades permit shorter turns than wider blades. Blades are available for cutting wood, metal, cardboard, and other material. Follow the manufacturer's instruction for the selection and use of blades.
3. Insert the blade and tighten the blade retention screw firmly. Do not apply so much pressure that the threads on the screw are stripped.
4. Adjust the base for the work being done. Some

saws have bases that tilt to make angle cuts. Almost all saws have a depth control.
5. Carefully secure all material being cut.
6. Start the cut at the edge of the board. When making inside cuts, bore a hole inside the circle, insert the blade and proceed.
7. A *blind cut* is a cut made by piercing a hole with the saw blade. To start blind cuts, place the toe of the base against the material with the blade free to start without striking the material. Start the saw, and carefully roll the saw backward to permit the blade to saw a hole gradually into the material.
8. Use firm pressure and uniform forward movement when operating the saw.
9. Make turns slowly and carefully and give the saw an opportunity to cut as it is directed into the curve.
10. Near the end of the cut take extra care to support the work. Also, reduce the rate of speed and pressure on the saw.
11. When finished, remove the blade and lightly retighten the blade retention screw to prevent accidental loss of the screw.
12. Store the tool properly.

Safe Use of Reciprocating Saws

The *reciprocating saw*, also referred to as *tiger saw*, is held and operated much like a portable drill. The action is at the end rather than underneath the tool. The main parts of the reciprocating saw are the

- motor housing
- blade retention screw

- blade
- shoe
- vents
- handles
- power cord
- trigger switch
- switch lock

These parts are shown in figure 14–13. Safe operating procedures for using the reciprocal saw include the following.

Procedure

1. Wear eye protection and protective clothing.
2. Select the correct blade for the job. Blades are available for cutting wood, plastic, metal, and other materials. Follow the manufacturer's recommendations for selecting blades for the reciprocal saw.
3. Select the speed for the job. Some saws have high speeds for woodworking and low speeds for metal working.
4. To make a blind or plunge cut, rest the saw on its shoe and gradually tilt the blade forward into the work.
5. Handle the saw as needed to make cuts in a fashion similar to the sabre saw.
6. Be careful not to bind, pinch, or crowd the blade when sawing.
7. Operate the saw with the shoe against the work at all times.
8. Remove the blade for storage.
9. Keep extra blades on hand for a variety of jobs.
10. Be especially careful not to wear loose fitting clothing. Observe other general safety precautions when using the reciprocating saw.

11. Always use the manufacturer's recommendations whenever there are questions of use and safety.

Safe Use of Circular Saws

The *portable circular saw*, also called a power hand saw, is a lightweight, motor-driven, round-bladed saw. It is perhaps the most popular saw used by people doing woodworking in agricultural mechanics. This saw is useful for many building and repair jobs on the farm and in other agricultural settings, figure 14–14. The major parts of the portable circular saw are the

- motor housing
- handles
- power cord
- trigger switch
- switch lock
- guard
- guard lift lever
- retractable guard
- blade
- spindle
- base
- angle adjustment lock
- angle scale

These parts are shown in figure 14–15.

The portable circular saw can be a very useful tool. However, its high speed and tendency to kick back make it a very dangerous tool. The user is cautioned to pay attention to detail and hold the tool very care-

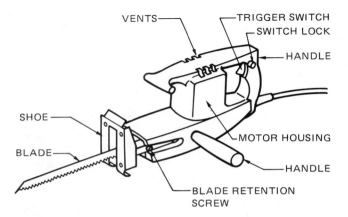

Figure 14–13. Major parts of a reciprocating saw

Figure 14–14. The circular saw is used extensively for ripping and cut-off work. Special blades are available for cutting concrete, brick, and stone.

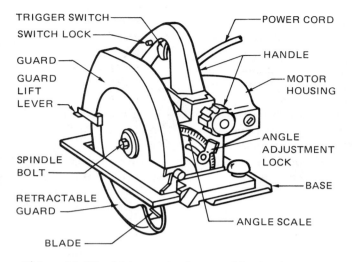

TRIGGER SWITCH
SWITCH LOCK
GUARD
GUARD
LIFT
LEVER
SPINDLE
BOLT
RETRACTABLE
GUARD
BLADE

POWER CORD
HANDLE
MOTOR
HOUSING
ANGLE
ADJUSTMENT
LOCK
BASE
ANGLE SCALE

Figure 14–15. Major parts of a portable circular saw

fully when using it. A safe procedure for using the power hand saw is as follows.

Procedure

1. Always use a face shield and protective clothing.
2. Choose an appropriate blade for the work being done. Special blades include crosscut, rip, combination, hollow ground, and safety blades.
3. Install the blade with care. Use the securing washers in the appropriate sequence. Use the correct wrench to tighten the spindle bolt. If a *bushing* is used for the blade to make it fit the spindle, be sure the bushing is in place. Install the blade so the bottom saw teeth point toward the front of the saw.
4. Support the work with trestles or other solid materials to avoid pinching or binding of the blade when sawing.

Caution:

Do not attempt to saw a board between two trestles. The board should be placed so the waste piece can drop off without binding the saw.

---

5. Adjust the blade for depth so that only one-quarter of an inch, or the length of a saw tooth, extends below the material.

6. Hold the saw securely with both hands at all times.
7. Start the saw while it is positioned near the work. Be careful that the blade is not touching another object or clothing.
8. Move the saw steadily into the work. If the saw stalls, back it up to clear its teeth from the material and correct the cause of the stalling.
9. When sawing, keep one eye on the line ahead of the saw and move the saw so its reference mark stays on the line.
10. Near the end of the cut reduce the pressure and release the switch as the cut is being finished.
11. When ripping, use the ripping guide if provided on the saw and if it is long enough for the cut being made.
12. For bevel cuts, adjust the angle of the base to the angle of the cut desired. Handle the saw the same as for other crosscuts or rip cuts.
13. When making a pocket cut, place the front of the saw base on the board, turn on the saw, and gradually lower the saw into the work.
14. When cuts are finished, be sure the saw blade stops completely before placing the saw at rest.
15. When the saw is not in use, unplug it to avoid accidental starting. Later model saws have a special switch or locking device which must be pushed before the switch will engage.
16. If the blade is gummy, remove the gum with alcohol or other solvent. If the blade is becoming dull, sharpen or replace it. If the retractable guard does not move freely and cover the blade at all times except when in use, correct the problem.
17. Properly store the tool.

□ STUDENT ACTIVITIES

1. Identify and correctly spell the names and parts of all power tools described in this unit.
2. Examine each portable power tool in the school agricultural mechanics shop. Determine its size classification, general condition, and the accessories available for use.
3. Study the operator's manual for each power tool in the shop.
4. Use each power tool described in this unit, under the close supervision of the instructor.
5. Examine all portable power tools which you have at home. Work with your parent(s) to arrange for any repairs that are needed on any of the tools.
6. Obtain a three-wire extension cord for the power tools at home if needed.

7. Arrange to have all bits, blades, and other accessories at home sharpened as needed.

☐ SELF-EVALUATION

A. Multiple Choice. Select the best answer.
1. A power tool that is double insulated has parts internally insulated in two locations or has a
   a. plastic motor housing
   b. three-wire cord
   c. special carrying case
   d. continuous duty cycle
2. Turning the first wrap of a power cord tightly around a power tool can cause the
   a. housing to break
   b. tool to use more electricity
   c. warranty to be voided
   d. insulation in the cord to break
3. When using power tools, electrical shock hazards may be reduced by wearing
   a. gloves
   b. rubber-soled shoes
   c. safety glasses
   d. coveralls
4. Compressed air is especially dangerous when
   a. coupling and uncoupling hoses
   b. room temperatures are high
   c. portable tools are used
   d. air lines are long
5. Most portable power drills have chuck sizes of
   a. ⅛, ¼, or ½ inch
   b. ⅛, ³⁄₁₆, or ½ inch
   c. ¼, ⅜, or ½ inch
   d. ¼, ⅜, or ⁹⁄₁₆ inch
6. A variable speed tool is one that will run
   a. a long time without damage
   b. backwards as well as forward
   c. in hot locations
   d. at different speeds
7. A feature that makes some drills especially good for drilling in masonry materials is its
   a. reversible action
   b. variable speed
   c. capacity to hammer
   d. continuous duty cycle
8. When drilling metal with a power drill
   a. use a center punch
   b. make a pilot hole
   c. use a cordless drill if appropriate
   d. all of these

B. Matching. Select the word or phrase in column I that best matches the word or phrase in column II.

Column I	Column II
1. circular pattern	a. good for sanding only wood
2. finishing sandpaper	b. good for sanding metal
3. flint paper	c. three-wire, grounded cord
4. aluminum oxide	d. type of saw blade
5. sabre saw	e. tiger saw
6. reciprocating saw	f. fine or very fine
7. power hand saw	g. ventilation
8. metal housing	h. battery pack
9. bushing	i. bayonet saw
10. cordless	j. circular saw
11. vent holes	k. blade insert
12. crosscut	l. orbital sander

C. Completion. Fill in the blanks with the word or words that will make the following statements correct.
1. When using portable power tools, it is important to have the work well _____ .
2. Another word for saw horse is _____ .
3. Under no circumstance should one use power hand tools without protecting the eyes with safety _____ or _____ .
4. A three-wire power cord and plug with a ground prong is used on power tools with _____ bodies.
5. The type of saw with the blade extending from its end is called a _____ saw.
6. Sanding belts should run _____ the grain of the wood.
7. A good accessory to use on portable grinders to remove rust from metal is the _____ _____ .
8. Grinding wheels that are _____ should be discarded.
9. When sawing large panels, _____ are used on trestles to provide extra support to the material.
10. The portable power tool that has a retractable guard is the _____ _____ .

D. Identification. Identify the parts of the following power tools.

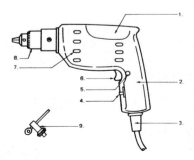

1. Portable drill.

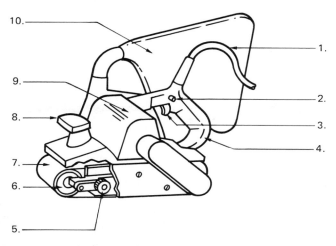

2. Belt sander.

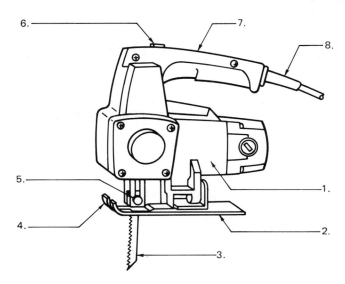

5. Sabre saw.

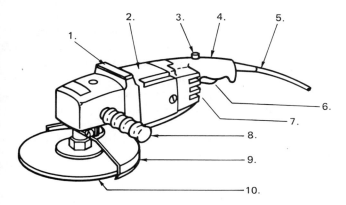

3. Portable grinder.

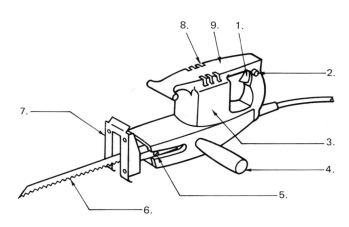

6. Reciprocating saw.

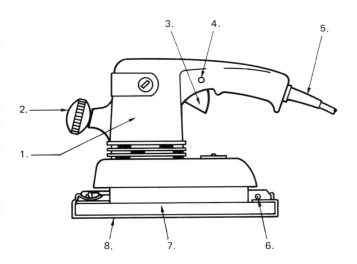

4. Finishing sander.

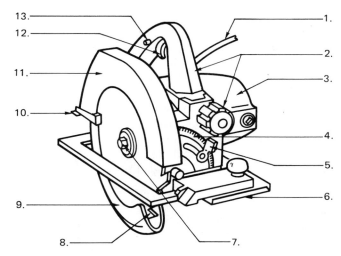

7. Circular saw.

Unit 15
Woodworking with Power Machines

To use stationary power woodworking machines in a safe manner.

After studying this unit, you should be able to:
* State basic procedures for using stationary power woodworking machines.
* Identify and properly spell major parts of specified machines.
* Operate a band saw.
* Operate a jig saw.
* Operate a table saw.
* Operate a radial arm saw.
* Operate a jointer.
* Operate a planer.
* Operate a sander.

power machine	bench saw
stationary	tilting arbor
band saw	parallel
blade guide	push stick
tilting table	dado heads
miter gage	molding heads
rip fence	radial arm saw
jig saw	cutoff saw
scroll saw	pivot
reciprocal	jointer
pulley	dress
table saw	rough lumber

Band saw
Jig saw
Radial arm saw
Table saw
Jointer
Planer
Bench brush and/or shop vacuum
Roller stand or helper
1 piece full dimension (rough) lumber, 2" × 4" × 14"

Note: Before proceeding with this unit, review the material in Unit 4 on personal safety.

The use of electricity has improved the quality of life for all who use it wisely. Electricity provides a way of moving energy from place to place. Electrical energy is generated from water power or fuels and is then sent over long distances by power lines to be received in homes, shops, and industries where it is put to use. In the agricultural mechanics shop, electrical energy is converted by relatively small motors to power in other forms used by shop machines. It is this power that makes machines hazardous to the careless operator. However, the careful operator uses the energy to get work done safely, quickly, and with ease.

☐ SAFETY WITH POWER MACHINES

A *power machine* is a tool driven by an electric motor, hydraulics, air, gas engine, or some force other than, or in addition to, human power. Some tools such as metal shears may use levers and cams to increase human power. Such machines develop so much force that they, too, should be regarded as power machines.

Large power tools are stationary and should be placed in permanent locations in the shop. The word *stationary* means having a fixed position.

To reduce the likelihood of injury when working with stationary power machines and to ensure efficient use of the machines, the following are recommended:

* Plan the location of each machine carefully.
* Firmly anchor each machine to the floor.
* Have a licensed electrician provide electrical hookups if needed.
* Use a stripe or narrow line to mark the safety zone around each machine.
* Follow the manufacturer's recommendations for the

installation, use, adjustment, and repair of each machine.
- Keep guards and shields in place on each machine at all times.
- Keep blades, knives, and bits sharp.

Safety Precautions

Safety rules are important and should be followed when using all shop machines.

1. Wear goggles and/or face shield.
2. Wear protective clothing that is not loose or baggy.
3. Walk—do not run around machines.
4. Only the operator is to be in the safety zone around a machine.
5. Do not use a machine without the instructor's permission.
6. Do not use a machine unless it is in good working order.
7. Perform only the procedures on a machine for which you have had instruction.
8. Do all operations slowly and cautiously.
9. Use a push stick to help push or guide small pieces.
10. Get help with large pieces of stock.
11. If the machine is worked too hard, an overload protector should stop the motor. If this happens, notify the instructor and the instructor will correct the problem.
12. Turn off the machine before leaving it.
13. Unplug the machine or switch off the circuit breaker when changing blades or doing repairs.
14. Clean woodworking machines with a brush or vacuum cleaner. Never clean the machines directly with the hand.
    Note: Before continuing with this unit, learn these safety precautions.

□ BAND SAW

The *band saw* is a power tool with saw teeth on a continuous blade or band, figure 15–1. The band saw will cut straight or curved lines and different kinds of materials. It is fast cutting and versatile when the proper blade and speed are used. Careful operation is absolutely necessary.

The blade of a band saw is flexible and is stretched over flat wheels. The face of each wheel is covered with flat rubber called a tire. The rubber protects the teeth of the saw blade. Generally, the wheel diameter determines the size of the saw. A band saw with 16-inch wheels is classified as a 16-inch band saw. This

Figure 15–1. The band saw gets its name from the fact that cutting is done by a thin, continuous band of steel that has teeth along its entire length (*courtesy of Delta International Machinery Corporation, formerly Rockwell Power Tool Division*).

means the saw can be used to cut to the center of stock up to 32 inches wide.

A *blade guide*, with carefully adjusted rollers, supports the blade. The blade guide and other parts of the guide and table assembly of a band saw are shown in figure 15–2. The rollers allow the operator to push material into the teeth of the blade. In addition, material may be turned gradually to make curved cuts. The guides keep the blade from bending, breaking, or moving off the wheels during normal use. Blades are available in various types and widths. The more narrow the blade, the shorter the curve which may be cut. Blades used in school shops are generally ¼ inch to ½ inch in width.

Most band saws have a *tilting table*. The table can be set at various angles so that cuts of 45° to 90° may be made.

Most band saw tables also are equipped with a miter gage. A *miter gage* is an adjustable, sliding device to guide stock into a saw at the desired angle. Some

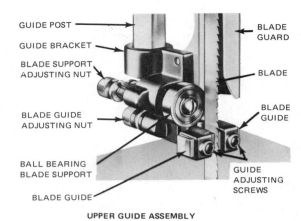

GUIDE POST
GUIDE BRACKET
BLADE SUPPORT ADJUSTING NUT
BLADE GUIDE ADJUSTING NUT
BALL BEARING BLADE SUPPORT
BLADE GUIDE
BLADE GUARD
BLADE
BLADE GUIDE
GUIDE ADJUSTING SCREWS

UPPER GUIDE ASSEMBLY

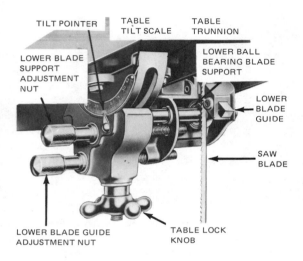

TILT POINTER
TABLE TILT SCALE
TABLE TRUNNION
LOWER BLADE SUPPORT ADJUSTMENT NUT
LOWER BALL BEARING BLADE SUPPORT
LOWER BLADE GUIDE
SAW BLADE
LOWER BLADE GUIDE ADJUSTMENT NUT
TABLE LOCK KNOB

LOWER GUIDE AND TABLE ASSEMBLY

Figure 15–2. Guide and table assemblies for a typical band saw (*courtesy of Delta International Machinery Corporation, formerly Rockwell Power Tool Division*)

band saws have a rip fence. A *rip fence* is a guide that helps keep work in a straight line with the saw blade. It can be adjusted to a distance as close to, or as far from, the blade as the table size permits.

Band saws may have a speed control. This permits the saw to be used for cutting metal as well as wood. A slow-running saw with an appropriate blade is needed to cut metal.

Safe Operation of Band Saws

Band saws are adaptable to many cutting jobs. Therefore, specific instructions are needed for various operations. However, the following procedure is recommended for use of all band saws.

Procedure

1. Obtain the instructor's permission to use the band saw.
2. Put on a face shield and protective clothing.
3. Check to see if the saw is equipped with a suitable blade for the job.
4. Be sure the blade is tight and the guide is properly adjusted.
5. Move the upper guide assembly down until it is within ⅛ inch of the top of the stock.
6. Place the miter gage or the rip fence in position and adjust as needed.

Caution:

Never use both the miter gage and the rip fence at one time as the stock will bind and create a serious hazard.

7. Clear the table and blade of materials. Keep hands out of the work area. Switch the machine on and off again.
8. If the blade seems to be tracking correctly, stand back from the blade and turn the machine on and let it run.
9. Slowly and carefully push the stock to be cut into the teeth of the blade.

Caution:

Keep fingers and hands away from the front of the blade.

10. When straight cuts are made, move the stock in a straight line.
11. When curved cuts are made, rotate the stock slowly as the saw cuts.
12. Try to arrange all cuts so you can cut all the way through the stock.

Caution:

If you must back out of a cut, turn the machine off to do so.

13. Make sure the blade has stopped before leaving the machine.

□ JIG SAW

The *jig saw* is also known as a *scroll saw* and is designed for sawing curves, figure 15–3. It is a poor tool for straight cutting. The saw cuts by means of the reciprocal action of the blade. *Reciprocal* means back and forth. The reciprocal action makes it a relatively safe tool if proper procedures are followed. However, the operator may receive serious injuries if the machine is not properly adjusted and if the work is not moved carefully into the blade. The major parts of a jig saw are shown in figure 15–4. Jig saws may be mounted on a bench or on a stand.

Blades for the jig saw are generally 6½ inches long. They resemble a coping saw blade except they do not have pins in the ends. Blades range in thickness from 0.05 inch to 0.25 inch. When buying blades, the specifications are stated by length, thickness, and width. For example, a 6½″ × 0.020″ × 0.110″ 20T blade means that it is 6½ inches long, 0.02 inch thick, 0.11 inch wide and has 20 teeth per inch. The number of teeth per inch indicates the size of the teeth and the speed at which the blade will cut. The narrower the blade, the shorter the turn the blade can make.

The tension assembly of a jig saw consists of a sliding sleeve, a spring, and a pistonlike rod with a chuck. This upper chuck holds the upper end of the blade; a lower chuck holds the other end. The lower chuck is driven by a crank mechanism in the housing of the base. The crank pulls the blade down and a spring

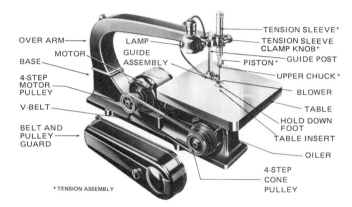

Figure 15–4. Major parts of the jig saw (*courtesy of Delta International Machinery Corporation, formerly Rockwell Power Tool Division*)

pulls it up. Blades must be installed with their teeth pointing down towards the table.

The jig saw's hold-down foot puts spring tension on the material being cut. This tension prevents the material from being lifted from the table as the blade rises. Since the teeth point downward, the blade cuts on the downward stroke.

Many jig saws have special pulleys. A *pulley* is a round device attached to a shaft and designed to fit a drive belt. Many jig saws have pulleys with two or more steps. These permit the speed of the machine to be changed by placing the belt on the various pulley steps. A motor with a large pulley drives a machine with a small pulley faster than the speed of the motor. Conversely, if the motor has a pulley smaller than the machine, the machine runs slower than the motor.

Safe Operation of Jig Saws

Jig saws may be used to cut wood, plastic, or metal. However, the correct blade and machine speed must be used. The following procedure is recommended when installing a blade and using a jig saw.

Procedure

1. Obtain the instructor's permission to use the machine.
2. Put on a face shield and protective clothing.
3. Unplug the motor.
4. Select an appropriate blade.
5. Attach the blade in the lower chuck.
6. Rotate the motor pulley until the blade is at its highest point.
7. Attach the blade in the upper chuck.
8. Loosen the tension sleeve clamp knob.

Figure 15–3. The jig saw will cut very short curves (*courtesy of Delta International Machinery Corporation, formerly Rockwell Power Tool Division*).

9. Lift the tension sleeve until moderate spring tension is placed on the blade. Retighten the tension sleeve clamp knob.
10. Rotate the pulley by hand to be sure the blade will go all the way up without buckling. Readjust, if needed.
11. Adjust the hold-down foot so it puts slight pressure on the stock.
12. Plug in and turn on the machine.
13. Push the work slowly into the front of the blade. Keep fingers to the side.
14. Make the saw kerf on the waste side of the line.
15. Rotate the stock as the saw cuts so it follows the curves.
16. Avoid backing out of cuts. Saw out across waste wood to the edge if necessary.
17. To make a cut inside a circle, first bore a hole in the stock inside the area to be cut. Remove the saw blade and reinstall it through the hole in the stock. Saw inside the circle.
18. Turn off the machine when finished.

□ TABLE SAW

The table saw is known also as a bench saw. A *table saw* or *bench saw* is a type of circular saw with either a tilting arbor or a tilting table for bevel cuts. A *tilting arbor* is a motor, belt, pulley, shaft, and blade assembly that moves as a unit, figure 15–5. The operator can adjust the arbor for depth or angle of cut.

Table saws are classified by the diameter of the blade. Most saws are 8 inches, 10 inches, or 12 inches. Saws with 8-inch blades make beveled cuts in wood up to 2 inches thick. The 8-inch saw is recommended for sawing small pieces. Its blade, guides, and guards work well for fine cutting and are easier to adjust than is the case with larger saws. Agricultural mechanics shops are often equipped with the larger, more powerful 10-inch or 12-inch saws. These are more suitable for cutting large boards and rough-sawed lumber.

Most saws have a hand wheel on the left side which is used to adjust the tilt of the blade from 45° to 90°. In addition, there are slots in the table top for a miter gage. The miter gage is also adjustable from 45° to 90°. This permits cross cutting of boards at the desired angle.

A hand wheel on the front of the saw controls depth of cut. The wheel is near the operator and is easy to reach. The saw blade should be adjusted so it is never higher above the board than a distance equal to the depth of the teeth.

Table saws are equipped with a rip fence. The rip fence slides on the front and back edges of the table. It has a scale to indicate the width of cut that will be made at any given setting. When the fence clamp is tightened, the fence guides lumber through the saw for cuts with parallel edges. *Parallel* means two edges or lines are the same distance apart at all points along the length of the object. Saws should be equipped with antikickback devices for ripping, figure 15–6.

All bench saws are provided with a blade guard, figure 15–7. The guard shields the hands of the operator from the dangerous cutting blade. It also provides some protection from splinters, saw dust, sparks, or metal bits that can be thrown by the blade.

The blade guard should always be kept in place.

Figure 15–5. A tilting arbor table saw (*courtesy of Delta International Machinery Corporation, formerly Rockwell Power Tool Division*)

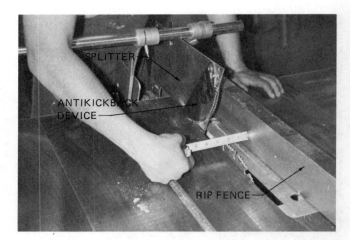

Figure 15–6. Setting the position of the rip fence on a table saw. For an exact measurement, the distance is checked between the ripping fence and a tooth that is pointing toward the ripping fence. Notice the antikickback device on the splitter located behind the saw blade.

Figure 15–7. Blade guards of various types are available for table saws. Here a special see-through design permits the operator to see the blade, yet is protected from direct contact with it (*courtesy Brett Guard Division, Foredom Electric*).

The guard will be lifted by approaching stock which glides under the guard. This provides maximum safety for the operator.

Sawing Large or Small Pieces

Two people are needed to safely manage pieces of material over three feet long or two feet wide. If a helper is not available, adjustable roller stands to help support the material should be used.

Pieces of wood being ripped to 3 inches or less should be pushed with a push stick. A *push stick* is a wooden device with a notch in the end to push or guide stock on the table of a power tool. Every effort should be made to keep hands several inches from the sides of a saw blade. The operator should never place the fingers or hands in front of the blade.

Selecting the Correct Saw Blade

Saw blades are classified generally by their diameter. Hence, a blade that has a diameter of 8 inches is called an 8-inch blade.

Figure 15–8. Many types of blades are available for table saws. The carbide tip blade has become very popular because it stays sharp longer than regular steel teeth (*courtesy of Sears, Roebuck and Co.*)

Blades are also classified by the type of teeth. A blade may be a rip blade, crosscut blade, or combination blade. It may have large teeth for fast, aggressive cutting or small teeth for smooth cutting. All blades must be sharpened according to their design. Blades with carbide tips on their teeth stay sharp longer because of their extra hardness, figure 15–8.

Blades are also designed and classified according to their unique function. For instance, *dado heads* are special blades that can be adjusted to cut kerfs from ⅛ inch to ¾ inch in a pass. Similarly, *molding heads* are blades that hold knives to shape wood into moldings of various types.

Care is needed when installing a blade on a saw. The hole in the blade must be the exact side of the shaft. A suitable bushing can be used if the hole is too large. The slot in the plate in the saw table must be suitable for the blade being used.

Safe Operation of Table Saws

Table saws are used for many sawing jobs. Instruction is needed for each special use. However, the following procedure is advisable when using table saws.

Procedure

1. Obtain the instructor's permission to use the saw.
2. Put on a face shield and protective clothing.
3. Unplug the motor.

4. Install the correct blade for the job with the teeth pointing toward the direction of rotation.

5. Adjust the blade to the correct angle. Use the degree scale on the saw or use a sliding T-bevel to determine the desired angle.

6. Adjust the height of the blade until only the teeth extend above the board to be sawed.

7. If ripping, move the miter gage off the saw. Always use the miter gage or the rip fence, but never both at the same time. Set the rip fence for the desired with of cut. This is checked by measuring from the fence to the tip of a tooth closest to the fence.

8. If crosscutting, move the rip fence to the right edge of the table or remove it. Adjust the miter gage to the desired angle. The correctness of the angle is determined by using the degree scale on the gage or by using a sliding T-bevel.

9. Check to see that the guard assembly is in place and the saw is ready.

10. Arrange for a helper or set up stands if a long or large piece of lumber is to be sawed.

11. Plug in the motor.

12. Stand to the side of the blade's path and turn on the motor.

13. Push the work toward the blade with a slow and even movement.

14. Hold the work firmly so as to support both pieces on both sides of the blade.

15. Push the final work through with a push stick.

16. Turn off the machine.

□ RADIAL ARM SAW

The radial arm saw is also known as a cutoff saw. The *radial arm saw* or *cutoff saw* is a power circular saw which rolls along a horizontal arm. The saw can be raised or lowered. The arm will also pivot up to 45° to the left and right.

The many movements of the radial arm saw make it capable of numerous cutting operations. These same movements make it a very dangerous tool. Whole books are written on using the radial arm saw. Only the simpler and more popular uses are described in this unit.

The radial arm saw has many moving parts and major assemblies, figure 15–9. To *pivot* means to turn or swing on. The arm pivots on the column, the yoke pivots under the arm, and the motor and blade assembly pivot at the bottom of the yoke. Each pivot point has a scale to indicate the position of the assembly. Each point also has a lock to hold the assembly once it is set. The major parts of the radial arm saw are shown in figure 15–10.

Safe Operation of Radial Arm Saws

The most popular use of the radial arm saw is for cutoff work, including squaring boards, cutting them to length, cutting them at angles and bevels, cutting dados, and cutting rabbets.

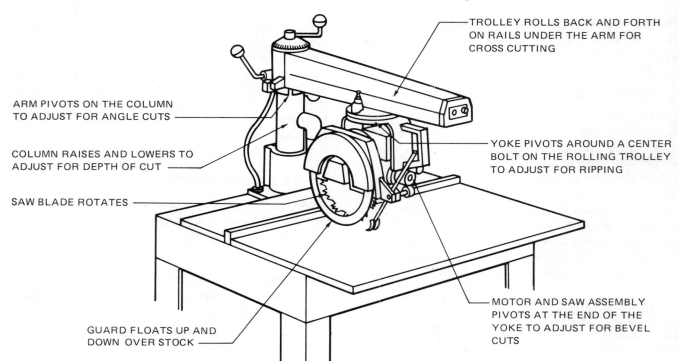

ARM PIVOTS ON THE COLUMN TO ADJUST FOR ANGLE CUTS

COLUMN RAISES AND LOWERS TO ADJUST FOR DEPTH OF CUT

SAW BLADE ROTATES

GUARD FLOATS UP AND DOWN OVER STOCK

TROLLEY ROLLS BACK AND FORTH ON RAILS UNDER THE ARM FOR CROSS CUTTING

YOKE PIVOTS AROUND A CENTER BOLT ON THE ROLLING TROLLEY TO ADJUST FOR RIPPING

MOTOR AND SAW ASSEMBLY PIVOTS AT THE END OF THE YOKE TO ADJUST FOR BEVEL CUTS

Figure 15–9. Moving assemblies on a radial arm saw

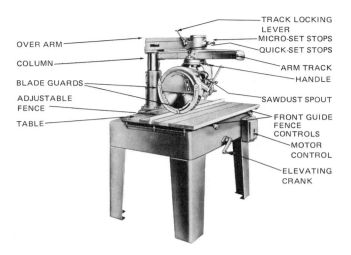

OVER ARM
COLUMN
BLADE GUARDS
ADJUSTABLE FENCE
TABLE

TRACK LOCKING LEVER
MICRO-SET STOPS
QUICK-SET STOPS
ARM TRACK
HANDLE
SAWDUST SPOUT
FRONT GUIDE FENCE CONTROLS
MOTOR CONTROL
ELEVATING CRANK

Figure 15–10. Major parts of a radial arm saw (*courtesy of Delta International Machinery Corporation, formerly Rockwell Power Tool Division*)

Checking the Table for Levelness. □ The radial arm saw cannot make accurate cuts unless the table is level and parallel to the arm. The following procedure may be used to check the table.

Procedure

1. Obtain the instructor's permission to check the table.
2. Put on a face shield and protective clothing.
3. Unplug the machine or lock the switch in the "off" position.
4. Loosen the column clamp handle.
5. Turn the elevating crank until the points of the teeth on the blade lightly touch the table.
6. Tighten the column clamp handle.
7. Loosen the trolley lock.
8. With the saw turned off, grasp the saw handle and slowly pull the assembly across the table. The teeth should just touch the table at all points. If they do not, report the condition to the instructor. Do not use the saw until the problem is corrected.

Crosscutting Boards. □ The following procedure is recommended for crosscutting.

Procedure

1. Obtain the instructor's permission to use the saw.

2. Put on a face shield and protective clothing.
3. Mark the board where the saw will start the cut.
4. Elevate the saw until the teeth will just reach through the board and touch the table top.
5. Position the board on the table tightly against the fence so the saw will cut on the waste side of the mark.
6. Grasp the saw handle with one hand.
7. Turn on the saw with the free hand.
8. Hold the board with the free hand, being careful to position the hand so the hand and arm are not in line with the blade.
9. Stand out of line of the path of the blade.
10. Very slowly and firmly pull the saw forward until the cut is complete, figure 15–11.
11. Push the saw back to the column.
12. Maintain the hold on the handle of the saw while you turn the saw off and lock the switch with the other hand.
13. Do not touch the lumber until the blade stops.

Figure 15–11. When using the radial arm saw for cutoff work, it is important to hold the work securely and pull the saw slowly into the wood (*courtesy of Delta International Machinery Corporation, formerly Rockwell Power Tool Division*).

Crosscutting Bevels. □ The saw can be swung to the left or right to crosscut at angles. A beveled cut may also be made at any angle within the swing radius of the arm. When cutting a bevel, the following additional steps are required.

Procedure

1. Elevate the saw about 2 inches above the table.
2. Loosen the bevel latch.
3. Swing the saw assembly to the desired angle as shown on the bevel scale.
4. Tighten the bevel latch.
5. Lower the saw to the table.
6. Proceed to crosscut.

Making Dados and Rabbets. □ Dados and rabbets are wide grooves cut only part way through the board. To cut dados and rabbets, the following procedure is used.

Procedure

1. Mark the board with a sharp pencil to show exactly where the grooves start and stop.
2. Mark the end of the board to indicate how deep the cut should be.
3. Install a dado blade if available.
4. Place the board on the saw table and pull the saw out so the blade is against the end of the board.
5. Elevate the blade until the points of the teeth touch the mark which indicates the depth of cut on the board.
6. Push the saw back and position the board so the cut will remove the desired material.
7. Turn on the saw.
8. Make the pass.
9. Push the saw back.
10. If a dado blade is not being used, move the board ⅛ inch and make another pass.
11. Repeat until the groove is finished.
12. Turn off the saw.

Many other types of cuts can be made with the radial arm saw. It is recommended that other books or bulletins on the subject be read if additional operations are desired.

Special Safety Precautions

When using a radial arm saw it is important to:

- Ask the instructor before attempting any operation.
- Always wear a face shield.

- Unplug or lock the switch in the off position when making adjustments.
- Always pull the saw toward the operator when sawing—never push the saw back into the wood.
- Always have all guards in place.
- Always have one hand on the saw handle when the saw is turning or about to be turned on (it is easy for the saw to accidentally roll forward).
- Never turn the saw on when the blade is touching wood.
- Never stop a turning blade by pushing a piece of wood into it.

□ JOINTER

The *jointer* is a machine with rotating knives used to straighten and smooth edges of boards and to cut bevels, figure 15–12. The jointer is potentially a very dangerous tool. The knives can inflict severe cuts, and lumber may be thrown if not handled properly.

The size of a jointer is determined by the length of the knives. Most school shops have either 6-inch or 8-inch jointers. However, jointers are available up to 15¾ inches or more.

The main parts of the jointer are shown in figure 15–13 and include the base, front infeed table, front table adjusting hand wheel, rear outfeed table, rear table adjusting hand wheel, rabbeting ledge, depth scale, knife assembly or cutterhead, guard, fence, fence clamp, and tilt scale.

The knives are installed so their cutting edges extend to the same height. The rear outfeed table is adjusted so it is level with the cutting edges of the knives; the table is locked at this level. Any adjustments of the rear outfeed table should be made by the instructor. The fence may be moved across the table to any point. Its position determines the maximum possible width of cut. In addition, the fence can be set at any angle from 45° to 90°. This position determines the angle of the edge which is created.

The height of the front infeed table determines the depth of cut. If the front infeed table is exactly level with the knives, the depth of cut will be 0 inches. The table is lowered to increase the depth of cut. The depth of cut is indicated by a scale on the side of the jointer.

Safe Operation of Jointers

The jointer is designed to remove a small amount of wood (1/16 inch or less) at a time. Generally, the operator should joint one edge smooth, then use a table saw to cut the board to ⅛ inch wider than

A. Using a jointer to dress the edge of a board

B. Using a jointer to bevel the edge of a board

Figure 15–12.   A jointer is used to dress and bevel the edges of boards.

needed. The board is then jointed down to the final width. This leaves the board with both edges smooth.

Strict attention must be given to the correct procedure to be used with the jointer.

Procedure

1. Obtain the instructor's permission to use the machine.
2. Put on a face shield and protective clothing.
3. Check to see that the guard is covering the knives.
4. Adjust the fence to the proper angle (usually 90°).
5. Adjust the front infeed table for a $\frac{1}{16}$-inch cut. Only $\frac{1}{16}$ inch per cut should be removed. Never cut deeper than $\frac{1}{8}$ inch in one pass.
6. Be sure the board is 12 inches or more in length. Arrange for a helper if the board is longer than 4 feet.

7. Stand aside and turn on the machine.
8. Place the board on its edge against the fence on the front infeed table.
9. Grip the board by placing the hands over the top edge.

> **Caution:**
>
> **Keep fingers high on the board and not over the cutterhead.**

10. Advance the board forward against the fence with moderate downward pressure.
11. When the leading hand is nearly over the knives, reposition it on the board over the rear outfeed table.

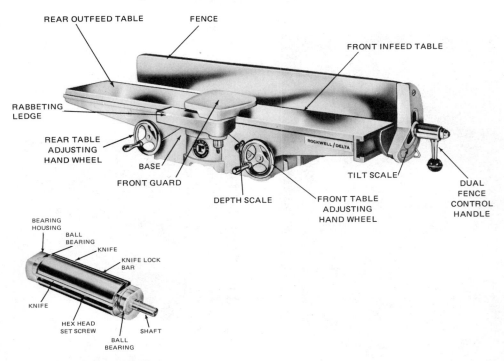

3-KNIFE CUTTERHEAD

Figure 15–13. Major parts of a jointer (*courtesy of Delta International Machinery Corporation, formerly Rockwell Power Tool Division*)

Caution:

Keep hands away from the cutter knives.

12. When most of the board has passed over the knives, transfer the second hand so it is also over the rear table.
13. Continue moving the board to finish the pass with decreasing pressure near the end. Use a push stick to complete the pass for boards that are less than 4 inches wide. Do not use the jointer on boards less than 2 inches wide.
14. Reverse the board from end to end each time another pass is made. This decreases the tendency to create a wedge-shaped board after many passes.
15. Make the last pass with the grain to get a smooth cut.

□ PLANERS

The planer (also known as a thickness planer or surface planer) is a machine with turning knives that dress the sides of boards to a uniform thickness. To *dress* means to remove material and leave a clean surface.

The planer is an excellent tool to convert low-cost, locally grown lumber into smooth, uniform materials. Planers are used to dress *rough lumber* to a desired thickness and leave it smooth. Lumber is rough when it comes from a sawmill. Rough lumber may be thicker on one end than on the other. By making successive passes through the planer, a board can be dressed down to the desired thickness, figure 15–14.

The planer is also an excellent tool to level and smooth wide pieces made by gluing boards together. Agriculture students can make attractive projects such as gates, wagon bodies, and bench tops from rough lumber at a reasonable cost if a planer is available.

The planer is a massive machine with many parts. It includes an adjustable bed with smooth rollers to support lumber, corrugated rollers to draw lumber into the machine, and a rotating cutter head with knives that dress the lumber. Once a board is started in the machine, the machine is self-feeding. The parts of the planer are shown in figure 15–15. However, most parts of the planer are shielded and normally hidden from view.

Figure 15–14. The planer dresses surfaces and leaves the board at the desired thickness (*courtesy of Delta International Machinery Corporation, formerly Rockwell Power Tool Division*).

Safe Operation of Planers

Several critical adjustments on the planer must be made by the instructor or power tool specialist. These include installing knives, adjusting the feed rollers, and adjusting the smooth rollers.

Caution:

Students are not to make these adjustments.

If used carefully, the planer will safely dress large volumes of lumber. If used carelessly, the knives can be ruined by the first board planed after they are installed. Careless use can also cause the operator to be seriously injured. The planer has a large motor capable of throwing massive pieces of lumber. Great care must be exercised with the use of the planer.

When using a planer, it is important that no metal or loose knots be in the lumber put through the planer. The cutter heads may throw these items, or the planer may be damaged by them.

The planer generates large volumes of wood chips. Therefore, a vacuum chip removal system should be used to reduce the dust created by planing.

The following procedure is recommended for dressing rough lumber with a planer. This same procedure is used for edge planing materials that are 2 inches

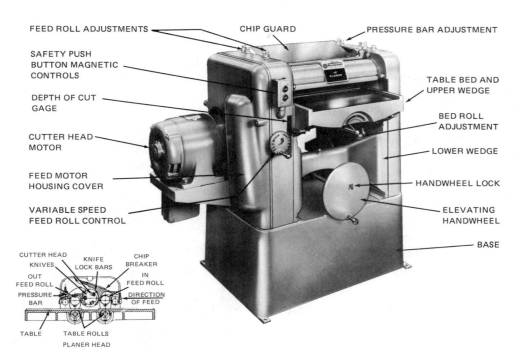

Figure 15–15. Major parts of the planer (*courtesy of Delta International Machinery Corporation, formerly Rockwell Power Tool Division*)

thick or more. Lumber should be reasonably straight for edge planing.

Procedure

1. Obtain the instructor's permission to use the planer.
2. Put on a face shield and protective clothing.
3. Check the machine and be sure the chip guard and other shields are in place.
4. Sort the lumber according to its general thickness.

> **Caution:**
>
> **Never plane lumber that has grit, staples, or other metal in it.**

5. Identify the thickness of the thickest end of the thickest piece of lumber.
6. Measure the thickest part of the thickest board.
7. Turn the thickness adjustment wheel until the thickness scale reads ⅛ inch less than the thickest piece to be planed. For instance, if the board is 2 inches thick, the planer should be set at 1⅞ inches.
8. Arrange for a helper.
9. Clear bystanders out of the path in front and back of the planer.
10. Set the feed control at its slowest rate.
11. Stand to the side and turn on the planer motor.
12. Have the helper place the thinner end of the board on the front end of the bed.
13. Lift the other end of the board until it is slightly higher than the end on the planer bed.
14. While staying beside the board and keeping the end high, push the board slowly forward into the planer.

> **Caution:**
>
> **When the thick part of the lumber is in the planer, the feed rollers will pull the lumber forward. Otherwise, the operator must push the lumber.**

15. When the lumber is halfway through the planer, the helper should move to the back of the machine to pull the lumber through and complete the first pass.
16. Run additional boards through the planer.

17. Raise the bed ⅛ inch with the thickness adjustment wheel.
18. Repeat steps 12 through 17 until both sides of all pieces are smooth and within ⅛ inch of the desired thickness.
19. Make one final pass on each side of the boards, going with the grain, and removing ¹⁄₁₆ inch per pass.
20. Turn off the machine and wait for the machine to stop.

Note: The last pass on each side of the board should be made so it is planed with the grain and removes only ¹⁄₁₆ inch per pass. This technique results in a smooth finish. The depth of each cut may vary from as little as ¹⁄₃₂ inch to ⅛ inch. The adjustments should be planned so the final pass leaves the board at the desired thickness.

□ SANDERS

Some agricultural mechanics shops may have stationary sanders, figure 15–16. For procedures in sanding, refer to "Portable Sanders" in Unit 14. When using portable sanders, the machine is moved to the material. To use stationary sanders the material is moved to the machine.

Sanders may be equipped with belts or discs suitable for use on wood or metal. The procedure for using a stationary sander is as follows.

Procedure

1. Obtain the instructor's permission to use the sander.
2. Wear safety glasses or other eye protection.
3. Wear a dust mask.
4. Use a belt or disc that is not clogged with particles.

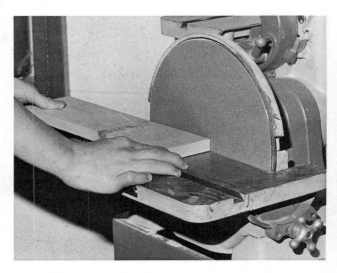

Figure 15–16. Using a stationary sander

5. Turn on the dust collector (if so equipped) and the sander motor.

6. Grip the material firmly with fingers well away from the surface to be sanded.

7. On belt sanders, place the work on the belt lightly and move it against the stop.

8. On belts, move the material back and forth in a sideways motion. This helps prevent the belt from over heating and becoming clogged with particles.

9. On discs, use the half of the disc moving downward toward the table. Again, keep the stock moving on the table.

10. Use only moderate pressure to sand.

11. Reduce pressure before lifting or pulling the work from the belt or disc.

12. When doing freehand sanding of faces, ends, edges, and corners, use light pressure and keep the work moving at all times.

13. Turn off the machine when finished.

□ STUDENT ACTIVITIES

1. Identify and correctly spell the names of major parts of each machine in this unit.

2. Examine each stationary power woodworking machine in your shop and make a list of items that need maintenance or repair. Submit the list to your instructor for review and appropriate action.

3. Copy the italicized terms in this unit and write a definition for each.

4. Properly clean a band saw, table saw, radial arm saw, jointer, planer, and other woodworking machines in the shop.

5. Do the following Power Workworking Machine Exercise.

POWER WOODWORKING MACHINE EXERCISE
(Courtesy of Douglas Hering, Brunswick High School, Brunswick, MD)

Purpose: The purpose of the Power Woodworking Machine Exercise is to obtain experience quickly. It requires the use of many power tools yet the material used is generally scrap material in an agricultural mechanics shop.

Note to the Teacher: Students should be instructed on the use of each machine before operations are performed.

Power Machine to Use	Procedures
Radial Arm Saw or Portable Circular Saw	1. Obtain a rough 2 x 4 that is at least 2 inches thick and 4 inches wide. Cut it to a length of 14 inches, figure 15–17.
Planer	2. Plane the 2 x 4 to a thickness of 1½ inches.
Jointer	3. Joint one edge of the 2 x 4 one time or until smooth. (Jointer should be set at a depth of $\frac{1}{16}$ inch.)
Table Saw	4. Place the jointed edge of the 2 x 4 against the fence. Rip the 2 x 4 to a width of 3⅝ inches.
Jointer	5. Joint the rough side of the 2 x 4 to a width of 3½ inches.
Jointer	6. Joint the top edges of the 2 x 4 at 45° angles, ⅛ inch deep. This is done by setting the jointer for a $\frac{1}{16}$ inch cut and making 2 passes.
Radial Arm Saw	7. Cut the 2 x 4 to a length of 12 inches.
	8. On the bottom side of the 2 x 4, mark and label the angles shown in figure 15–18 using a try square and pencil.
	9. Cut the pieces labeled "A" from the 2 x 4.
Band Saw	10. Cut the piece labeled "B" from the 2 x 4.
	11. On the bottom side of the 2 x 4, starting at the pointed end, make pencil marks at 3, 6, and 9 inches down the center. Locate the exact center so that cross marks can be drawn, figure 15–19.
Portable Hand Drill	12. Drill a ½-inch hole through the 2 x 4 at the cross mark closest to the point.
Drill Press or Portable Power Drill	13. Drill two 1-inch holes through the other two cross marks with a spade bit.
	14. Connect the edges of the two 1-inch holes with a straight edge and pencil. Label the area inside as "C," figure 15–20.
Sabre Saw or Jig Saw	15. Cut "C" out by sawing on the two pencil lines.
Router (optional)	16. Shape the inside edge of the top side of the 2 x 4, moving the router clockwise around the hole.
Disc Sander or Belt Sander	17. Sand the point of the 2 x 4 project lightly.
	18. Put your name on the bottom side of the 2 x 4 project in pencil.
	19. Turn your project in to your instructor to be graded, figure 15–21.

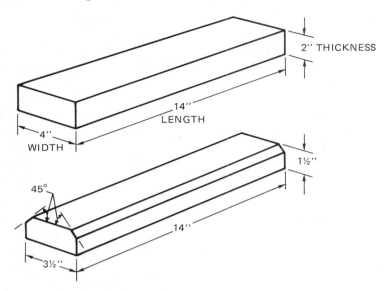

Figure 15-17.

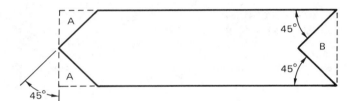

Figure 15-18.

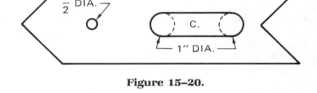

Figure 15-20.

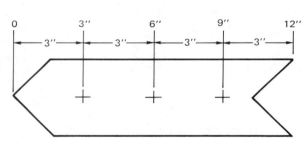

Figure 15-19.

Figure 15-21. The completed project for the exercise should look like the piece shown in this illustration.

□ SELF-EVALUATION

A. Multiple Choice. Select the best answer.
1. The number of people permitted in the safety zone around a machine is
   a. one
   b. two
   c. three
   d. any number
2. Woodworking machines should be cleaned with
   a. a brush
   b. the hand
   c. a rag
   d. an air gun
3. If a machine is worked too hard, the electric motor should stop because of
   a. burnout
   b. general fatigue
   c. overload protection
   d. voltage drop
4. When using a band saw, the operator should avoid
   a. backing out of cuts
   b. crosscutting

c. cutting metal

d. sawing curved lines

5. The band saw blade is held in position when cutting by

a. wheels

b. tires

c. levers

d. guides

6. Jig saws are best for cutting

a. very short curves

b. straight lines

c. rabbets

d. dados

7. The blade on a table saw should extend how far above the work:

a. 1 inch

b. 2 inches

c. 3 inches

d. none of these

8. Small pieces of wood should be moved on a saw table by a

a. bare hand

b. gloved hand

c. push stick

d. hammer handle

9. Table saw blades are classified by the

a. type of teeth

b. size in diameter

c. unique function

d. all of these

10. The most popular use of the radial arm saw is

a. ripping

b. cutoff work

c. curve cutting

d. dado cutting

11. Minimum protection when using power machines starts with

a. steel-toed shoes

b. leather apron

c. finger guards

d. face shield

12. When cutting with the radial arm saw, the operator should

a. move the wood into the saw

b. pull the saw into the wood

c. push the saw into the wood

d. any of the above are safe

13. An adjustment that should be made on a jointer by the instructor is the

a. rear outfeed table

b. front infeed table

c. fence

d. miter gage

14. The maximum safe depth per cut by a jointer is

a. ½ inch

b. ¼ inch

c. ⅛ inch

d. $\frac{1}{16}$ inch

15. The power machine that generates large volumes of wood chips is the

a. band saw

b. bench saw

c. jointer

d. planer

16. For a good job when planing lumber, the operator should

a. make each final pass while planing with the grain

b. make each final pass with a shallow cut

c. end up with the desired thickness

d. all of these

B. Matching. Match the word or phrase in column I with the correct word or phrase in column II.

Column I	Column II
1. miter gage	a. table saw
2. rip fence	b. parallel to the blade
3. radial arm saw	c. guides angle cuts
4. many moving assemblies	d. saw kerf ⅛ inch to ¾ inch wide
5. tilting arbor	e. jointer
6. rotating cutterhead	f. dress sides of lumber to a uniform thickness
7. dado head	g. also called a cutoff saw
8. planer	h. radial arm saw
9. sander	i. special need for dust mask
10. jig saw	j. reciprocal action

C. Completion. Fill in the blanks with the word or words that will make the following statements correct.

1. Electrical wiring for shop machines should be installed by a _____ _____.

2. The safety zone around a machine should be marked with a _____.

3. Bits, knives, and blades should be kept _____.

4. Do only the procedures on a power machine for which you have had _____.

5. The table saw is also known as a _____ _____.

Unit 16
Metalworking with Power Machines

Objective

To safely use stationary power machines for metalworking in agricultural mechanics.

Competencies to Be Developed

After studying this unit, you will be able to:
- State basic procedures for using stationary machines for metalworking.
- Identify and properly spell major parts of specified machines.
- Operate a drill press.
- Operate a grinder.
- Operate power metal-cutting saws.
- Operate a power shear.
- Operate a metal bender.

Terms to Know

drill press	coolant
grinding wheel	horizontal
grit	horizontal band saw
face	shear
dress	

Materials List

Drill press
Grinder with grinding and wire wheels
Power hacksaw and/or metal-cutting band saw
Metal shear
Metal bender
Drill bits, ⅛ inch and ½ inch
Length of ⅛" × 1" flat steel (longer than 6 inches)

Note: Before proceeding with this unit, review Unit 4, "Personal Safety in Agricultural Mechanics" and the topic "Safety with Power Machines" in Unit 15.

Power is used in a number of ways for metalworking in agricultural mechanics. It is used to cut, make holes, shape, and sharpen metal. These functions will be discussed under the following headings of drill press, grinder, metal-cutting saws, power shears, and metal bender.

☐ DRILL PRESS

The *drill press* is a stationary tool used to make holes in metal and other materials. Its design and structure permit it to drive large drills and apply heavy pressure on bits. When material can be placed on a drill press, this tool is preferred over the portable power drill for drilling large holes. The drill press is also preferred for precision drilling or when many holes are needed.

Drill presses are available as floor models or bench models. The parts are similar for both. The drill press consists of a base, column, table, and head. The head is an assembly consisting of the motor, switch, drive belt, speed control, shaft, quill, chuck, and feed. Parts of the drill press are shown in figure 16–1.

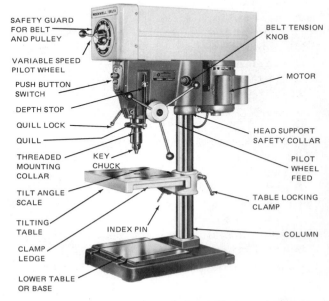

SAFETY GUARD FOR BELT AND PULLEY
VARIABLE SPEED PILOT WHEEL
PUSH BUTTON SWITCH
DEPTH STOP
QUILL LOCK
QUILL
THREADED MOUNTING COLLAR
KEY
CHUCK
TILT ANGLE SCALE
TILTING TABLE
INDEX PIN
CLAMP LEDGE
LOWER TABLE OR BASE
BELT TENSION KNOB
MOTOR
HEAD SUPPORT SAFETY COLLAR
PILOT WHEEL FEED
TABLE LOCKING CLAMP
COLUMN

Figure 16–1. Major parts of the drill press (*courtesy of Delta International Machinery Corporation, formerly Rockwell Power Tool Division*)

Safe Operation of the Drill Press

The motor of the drill press drives a shaft which, in turn, drives the chuck. The speed of the chuck is controlled by the use of step pulleys, variable speed pulleys, or a variable speed motor. When the operator moves the feed handle, a gear moves the quill up and down. This movement raises and lowers the chuck to permit drilling, figure 16–2. The movement of the quill may be limited by the depth stop, or the quill may be locked for special functions.

The table slides up or down on the column. This permits thick or thin material to fit on the machine. The table may be swung to one side so that large objects can rest on the base in a drilling position. Some drill presses have tables that tilt. Tilting tables permit materials to be supported for angle drilling.

The following procedure is recommended for the safe use of the drill press.

Procedure

1. Obtain the instructor's permission to use the machine.
2. Wear a face shield and protective clothing. Leather gloves are recommended when drilling sharp or irregular-shaped metal.

Figure 16–2. When the operator moves the feed handle, a gear moves the quill up and down. This action raises and lowers the chuck allowing drilling to take place (*courtesy of Delta International Machinery Corporation, formerly Rockwell Power Tool Division*)

3. When drilling wood, use a sharp pencil to mark where holes are to be drilled. A cross (+) or caret (^) is useful to mark an exact location.
4. When drilling metal, use a center punch to aid in starting the drill.
5. Use only straight shank drills in gear chucks, and taper shank drills in taper chucks.
6. Select the correct drill by observing the size stamped on the shank, or use a drill gage to determine drill size. Use a pilot hole for holes larger than ⅜ inch.
7. Tighten gear chucks securely by inserting the chuck wrench into at least two of the three holes, twisting the wrench as tightly as possible in each.

Caution:

Be sure to remove the chuck wrench after tightening.

8. Check table alignment. To do this, move the feed handle until the drill passes down through a hole in the table and then return it. A flat piece of wood may be used to protect the table.
9. Clamp flat material to the table; clamp round material in a V-block.

Caution:

The long end of the material should be on the operator's left and against the drill press column to prevent any rotation.

10. Turn on the motor.
11. Grasp the feed handle and lower the drill to start the hole.
12. Apply steady pressure while drilling.
13. Add cutting oil to cool the drill when drilling steel.
14. Ease off and use very little pressure as the drill breaks through the metal.
15. Raise the drill.
16. Turn off the machine.
17. Use a file to remove burrs from the metal.
18. Use a bench brush to remove metal chips from the project and work area.

19. Use a cloth to clean oil from the project and work area.

(**Note:** Refer to Unit 10, "Fastening Wood," and Unit 13, "Fastening Metal," for applications where drilling is needed.)

□ GRINDER

A grinder removes metal by abrasive action. Grinders are available in many types and sizes. They are used to sharpen tools, shape metal, prepare metal for welding, and remove undesirable metal.

Most shop grinders consist of a pedestal, a double-shafted motor, switch, wheels, guards, tool rests, and safety shields, figure 16–3. Some grinders have a light to improve vision and a water pot for cooling metal. Small grinders have wheels that are 6 inches or 7 inches in diameter and 1 inch wide. Large grinders generally have wheels that are 12 inches in diameter and 2 inches wide.

Grinders may be equipped with grinding wheels or wire brush wheels. A *grinding wheel* is a wheel made of abrasive cutting particles formed into a wheel by a bonding agent. The cutting particles are called *grit.*

When selecting a grinding wheel it is important to use the correct wheel for the job. The wheel must have the correct speed rating for the motor. Failure to select the proper wheel can result in a serious accident. The wheel may fly into pieces, hurling rocklike fragments into the face and body of the operator and bystanders.

When replacing wheels, obtain specifications from the operator's manual or from the wheel on the machine. Bushings that fit the shaft must be used when a bushing is required.

A coarse or very coarse texture is recommended for wheels used to shape metal for sharpening, welding, or construction. A medium texture wheel is good for sharpening axes, rotary lawn mowers, knives, and other tools. Fine textured wheels are not generally recommended for agricultural mechanics except for power oil stone or wet stone grinders. Fine wheels cut slowly and overheat the metal easily.

Grinding wheels are manufactured to be used on one surface only. The surface intended for use is called the *face.*

Wire wheels are useful for removing rust and dirt from machinery parts, figure 16–4. Such cleaning is useful before welding or painting. Wheels with large, stiff wire are called coarse wire wheels. Coarse wheels cut aggressively and can be dangerous. Care must be exercised to use a tool rest to support metal being brushed. Otherwise, the wheel may drag the metal down between the wheel and the guard. Such accidents may result in injury to the operator, a damaged object, or a broken machine. Fine wire wheels are good for moderate cleaning jobs. They are softer than coarse wheels and safer for the operator.

TWIN-LITE SAFETY SHIELD — MOTOR — ADJUSTABLE SPARK DEFLECTOR — 7" x ¾" GRINDING WHEEL — WATER POT — WHEEL GUARD — ADJUSTABLE TOOL REST — PLANE BLADE GRINDING ATTACHMENT — SWITCH — DUST CHUTE — PEDESTAL

Figure 16–3. Major parts of the grinder (*courtesy of Delta International Machinery Corporation, formerly Rockwell Power Tool Division*)

Figure 16–4. Grinder equipped with wire wheel. Wire wheels on grinders are useful for removing rust and welding slag.

Safe Operation of Grinders

In addition to the points made above, the following procedure is recommended for the safe operation of a grinder:

Procedure

1. Obtain the instructor's permission to use the grinder.
2. Wear a face shield.
3. Wear close-fitting leather gloves and a leather apron. Small work pieces should be held with lever lock pliers.
4. Check the wheels. Use the machine only if the wheels are clean and free of nicks, chips, and cracks.

Caution:

If any of these conditions are not met, have the instructor inspect the wheels and correct the problem(s).

---

5. Adjust the tool rest so its top surface is level with the center of the motor shaft and within $\frac{1}{16}$ inch of the wheel. For bevel grinding, set the tool rest at the appropriate angle.
6. Hold the metal to be ground firmly on the tool rest, figure 16–5.
7. Move the metal back and forth or in a curved motion as needed. The movement will help keep the stone clean and avoid overheating of the metal.

Cautions:

- **Do not use excessive pressure on the metal. The speed of grinding is determined by the grit and speed of the wheel, not by pressure.**
- **Never grind with the side of the wheel.**
- **Grinding creates heat, therefore, avoid handling the metal with bare hands.**

8. Turn off the machine when finished and wait for it to stop running.

Dressing the Grinding Wheel. ☐ After proper instruction, the student may dress a grinding wheel. To *dress* a wheel means to remove material so the wheel is perfectly round with the face square to the sides and sharp abrasive particles exposed. Dressing a wheel removes clogged abrasives and slight bulges that cause the wheel to get out of balance, figure 16–6. The procedure for dressing a wheel is as follows.

Procedure

1. Obtain the instructor's permission.
2. Wear a face shield. Leather gloves and leather apron are also recommended.
3. Wear a dust-type filter respirator.
4. Obtain a wheel-type grinding wheel dresser.
5. Turn on the grinder.
6. Place the dresser on the tool rest of the grinder.

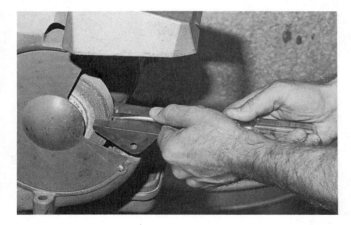

Figure 16–5. The tool rest helps position and guide small objects being ground. The tool rest should be adjusted so it is $\frac{1}{16}$ inch or less from the wheel and level with the motor shaft.

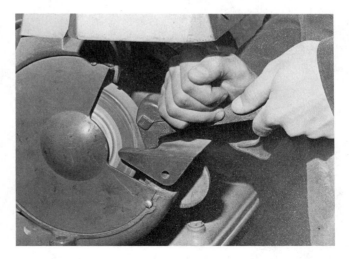

Figure 16–6. When dressing a grinding wheel, material is removed to leave the wheel perfectly round. The face is left square with the sides of the wheel.

7. Slowly rock the dresser wheels forward until they touch the grinding wheel.

8. Apply firm and even pressure on the dresser and move it back and forth across the wheel for about 30 seconds.

Caution:

Be prepared for abrasive particles to fly off the wheel. The wheel is cleaned, balanced, and squared by removing particles.

A. Checking a wheel for roundness using the tool rest.

B. Checking the wheel for squareness of the face using a combination square. The tool rest may also be used if it has not been damaged.

Figure 16–7. Checking the grinding wheel

9. Remove the dresser and turn off the grinder.
10. Readjust the tool rest so it is within 1/16 inch of the wheel. Then rotate the wheel by hand to check for roundness, figure 16–7A.
11. Use a combination square or the tool rest to determine if the face of the wheel is square to the side, figure 16–7B.
12. Repeat steps 5 through 11 if additional dressing is needed.

Applications for grinders will be found in units on metalworking, welding, tool fitting, and others. The grinder is a basic tool and is found in most shops.

□ METAL-CUTTING POWER SAWS

Metal-cutting power saws are reciprocating hacksaws, band saws or thin grinding-type wheels that cut metal. The hacksaw and band saw are used most often in agricultural mechanics.

Hacksaws

The power hacksaw has a reciprocating movement which operates like a hand hacksaw, figure 16–8. However, it is motor driven and cuts much faster. Fast cutting saws have a system to pump a coolant onto the cutting area. A *coolant* is a liquid used to cool parts or assemblies.

The power hacksaw consists of a motor driven frame mounted on a stand. The frame holds a rigid blade which is ¾ to 1 inch wide and 12 inches to 18 inches long. Blades are available with fine to coarse

Figure 16–8. The power hacksaw has a reciprocating movement which works like a hand hacksaw. The blade is hard and, therefore, brittle and easily broken. The metal being cut must be clamped carefully to eliminate blade breakage due to binding.

teeth. Very hard blades are needed to cut hard steel. Such blades are brittle and break easily if the frame is dropped or if the saw binds. Extra care is needed to prevent blade breakage.

The power hacksaw stand is narrow. Therefore, metal must be supported to avoid tipping the machine. Sturdy models can support the weight of heavy steel. However, long stock must be supported by stands or a helper.

The power hacksaw is very useful in the agricultural mechanics shop. However, an expensive blade can be broken in an instant if the machine is not used properly. The following procedure is recommended for using a power hacksaw.

Procedure

1. Obtain the instructor's permission to use the saw.
2. Wear a face shield, leather gloves, and a leather apron.
3. Check the machine to be certain it has the proper blade and that the blade is tight.
4. Place the frame in the raised position.
5. Adjust the vise on the machine to hold the metal at the desired angle.
6. Position the metal in the vise on the machine and tighten securely.

> **Caution:**
>
> **If the vise does not hold the metal so that it is rigid, the blade will break.**

7. Turn on the machine.
8. Lower the frame slowly and carefully until the blade is on the stock and starting to cut.
9. Turn on the coolant, if it does not turn on automatically.
10. Stay near the machine while it is cutting.

> **Caution:**
>
> **Do not put pressure on the blade or otherwise interfere with the machine while it is running.**

Figure 16–9. The horizontal band saw is especially useful for cutting large pieces of structural metal, such as round stock, flats, angles, I beams, and channels (*courtesy of Wells Manufacturing*).

11. Switch off the machine when the cut is finished if it does not turn off automatically.
12. Remove all scrap metal and clean up all metal dust and coolant.

Horizontal Band Saw

The word *horizontal* means flat or level. The *horizontal band saw* has a blade that saws parallel to the ground. Generally, however, it is constructed like an upright band saw. It has a band-type blade that travels on wheels and moves through rollers and guides, figure 16–9. Since blade movement is forward at all times, it cuts continuously. As a result, it cuts faster than a power hacksaw.

The procedure for cutting with a horizontal band saw is similar to the procedure described for a hacksaw. However, the manufacturer's instructions should be followed. Machines vary in minor details of design and function.

□ POWER SHEAR

Sheet metal is easy to cut with hand shears. Therefore, power sheet metal shears may not be seen in many agricultural mechanics shops. To *shear* means to cut at an angle. Generally, shearing is done with two movable blades.

Figure 16–10. Metal shear for cutting flat and round stock. Special knives for some shears make them capable of cutting angle iron.

The cutting of flat, angle, and other structural steel is difficult by hand, and slow by power saw. The cutting of these materials is fast and clean when a shear is used, figure 16–10. Some shears have features that permit them to bend, cut, and punch metal. To operate a hand power shear, the following procedure is recommended.

Procedure

1. Obtain the instructor's permission to use the machine.
2. Put on a face shield, leather gloves, and a leather apron.
3. Raise the handle of the shear.
4. Insert the metal to be cut or punched.
5. Carefully align the cut mark with the cutting edge of the stationary blade.
6. Support the metal so it is level.
7. Lower the handle until the cut is complete.

Caution:

Keep hands away from the cutting shears or punch.

8. Store the handle so it cannot accidentally fall and operate the shear.

Caution:

The weight of the handle alone is sufficient to close the shears with enough power to cut off fingers.

The procedure described for use of a hand power shear also applies to hydraulic shears. In the case of hydraulic shears and punches, power is delivered by cylinders driven by hydraulic fluid.

□ METAL BENDERS

Bending metal by hand is discussed in Unit 12. However, some shops have hand operated benders. These are useful for bending small round and bar stock. By using cams, pins, and levers, such tools permit fast and accurate bends, figure 16–11.

Figure 16–11. The hand metal bender is useful for bending round, flat, and angle stock (*courtesy of Strippit/Di-Acro Houdaille*).

Figure 16–12. The hand operated sheet metal bender is useful for making angle bends and seams.

Sheet metal benders are available for making rain gutters, heating and air conditioning ducts and other commercial uses, figure 16–12. However, sheet metal bending in agricultural mechanics is generally limited to repair work. Such occasional work may be done with hand tools.

□ STUDENT ACTIVITIES

1. Copy the italicized terms in this unit and write the definition of each.
2. Examine the grinders in the shop to determine if the wheels need dressing. Ask the instructor to demonstrate how to dress a wheel.
3. Check the wheels on any grinders you have at home. Dress them if needed.
4. Do the following sawing, brushing, grinding, drilling exercise.
   a. Use a metal-cutting power saw to cut a 6 inch piece of ⅛ × 1″ flat steel.
   b. Clean the ⅛ × 1″ × 6″ piece with a power wire brush.
   c. Round the corners and edges of each end slightly with a power grinder.
   d. Measure in one inch from one end and center punch on the center line.
   e. Drill the center punched hole with a ⅛-inch drill bit.
   f. Redrill the hole with a ½-inch drill bit.
   g. Remove all burrs or rough spots with a power wire wheel.
   h. Measure in 1½ inches from the end without a hole.
   i. Make a 90° bend at the line with a metal bender.
   j. Submit your project to the instructor.

□ SELF-EVALUATION

A. Multiple Choice. Select the best answer.
1. After installing a drill in a gear chuck, the next important thing is to
   a. start the motor
   b. remove the chuck wrench
   c. place the table off center
   d. check the belt for tightness
2. Round stock is best held for drilling by a
   a. C-clamp
   b. helper
   c. vise
   d. V-block
3. A grinding wheel may fly apart when running if the wheel does not have
   a. a coarse texture
   b. a coolant device
   c. a clean surface
   d. adequate speed rating for the motor
4. Grinding wheels are cleaned and restored to roundness with a
   a. grit cutter
   b. screwdriver
   c. wheel dresser
   d. any one of these
5. A dangerous act is to grind
   a. with heavy pressure on the metal
   b. using the side of the wheel
   c. without wearing a face shield
   d. all of these
6. Blades for power hacksaws are generally
   a. Less than 12 inches long
   b. 12 inches to 18 inches long
   c. ½ inch to ⅝ inch wide
   d. none of these
7. A critical step to prevent breaking of power hacksaw blades is
   a. buy brittle blades
   b. clamp the work securely
   c. cool the blade frequently
   d. provide some slack in the blade
8. Gloves that are worn while doing metalwork should be made of
   a. asbestos
   b. cotton
   c. leather
   d. all are recommended
9. Power metal shears can cut
   a. angle stock
   b. flat stock
   c. round stock
   d. all of these

B. Matching. Match the word or phrase in column I with the correct word or phrase in column II.

Column I	Column II
1. chuck	a. type of drill bit
2. quill	b. bonded grit
3. center punch	c. coarse grinding wheel
4. taper shank	d. medium grinding wheel
5. shear	e. power for some shears
6. grinder wheel	f. tighten in two holes
7. used to shape metal	g. helps start drill bit
8. used to sharpen small tools	h. to cut at an angle
9. hydraulic	i. controlled by the feed handle

C. Completion. Fill in the blanks with the correct word or words that will make the following statements correct.

1. The _____ _____ of the drill press permits stock to be supported for angle drilling.

2. Grinding wheels may be checked for squareness by using a _____ or a _____ _____.

3. The liquid used to cool metal-cutting saw blades is called a _____.

4. Removing material so the grinding wheel is perfectly round with the face square to the sides and the abrasive particles exposed is called _____ the wheel.

Section 5
PROJECT PLANNING

Unit 17
Sketching and Drawing Projects

Objective

To use simple drawing techniques to create plans for personal projects.

Competencies to Be Developed

After studying this unit, you will be able to:

- Identify common drawing equipment.
- Match basic drawing symbols with their definitions.
- Distinguish between pictorial and three-view drawings.
- Use common drawing techniques to represent ideas.
- Read and interpret a drawing.
- Make a three-view drawing of a given object.

Terms to Know

represent
sketch
drawing
dimension
pictorial drawing
three-view drawing
protractor
border line
object line
hidden line
dimension line
extension line
break line
center line
leader line
border
title block
full scale
scale
show box

Materials List

Unlined paper, 8½″ × 11″
Lined paper, 8½″ × 11″
Ruler, 12-inch
Pencil
Soft eraser
Sample drawing blocks
Optional:
 a. Drawing board
 b. T-square
 c. Right triangle
 d. Tape

Many objects in the environment are represented by other things. To *represent* means to stand for or to be a sign or symbol of. For instance, a map uses lines to represent roads in a geographic area such as the community or state, figure 17–1. Books and bulletins use words to describe and create images in the mind that stand for ideas or objects. A photograph represents a person or object.

□ REPRESENTING BY SKETCHING AND DRAWING

A *sketch* is a rough drawing of an idea, object, or procedure. A *drawing* is a picture or likeness made with a pencil, pen, chalk, crayon, or other instrument.

Sketching and drawing are used in agricultural mechanics to put ideas on paper. For example, in a building project an accurate sketch records the details of

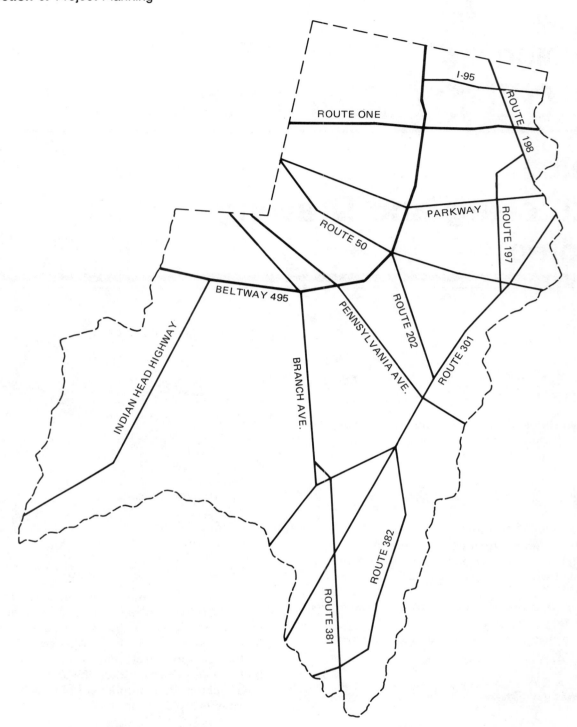

Figure 17–1. On maps, lines are used to represent roads and other features.

construction and indicates how the finished product should look. A sketch is needed to determine the amount of lumber, nails, bolts, and other materials to be used. Any construction project will be more efficient if there is an accurate sketch or carefully drawn plan to follow.

The ability to make sketches and simple drawings is a valuable skill. It is a way to record ideas for present and future use. In this way, a project can be planned in every detail before investing time and materials in construction. As the plan develops, some ideas may prove unworkable. Since the discovery is made before lumber or metal are cut, other workable ideas can be substituted.

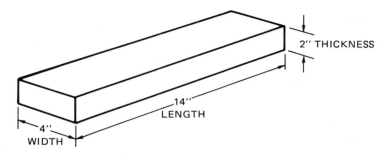

Figure 17–2.   On drawings, lines show edges and corners of objects. Dimensions indicate length, width, and thickness.

A sketch or plan can be simple; it does not have to be done with specialized drawing equipment. However, a sketch must be complete. It must include lines to represent edges and corners of sections or parts of an object. Dimensions must be included to indicate the size of each part of the object. A *dimension* is a measurement of length, width, or thickness, figure 17–2.

Perspective

Figure 17–2 is a pictorial drawing. A *pictorial drawing* shows all three dimensions at once. That is, it shows the item turned so that the front, side, and top are in view. Pictorial drawings are very useful, but it takes a good deal of training and experience to draw them accurately.

Almost anyone can make clear drawings by showing one view at a time. This method is called *three-view drawing.* It will be discussed in detail later in this unit.

Drawing Instruments

Sketches and simple drawings can be prepared with instruments found in most homes. These instruments include a sharp lead pencil, eraser, 12-inch ruler, compass, and protactor, figure 17–3. A *protractor* is an instrument for drawing or measuring angles.

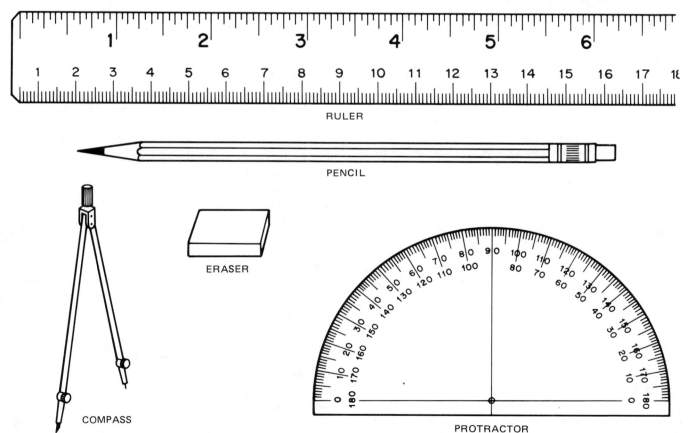

Figure 17–3.   Instruments for making sketches and simple drawings

If much drawing is to be done, several additional items will help save time. These items include a small drawing board with T-square, a 30° × 60° × 90° plastic triangle, a plastic scale, and some masking tape, figure 17–4.

To use these tools, place the T-square across the bottom of the drawing board. Then place an 8½″ × 11″ piece of paper on the board above the T-square. Adjust the paper so that it rests on the T-square with its base parallel to the base of the board. Using masking tape, attach the paper to the board as shown. The T-square can now be used to make horizontal lines. Place the base of the triangle on the T-square. Using the triangle, lines can be drawn at 30°, 60°, or 90° angles to the T-square. Use of the scale will be discussed later.

Symbols Used in Drawing

Hundreds of figures have been adopted by architects and engineers for use as symbols in drawing. Only a few are needed for preparing sketches and simple drawings.

Several types of lines have specific meanings, figure 17–5. They are as follows:

- *border line*—a heavy, solid line drawn parallel to the edges of the drawing paper
- *object line*—a solid line showing visible edges and form of an object
- *hidden line*—a series of dashes which indicates the presence of unseen edges
- *dimension line*—a solid line with arrowheads at the ends to indicate the length, width, or height of an object or part
- *extension line*—a solid line showing the exact area specified by a dimension
- *break line*—a solid, zigzag line used to show that the illustration stops but the object does not
- *center line*—a long-short-long line used to indicate the center of a round object
- *leader line*—a solid line with an arrow used with an explanatory note to point to a specific feature of an object

These lines and symbols help the person who is drawing to communicate with the person who will use the plan. Learning to recognize these lines and symbols is the first step in learning to read plans. Learning to use them is the first step in learning to draw plans.

☐ ELEMENTS OF A PLAN

Paper of any size may be used for drawing. However, it is advisable to start with 8½″ × 11″ paper. This is standard notebook size and is readily available.

Border. ☐ When making a drawing, it is useful to draw a heavy line all around and close to the outer edges of the paper. This is called a *border*. The lines used to make a border are called border lines. It is suggested that borders be made ½ inch in from the edge of the paper.

Title Block. ☐ A few items of information about the total drawing are necessary. These include:

- the name of the person who prepared the drawing
- the date when the drawing was completed
- the name of the drawing
- the scale of the drawing

The section of a drawing reserved for information about the drawing in general is called the *title block*. For simple drawings, a ½-inch line drawn above the border line at the bottom of the paper works well. The information can be printed within these lines. The addition of very light guidelines drawn ⅛ inch from the top and bottom lines of the title block makes lettering easier.

Refer to figure 17–6. Note that a border line has been drawn around the edges of the paper. The title block indicates that the drawing is a sample block, drawn to full scale by Bill Brown on January 1, 1986. *Full scale* means that the drawing is the same size as the object it represents. In other words, 1 inch on the paper represents 1 inch on the block.

Views. ☐ The sample block is represented in this drawing by two views. The upper left-hand drawing is a top view of the block. In other words, the viewer is looking down on the block from above. The dimensions indicate that the block is 2 inches long and 2 inches wide. The inner circle indicates that there is a hole in the center of the block. The dimension indicates that it is 1 inch in diameter. The outer circle shows that a larger hole goes part way through the block. The size of the outer hole is not shown. However, remember that the object is drawn to full scale. This means the drawing is the same size as the block. Measuring the outer circle on the drawing shows it to be 1¼ inches in diameter. Therefore, the hole in the block is the same size.

The lower drawing in figure 17–6 shows a front view of the block. This view shows that the height of the block is 1½ inches (¾ inch + ¾ inch = 1½ inches). It also shows that the larger hole is ¾ inch deep and the smaller hole goes through the remaining ¾ inch of the block. A note is provided just above the title block on the right to specify the material from which the block is to be made. Enough informa-

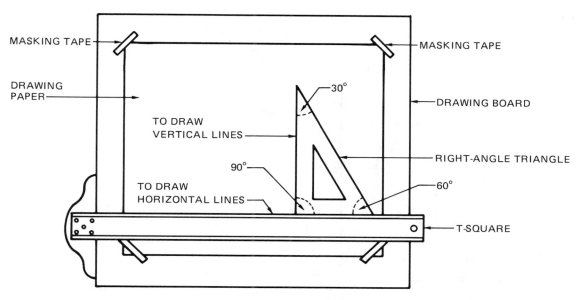

Figure 17–4. A drawing board with paper taped in place. The T-square is helpful for drawing horizontal lines. A right angle triangle is useful for drawing vertical lines.

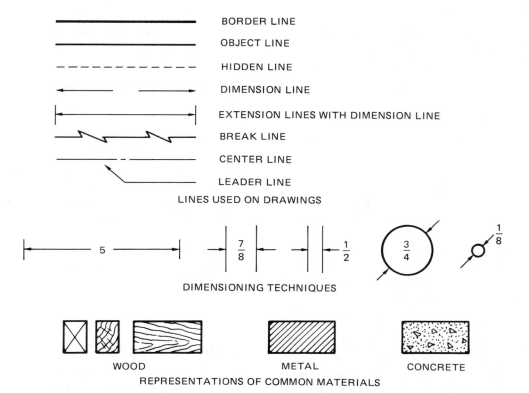

Figure 17–5. Some symbols used on drawings

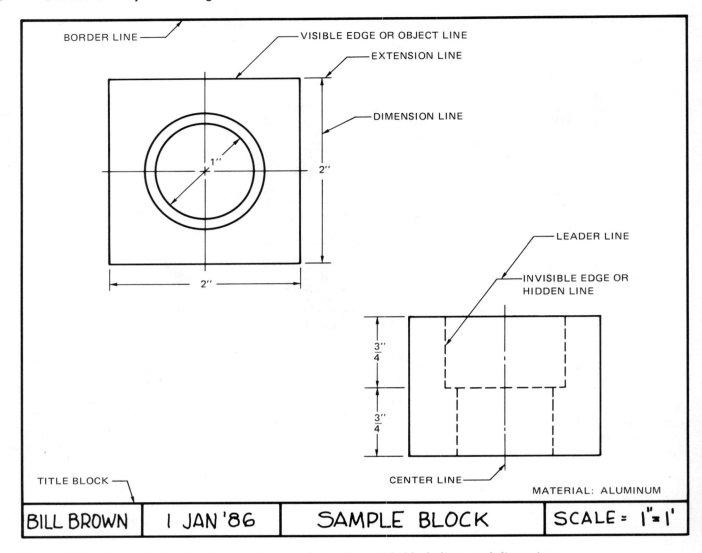

Figure 17–6. A sample drawing showing the use of a border, a title block, lines, and dimensions

tion is provided in figure 17–6 to make the block correctly. Therefore, it is a good drawing.

Most objects require three views in a drawing to provide complete information. For example, consider the bookend shown in figure 17–7. The top view is at the upper left. The front view is at the lower left. However, the object is not recognized as a common bookend until the end view is seen. In this case, the end view gives the best indication of the shape of the object. End views are placed to the left or right of the front view and in line with it.

The views are placed in the drawing with the front view in the lower left, the top view above it, and the end view to the right of the front view. This arrangement allows dimensions placed on the drawing to represent more than one view, figure 17–8. By using this technique, dimensions are not repeated and the drawing is less cluttered, figure 17–9.

☐ SCALE DRAWING

The word *scale* as used here means the size of a plan or drawing as compared to that of the object it represents. In figure 17–6, the drawing was made as large as the object. Since the size of the object was rather small, it fit on the paper. However, most objects are much larger than a sheet of 8½″ × 11″ paper. Therefore, the scale to use in preparing the drawing must be determined.

Determining the Scale

A piece of 8½″ × 11″ paper, turned sideways, with borders and title block drawn, has 10 inches of horizontal drawing space and 7 inches of vertical drawing

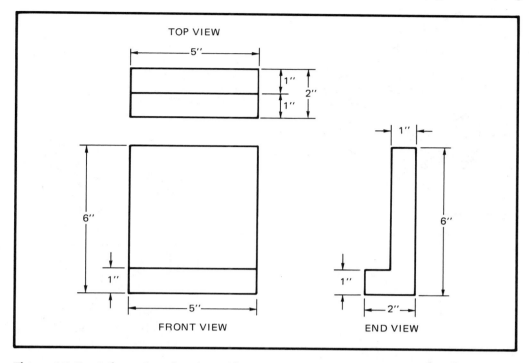

Figure 17–7. A three-view drawing with repeated dimensions. The names of the views do not appear on a final drawing.

space. The three views must be planned to fit into this space without crowding.

A good place to start in deciding upon the scale is to assume that 1 inch equals 1 foot (1″ = 1′). It is

recommended that 1 inch of space for dimensions be allowed on every side of each view. Therefore, 1 inch on the left of the front view, 1 inch between views, and 1 inch on the right of the end view is

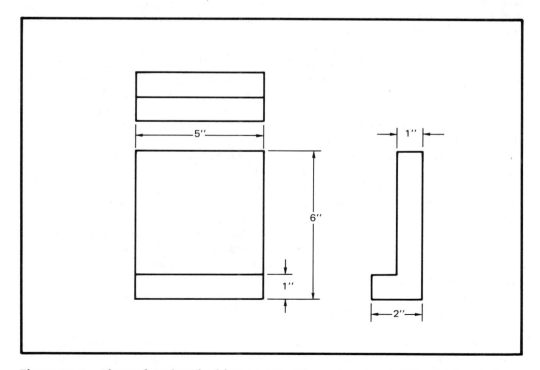

Figure 17–8. The preferred method for presenting three-view drawings in which each dimension is shown only once

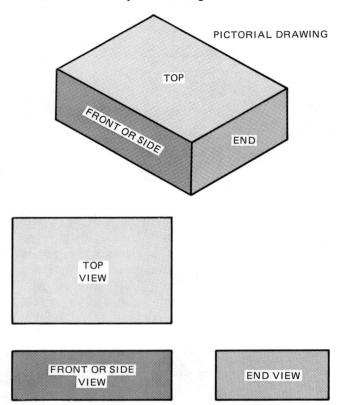

PICTORIAL DRAWING

TOP

FRONT OR SIDE

END

TOP VIEW

FRONT OR SIDE VIEW

END VIEW

Figure 17–9. Placement of views in a three-view drawing

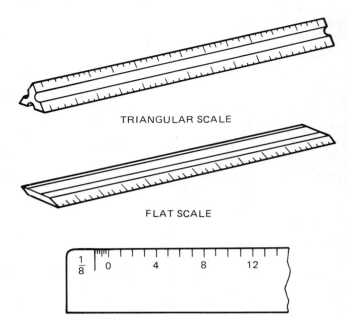

TRIANGULAR SCALE

FLAT SCALE

1/8 | 0   4   8   12

Figure 17–10. Triangular and flat scales. Notice that ⅛ inch is used to represent 1 inch in the bottom example.

reserved. The original 10 inches of horizontal space, reduced by 3 inches for dimensions, leaves 7 inches of horizontal space for the two views.

A three-view drawing has a front view and a top view, one over the other. Three spaces of 1 inch each are allowed for dimensions. This reduces the original 7 inches of vertical space by 3 inches, leaving 4 inches of vertical drawing space.

This procedure indicates that a scale of $1'' = 1'$ is adequate to draw objects whose front and end dimensions (horizontal) add up to 7 feet or less and whose front and top dimensions (vertical) total no more than 4 feet. Larger objects require a larger scale such as $1'' = 2'$ or $1'' = 3'$, and so on. For example, consider a three-view drawing of a wagon body with hay racks. The dimensions are:

Length = 16′
Width = 8′
Height = 8′

Remember that there are 7 inches of horizontal drawing space and 4 inches of vertical drawing space. With a scale of $1'' = 4'$, 1 inch of drawing represents 4 feet of wagon:

L = 16′ actual dimension, 4″ scale value
W = 8′ actual dimension, 2″ scale value
H = 8′ actual dimension, 2″ scale value

For the horizontal views, the dimensions would be:
4″ (front view) + 2″ (end view) = 6″
Since there are 7 inches of drawing space, the scale is usable for the horizontal views. The extra inch will simply be extra space.

For the vertical views, the dimensions would be:
2″ (front view) + 2″ (top view) = 4″
Since there are 4 inches of vertical drawing space, the scale will work.

Using a Scale

A *scale* is an instrument with all increments shortened according to proportion, figure 17–10. On a ½ scale, 1-inch marks are only ½-inch long. It can be used to let ½ inch equal 1 inch, 1 foot, 1 yard, 1 mile, and so on.

Some instruments have a different scale on each edge. A triangular scale has three sides but six scales, with each side showing two scales. Some instruments show different scales at each end for a total of four scales per side, or 12 scales per instrument.

An expensive scale should never be used as a straightedge for drawing. However, low-cost plastic scales may be used for this purpose.

☐ MAKING A THREE-VIEW DRAWING

To demonstrate the procedure for making a three-view drawing, a drawing of a show box will be made. A *show box* is used to carry livestock equipment and supplies. It can also be used to store and transport tools and materials for other purposes.

The show box will be 4 feet long, 2 feet wide, and 1½ feet high. The scale will be 1″ = 1′ or ¹⁄₁₂″.

Blocking in the Views

Procedure

1. Select a plain 8½″ × 11″ piece of heavy paper. Regular drawing paper is preferred, but mimeograph paper will do. Turn the paper lengthwise in front of you, figure 17–11. Draw a ½-inch border around the paper. Add a ½-inch title block.

2. In the title block, letter your name and the date. For the name of the project print "Show Box"; for scale, Scale = 1″ = 1′.

3. Make a light mark 1 inch in from the left margin and draw a very light line through it from top to bottom and parallel to the left border. This becomes the object line for the left side of the top and front views.

4. Make a light mark 1 inch up from the title block and draw a very light line through it from border to border and parallel to the title block. This becomes the bottom object line.

5. The length of the box is 4 feet. The scale is 1″ = 12″, or 1″ = 1′. Measure across 4 inches from the left object line and draw a very light vertical line from top to bottom. This becomes the object line on the right side of the top and front views.

6. Measure 1 inch to the right and draw a very light vertical line from top to bottom. This becomes the left side of the end view.

7. Measure another 2 inches to the right and draw another very light vertical line. This becomes the right side of the end view.

8. The chest is 1½ feet high. Start at the bottom object line and measure up 1½ inches. Draw a very light line to form the top object lines of the front and end views. **Note:** Do not extend the lines into the spaces around the views.

9. Measure up 1 inch from the top object line and draw a light line to form the bottom object line of the top view. **Note:** Do not extend the line into the spaces around the views.

10. The chest is 2 feet wide. Measure up another 2 inches and draw a light line to form the upper object line of the top view.

11. Now all three views are blocked in and drawn to scale, figure 17–12. Erase excess lines that extend beyond the views, figure 17–13. **Note:** Remember that you have planned for 1 inch or

more of space around views. Therefore, when blocking in views, lines may be stopped before they touch the borders. This saves erasing.

All three views are now blocked in. This means the outside object lines are drawn. Next the details of each view must be added.

Completing the Front View

The most prominent view is generally the front view. Therefore, the front view should be drawn first. It will probably tell the viewer more about the project than either the top or the end view.

Assume that the show box will be made of ¾-inch plywood. The sides will be rabbetted and the ends set in. The bottom will be hidden by the sides and ends. The lid will fit on top of the sides and ends.

Before drawing, it must be decided how far apart two lines should be on paper to represent the thickness of ¾-inch plywood at ¹⁄₁₂ scale. For convenience, round off the ¾-inch thickness to 1 inch. Therefore ¹⁄₁₂ of an inch on the paper represents lumber that is ¾ inch thick. Rulers are laid out in eighths and sixteenths, not twelfths. Therefore, either ¹⁄₁₆ or ⅛ of an inch must be chosen to represent an inch or a special ¹⁄₁₂ drawing scale must be obtained. For sketching purposes, ⅛ inch will be used to represent the ¾-inch plywood to better show the detail of the joints.

To draw in the details of the front view, the steps are as follows.

Procedure

1. Measure in ⅛ inch from the left side of the front view and draw a broken vertical line from top to bottom. This line is a hidden line because the edge of the end panel cannot be seen, figure 17–14.

2. Measure in ⅛ inch from the right side of the front view and draw a broken vertical line from top to bottom. This line represents the edge of the right-hand end panel.

3. Measure up from the bottom ⅛ inch and draw a broken horizontal line between the two vertical broken lines. This line represents the hidden top edge of the bottom panel.

4. Measure down ⅛ inch from the top of the front view and draw a horizontal solid line from solid line to solid line to represent the thickness of the lid.

5. Erase any parts of the broken vertical lines that show inside the two lines representing the edge of the closed lid.

6. Show where handles will be attached, 12 inches up from the bottom of the chest.

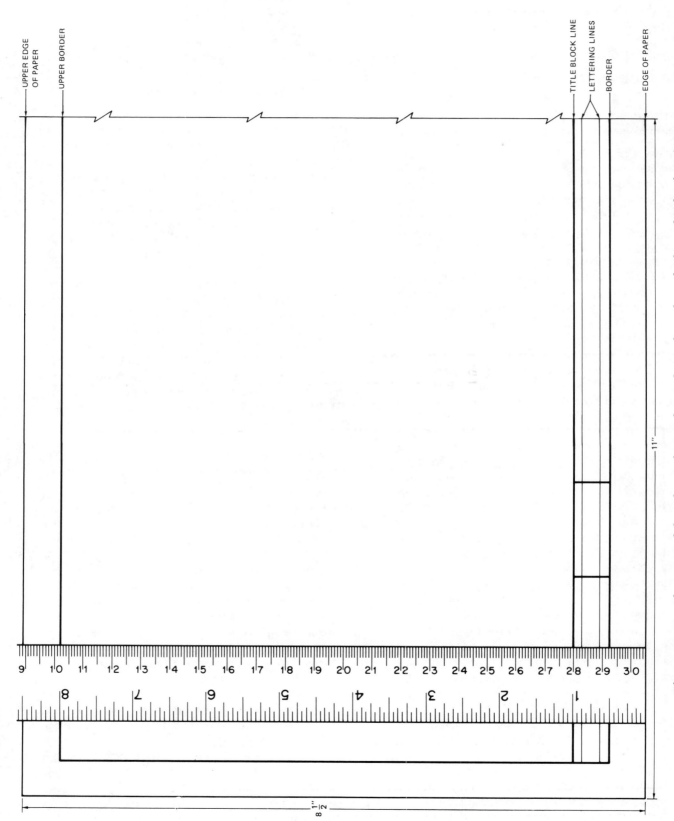

Figure 17-11. Placement of the ruler so that marks can be made to draw the lower border, lettering guidelines, title block line, and upper border line at the same time

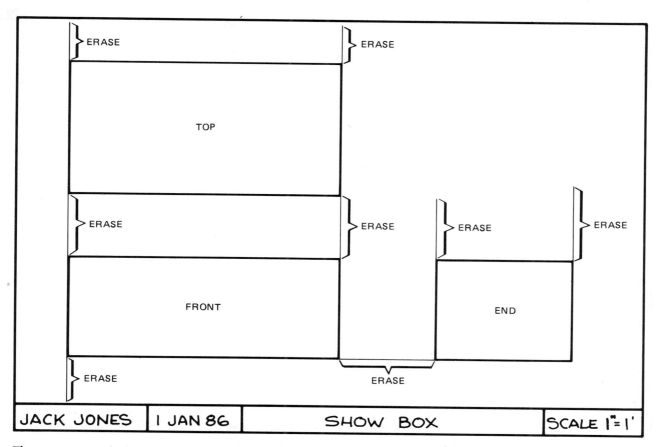

Figure 17-12. Blocked-in, three-view drawing of the show box before excess lines are erased

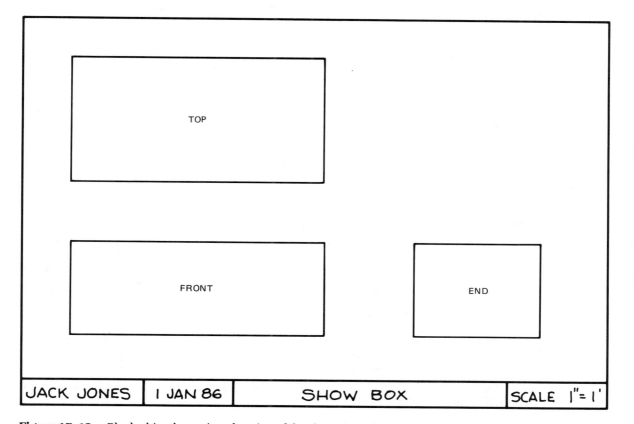

Figure 17-13. Blocked-in, three-view drawing of the show box after excess lines are erased

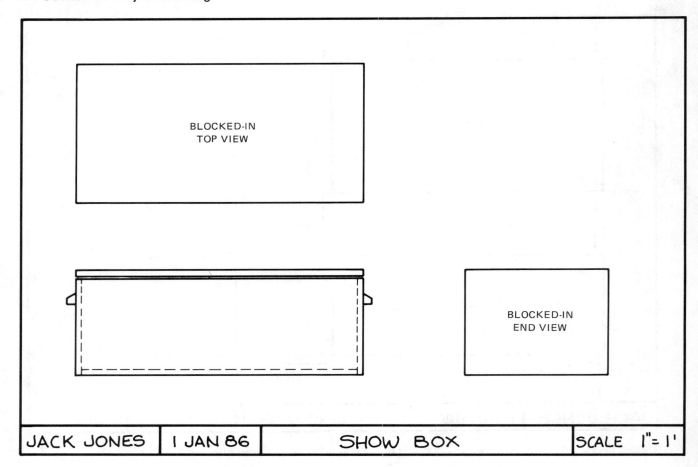

Figure 17–14. Completed front view of the show box

This completes the front view unless the exposed heads of nails or screws used to attach the front to the ends and bottom are to be shown. A hasp or other locking device could also be pictured.

Completing the End View

With experience in drawing, some of the parts of the top and end views can be drawn at the same time the front view is drawn. Many of the lines can be drawn by placing the ruler on the measurement and drawing the lines in two views without moving the ruler.

Refer to figure 17–15. The steps in drawing the end view are as follows:

Procedure

1. Place the ruler on the bottom line of the lid in the front view. The ruler should extend across the end view also. Draw the bottom line of the lid in the end view (use a solid line).
2. The end view of the sides needs special treatment to show the rabbet joints. That is, ⅜ inch of the plywood end grain is seen; the other ⅜ inch is hidden. To show this, measure in 1/16 inch from the left-hand side of the end view and draw a solid vertical line. Measure another 1/16 inch and draw a vertical broken line. Repeat the procedure on the right-hand side of the end view.
3. Place the ruler on the hidden top edge of the bottom panel in the front view. Draw in the hidden top edge of the bottom panel in the end view (use a broken line).
4. Draw in the handle. Notice that the handle is on the same level in both the front and end views. The handle consists of a 1″ × 2″ × 6″ hollowed strip of wood. Since the scale is 1″ = 12″, the handle is ½″ long on the drawing. Measure to the center of the end and then outward ¼ inch in each direction.

Completing the Top View

Refer to figure 17–16. The steps in drawing the top view are as follows.

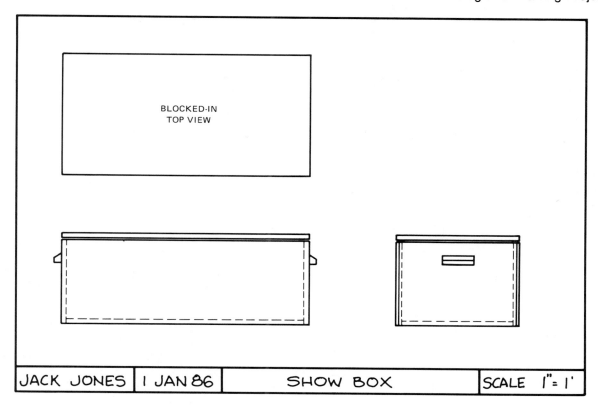

Figure 17–15. Completed front and end views of the show box

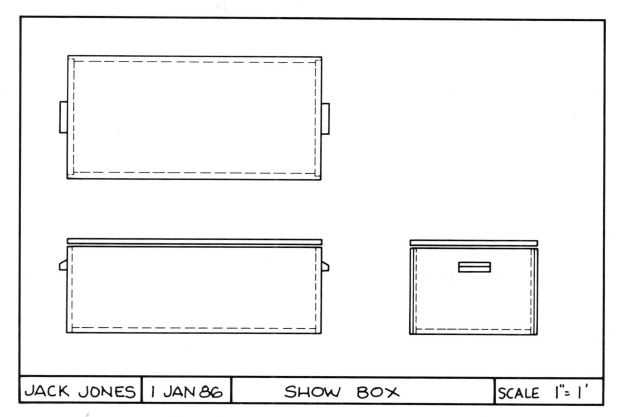

Figure 17–16. All three views of the show box completed without dimensions

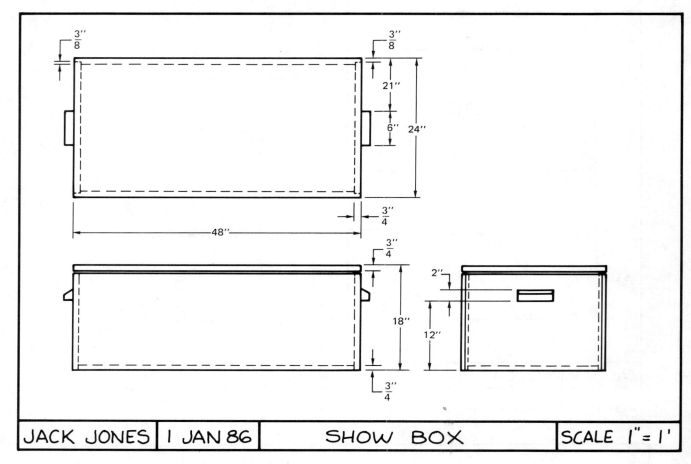

JACK JONES | 1 JAN 86 | SHOW BOX | SCALE 1"= 1'

Figure 17–17. Completed drawing of the show box with dimensions

Procedure

1. Place the ruler on the hidden inner edge of the left end panel of the front view and extend it over the top view (use a broken line).
2. Repeat step one on the right side of the view.
3. Measure ⅛ inch up from the bottom and ⅛ inch down from the top lines and draw the inner edges of the sides with broken lines.
4. Show the rabbet joints correctly in each corner. That is, show how the side panels are sawed and the end panels fit into the rabbet grooves.
5. Add the handles on each end by locating the center and measuring out ¼ inch in each direction. Assume a thickness of 1-inch wood or ⅛ inch on the paper.

Dimensioning the Drawing

The final step to complete the drawing is to add dimensions, figure 17–17. The dimensions are the exact measurements, more exact than the drawing itself.

The drawing focuses on details of the assembly. The dimensions tell the exact length, width, and height of the object. The dimensions also indicate the thickness of materials and other construction details as needed.

□ STUDENT ACTIVITIES

1. Study agricultural mechanics project plans to visualize how individual parts look and how they are fastened together.
2. Practice printing the alphabet in capitals and lower-case letters. Follow the form provided in figure 17–5. Use lined paper.
3. Using a ruler, make 20 arrows following the form used in figure 17–5 for dimension lines.
4. Practice making neat figures for dimensions by numbering from 0 through 9. Follow the form provided in figure 17–5.
5. Practice measuring accurately by drawing lines on a sheet of paper the length of each figure stated below. Make some lines solid and some

broken. Submit the paper to your instructor for evaluation.

a. 1 inch
b. 1½ inches
c. 1¾ inches
d. 3⅛ inches
e. 3⅝ inches
f. 1¹⁄₁₆ inches
g. 1³⁄₁₆ inches
h. 1¼ inches
i. 1⁹⁄₁₆ inches
j. 1⅜ inches

6. Ask the instructor to provide blocks or other objects for practice drawing. Draw one or more of such objects.

7. Convert the following project dimensions to the measurements they would be on paper, first using a $\frac{1}{12}$ scale (1″ = 12″); then using a $\frac{1}{2}$ scale (1″ = 2″).

a. 12 inches = _____ _____
b. 18 inches = _____ _____
c. 36 inches = _____ _____
d. 6 inches = _____ _____
e. 3 inches = _____ _____

8. If a drawing instrument with $\frac{1}{12}$ and $\frac{1}{2}$ scales is available, draw lines on a sheet of paper representing the length of each item in activity 7.

□ SELF-EVALUATION

A. Multiple Choice. Select the best answer.

1. A simple plan must
   a. be done with specialized equipment
   b. be done by a draftsman or engineer
   c. have dimensions
   d. all of these

2. The triangle recommended as an aid for simple drawing is
   a. 30°, 60°, 90°
   b. plastic
   c. three sided
   d. all of these

3. Which scale is best for making a three-view drawing on a sheet of 8½″ × 11″ paper if the project is 4 feet long, 2 feet wide, and 1½ feet high?
   a. 1″ = 12″
   b. 1″ = 6″
   c. 1″ = 3″
   d. 1″ = 1″

4. Which scale is best for making a drawing of the front view only on an 8½″ × 11″ sheet of paper if the project is 4 feet L, 2 feet W, and 1½ feet H?
   a. 1″ = 12″
   b. 1″ = 6″
   c. 1″ = 3″
   d. 1″ = 1″

5. The process of first drawing outside object lines for all three views is called
   a. blocking in
   b. pictorial drawing
   c. scale drawing
   d. sketching

B. Matching. Match the terms in column I with those in column II.

Column I	Column II
1. dimension	a. ½″ from edge of paper
2. pictorial	b. use with dimensions
3. object line	c. round objects
4. border line	d. three views in one
5. hidden line	e. targets special information
6. extension line	f. measurement
7. leader line	g. name, date, project, scale
8. center line	h. proportion
9. title block	i. broken line
10. scale	j. visible edges

C. Completion. Fill in the blanks with the word or words that make the following statements correct.

1. The view placed in the upper left corner of a plan is the _____.

2. The view placed at the lower left corner of a plan is the _____.

3. The view placed in the lower right corner of the plan is the _____.

D. Brief Answers. Briefly answer the following questions.

1. What is a drawing?
2. What is a sketch?
3. What is a protractor?
4. Using a scale of 1″ = 12″, convert the following measurements.
   a. 48 inches = _____
   b. 6 inches = _____
   c. 9 inches = _____
   d. 1 inch   = _____

Unit 18
Figuring a Bill of Materials

Objective

To state the use and format of a bill of materials and make all calculations needed to develop a bill of materials.

Competencies to Be Developed

After studying this unit, you will be able to:
- Define terms associated with a bill of materials.
- State the components of a bill of materials.
- Record dimensions of structural metals and lumber.
- Calculate board feet.
- Calculate costs included in a bill of materials.
- Prepare a written bill of materials.

Terms to Know

bill of materials	galvanize
item	cadmium
rounded up	board foot
tongue and groove	

Materials List

Paper
Pencil

☐ COMPONENTS OF A BILL OF MATERIALS

A *bill of materials* is a list and description of all the materials to be used in constructing a project. For each component, the bill of materials includes the following:

- item or part name
- number of pieces
- type of material
- size of pieces
- description of parts
- total feet
- unit cost
- cost

A bill of materials is used to determine the material that must be assembled for a project and its estimated cost. To determine the amount of lumber needed, the student must be able to compute board feet. The student must also be able to combine the costs of numerous pieces and types of materials to arrive at a total estimate of the cost of the project.

Several formats are used for bills of materials. These vary with the project or building being planned. One format for listing items is shown in figure 18–1.

The term *item* means a separate object. In a bill of materials, the term may refer to a part of a structure such as the side, end, bottom, or top. In a trailer body, an item may be "brackets" for attaching cross members to beams, "leg screws" for attaching pockets for side boards, "boards" for the bottom, "steel" for the sides, or "paint" for the whole project.

An item in a bill of materials may be "paint," "rafters," "doors," "nails," "hinges," or "2 × 4s." The important thing to remember in making a bill of materials is that items of different types, materials, sizes, or costs must *not* be mixed and listed as one item.

The number of pieces is helpful since it can be used to buy materials efficiently. It is also needed in the calculation of total feet and total cost. For example, consider a plan to build a small storage building. The door will be made from boards that are 1" × 6". These boards can be bought in 6-, 8-, 10-, 12-, 14-, and 16-foot lengths.

The plan indicates that the door is 6 feet high and 3 feet wide, figure 18–2. This knowledge enables the worker to determine the number of pieces as follows: 36 inches (width of door) divided by 5½ inches (width of a finished 1" × 6") = 6.5 boards. This calculation indicates that 7 boards, each 6 feet long, are needed to make the 36-inch door. The number 6.5 is rounded up to 7, because boards are not sold in fractional sizes. When a number is *rounded up*, it means the next higher whole number is used.

The plan also shows cross boards at the top and bottom of the door to hold the vertical boards together. Each cross piece is 36 inches (or 3 feet) long or equal to the distance across the 36-inch door. Finally, the plan calls for a 1" × 6" diagonal brace. Assume this brace needs to be 5 feet 2 inches long.

The total number of 1" × 6" pieces required to make the door can now be stated as follows:

7 pieces—6 feet long

ITEM	NO. OF PIECES	TYPE OF MATERIALS	SIZE	DESCRIPTION	TOTAL FEET	UNIT COST	COST

Figure 18–1. Format for a bill of materials

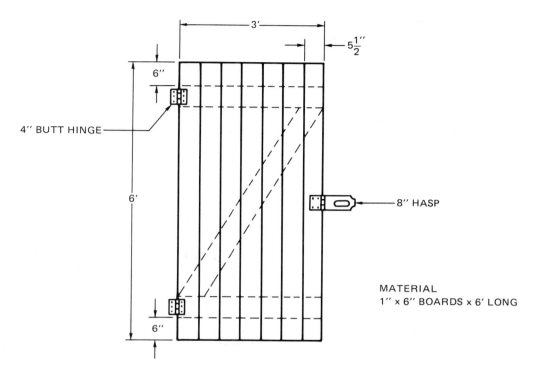

Figure 18–2. Plan of a storage shed door

2 pieces—3 feet long

1 piece—5 feet 2 inches long

Since the pieces are all 1″ × 6″ boards, the items listed can be combined. In order to combine the three items into one item in the bill of materials, the two 3-foot pieces are cut from one 6-foot piece. The 5 foot 2 inch piece can also be cut from a 6-foot piece. Therefore, to make the door, 9 pieces of 1″ × 6″ × 6′ lumber are needed.

To make the door weathertight and strong, tongue and groove (T&G) lumber should be used. *Tongue and groove* boards have a tonguelike edge sticking out on one side and a groove cut into the other which permits them to lock together.

Assume that the cost of the boards will be $.30 per linear foot. The lumber requirement would be included in the bill of materials as shown for "door" in figure 18–3. The plan also calls for a pair of 4-inch

ITEM	NO. OF PIECES	TYPE OF MATERIALS	SIZE	DESCRIPTION	TOTAL FEET	UNIT COST	TOTAL COST
Door	9	Boards	1″ × 6″ × 6′	S4S T&G Pine	54	$0.30	$16.20
Hinge	1 pr	Steel	4″	Butt	N/A	$4/pr	4.00
Hasp	1	Steel	8″	Galvanized	N/A	$3.50	3.50
Nails	1 lb	Steel	6d	Common	N/A	$0.75/lb	0.75
Screws	24	Steel	No. 10 × ¾	Cd finish	N/A	$0.06 ea	1.44
Paint	1	Oil Base	Quart	Red	N/A	$6.50	6.50
							$32.39

Figure 18–3. Lumber, hardware, and paint requirements in a bill of materials

butt hinges and an 8-inch hasp. These items of hardware, along with nails and screws, are also included on the bill of materials shown in figure 18–3.

Some additional explanation is in order regarding the sample bill of materials. The lumber was priced by the linear foot since tongue and grooved boards are used. This means that each running foot, or each foot of length in the board, costs $.30. The amount of cost of a linear foot is influenced by the width and thickness. The total number of linear feet is 54; thus, 54 × $.30 = $16.20.

Rough lumber from the sawmill is sold by the board foot. Calculating board feet requires computation of the volume of the board. Calculating board feet will be discussed later. The *S4S* under "Description" is a symbol designating which surfaces of the board have been planed. S4S means the boards are planed or *surfaced* on four sides. The symbol *S2S* would mean the boards were surfaced on two sides.

Since the hasp in the example is galvanized, the finish on the screws is specified as cadmium (Cd). To *galvanize* is to coat a metal with zinc. *Cadmium* is a similar metal used for rust-resistant plating of fasteners and other steel products. "No. 10 × ¾" size screws means the screws are number 10 diameter and ¾" inch long. "N/A" means "not applicable" and is used when requested information does not apply to the situation.

If the door is a single project, the paint should be added to the bill of materials. The door could be part of a whole building under construction. If so, paint would appear as a single item for the entire building unless the door was painted a different color.

□ UNITS OF MEASURE

A bill of materials provides much information in a small space. Therefore, standard abbreviations are helpful, figure 18–4.

It is easier to plan and build projects with a knowledge of the sizes and weights of lumber and steel products. These are given in figures 18–5 through 18–8.

□ CALCULATING BOARD FEET

Lumber in the form of trees, logs, or rough lumber is measured and sold by board feet. Sometimes dressed lumber is sold by the board foot as well.

A *board foot* is an amount of wood equal to a board 1 inch thick, 1 foot wide, and 1 foot long; or 1 board foot is $1'' \times 12'' \times 12''$. A board foot is also 144 cubic inches (cu in or in³). That is, $1'' \times 12'' \times 12'' = 144$ cu in), figure 18–9. Cubic measures are determined by multiplying thickness by width by length. To convert cubic inches back to board feet, divide by 144. Hence, 144 cubic inches divided by 144 = 1 board foot.

Two Methods for Calculating Board Feet

Two methods or formulas can be used to calculate board feet. Each method is useful for certain sizes of lumber.

Small Pieces of Lumber. □ Remember that 1 board foot equals 144 cubic inches. With this in mind, the board feet in small or odd-sized pieces of lumber can be calculated. The formula is:

$$BF = \frac{\text{Thickness (in)} \times \text{Width (in)} \times \text{Length (in)}}{144}$$

" or in = inch	S3S = surface 3 sides
' or ft = foot	S4S = surface 4 sides
yd = yard	no. or # = number
mi = mile	in² or sq in = square inch
ea = each	ft² or sq ft = square foot
@ = at	yd² or sq yd = square yard
N/A = not applicable	square = 10' × 10' or 100 square feet
pt = pint	NC = national coarse threads
qt = quart	NF = national fine threads
gal = gallon	NPT = national pipe threads
LF = linear foot	d = penny (nails)
BF = board foot	lb = pound
SIS = surface 1 side	oz = ounce
S2S = surface 2 sides	Cwt = hundredweight (100 pounds)

Figure 18–4. Useful abbreviations for making a bill of materials

LUMBER (All available in 6', 8', 10', 12', 14', and 16' lengths)

1" × 6"	2" × 6"
1" × 8"	2" × 8"
1" × 10⁺	2" × 10"
1" × 12"	2" × 12"
2" × 4"	4" × 6"

PLYWOOD (Interior and Exterior Grades)

¼" × 4' × 8'	⅝" × 4' × 8'
⅜" × 4' × 8'	¾" × 4' × 8'
½" × 4' × 8'	1" × 4' × 8'

STRUCTURAL STEEL (Standard length = 20')

FLAT BAND IRON—⅛", ³⁄₁₆", ¼", ⁵⁄₁₆", ⅜", and ½" thick
—½", ¾", 1", 1½", 2", 3", and 4" wide

ANGLE IRON—⅛", ³⁄₁₆", ¼", ⁵⁄₁₆", and ⅜" thick
—½" × ½", ¾" × ¾", 1" × 1", 1½" × 1½",
2" × 2", and 3" × 3" (width of legs)

ROUND (Hot Rolled, Cold Rolled, or Tool Steel)
—¼", ⁵⁄₁₆", ⅜", ½", ⅝", ¾", 1", 1½", and 2"
(diameter)

BLACK OR GALVANIZED STEEL PIPE (Standard Length = 21') — ¼", ⅜", ½", ¾", 1", 1¼", 1½", 2", 2½", and 3" (inside diameter)

Figure 18–5. Common sizes and units of lumber and structural metals

What is the number of board feet in a piece of wood 1 inch thick, 2 inches wide, and 24 inches long? The answer is calculated as follows:

$$BF = \frac{T'' \times W'' \times L''}{144}$$

$$BF = \frac{1 \times 2 \times 24}{144}$$

$$BF = \frac{48}{144}$$

$$BF = .33$$

Consider a wooden nail box with two sides that are each 1" × 4" × 14", two ends that are each 1" × 4" × 9", and a bottom that is 1" × 9" × 12½". How many board feet are needed for the project? The answer is obtained by the steps shown in figure 18–10.

Large Pieces of Lumber. □ There is a better formula for large pieces of lumber, such as those used in wagon bodies or truck racks. The formula is:

$$BF = \frac{T'' \times W'' \times L'}{12}$$

In this formula the length is expressed in feet instead of inches and divided by 12 instead of 144. The formula is derived by dividing both the top and bottom by 12. Hence

$$BF = \frac{1'' \times 6'' \times 24''}{144} = \frac{1'' \times 6'' \times 2'}{12} = 1$$

This formula can be used to determine the number of board feet in the door shown in figure 18–11. The math is easier in the second formula because expressing the length in feet makes the figures smaller. Remember that lumber must be bought in lengths of even numbers of feet. Calculations must be based on the length of the lumber to be purchased, not just on the length to be used in the project. To ignore the waste lumber means that the final cost of the project will be higher than calculated.

□ PRICING MATERIALS

Lumber is priced by the thousand board feet. The Roman numeral for one thousand is a capital M. For example, lumber dealers may describe the price of lumber as $300/M. This means $300 for 1000 board feet. In this example, to determine the price of 1 board foot, simply divide $300 by 1000:

$$\frac{\$300}{1000} = \$.30$$

ANGLES—BAR SIZES

SIZE IN INCHES	WEIGHT PER FT (LBS)	SIZE IN INCHES	WEIGHT PER FT (LBS)
1 × 1 × 1/8	.80	2 × 1 1/2 × 1/8	1.44
3/16	1.16	3/16	2.12
1/4	1.49	1/4	2.77
1 1/8 × 1 1/8 × 1/8	.90	2 × 2 × 1/8	1.65
1 1/4 × 1 1/4 × 1/8	1.01	3/16	2.44
3/16	1.48	1/4	3.19
1/4	1.92	5/16	3.92
1 3/8 × 7/8 × 1/8	.91	3/8	4.70
3/16	1.32	2 1/4 × 1 1/2 × 3/16	2.28
1 1/2 × 1 1/4 × 3/16	1.64	2 1/2 × 1 1/2 × 3/16	2.44
1 1/2 × 1 1/2 × 1/8	1.23	1/4	3.19
3/16	1.80	5/16	3.92
1/4	2.34	2 1/2 × 2 × 3/16	2.75
1 3/4 × 1 1/4 × 1/8	1.23	1/4	3.62
1 3/4 × 1 3/4 × 1/8	1.44	5/16	4.50
3/16	2.12	3/8	5.30
1/4	2.77	2 1/2 × 2 1/2 × 3/16	3.07
2 × 1 1/4 × 3/16	1.96	1/4	4.10
1/4	2.55	5/16	5.00
		3/8	5.90
		1/2	7.70

ANGLES—STRUCTURAL

SIZE IN INCHES	WEIGHT PER FT (LBS)	SIZE IN INCHES	WEIGHT PER FT (LBS)
3 × 2 × 3/16	3.07	3 × 3 × 3/16	3.71
1/4	4.10	1/4	4.90
5/16	5.00	5/16	6.10
3/8	5.90	3/8	7.20
1/2	7.70	7/16	8.30
3 × 2 1/2 × 1/4	4.50	1/2	9.40
5/16	5.60	3 1/2 × 2 1/2 × 1/4	4.90
3/8	6.60	5/16	6.10
1/2	8.50	3/8	7.20
3 1/2 × 3 × 1/4	5.40	1/2	9.40
5/16	6.60	5 × 5 × 3/8	12.3
3/8	7.90	1/2	16.2
1/2	10.20	5/8	20.0
3 1/2 × 3 1/2 × 1/4	5.80	3/4	23.6
5/16	7.20	6 × 3 1/2 × 5/16	9.8
3/8	8.50	3/8	11.7
7/16	9.80	1/2	15.3
1/2	11.10	6 × 4 × 5/16	10.3
4 × 3 × 1/4	5.80	3/8	12.3
5/16	7.20	7/16	14.3
3/8	8.50	1/2	16.2
7/16	9.80	5/8	20.0
1/2	11.10	3/4	23.6
5/8	13.60	7/8	27.2
4 × 3 1/2 × 1/4	6.2	6 × 6 × 3/8	14.9
5/16	7.7	7/16	17.2
3/8	9.1	1/2	19.6
7/16	10.6	5/8	24.2
1/2	11.9	3/4	28.7
4 × 4 × 1/4	6.6	7/8	33.1
5/16	8.2	1	37.4
3/8	9.8	7 × 4 × 3/8	13.6
7/16	11.3	1/2	17.9
1/2	12.8	5/8	22.1
5/8	15.7	8 × 4 × 1/2	19.6
3/4	18.5	8 × 6 × 1/2	23.0
5 × 3 × 1/4	6.6	5/8	28.5
5/16	8.2	3/4	33.8
3/8	9.8	1	44.2
7/16	11.3	8 × 8 × 1/2	26.4
1/2	12.8	5/8	32.7
5 × 3 1/2 × 3/16	8.7	3/4	38.9
3/8	10.4	7/8	45.0
7/16	12.0	1	51.0
1/2	13.6		
5/8	16.8		
3/4	19.8		

Figure 18–6. Weight per foot of angle steel. To determine the cost per foot, multiply the weight in pounds per foot by the price per pound.

CONCRETE REINFORCING BARS

BAR SIZES OLD (INCHES)	NEW (NUMBERS)	WEIGHT PER FT (LBS)	DIAMETER (INCHES)
1/4	2	.167	.250
3/8	3	.376	.375
1/2	4	.668	.500
5/8	5	1.043	.625
3/4	6	1.502	.750
7/8	7	2.044	.875
1	8	2.670	1.000
1	9	3.400	1.128
1 1/8	10	4.303	1.270
1 1/4	11	5.313	1.410

ROUNDS

SIZE IN INCHES	WEIGHT PER FT (LBS)	SIZE IN INCHES	WEIGHT PER FT (LBS)
1/4	0.167	2 1/4	13.52
5/16	0.261	2 3/8	15.06
3/8	0.376	2 1/2	16.69
7/16	0.511	2 5/8	18.40
1/2	0.668	2 3/4	20.20
9/16	0.845	3	24.03
5/8	1.043	3 1/4	28.21
3/4	1.502	3 1/2	32.71
7/8	2.044	3 3/4	37.55
1	2.670	4	42.73
1 1/8	3.379	4 1/4	48.23
1 1/4	4.173	4 1/2	54.08
1 3/8	5.05	4 3/4	60.25
1 1/2	6.01	5	66.76
1 5/8	7.051	5 1/4	73.60
1 3/4	8.18	5 1/2	80.78
1 7/8	9.39	5 3/4	88.29
2	10.68	6	96.13
2 1/8	12.06		

CHANNELS—STRUCTURAL

DEPTH OF CHANNEL INCHES	WEIGHT PER FT (LBS)	THICKNESS OF WEB	WIDTH FLANGE (INCHES)
3	4.1	0.170	1.410
	5.0	0.258	1.498
	6.0	0.356	1.596
4	5.4	0.180	1.580
	7.25	0.320	1.720
5	6.7	0.190	1.750
	9.0	0.325	1.885
6	8.2	0.200	1.920
	10.5	0.314	2.034
	13.0	0.437	2.157
7	9.8	0.210	2.090
	12.25	0.314	2.194
	14.75	0.419	2.299
8	11.5	0.220	2.260
	13.75	0.303	2.343
	18.75	0.487	2.527
9	13.4	0.230	2.430
	15.0	0.285	2.485
	20.0	0.448	2.648
10	15.3	0.240	2.600
	20.0	0.379	2.739
	25.0	0.526	2.886
	30.0	0.673	3.033
12	20.7	0.280	2.940
	25.0	0.387	3.047
	30.0	0.510	3.170
13	31.8	0.375	4.000
	50.0	0.787	4.412
15	33.9	0.400	3.400
	40.0	0.520	3.520
	50.0	0.716	3.716
18	45.8	0.500	4.000

Figure 18–7. Weight per foot of round and channel steel. Multiply the weight in pounds per foot by the price per pound to get the cost per foot.

GAUGE	THICKNESS (INCHES)	WEIGHT/SQ. FT (POUNDS)
No. 18	0.0478	2.00
No. 16	0.0598	2.50
No. 14	0.0747	3.125
No. 12	0.1046	4.375
No. 11	0.1196	5.000
No. 10	0.1345	5.625
No. 8	0.1644	6.875
No. 7	3/16	7.500

Figure 18–8. Weight per square foot of sheet metal. Multiply the weight in pounds per square foot by the price per pound to determine the cost per square foot.

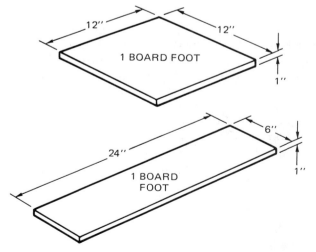

Figure 18–9. A board foot is any volume of wood that equals 144 cubic inches.

ITEM	NO. OF PIECES	SIZE	BF PER PIECE	TOTAL BF
1. Sides	2	$1'' \times 4'' \times 14''$	0.39	0.78
2. Ends	2	$1'' \times 4'' \times 9''$	0.25	0.50
3. Bottom	1	$1'' \times 9'' \times 12\frac{1}{2}''$	0.78	0.78
			Grand Total	2.06

Step 1. Sides: $\dfrac{1'' \times 4'' \times 14''}{144} = \dfrac{56}{144} = 0.39 \times 2 \text{ pcs.} = 0.78$

Step 2. Ends: $\dfrac{1'' \times 4'' \times 9''}{144} = \dfrac{36}{144} = 0.25 \times 2 \text{ pcs.} = 0.50$

Step 3. Bottom: $\dfrac{1'' \times 9'' \times 12\frac{1}{2}''}{144} = \dfrac{112.5}{144} = 0.78$

Figure 18–10. Calculation of board feet for small pieces

ITEM	NO. OF PIECES	SIZE	BF PER PIECE	TOTAL BF
1. Vertical	7	$1'' \times 6'' \times 6'$	3	21
2. Cross	2	$1'' \times 6'' \times 3'$	1.5	3
3. Diagonal	1	$1'' \times 6'' \times 6'$	3	3
			Grand Total	27

Step 1. Vertical: $\dfrac{1'' \times 6'' \times 6'}{12} = \dfrac{6 \times 6}{12} = \dfrac{36}{12} = 3 \quad \times 7 \text{ pcs.} = 21 \text{ BF}$

Step 2. Cross: $\dfrac{1'' \times 6'' \times 3'}{12} = \dfrac{6 \times 3}{12} = \dfrac{18}{12} = 1.5 \times 2 \text{ pcs.} = 3 \text{ BF}$

Step 3. Diagonal: $\dfrac{1'' \times 6'' \times 6'}{12} = \dfrac{6 \times 6}{12} = \dfrac{36}{12} = 3 \quad \times 1 \text{ pc.} = 3 \text{ BF}$

Step 4 $\hspace{4cm}$ Grand Total $= 27$ BF

Figure 18–11. Calculation of board feet for large pieces

Wood products that are sold in sheets are sold by the square foot or panel. Examples of such products are plywood, pressed wood, chip board, wall paneling, and insulation board. These panels generally are sold in $4' \times 8'$ sheets of various thicknesses. A ¾″ × $4' \times 8'$ piece of exterior grade plywood may sell for $.50 per square foot. However, an entire $4' \times 8'$ sheet will probably have to be purchased for $16 ($4' \times 8'$ = 32 sq ft $\times$ $.50/sq ft = $16).

☐ **STUDENT ACTIVITIES**

1. List the eight components that are recommended for a bill of materials.
2. Calculate the number of board feet in each item below.

   a. 1 piece $1'' \times 6'' \times 12'$
   b. 1 piece $1'' \times 12'' \times 12'$
   c. 1 piece $2'' \times 6'' \times 8'$
   d. 5 pieces $2'' \times 4'' \times 12'$
   e. 1 piece $1'' \times 3'' \times 18'$
   f. 3 pieces $1'' \times 12'' \times 14''$

3. Calculate the cost of materials in each item below.

   a. 4 pieces $2'' \times 6'' \times 12'$ at a cost of $.50/BF
   b. 20 pieces $2'' \times 4'' \times 16'$ at a cost of $.40/BF
   c. 12 pieces $1'' \times 10'' \times 12'$ at a cost of $.55 per LF
   d. 1 piece ¼″ × $1'' \times 1''$ angle steel 20 feet long at a cost of $.30 per pound (use Figure 18–6)
   e. 1 piece ¾-inch black pipe 21 feet long at $.45 per foot

□ **SELF-EVALUATION**

A. Multiple Choice. Select the best answer.
1. A board with all four surfaces planed is specified by writing
    a. S2S
    b. S4S
    c. S2E
    d. S2+2
2. Wood products for building may be sold by the
    a. board foot
    b. linear foot
    c. square foot
    d. all of these
3. Boards that lock together are
    a. T&G
    b. BF
    c. LF
    d. S4S
4. Rough lumber is generally sold by the
    a. T&G
    b. BF
    c. LF
    d. piece
5. A board foot is based on the
    a. number of cubic inches
    b. length and width only
    c. length × width × thickness divided by 50
    d. none of these
6. Sheet steel is sold by the pound. To determine the cost per square foot
    a. divide the cost per sheet by 12
    b. multiply length by width of sheet
    c. multiply price per pound by weight per square foot
    d. none of these

B. Matching. Match the terms in column I with those in column II.

Column I	Column II
1. steel, angle	a. priced by the square foot
2. dressed lumber	b. priced by BF and LF
3. logs	c. priced by the pound
4. pipe	d. priced by board feet
5. $500/M	e. priced by linear foot
6. plywood	f. price per 1000 board feet

C. Completion. Fill in the blanks with the word or words that make the following statement correct.
1. A bill of materials is a _____ .
2. The components of a bill of materials for agricultural mechanics projects are
    a. _____
    b. _____
    c. _____
    d. _____
    e. _____
    f. _____
    g. _____
    h. _____
3. In a board 1″ × 4″ × 12′
    a. the 1″ refers to the _____
    b. the 4″ refers to the _____
    c. the 12′ refers to the _____
4. Five items that might be in a bill of materials for a trailer are
    a. _____
    b. _____
    c. _____
    d. _____
    e. _____
5. The two formulas used to calculate board feet are
    a. _____
    b. _____

D. Brief Answers. Briefly answer the following questions.
1. Which of the two formulas in question 5, Completion, is best for small boards?
2. Calculate board feet.
    a. 1 piece 1″ × 10″ × 16′ = _____ BF
    b. 1 piece 2″ × 8″ × 6′ = _____ BF
    c. 1 piece 2″ × 6″ × 8′ = _____ BF
    d. 2 pieces 2″ × 4″ × 8′ = _____ BF
3. Calculate board feet and costs of a project with the following parts if lumber costs $.40 per board foot.

	Board Feet	Cost
2 sides, each 1″ × 6″ × 18″ =	_____	____
2 ends, each 1″ × 6″ × 10″ =	_____	____
1 bottom, 1″ × 10″ × 16½″ =	_____	____
8 screws, $.07 each =		____
1 pint of paint, $4.50 =		____
Total =		____

4. Calculate the cost of each item if the price of steel is $0.50 per pound.
    a. 1 piece ⅛″ × 1″ × 1″ × 20′ angle steel (pounds per foot = .80)
       Cost = _____
    b. 4 pieces ¼″ × 2″ × 2″ × 1′ (pounds per foot = 3.19)
       Cost = _____
    c. 2 pieces ½″ × 10′ concrete reinforcing bar (pounds per foot = .67)
       Cost = _____

Unit 19
Selecting, Planning, and Building a Project

Objective

To select and plan projects that develop the woodworking and metalworking skills needed in agricultural jobs.

Competencies to Be Developed

After studying this unit, you will be able to:
- Determine the skills you now have in agricultural mechanics.
- Select projects that require the use of woodworking tools.
- Select projects that require the use of metalworking tools.
- Select projects that are appropriate for developing basic skills.
- Modify plans of projects to meet personal needs.

Terms to Know

project
scope
hands-on
graph paper

Materials List

Lead pencil
Eraser
Ruler
Graph paper, ¼-inch

This unit will show how to plan a project for construction in the agricultural mechanics shop. A *project* is a special activity planned and conducted to aid learning. All school project work is supervised by an instructor. Shop construction projects are fun. They develop eye-hand coordination and skills for jobs. In addition to personal development, there is a usable product when the project is completed.

☐ PURPOSES OF PROJECTS

The project method of learning has been used in American schools for about a century. It stimulates students to think and learn. It helps students develop good and safe work habits. Any project a student builds for himself or herself stimulates interest and a desire to do a good job.

Students may design their own projects or they may modify plans from another source. Either way, the design phase requires thinking—to seek solutions and to solve problems. The project may stimulate the student to buy materials and get to know local suppliers. It can provide experience in shopping for the best prices and in looking for other ways to cut costs. These are good business skills.

School projects help to develop construction skills. These include measuring, cutting, drilling, assembling, and finishing. A good project will require the use of a wide variety of shop skills.

A good project will also help students judge their own achievement. When students are finished, they can decide what they did well. They can also see where they made mistakes and decide how to avoid them the next time.

Finally, the project will leave a visible reminder of students' efforts, figure 19–1. If they did well, the project will show it. If they were careless, lacked skill,

Figure 19–1. A properly constructed shop project can be rewarding. This vocational agriculture student won a $100 welding award for designing and building this utility trailer.

or did not complete the project, this too will show. Students like to take home well-made projects. These are a source of pride for students and their families as well as being useful objects.

□ SCOPE OF PROJECTS

The *scope* of a project refers to its size and complexity. Students in vocational agriculture should select and plan projects that help them learn. Learning is the most important reason for doing a shop project. Therefore, woodworking projects are appropriate when woodworking is being studied. Similarly, metalworking projects are appropriate when metalworking is being studied. The project provides an opportunity for hands-on experience. *Hands-on* means actually doing something, rather than reading or hearing others tell about it. It means practicing an activity to develop a skill.

Shop projects should be relatively simple during the first course in vocational agriculture. At this time the student is learning basic skills. Many tools and equipment will be used for the first time. Work may progress slowly. In addition, there may be competition for tools in the class. However, in completing these projects, the students will develop skills to help them move on to more complex projects in the future.

Projects such as nail boxes, tool boxes, feed scoops, bird houses, stools, funnels, bench stops, foot scrapers, and gate staples have been favorites of first-year students and instructors for years, figures 19–2 and 19–3. They have been popular because they require skills appropriate to the experience of beginning students.

For second-year students, more complex projects are desired. Such projects typically are larger, require more skills, and take longer to complete. Examples are jack stands, car ramps, sawhorses, shop stands, storage units, yard gates, feed carts, pickup truck racks, and picnic tables, figures 19–4 through 19–9.

Advanced students should choose more challenging projects. Third- and fourth-year classes generally are smaller. This permits larger projects to be built. Such projects should be designed to provide additional skill development in farm carpentry, metal shaping and bending, welding, pipe cutting and bending, nailing, bolting, and painting, figure 19–10. Good projects for developing such skills include loading chutes, farm and utility trailers, wagon bodies, truck bodies, horse trailers, cattle guards, cattle feeders, post hole diggers, and portable buildings.

If farm tractor and machinery maintenance are be-

Figure 19–2. Tool boxes make excellent projects for beginning agriculture students. They provide practice in appropriate skills in woodworking and encourage the student to acquire and organize personal tools.

ing taught, farm machinery and tractors that need repair should be brought to the school shop from home or work. Appropriate repair and maintenance operations include engine tuneup, horsepower checks, minor engine repair, clutch and brake adjustments, servicing wheel bearings, replacing electrical wiring, repairing frames, replacing broken parts and painting.

Whether first or fourth year, projects must be matched with level of experience. Before proceeding too far with a project, students should discuss their ideas with their instructor.

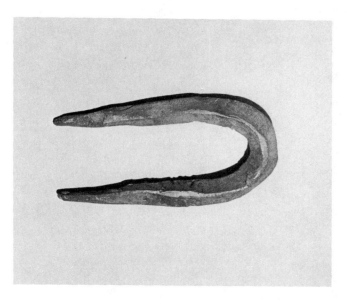

Figure 19–3. The gate staple is a good project for learning hot metalworking skills.

A. AUTOMOBILE JACK STAND

B. HEAVY-DUTY TRACTOR AND MACHINERY JACK STAND

Figure 19–4. Jack stands for automobiles and machinery develop skill in metalworking and welding.

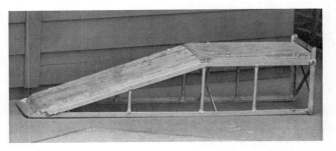

Figure 19–5. A car ramp project combines experiences in metalworking and woodworking. When making projects such as jack stands and ramps, special care must be taken to use designs that provide for extra strength. Careful design and quality workmanship are needed to assure the safety of users.

Figure 19–6. This lightweight (spider) trestle was made in an agricultural mechanics shop and has been in service for over 20 years. It features long, slender legs and light-weight construction.

□ SELECTING A PROJECT

Projects should be selected carefully. Projects that are unsuitable may not be completed or may be poorly constructed. Either way, they do not achieve their intended purpose. The following steps are recommended for selecting a shop project.

Procedure

1. Take a personal inventory of agricultural mechanics skills (competencies), figure 19–11.
2. Discuss the results of the inventory with your instructor.

Figure 19–7. Storage units may be built for the home agricultural library or other areas of the home or farm.

Figure 19–8. A rack made from pipe is useful for lumber storage in shops at home or school.

3. Ask your instructor to suggest projects that will help you develop appropriate skills.
4. Search books, project bulletins, and departmental files for project ideas and plans.
5. Examine sample projects and look for details of design.
6. Ask parents, guardians, friends, and employers for ideas and suggestions.

Figure 19–9. Picnic tables are popular home improvement projects. Good plans are available for picnic tables made of lumber, pipe, or steel.

Figure 19–10. Wagons and trailers can be relatively simple if only the low sideboards are constructed. However, if the entire running gear and body are constructed, it is a challenging project for the advanced agricultural mechanics student.

7. Make a tentative decision and obtain approval from your instructor.

□ PREPARING A PLAN

Good plans for projects are often available in textbooks, special plan books, magazines, and agricultural mechanics shop files. A good plan with detailed drawings should be used. Such plans are generally well thought out and tested through actual construction.

If an exact plan is not available, a plan that is close to the need can be modified. If nothing is available, then an original design can be made.

Detailed information on how to make a simple project plan and a bill of materials is presented in previous units. However, project ideas can be tried out quickly on graph paper. *Graph paper* is paper laid out in squares of equal size. Paper with ¼-inch squares is recommended for preparing views of projects, figure 19–12A. With graph paper, the blocks can be counted

AGRICULTURAL MECHANICS COMPETENCY INVENTORY

INSTRUCTION: **Check the column that best describes your personal level of experience in each area of competency.**

	Level of Experience			
	1 Have Done	2 Am Familiar With	3 Will Learn This Year	4 No Skill; Will Not Learn This Year

1. Practice Safety and Shop Organization
   a. Work safely with hand tools
   b. Work safely with power tools
   c. Follow safety rules in shop
   d. Maintain safe work areas
   e. Eliminate fire hazards
   f. Select and use fire extinguishers
   g. Interpret product labels
   h. Act correctly in an emergency
   i. Store materials correctly
   j. Clean the shop effectively

2. Identify and Fit Agricultural Tools
   a. Identify tools and equipment
   b. Sharpen hand tools
   c. Sharpen twist drills
   d. Fit handles for hammers and axes
   e. Fit handles for hoes and shovels
   f. Dress grinding wheels

3. Maintain and Service a Home shop
   a. Clean and maintain an orderly shop
   b. Observe safety regulations
   c. Place and use fire fighting equipment
   d. Inventory tools and equipment
   e. Plan shop layout
   f. Select shop site

4. Apply Farm Carpentry Skills
   a. Select lumber for a job
   b. Operate power saws
   c. Operate a jointer
   d. Operate a planer
   e. Saw to dimension
   f. Bore holes
   g. Glue wood
   h. Drive and remove nails
   i. Set screws and install bolts

5. Properly Use Paint and Paint Equipment
   a. Select paints or preservatives
   b. Compute area for painting
   c. Prepare wood surfaces

Figure 19–11. Agricultural mechanics competency inventory

	Level of Experience			
	1 **Have** **Done**	**2** **Am** **Familiar** **With**	**3** **Will** **Learn** **This Year**	**4** **No Skill;** **Will Not** **Learn This Year**
d. Prepare metal surfaces				
e. Apply paint with a brush or roller				
f. Clean and store paint brushes				
g. Mask areas prior to painting				
h. Apply paint with a spray can				
i. Clean and store spray gun				
6. Operate an Arc Welder				
a. Practice safety in arc welding				
b. Prepare metal for welding				
c. Determine welder settings				
d. Select electrodes				
e. Operate AC and DC welders				
f. Strike an arc				
g. Run flat beads				
h. Make weld joints				
i. Weld horizontally				
j. Weld vertically				
k. Weld overhead				
l. Cut with arc				
m. Weld cast iron				
n. Hard surface steel				
7. Operate an Oxyacetylene Welder				
a. Check for leaks				
b. Turn equipment on and off				
c. Adjust equipment				
d. Change cylinders				
e. Cut with cutting torch				
f. Choose proper tips				
g. Braze thin metal				
h. Run beads				
i. Make butt welds				
8. Perform Skills in Hot and Cold Metal Work				
a. Identify metals				
b. Cut with hand hacksaw				
c. Cut with cold chisel				
d. Bend metal				
e. Cut with tinsnips				
f. Drill holes				
g. Cut threads				
h. Use files				
i. Solder metal				
j. Use power hacksaw				
k. Heat metal with torch				
l. Operate gas forge				

Figure 19–11. Cont.

	Level of Experience			
	1 **Have** **Done**	**2** **Am** **Familiar** **With**	**3** **Will** **Learn** **This Year**	**4** **No Skill;** **Will Not** **Learn This Year**
9. Perform Skills in Concrete and Masonry Work				
a. Build and prepare forms				
b. Treat forms				
c. Reinforce concrete				
d. Test aggregates for impurities				
e. Mix concrete				
f. Pour concrete				
g. Embed bolts				
h. Finish concrete				
i. Protect concrete while curing				
j. Trowel concrete				
k. Remove forms				
l. Lay concrete blocks				
m. Drill holes in concrete				
10. Operate and Maintain Farm Engines				
a. Read and follow operator's manual				
b. Start and operate engines				
c. Change oil and oil filters				
d. Service air and fuel filters				
e. Maintain battery water				
f. Maintain and operate small gas engines				
g. Identify engine components and systems				
h. Remove and connect battery cables				
i. Charge batteries				
j. Read battery hydrometer				
k. Test batteries				
l. Troubleshoot fuel problems				
m. Troubleshoot ignition problems				
11. Perform Minor Tuneup and Repair of Farm Engines				
a. Clean, gap, and replace spark plugs				
b. Adjust carburetor mixture and speed screws				
c. Install gap breaker points				
d. Set dwell with meter				
e. Install condensers				
f. Time engines using timing light				
g. Disassemble and reassemble distributors				
h. Measure compression				
i. Adjust valves				
j. Set float levels in carburetors				
k. Rebuild carburetors				
l. Operate engine analyzers				
m. Test and replace coils				
n. Clean and install brushes				
o. Clean commutator bars				

Figure 19–11. Cont.

	Level of Experience			
	1 **Have Done**	**2** **Am Familiar With**	**3** **Will Learn This Year**	**4** **No Skill; Will Not Learn This Year**
12. Perform Electrification Skills				
a. Use safety measures in electrical wiring				
b. Select correct fuse sizes				
c. Replace fuses				
d. Make splices				
e. Repair electrical cords				
f. Wire on-off switches				
g. Select wire sizes				
h. Know electrical terminology such as: volts, amps, watts, ohms				
i. Attach wires to terminals				
j. Solder splices				
k. Install wire nut connectors				
l. Install light fixtures				
m. Install electric motors				
n. Wire buildings and structures				
o. Wire three-way switches				
p. Wire four-way switches				
13. Learn Plumbing Skills				
a. Repair leaky faucets				
b. Assemble pipe and pipe fittings				
c. Thread pipe				
d. Measure and cut pipe				
e. Cut plastic tubing/pipe				
f. Install plastic tubing/pipe				
g. Flare copper tubing				
h. Cut copper tubing				
i. Ream pipe				
j. Install fixtures on plastic pipe				
k. Sweat joints on copper pipe				
l. Adjust air control mechanism on pressure system				
m. Install pressure pumps				
n. Repair pumps				
o. Install water heaters				
p. Cut soil pipe				
q. Caulk joints on soil pipe				
r. Lay soil pipe				

Figure 19–11. Cont.

as a means of measuring. A ruler can then be used as a straightedge to draw lines. On plain paper, every line requires two or more measurements to establish length and position, figure 19–12B.

To design on paper, a tentative scale is first tried. If it doesn't work, another scale is tried. Graph paper with ¼-inch squares has 34 blocks one way and 44 blocks the other. To draw a project view on a sheet of this graph paper, follow these steps.

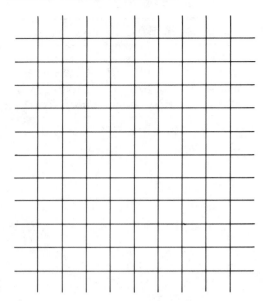

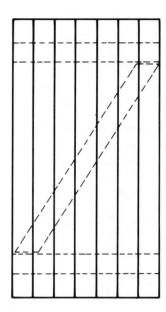

A. GRAPH PAPER WITH 1/4-INCH SQUARES IS EXCELLENT FOR PROJECT PLANNING. ON GRAPH PAPER, DISTANCES ARE DETERMINED BY COUNTING SQUARES AND USING A RULER TO CONNECT POINTS.

B. A PROJECT DRAWN ON PLAIN PAPER REQUIRES MORE TIME AND EFFORT THAN ONE DRAWN ON GRAPH PAPER. ON PLAIN PAPER, EVERY LINE REQUIRES TWO OR MORE MEASUREMENTS TO ESTABLISH LENGTH AND POSITION.

Figure 19–12. Graph paper makes plan drawing easier.

Procedure

1. Determine the longest dimension of the project. This may be the length or the height.
2. Turn the graph paper horizontally for long projects, vertically for high ones.
3. Divide 44 (maximum number of squares in the long direction) by the longest dimension of the project. The result tells how many feet or inches one square must represent.
4. Choose a scale that permits a one-view drawing on the paper with some room for dimensions and notes.
5. Count squares and draw lines to complete one view. Include dimensions.
6. Draw additional views on additional sheets of paper.
7. Modify the plan until it is what you want.
8. Review the plan with the instructor.
9. Draw better plans after obtaining the instructor's approval.

After the plan is drawn, a bill of materials can be prepared and the materials purchased. It is best not to change the design after construction has started. However, if errors in design become obvious during construction, design modifications must be made.

□ DESIGN

Design work is not easy. It generally requires a lot of experience to be done well. When designing projects, the following principles and facts should be kept in mind:

- Base designs on engineered plans—that is, start with tested plans.
- Use materials of extra strength and size if safety is a factor. For example, car ramps and jack stands should be extra sturdy. Failure of these objects could cause injury or death.
- Structural steel generally holds up better than wood.
- Welding is stronger than riveting.
- Gluing is generally the most secure method of fastening wood.
- Bolts are more secure than screws; screws are more secure than nails.
- Crimps and folds increase the strength of sheet metal panels.
- Tongue and groove, dowels, and splines all increase the strength of joints in wood.
- Both wood and metal must be protected by paint if they are to be exposed to weather.

Examples of project plans are presented in Appendix A.

□ STUDENT ACTIVITIES

1. Take a personal agricultural mechanics skills inventory.
2. Write down the skills that you need to develop through project construction.
3. Discuss project ideas with your instructor.
4. Search books, bulletins, files, and other sources for project ideas.
5. Select a project that fits your shop experience.
6. Make or modify plans as needed.

□ SELF-EVALUATION

A. Multiple Choice. Select the best answer.
   1. The project method of learning is used because it
      a. develops body coordination
      b. develops skills for jobs
      c. provides a product when finished
      d. all of these
   2. The project method has been used in the public schools for about
      a. ten years
      b. two decades
      c. thirty years
      d. one century
   3. The project method of learning usually
      a. causes students to think
      b. causes students to work carefully
      c. creates personal interest
      d. all of these
   4. A suitable project must
      a. provide practice in skills taught in class
      b. earn the student money
      c. be completed in four weeks
      d. be paid for totally by student funds
   5. Projects that are not carefully chosen will
      a. result in student failure
      b. never be completed
      c. not achieve the educational purposes
      d. be too costly
   6. Good project ideas may come from
      a. books
      b. teachers
      c. parents or guardians
      d. all of these
   7. Good sample projects have the benefit of
      a. showing details of design
      b. showing mistakes to avoid
      c. eliminating the need for a bill of materials
      d. all of these
   8. In general, preferred projects for first-year students are
      a. those already designed with plans published
      b. those designed by the student
      c. those modified by the student
      d. all of these
   9. Graph paper for drawing permits
      a. easy drawing of angles
      b. quick measurement
      c. easy three-view drawing
      d. economy of cost

B. Matching. Match the terms in column I with the terms in column II.

Column I	Column II
1. hands-on	a. second- or third-year project
2. nail box	b. third- or fourth-year project
3. feed cart	c. first-year project
4. farm trailer	d. doing
5. graph paper	e. strongest wood joint
6. nailing	f. weakest wood joint
7. gluing	g. strongest metal joint
8. steel	h. stronger than wood
9. paint	i. ¼-inch squares
10. weld	j. protects from weather

C. Completion. Fill in the blanks with the word or words that make the following statement correct.
   1. Building shop projects helps students develop buying skills because _____.
   2. Shop skills developed by building suitable projects include _____, _____, _____, _____ and _____.
   3. Students in vocational agriculture should select and plan projects that _____.

Section 6
TOOL FITTING

Unit 20
Repairing and Reconditioning Tools

Objective

To restore worn, damaged, or abused tools to good working condition.

Competencies to Be Developed

After studying this unit, you will be able to:
- Remove rust from metal tools.
- Apply rust-inhibiting materials to metal surfaces.
- Repair split wooden handles.
- Replace broken wooden handles.
- Reshape screwdriver tips.
- Reshape the heads of driving and driven tools.

Terms to Know

abuse	neat's-foot oil
rust	ferrule
mushroomed	tang
tool fitting	crown
recondition	torque
saddle soap	

Materials List

Tools that need repair or reconditioning:
  a. Tools with rust showing
  b. Driving tool with a split wooden handle
  c. Driving tool with a broken wooden handle
  d. Fork, rake, hoe, or shovel with broken handle
  e. Screwdriver that needs fitting
  f. Leather tools or protective clothing that need reconditioning
Replacement wooden handles for previous items
Saddle soap or neat's-foot oil
Stiff bristle scrub brush
Soft cloths
Paste finishing wax
Varsol or kerosene
Wire brush or wire wheel
Waterproof silicon carbide paper, 220 and 400 grit
Lightweight lubricating oil, No. 10 or lighter
Mediumweight lubricating oil, No. 20 or 30
Wood glue
Vise, hand tools, and electric drill
Assorted steel rivets
Wood and metal handle wedges

Modern farms and agribusinesses make large investments in tools. The most obvious use of tools is to build, repair, or adjust objects and machines. However, tools need repair too.

This unit provides instruction in repairing tools. The student is advised to review units 14 and 16 on the use of power drills and grinders. These tools are useful in performing the tasks described in this unit.

A high-quality tool will last for years if used and maintained properly. Normal use does cause tools to lose their shape, however. Also, tools are frequently abused. To *abuse* is to use wrongly, make bad use of, or misuse. Such practices necessitate tool repair.

Metal tools rust if they get wet and are not dried off quickly. *Rust* is a reddish brown or orange coating that results when steel reacts with air and moisture. It creates pits in the surface of steel.

Wood handles may split or break, and metal handles bend out of shape leaving tools useless until repairs are made. Tools such as punches, chisels, wedges, and sledges get mushroomed faces. *Mushroomed* means a spread or pushed-over condition caused by being struck many times and occurs over a long period of use. After much use, screwdrivers get rounded tips and cannot grip screws properly. Screw heads are then damaged by attempts to use such screwdrivers.

The term *tool fitting* means to clean, reshape, repair, or resharpen a tool. To *recondition* means to do what is needed to put the tool into good condition. This unit will cover cleaning and protecting tool surfaces as well as reshaping and otherwise repairing tools. The next unit will cover tool sharpening. Many students in agricultural mechanics start or add to their tool holdings by restoring tools that others have discarded.

□ PROTECTING TOOL SURFACES

Most tool parts are made of wood, plastic, leather, or steel. Plastic is tough and is not affected by moisture. It is too soft to restore by grinding or buffing. Therefore, plastic parts must be handled carefully at all times. It is important to avoid cuts, nicks, or other damage to plastic.

Restoring Leather Parts

Leather is used in a few tools. It may be rolled very tightly to form a mallet head. Layers of leather may be stacked and glued on wooden chisel handles. This cushions the rest of the tool when it is struck with a mallet. Leather is also used in protective clothes such as gloves and aprons.

When leather becomes stiff or dry, it can be restored by rubbing with saddle soap. *Saddle soap,* when mixed with water, will clean, soften, and preserve leather. The procedure is as follows:

Procedure

1. Brush off all dirt with a stiff bristle brush.
2. Moisten a sponge or soft cloth with water.
3. Rub the cloth onto the surface of the saddle soap until suds develop into a lather.
4. Rub the sudsy mixture into all leather surfaces until the leather is clean and soft.
5. Remove suds with a damp cloth.
6. Let leather dry.
7. Rub with a soft, dry cloth or brush with a soft bristle brush.

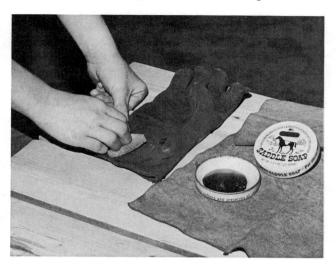

Figure 20–1. Leather may be restored with saddle soap, neat's-foot oil, or other commercial products containing silicone or lanolin.

Another product used to soften, restore, and protect leather is neat's-foot oil. *Neat's-foot oil* is a light yellow oil obtained by boiling the feet and shinbones of cattle. Simply apply neat's-foot oil to clean leather surfaces with a lamb's wool pad or soft cloth.

Other products may be used to recondition leather. They contain various mixtures of silicone, lanolin, or other oils, figure 20–1.

Restoring Wooden Surfaces

Wooden handles frequently become dried out, porous, and rough. Repeated use of such handles results in sore hands and sometimes splinters. Handles that dry out become loose and dangerous.

Wooden handles and other parts are easy to restore if treated as soon as drying is noticed. The procedure is as follows.

Procedure

1. Sand wooden parts with very fine sandpaper to remove all rough areas.
2. Rub all wood with a soft cloth dipped in boiled linseed oil, figure 20–2.
3. After the oil dries, rub the wood briskly with a soft, dry cloth.
4. Repeat steps 2 and 3 several times if the wood is very porous.

Another product that may be used to restore wooden tool handles is paste finishing wax. The procedure is similar to that for linseed oil.

Figure 20–2. Linseed oil or paste finishing wax will restore and protect wooden parts.

Procedure

1. Before applying the wax, sand the wood thoroughly.
2. Rub on a thin coat of wax and let it dry.
3. Once the wax is dry, polish it with a soft cloth.
4. Repeat the process if needed to obtain a smooth surface that will prevent moisture from entering the wood.

Restoring Metal Surfaces

Metal surfaces may be quite difficult to restore if they are badly neglected. However, a few standard procedures can stop rust and restore a neglected tool to service.

Removing Grease and Oil. □ Grease and oil on tool handles are dangerous as well as messy. Tools are likely to slip out of the hand or be difficult to control when greasy.

To remove grease or oil, wipe the tool with a clean cloth dipped in kerosene, varsol, or other commercial solvent for grease and oil. After cleaning with a solvent, dry the tool with a clean cloth or paper towel.

Caution:

Do not use gasoline to remove grease and oil from tools. Gasoline is very volatile. Its explosive fumes will travel to sparks, flames, or other sources of fire.

Removing Dirt. □ Tools such as shovels, hoes, and rakes may become caked with mud or concrete. When dry, these materials can be removed by tapping the tool with a metal object. The vibration generally causes the material to drop off. Any remaining material should be removed with a scraper, wire brush, or wire wheel.

Removing Rust. □ Light rust can be removed with varsol, kerosene, or other solvents. However, once pitting starts, a wire brush, wire wheel, steel wool, emery cloth, or aluminum oxide paper must be used, figure 20–3. These materials will dislodge deep rust but may scratch polished surfaces. A smooth final surface may be obtained by using 400 grit silicon carbide paper. Use water to help keep the paper clean while sanding.

Immediately after sanding, dry the tool thoroughly and coat all smooth metal surfaces with a light oil such as No. 10. Rough surfaces should be coated with mediumweight oil such as No. 20 or 30. If the tools are not treated in this manner, the water will cause rusting to start again within a few hours.

Figure 20–3. When pitting is evident on steel, rust must be removed by using a wire brush, wire wheel, steel wool, emery cloth, silicon carbide paper, or other abrasives.

Protecting Steel from Rust. □ As noted before, rust occurs when steel is exposed to moisture and air. Therefore, rust is prevented by keeping air or moisture away from steel. Special rust-resisting primers and paints are the products most frequently used to protect steel from rust. However, many tool surfaces must remain smooth and unpainted in order to cut well. In addition, smooth, unpainted surfaces permit material to slide over metal more easily. Table tops on power machines are examples. Unpainted steel edges and surfaces are protected by rubbing with a good grade of lubricating oil. Paste wax also works satisfactorily on unpainted steel. However, wax is useful only if the steel is free of rust when it is applied.

□ REPAIRING DAMAGED TOOLS

Tools may be repaired by welding or gluing of broken parts. Wooden handles on files, wood chisels, hammers, axes, shovels, and the like can be replaced when broken. Sometimes it is cheaper to replace the tool than it is to replace the handle. However, for most tools, replacing handles is the most economical step.

Repairing Wooden Handles

Split Handles. □ Wooden handles may split when exposed to excessive strain. If the split is clean, it can be repaired with wood glue, figure 20–4. The procedure is as follows.

Procedure

1. Position the tool in the vise so the split can be forced open.
2. Force open the split and spread glue on both surfaces.
3. Close the split and clamp the handle in several places.

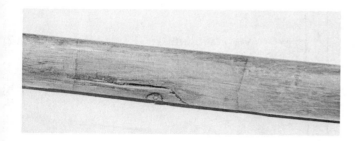

Figure 20–4. This split handle is strong, but may cause splinters if the split area is not taped. Split wooden handles may be repaired by gluing.

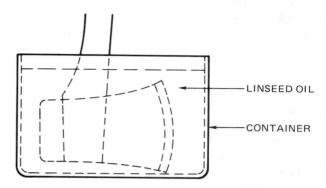

Figure 20–5. Treatment with linseed oil for 12 to 24 hours prevents drying and loosening of wooden handles.

4. Wipe off all excess glue.
5. After the glue dries, thoroughly sand the handle.
6. Treat the handle with boiled linseed oil and polish with a clean cloth when dry.

Loose Handles. □ Loose handles on hammers, hatchets, sledges, and axes are very dangerous. Handles have a wooden wedge and one or two metal wedges driven into the end. A *wedge* is a piece of wood or metal that is thick on one end and tapers down to a thin edge on the other. Check frequently to ensure that wedges are securely in place. If they are not, drive them into place. Even when handles are properly attached, they can still loosen due to drying. To reduce this problem, soak the end of the handle which is in the head of the tool with linseed oil about once a year, figure 20–5. An alternate method for treating tool heads with linseed oil follows.

Procedure

1. Clamp the head of the tool in a vise with the end of the handle pointing upward. Brush enough boiled linseed oil over the end of the handle and wedges to saturate all parts. Keep them saturated for 12–24 hours if possible.
2. Permit the oil to dry while the head of the tool is still clamped in the vise.
3. Remove the tool from the vise and wipe a coat of boiled linseed oil on the rest of the handle.
4. Polish the handle with a dry cloth after the oil dries.

Replacing Wooden Handles

As mentioned before, most wooden handles are secured in the heads of tools by the use of wedges. The hole in the head of the tool is called the eye.

This hole is smaller on the side where the handle enters than on the opposite side. Therefore, once the handle is inserted, it can be wedged out to fill the larger portion of the hole. As long as the wedges stay in place, the head remains tight on the handle.

Removing Broken Handles. □ A wooden handle generally breaks off flush with the head of the tool. When this occurs, the wooden core and metal wedges are still lodged securely in the tool head. To remove the broken piece of handle, the following procedure is recommended.

Procedure

1. Place the head securely in a vise.
2. Use a ¼-inch or ⅜-inch metal cutting drill bit and drill numerous holes into the wooden core.
3. Drive the remaining honeycomb of wood out of the head using a large punch or rod.

> **Caution:**
>
> **Do not attempt to remove wooden pieces from metal tools by burning them out. The heat will cause the metal to lose its hardness.**

Attaching Handles to Hammers and Axes. □ When purchasing a replacement handle it is important to buy the correct size. The end of the handle that is shaped to go into the head should be slightly longer than the head is deep and should have the same shape as the hole in the head. It should also be slightly larger in cross section than the smaller part of the eye, or hole into which it is to be inserted.

Handles must be shaped to fit a head exactly. Start by placing the handle in a vise. The end to be worked on should be exposed so that it can be sawed, filed, and shaped. The procedure for attaching handles to hammers and axes follows and is also illustrated in figures 20–6 and 20–7.

Procedure

1. Place the head against the end of the handle to see if it will start on.
2. Observe where wood must be removed to shape the handle to enter the head.
3. Use a wood rasp or coarse file to shape the handle.

> **Caution:**
>
> **Do not remove too much wood from the end. This will help get the head on, but it will also damage the wedged fit.**

4. Try the head frequently as wood is removed to avoid a loose fit.

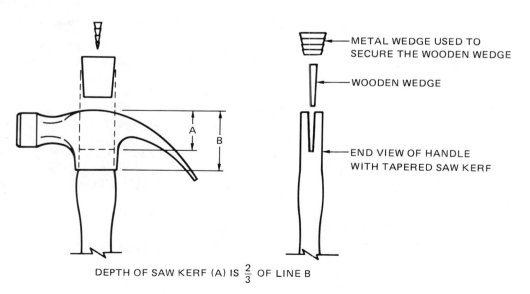

DEPTH OF SAW KERF (A) IS $\frac{2}{3}$ OF LINE B

Figure 20–6. A handle is attached to a tool by wedging.

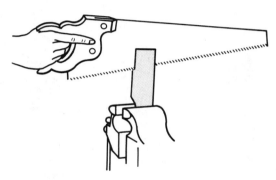

PLACE THE HANDLE SECURELY IN A
VISE AND USE A HAND SAW TO MAKE
A KERF ACROSS THE LONGEST CENTER
LINE OF THE HANDLE.

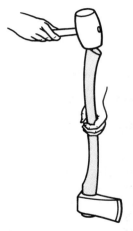

DRIVE THE HANDLE INTO THE
HEAD WITH A WOODEN MALLET.

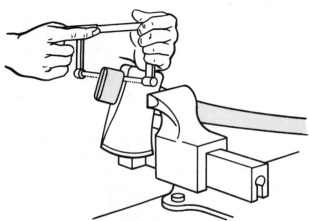

USE A HACKSAW TO SAW OFF THE
EXCESS HANDLE FLUSH WITH THE HEAD.

DRIVE IN A WOODEN WEDGE SO THE
HANDLE SPREADS AND FILLS THE HEAD.

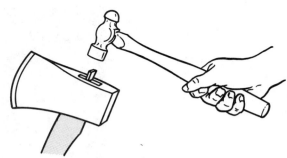

DRIVE ONE OR TWO METAL WEDGES IN AT RIGHT ANGLES TO
THE WOODEN WEDGE TO HOLD THE WOODEN WEDGE IN PLACE.

Figure 20–7. Procedure for attaching a wooden handle to an axe or hammer

Caution:

Do not drive the head. It will bind and be very hard to remove.

5. Work the handle down until the head slides on snugly. It should come to rest about ½ inch from the enlarged part of the handle.
6. Mark the handle on both sides of the head.
7. Remove the head.
8. Reposition the handle vertically in the vise.
9. Use a handsaw to make a kerf across the longest center line of the handle. The kerf should extend two-thirds of the distance between the two marks made in step 6.
10. Reposition the handle in the vise and squeeze the end until the saw kerf is completely closed.
11. Run the saw down through the kerf again. When released, the kerf will be wider at the end than further down in the handle.
12. Now make a wooden wedge as wide as the oval hole in the head and thick enough to spread the handle when driven in.
13. Slide the handle into the head and drive it in securely with a wooden mallet.

Caution:

A steel hammer will ruin the end of the handle.

14. Grip the handle with the vise just below the head.
15. Use a hacksaw to saw off the excess handle flush with the head.
16. Drive in the wooden wedge so the handle spreads and fills the head.

Caution:

Drive the wedge evenly or it will split.

17. Use a hacksaw to saw off the excess wooden wedge.
18. Drive one or two metal wedges in at right angles to the wooden wedge. These will hold the wooden wedge in place.

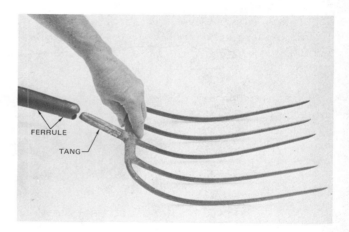

Figure 20–8. Forks, hoes, and rakes have a tang which is held in the handle with a nail or rivet or by friction between the wood and metal.

19. Place the tool, head down, into a metal or plastic container about the size of the head.
20. Add several inches of boiled linseed oil and brush oil around the handle and head.
21. The handle and head should soak for several days to seal the wood.
22. Remove the tool, rub oil on all parts of the handle, dry, and polish.

Attaching Handles to Rakes, Hoes, and Forks. □ Rakes, hoes, and forks are driven into their handles. The handles are fitted with metal collars to prevent splitting of the wood. This metal collar is called a *ferrule*. The tool has a metal finger called a *tang* which is driven into the ferrule, figure 20–8. In some handles, a hole is provided in the ferrule for a nail or rivet.

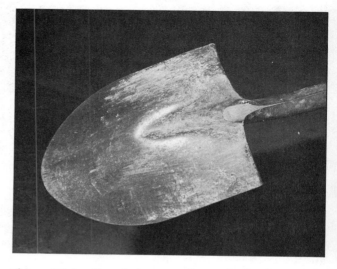

Figure 20–9. Shovels have a split metal tube that is tightened around the handle for support. A rivet is installed through the metal and handle to hold the handle in place.

The nail or rivet is driven through the hole into the handle and tang to help hold the parts together. Other handles rely on friction only to keep the parts together.

Attaching Handles to Shovels. □ Shovel handles are exposed to tremendous forces when digging. Therefore, a metal tube extends up the handle for extra support, figure 20–9. Some shovel handles have a single bend and others have a double bend. Shovel handles are made to fit specific shovels. Therefore, it is best to take the shovel to the store when selecting a replacement handle. To attach a shovel handle, the following procedure is suggested.

Procedure

1. Grind off the head of the rivet that holds the handle.
2. Remove the rivet with a drift punch.
3. Spread the metal tube and drive out the remains of the old handle.
4. Use a mallet to drive the new handle securely into place.

Caution:

If the handle drives very hard, consider removing some wood to improve the fit.

---

5. Use a vise to close the metal around the handle.
6. Insert a drill through the hole in the metal and drill through the wooden handle.
7. Install the replacement rivet securely.

Many other tools have replaceable handles. The procedures outlined for the tools in this unit will apply to other tools as well.

Reshaping Tools and Tool Heads

All metal tools that are driven will eventually flatten. If the tools are not reground, the surfaces will mushroom. A mushroomed tool head is dangerous. When struck, fragments of steel are likely to break off and fly into the face and body of the operator or bystanders.

Driven tools such as punches, chisels, and wedges are designed with a slight bevel around the crown. The *crown* is the end of a tool which receives the blow of a hammer. The bevel and the crown together make up the head of the tool, figure 20–10.

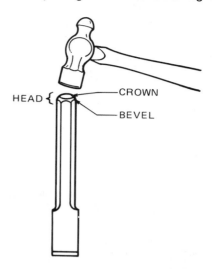

Figure 20–10. Driven tools such as punches, chisels, and wedges have a beveled edge around the crown to resist flattening when struck.

To reshape the head of a driven tool, the procedure is as follows.

Procedure

1. Examine the head of a similar tool that is in good condition.
2. With the good tool as a guide, use a medium grinding wheel to grind a taper from each flat surface to the crown.

Caution:

Use a face shield and leather gloves when grinding.

---

3. Finish the taper by twirling the tool slightly so that all corners are slightly rounded. Round tools will have an even taper in a circular pattern around the crown.

The heads of axes, sledges, and hammers can also be reshaped using the procedures just described. However, these tools do not have much of a taper around the crown. Instead, they have rounded edges. It is important to keep the edges slightly rounded to avoid mushrooming of the head.

Reshaping Screwdrivers. □ The tips of standard screwdrivers can be reshaped. Phillips and other crossheaded screwdrivers must be discarded when their tips become worn. Standard screwdriver tips

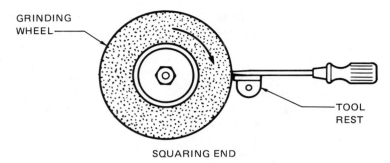

SQUARING END

A. TO SQUARE THE END OF A SCREWDRIVER, SLOWLY
PUSH THE TIP OF THE SCREWDRIVER ACROSS THE
TOOL REST AND INTO THE GRINDING WHEEL.

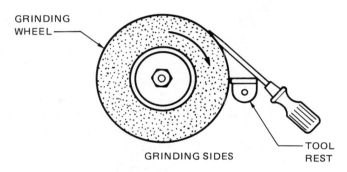

GRINDING SIDES

B. TO FLATTEN THE SIDES OF A SCREWDRIVER, HOLD
THE SCREWDRIVER DIAGONALLY ACROSS THE TOOL
REST WITH THE SIDE FLAT AGAINST THE WHEEL
WHILE GRINDING.

Figure 20–11. Shaping the tip of a standard screwdriver

must be reground often to maintain a shape that will grip screws securely. Properly shaped standard screwdriver tips have a flat end and parallel sides. The width and thickness of the tip are determined by the size of the slot in the screw to be used.

When reshaping a standard screwdriver tip, the following procedure is recommended.

Procedure

1. Obtain a screw of the size you wish to drive with the screwdriver.
2. Check to see that the screwdriver tip is as wide as the screw head. If it is not, obtain a wider screwdriver.
3. Check the face of the grinding wheel to be sure it is flat.
4. Check to see that the grinder's tool rest is properly adjusted.
5. Slowly push the flat end of the screwdriver tip across the tool rest and into the grinding wheel, figure 20–11A.
6. Grind the tip until it is flat, square, and even.

Caution:

Avoid overheating the tip of the screwdriver.

7. Place the tip into the screw slot. It should reach to the bottom of the slot and fit the sides snugly.
8. If the tip is too thick, grind the flat sides of the tip slightly to narrow the screwdriver blade, figure 20–11B.

Caution:

Be careful to maintain the flat end. The flat surfaces of screw and screwdriver must fit perfectly. Any taper in the tip will cause it to twist out of the screw slot when torque is applied. *Torque* is a twisting force.

9. Dress the edges of the tip so they are equal to the width of the screw head.

□ **STUDENT ACTIVITIES**

1. Bring hand tools from home that need cleaning, rust protection, or repairing.
2. Remove dirt and rust from tools.
3. Repair wooden handles on tools.
4. Replace handles on driving and digging tools.
5. Reshape the heads of driven tools such as chisels and punches.
6. Reshape screwdriver tips.

□ **SELF-EVALUATION**

A. Multiple Choice. Select the best answer.
1. A common problem with wooden handles is
   a. rotting
   b. splitting
   c. rusting
   d. fatigue
2. Leather parts should be reconditioned with
   a. neat's-foot oil
   b. saddle soap
   c. lanolin products
   d. any of these
3. Hammer handles become loose if
   a. the handle dries out
   b. linseed oil is not applied occasionally
   c. the handle is not properly installed
   d. all of these
4. Wooden handles should be treated with
   a. linseed oil
   b. shellac
   c. paint
   d. all of these
5. A solvent recommended for removing grease and light rust is
   a. gasoline
   b. varsol
   c. turpentine
   d. water
6. After water touches an unprotected steel surface, rusting starts within
   a. hours
   b. days
   c. weeks
   d. months

7. Rust may be prevented from forming on metal surfaces by applying
   a. oil
   b. water
   c. saddle soap
   d. any of these
8. Split wooden handles are best repaired with
   a. nails
   b. screws
   c. glue
   d. tape
9. Driving tool handles are held in place by
   a. bolts
   b. nails
   c. screws
   d. wedges
10. When standard screwdrivers have rounded tips they
    a. work better
    b. need replacing
    c. slip out of screw slots
    d. should be heated and reshaped

B. Matching. Match the terms in column I with those in column II.

Column I	Column II
1. mushroomed	a. damaged head
2. linseed oil	b. hammer
3. tool fitting	c. reconditioning
4. neat's-foot oil	d. treat handles
5. mallet	e. rakes and hoes
6. ferrule	f. drive handles
7. wedged handle	g. leather

C. Completion. Fill in the blanks with the word or words that will make the following statements correct.
1. When a tool head has mushroomed, it has _____.
2. The crown is the part of the tool that _____.
3. When reconditioning leather, a sponge may be used to apply _____.
4. Shovel handles are held in place with a _____.
5. A mushroomed condition is corrected by _____.
6. To obtain a smooth final surface, metal tools should be sanded with _____.
7. A good tool for removing rust is a _____ _____.

Unit 21
Sharpening Tools

Objective

To keep cutting tools sharp.

Competencies to Be Developed

After studying this unit, you will be able to:
- Examine tools and determine the design of the cutting edges.
- Select appropriate procedures for sharpening tools.
- Sharpen knives.
- Sharpen wood chisels and plane irons.
- Sharpen cold chisels and center punches.
- Sharpen axes and hatchets.
- Sharpen twist drills.
- Sharpen rotary mower blades.
- Sharpen digging tools.

Terms to Know

bench stone	temper
hand stone	draw the temper
inclined plane	whet
convex	serrated
concave	field use
hollow ground	high-speed drill
tool steel	clockwise
anneal	balance
true	arbor

Materials List

Grinder with medium and fine grit wheels
Eye or face protection and protective clothing
Heavy-duty portable sander with fine aluminum oxide discs
Large, flat file with medium or fine teeth
Machinist's vise
Assorted hand tools
Oil bench stone (one fine side and one coarse side)
Ax stone
Tool fitting gauge
Tools requiring repair and sharpening
  a. Pocket or pruning knife
  b. Plane iron and/or wood chisel
  c. Cold chisel
  d. Center punch
  e. Ax or hatchet
  f. Twist drill
  g. Rotary mower blade
  h. Digging tools

A properly sharpened tool of good quality can make cutting, drilling, and digging jobs relatively easy. Such tools result in work of excellent quality when used by skilled craftsmen. Sharp tools are safer than dull ones because they require less pressure. A craftsman should not attempt a job with a dull tool, because the resulting quality of work will be disappointing.

☐ TOOLS USED FOR SHARPENING

Tools in the agricultural mechanics shop are generally sharpened with files, bench stones, hand stones, sanders, and grinders. The use of files and grinders is covered elsewhere in this text. A *bench stone* is a sharpening stone designed to rest on a bench. A *hand stone* is one designed to be held in the hand when in use. Sharpening stones include the common 6″ × 2″ × 1″ bench stone, hand-held slip stones, and power-driven wheels. Bench stones are generally mounted in a wooden box which holds the stone during use and storage, figure 21–1. The box also helps keep the stone clean and oil filled. Many stones require either water or oil to keep them clean and to ensure that the abrasive materials cut during use.

Files needed for sharpening tools include flat, round, triangular, and special styles. Examples of special files are auger bit files and chain saw files. Coarse files are useful for sharpening large objects such as

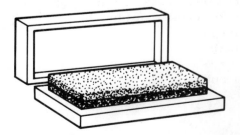

Figure 21–1. An oil-filled bench stone mounted in a wooden box

blades for rotary mowers. Fine-toothed files are used for the majority of sharpening jobs.

□ CUTTING EDGES

Always observe the design and shape of the edges of cutting tools when they are new. The tool is designed to do a specific job. The nature of that job determines the shape of the cutting edge. For example, shaving should remove all evidence of hair. Hair is relatively soft. Therefore, a razor blade has a very thin, sharp edge. This edge does a good job in shaving but it is too fragile for most other cutting jobs.

In contrast, steel is tough and hard. Therefore, tools used to cut steel must be tough and hard. Their edges must be thick and strong.

Wood is relatively soft, so wood chisels and plane irons can be ground to fine, sharp edges. However, this means that they nick, break or dull quickly when they are used on metal objects or come in contact with dirt or metal surfaces.

Tool edges are usually designed with either single or double inclined planes, figure 21–2. An *inclined plane* is a surface that is at an angle to another surface. It is one of the six simple machines.

Most tool blades consist of a single inclined plane. Such tools include wood chisels, plane irons, shears, drills, mower knives, digging tools, and saws. All are sharpened on only one side of the edge.

Other tools are designed with edges formed from two inclined planes. These tools must be sharpened from both sides. Some examples are knives, axes, hatchets, and cold chisels.

The term *convex* means curved out. An edge that is convex is strong because it has extra steel to back it up. However, the bulging nature of a convex edge means it requires more force to make it cut. Axes and similar tools with convex edges get their force from their speed and weight. Wedges get their force from a driving tool.

The term *concave* means hollow or curved in. Wood chisels and plane irons are ground to a concave edge. Their edges are very sharp and require little force to make them cut. Certain saws may be ground so the teeth are wider at the points than they are at their base. Such blades are said to be *hollow ground*. The teeth are wider at their points than the thickness of the blade behind the teeth. Therefore, the teeth have no set and the blade makes a very smooth cut.

Preserving Temper

Tools are made from tool steel. *Tool steel* is steel with a specific carbon content that allows the tool to be annealed and tempered. To *anneal* is to heat and then cool the steel slowly so as to make it soft and malleable. To *temper* is to heat the steel and then cool it more quickly so as to control the degree of hardness. After tools are properly tempered, excessive heating will *draw the temper,* or modify it to render the steel soft and useless until it is retempered. To *draw the temper* means to modify the temper.

□ TESTING FOR SHARPNESS

Knives, plane irons, and chisels should be sharpened to a very fine edge. To check for sharpness, place the tool blade in a flat position on your thumb nail. Then raise the back slightly and push the cutting edge across the nail. The edge should catch the nail and shave off a sliver. If this doesn't happen, the tool is not very sharp.

Caution:

Any test of sharpness using the hands or body must be done with extreme care.

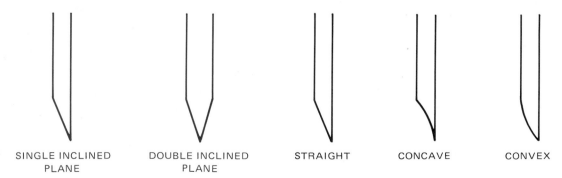

SINGLE INCLINED PLANE   DOUBLE INCLINED PLANE   STRAIGHT   CONCAVE   CONVEX

Figure 21–2. Tool edges usually have single or double inclined planes. These planes may also be straight, concave, or convex.

TOOL	SIDES TO SHARPEN	SHAPE OF THE EDGE	RECOMMENDED ANGLE
Ax/hatchet	Both	Convex	20°
Center punch	Point	Straight	60° to 75°
Cold chisel	Both	Straight	60° to 75°
Knife	Both	Straight	As original
Hoe/shovel	One	Straight	As original
Plane iron	One	Concave	29°
Rotary mower blade	One	Concave	As original or 45°
Scissors or snips*	One	Concave	80°
Twist drill	Both	Straight	118° to 120°
Wood chisel	One	Concave	25° to 29°

* Scissors and snips have two moving blades. Each blade is sharpened on one side only at an 80° angle.

Figure 21–3. A guide for sharpening tools

☐ SHARPENING TOOLS

When sharpening tools, first remove all metal burrs or rough spots. Burrs should be ground or filed away without changing the shape of the edge. If deep nicks are present, straighten the entire cutting edge. The edge should be ground to the proper shape and angle.

> **Caution:**
>
> **Grind slowly to avoid drawing the temper. A guide for sharpening tools is given in figure 21–3.**

When using grinders, be sure to wear a face shield and protective clothing. Check the grinding wheel(s) for soundness. *True* and clean the wheel, if needed, and then proceed to sharpen tools.

Tools that require a very sharp edge must be whetted. To *whet* means to sharpen by rubbing on a stone.

Knives

Knives with smooth edges may be sharpened in the shop. Those with serrated edges should be sharpened by the manufacturer or a business with special sharpening equipment. *Serrated* means notched, like the edge of a saw. Most schools have policies regarding students carrying knives to school. Therefore, it is recommended that students obtain permission from their instructor before bringing knives to school for fitting and sharpening.

To sharpen most knives, the following procedure is suggested.

Procedure

1. Select a fine grit wheel for the grinder.
2. Remove the nicks and grind away some of the thickness of the blade along the edge. Grind slowly to avoid overheating.

> **Caution:**
>
> **Hold the knife with the cutting edge up, figure 21–4. Move the knife back and forth to create an even grinding mark about ¼ inch wide starting at the edge.**

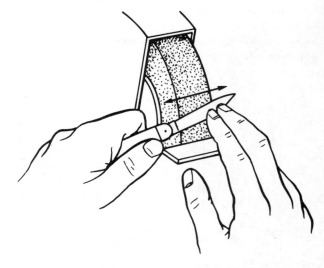

Figure 21–4. Hold the knife with the cutting edge up and move it back and forth across the grinding wheel.

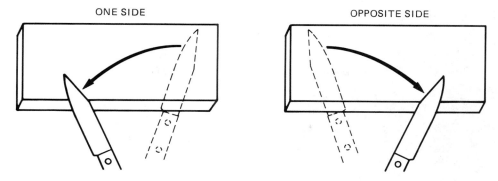

ONE SIDE   OPPOSITE SIDE

Figure 21–5.   Finish sharpening a knife by whetting it on an oil stone.

3. If available, use a grinder with a fine oil stone to remove grinding marks.

4. Use an oil stone to whet the blade to a sharp edge. Whetting is done by holding the knife flat to the stone with the back of the blade slightly raised. Draw the knife, edge first, across the oiled stone; then flip the knife over and push it in the opposite direction, figure 21–5. Continue to alternate sides until the edge is even and sharp.

5. Some stones have one coarse and one fine side. Always make the final strokes on the fine side.

6. When finished, wipe the oil slurry from the stone. Cover the stone with an even coat of light oil and store it in its wooden box or original carton.

Plane Irons and Wood Chisels

The edges of plane irons and wood chisels are identical in type and shape. Both are sharpened on one side only to a 29° concave angle. (A tool sharpening gauge, figure 21–6, is used to check the angles.)

To sharpen a plane iron or wood chisel, grind the

end until it is square and free of nicks; then use the following procedure.

Procedure

1. Loosen the tool rest on a grinder with a medium or fine wheel. Angle the tool rest so that the plane iron lies on the tool rest while its beveled edge matches the curve of the wheel, figure 21–7. Tighten the tool rest.

2. To grind the plane iron, move it back and forth on the tool rest. After grinding, the cutting edge should be square to the side of the plane iron.

3. Check the angle of the edge with a 29° gauge, figure 21–8. It can also be compared to another plane iron that is properly sharpened. Resharpen if necessary.

4. The next step is to put a fine edge on the plane iron by whetting. To whet a plane iron, place the ground section flat against the surface of an oiled stone, then raise the heel of the ground area

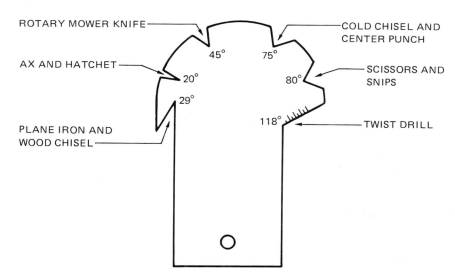

ROTARY MOWER KNIFE

AX AND HATCHET

45°

20°

29°

PLANE IRON AND
WOOD CHISEL

COLD CHISEL AND
CENTER PUNCH

75°

80°

SCISSORS AND
SNIPS

118°

TWIST DRILL

Figure 21–6.   A tool sharpening gauge greatly improves the accuracy of tool sharpening.

Figure 21–7. Sharpening a plane iron using an oil grinder

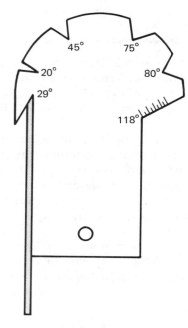

Figure 21–8. Checking the blade angle of a plane iron using a tool sharpening gauge

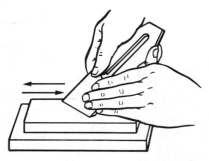

A. LIFT THE HEEL SLIGHTLY AND PUSH THE PLANE ACROSS THE STONE.

B. LAY THE PLANE IRON FLAT ON THE STONE AND PUSH IT FORWARD TO REMOVE THE WIRE EDGE.

Figure 21–9. Whetting a plane iron using an oil bench stone

When sharpening a cold chisel, select a medium grit grinding wheel. If the chisel edge has nicks, hold the chisel horizontally on the tool rest and ease the edge against the wheel. Move the chisel back and forth across the wheel and grind the end until all nicks are removed.

To resharpen, use the following procedure.

Procedure

1. Hold the chisel on the tool rest. The chisel should be pointed upward and leaning toward the wheel. Form the cutting edge by making one pass across the wheel.
2. Then rotate the chisel half a turn and make one pass on the other side.
3. Alternate passes until a sharp edge is formed, figure 21–10.
4. Check the edge for squareness.
5. Use a tool fitting gauge to check the edge for the correct angle.
6. Regrind the chisel until a perfect edge is formed.

Center punches are also sharpened at angles ranging from 60° to 75°. A center punch is sharpened by holding it at an angle to the wheel and twirling it as it is ground. The twirling action should create a round, even point.

until it just lifts from the stone, figure 21–9A. Push the iron across the stone to sharpen the edge.

5. Lay the blade on its back and push it across the stone to remove the wire edge, figure 21–9B. Alternate these two steps until the blade has a fine, sharp edge.

Cold Chisels and Center Punches

The end of a cold chisel is relatively thick, because it must withstand heavy blows from a hammer to cut steel. However, the thickness tapers at a second angle to a sharp edge. This secondary taper forms a 60° to 75° angle.

Figure 21–10. Sharpening a cold chisel using a grinder. Here the operator is grinding the edge to a 60° angle.

Axes and Hatchets

Axes are large and hard to grind on standard grinding machines equipped with the proper guards. For this reason some shops use grinding wheels mounted on a belt-driven *arbor*. Another method is to clamp the ax in a vise and use a portable disc sander, figure 21–11. The edge of an ax is ground to a 20° angle.

The ax is slowly ground away by many sharpenings.

Figure 21–11. A disc sander can be used to sharpen an ax.

This causes the angle to increase and the area behind the edge to get thicker. This problem is reduced by removing metal from as far back as an inch from the edge.

While cutting wood the ax may be touched up or resharpened slightly with an ax stone. An ax stone is a small round stone held in the hand for field use. *Field use* means the area where an operation takes place. In this case field means the forest or wood pile.

Hatchets may be sharpened with a disc sander or grinder. The common shop grinder with a medium wheel may also be used to sharpen a hatchet. First remove any small nicks by pushing the blade horizontally across the tool rest and into the wheel. Make one pass across the wheel. Then examine the edge to see if all small nicks are removed. Make another pass if needed.

In some cases, it may be better to leave large nicks in the blade than to grind away so much metal that the hatchet or ax becomes excessively thick. Generally, large nicks occur at the end of the blade. If so, the end of the blade may be rounded in order to grind the nick into a sharp edge. Needless to say, large nicks ruin an ax or hatchet. They should be avoided by never cutting wood that may contain metal. Another possible source of nicks is chopping through wood and into the ground.

To sharpen a hatchet, use the following procedure.

Procedure

1. Draw a line on each side of the blade parallel to and ½ to ⅝ of an inch from the edge to serve as a guide.
2. Hold the head, edge up, against the wheel. Use

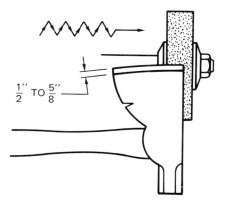

Figure 21–12. When using the grinder to sharpen a hatchet, move the hatchet across the wheel with quick up-and-down movements (shown by the arrows) to produce a long taper on the edge.

a gloved hand to move the head quickly up and down half an inch or more as the tool is moved slowly across the wheel, figure 21–12. The up-and-down motion creates a wide grinding band and maintains the correct angle.

3. Reverse the position of the tool and make a similar pass on the other side.
4. Repeat the process until a good edge is obtained.
5. Remove grinding marks with a hand stone or fine flat file.

A large, sharp, medium-cutting, flat file provides an alternate and equally effective way to sharpen a hatchet or ax, figure 21–13. The tool to be sharpened is first clamped securely in a machinist's vise, then the end of the file is placed on the flat of the tool. If wearing leather gloves, start each stroke at the cutting edge and file toward the handle. This produces a sharper edge.

Caution:

Use extreme care to prevent fingers or hand from slipping into the cutting edge. If not wearing leather gloves, direct the file away from the cutting edge for safety reasons.

Figure 21–13. When using a file to sharpen a hatchet, direct the file away from the cutting edge, if no gloves are worn. A sharper edge can be achieved by wearing leather gloves and directing the file toward the cutting edge.

Twist Drills

Twist drills stay sharp for a long time when drilling wood or plastic. However, when drilling metal they quickly become dull. Although aluminum is soft, it dulls drill bits rather quickly. Drilling steel has a moderate dulling effect on high-speed drills. A *high-speed drill* is a twist drill made and tempered specially to drill steel.

To sharpen a twist drill requires familiarity with its design. The cutting tip of a drill consists of a dead center and two cutting lips, figure 21–14.

The dead center must be exactly centered after the drill is sharpened. The dead center and two cutting lips must touch metal before the rest of the drill end. Otherwise the drill will ride on its rounded end rather than cut into the metal.

The two cutting lips start at the two ends of the dead center. They are ground at 59° to the center line of the drill shank and together form a 118° angle when new. When resharpening a twist drill, a 118° or 120° gauge is used to test the accuracy of the angle.

The metal recedes or falls back from the cutting tip to the heel. The resulting angle is 12° on a new drill and should be kept as close to that as possible when resharpening. This provides clearance for the cutting lips and also gives adequate support to keep the lips from breaking.

Twist drills should be sharpened with a special jig to obtain the correct angles. However, with practice, they can be sharpened freehand by using the following procedure.

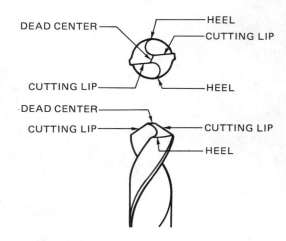

Figure 21–14. The cutting tip of a twist drill

Procedure

1. Redress a fine grinding wheel if needed to make it true.
2. Set the tool rest horizontal to the center of the wheel.
3. Study the shape of a new drill which is about the size of the one to be sharpened.
4. Hold the drill between your thumb and index finger with the tip of the drill exposed about 1 inch, figure 21–15.
5. Place the back of your index finger on the tool rest with your thumbnail up and the drill at an angle to the wheel. The lip on the left should be visible and parallel to the stone.
6. Touch the lip against the grinding wheel and lower the opposite end of the drill as you give it a slight clockwise twist.
7. Rotate the drill half a turn so the other lip is visible.
8. Repeat step 6 to sharpen the second lip.

Use a tool sharpening gauge to check the angle of the lips, position of the dead center, and clearance of the heels, figure 21–16. If the angle of the lips is less than 118°, the bit will cut too fast. If the angle is flatter (more than 118° to 120°), it will cut too slowly. If correct, test the tool by drilling a piece of mild steel. If the drill does not cut well, it must be resharpened.

Rotary Mower Blades

Blades for both lawn and field rotary mowers need frequent sharpening. Sharp blades require less power, do a better looking job, and damage the plants less.

Figure 21–15. Sharpening a twist drill on a grinder (*courtesy of Delta International Machinery Corporation, formerly Rockwell Power Tool Division*).

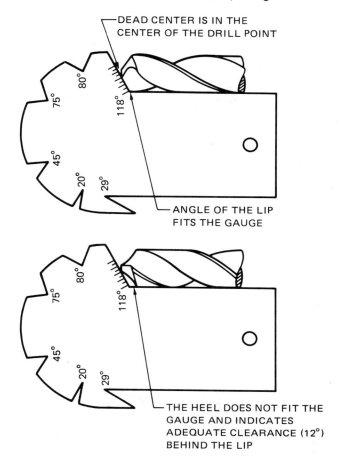

Figure 21–16. A tool sharpening gauge is used to check the angle of the lips, position of the dead center, and clearance of the heels of a resharpened twist drill.

When sharpening blades, try to maintain the original angle. If this is not known, grind the blades to a 45° angle. One side of the blade is kept perfectly flat; the other side is ground to a cutting edge.

When grinding mower blades, remove the same amount of metal from both ends. This keeps the blade in balance. *Balance* means the weight is equally distributed on both sides of the center. Therefore, if one end is badly nicked, grind both ends equally until the nick is removed. Failure to do so will cause the blade to be out of balance and the mower to vibrate. This causes damage to the shaft, bearing, and mower body.

To sharpen a rotary mower blade, first remove the nicks by grinding or filing the cutting edge, then clamp the blade in a machinist's vise. Restore the flat side by using a large, medium-cutting, flat file. Restore the cutting edge using a flat file or aluminum oxide disc on a heavy-duty portable sander.

Mower blades may be sharpened on a stationary grinder. The blades are hard to position properly on a grinder and so this is a less desirable method.

Caution:

Whenever sanders or grinders are used, care must be taken to avoid overheating, which can cause the tool to lose its hardness.

Digging Tools

The edges of digging tools such as hoes, shovels, spades, and spading forks get ragged and dull with use. The same is true for scoop shovels. To reshape these tools, first observe the side on which they were ground before. If metal is curled back, place the tool on an anvil and tap the bent metal back into place with a steel hammer. Restore the cutting lip with a file or grinder, figure 21–17. All shovels are sharpened on the inside. The outside or back remains straight.

Figure 21–17. Restoring the cutting edge of a digging tool

☐ STUDENT ACTIVITIES

1. Copy the data in figure 21–3 into your notes.
2. Use a grinder to sharpen a
   a. knife
   b. plane iron or wood chisel
   c. cold chisel
   d. center punch
   e. hatchet
   f. twist drill
3. Use a bench stone to whet a knife and plane iron.
4. Use a file to sharpen a hatchet, rotary mower blade, and shovel.
5. Use a portable power sander to sharpen an ax.
6. Do the tool sharpening gauge exercise. (A drawing of the tool sharpening gauge is shown in figure 21–18.)

TOOL SHARPENING GAUGE EXERCISE

1. Cut a piece of 26-gauge galvanized steel, 3″ × 4½″. All corners must be square.
2. Measure in 1½ inches from one end and mark the location.
3. Square a line across the steel through the 1½-inch location.
4. Measure 1½ inches along the line to its midpoint. Make a light center punch mark.
5. Set a divider at 1½ inches.
6. Set one leg of the divider in the punch mark and scribe a semicircle across the end of the plate.
7. Reset the divider to a radius of 1³⁄₁₆ inches.
8. Scribe a semicircle inside the first arc.
9. Center a protractor on the scribed line.
10. Mark the inner semicircle at 20°, 60°, 120°, and 160°. These points are located along lines AC, AD, AE, and AF.
11. Use a protractor to lay out and cut triangles from heavy paper to use as patterns to mark 20°, 45°, 75°, and 80° angles on the sheet metal.
12. Place the point of the 20° angle pattern on the inner semicircle and scribe the cutout C.
13. Scribe cutouts D, E, and F.
14. Use tinsnips and cut out C, D, E, and F.
15. Lay out and cut out the 29° chisel gauge.
16. Lay out and cut out the 118° drill gauge.
17. Measure in ⅜ inch from the end and mark for a ¼-inch hole centered from side to side.
18. Drill the ¼-inch hole.
19. Remove all sharp corners with a file.
20. Scribe ¹⁄₁₆-inch reference marks on the drill gauge.

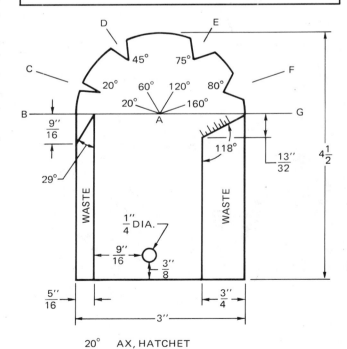

20°	AX, HATCHET
45°	ROTARY MOWER BLADE
75°	COLD CHISEL, CENTER PUNCH
80°	SCISSORS, SNIPS
29°	PLANE IRON, WOOD CHISEL
118°	TWIST DRILL

Figure 21–18. Plan for a tool sharpening gauge

□ **SELF-EVALUATION**

A. Multiple Choice. Select the best answer.
1. The reason for learning tool sharpening skills is that
   a. better work is possible with sharp tools
   b. sharp tools are easier to use
   c. sharp tools are safer
   d. all of these
2. A 6″ × 2″ × 1″ oil stone is called
   a. an ax stone
   b. a bench stone
   c. a common stone
   d. a slip stone
3. The best tool to sharpen an ax is a
   a. grinder
   b. belt sander
   c. portable disc sander
   d. all of these
4. When sharpening tools with a grinder, the tool rest should be
   a. removed
   b. placed at a 90° angle
   c. placed to support the tool
   d. pushed down out of the way

5. A properly ground drill must have the
   a. proper angles
   b. proper clearance
   c. proper centering
   d. all of these

B. Matching. Match the terms in column I with those in column II.

Column I	Column II
1. concave	a. make soft
2. convex	b. curved out
3. serrated	c. curved in
4. anneal	d. sharpen
5. temper	e. drill part
6. whet	f. notched
7. dead center	g. make hard

C. Brief Answers. Briefly answer the following questions.
1. List the number of sides to be sharpened and the recommended angle for grinding each of the following.
   a. ax
   b. cold chisel
   c. hatchet
   d. knife
   e. mower blade
   f. plane iron
   g. scissors
   h. shovel
   i. wood chisel

Section 7

GAS HEATING, CUTTING, BRAZING, AND WELDING

Unit 22
Using Propane and Oxyacetylene Equipment

Objective

To use a gas forge and gas welding equipment safely.

Competencies to Be Developed

After studying this unit, you will be able to:
- Identify major parts of propane and oxyacetylene welding equipment.
- Operate a gas forge.
- Change oxygen and acetylene cylinders.
- Turn on and adjust oxyacetylene controls.
- Light and adjust oxyacetylene torches.
- Shut off and bleed oxyacetylene equipment.
- Check for leaks in gas equipment.

Terms to Know

weld	gas forge
fusion	torch
gas	cylinder
compress	valve
flammable	regulator
apparatus	gauge
manifold	crack the cylinder
oxyacetylene	hose
rig	seat
gas furnace	purge the lines

carbonizing flame
neutral flame
oxidizing flame

tip cleaner
bleeding the lines

Materials List

Gas furnace and/or forge
Oxyacetylene welding outfit
Variety of welding tips
Welding gloves and apron
No. 5 shaded goggles
Tip cleaners
Spark lighter

Many gases will burn. This quality makes them both dangerous and beneficial. The burning qualities of gases are used in agricultural mechanics to heat, cut, and weld metals. To *weld* means to join by fusion. *Fusion* means melting together.

A *gas* is any fluid substance that can expand without limit. This expandable nature also means that a gas can be compressed. To *compress* means to apply pressure to reduce in volume. For example, an air compressor uses the force of a pump to compress great volumes of air into a small tank.

A compressed gas is dangerous simply because it is under pressure and is always trying to get free. It is like a spring in a compressed state. If a compressed gas is *flammable* (meaning that it burns easily) there is the additional danger that it may explode or burn out of control. Fortunately, suitable equipment and techniques are available for using compressed, combustible gases safely.

The most popular gases for heating, cutting, and welding metals are propane and acetylene. Before these gases will burn, they must be mixed with oxygen from the air or pure oxygen from tanks. The student is referred to the unit, "Reducing Hazards in Agricultural Mechanics," for information on how to prevent and control fires.

□ GENERAL SAFETY WITH COMPRESSED GASES

Some general precautions in handling compressed gases are:

- Wear safety goggles or a face shield at all times.
- Obtain the instructor's permission before using compressed gases.
- Store fuel gas cylinders separately from oxygen cylinders.
- Keep gas cylinders upright and chained securely at all times, figure 22–1. They should be stored outdoors or in well-ventilated, fire-safe areas.
- Do not bump or put pressure on pipes, connections, valves, gauges, or other equipment connected to compressed gas cylinders.
- When connections are opened or cylinders changed, check thoroughly for leaks before using the equipment.
- Never use equipment exposed to oil or grease. Spontaneous or instant fires may result.

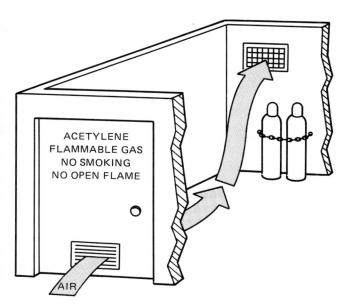

Figure 22–1. Cylinders must be fastened securely in place at all times.

- Follow specific procedures for turning systems on and off.
- Work only in areas that are free of materials that burn.
- Never use gas-burning equipment without approved fire extinguishers in the area.
- Always wear leather gloves and apron when using gas-burning equipment.
- Screw the caps on all cylinders that do not have regulators or other apparatus attached. *Apparatus* means objects necessary to carry out a function.
- All equipment or cylinders that may discharge gas should be pointed away from the operator, other people, and clothing. Fire will follow a gas stream.
- Never leave clothing where it can become saturated by oxygen or fuel gases.
- If gas equipment catches fire, immediately turn off the gas at the tanks. If this is not practical, or if this action does not extinguish the gas fire, evacuate the area and call for help.
- Learn to recognize the odors of combustible fuels.
- Protect gas cylinder storage areas with locked chain link fences or concrete enclosures.

□ GAS-BURNING EQUIPMENT FOR AGRICULTURAL MECHANICS

Shops generally use propane and/or acetylene and oxygen for heating, cutting, and welding. Propane and acetylene are fuels. Oxygen is not a fuel, but it must be present for other fuels to burn. Some shops have manifolds to which many cylinders and welding outfits may be attached. A *manifold* is a pipe with two or more outlets. Most agricultural mechanics shops have one or more portable oxyacetylene rigs. A *rig* is a self-contained piece of apparatus assembled to conduct an operation. *Oxyacetylene* is a shortened version of the words oxygen and acetylene. It refers to equipment and processes where the two gases are used together.

Shops may also have propane and oxygen torches and propane furnaces for soldering. Propane and oxygen torches are desirable as cutting units since propane gas is generally cheaper than acetylene.

The Gas Forge

A fast and efficient unit for heating small pieces of metal and tempering tools is the *gas furnace* or *gas forge*.

Major Parts. □ A forge consists of a: (1) firebox, (2) lid, (3) work rack, (4) air system, and (5) gas controls,

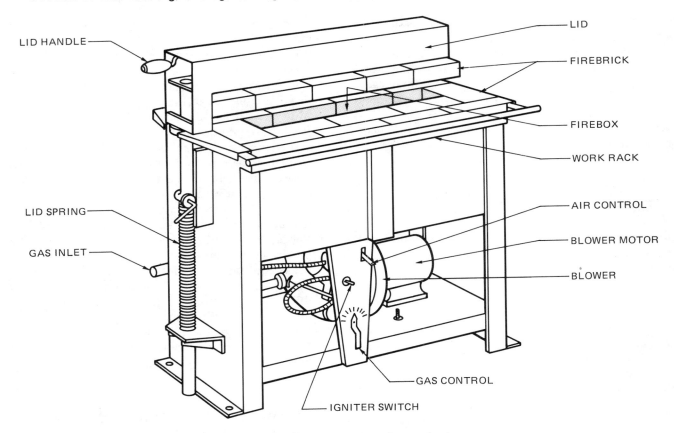

Figure 22–2. Major parts of a gas forge

figure 22–2. Individual controls used in operating the gas forge are the: (1) lid handle, (2) air control, (3) gas control, (4) igniter switch, and (5) gas inlet.

The gas forge is connected to a gas line and electrical outlet. The firebox is lined with firebrick. An electric arc is used to ignite the gas burner in the bottom. A fan driven by an electric motor provides air (which contains oxygen) to balance the flame.

The lid of the gas forge is heat resistant. This lid swings aside so that long metal objects can be placed in the forge, then swings back over the firebox while the object is being heated. The lid reduces the fire hazard and conserves fuel.

Operating Procedures. ☐ The forge must be anchored securely to the floor. It should be located away from combustible materials such as wood, sawdust, and painting supplies.

The forge is gas operated and all safety procedures regarding gas-operated equipment apply. Always follow the manufacturer's instructions. However, in the absence of manufacturer's instructions, general procedures for lighting and shutting down gas forges are as follows.

Procedure

A. Lighting the Forge
1. Wait 5 minutes or more if the forge was in operation but recently shut off.
2. Swing the lid aside so it is not covering the firebox.

Caution:

Keep face and other body parts away from the slot above the firebox. Wear face and body protection.

3. Open the air control part way.
4. Turn on the switch to start the blower and spark igniter.
5. Press and hold in the igniter button while slowly turning the gas control lever. Keep the igniter button pressed for about 20 seconds after the fire starts or the flame will go out due to a safety feature in the system.

6. As the burner lights, adjust the gas control until the flame roars with a steady sound.
7. Release the igniter button slowly.
8. Make final adjustments of gas and air until sharp-tailed flames are observed. Increase both gas and air to get a hotter flame. Rebalance the gas-air mixture as needed to keep the sharp-tailed flames.
B. Shutting Down the Forge
1. Turn the gas control to the closed position.
2. Turn off the stop button switch to stop the blower.

When using the forge, use blacksmith's tongs, leather gloves and goggles, an anvil, and other tools designed for hot metalworking. The unit, "Identifying, Marking, Cutting, and Bending Metal," gives tool lists and detailed procedures.

Oxyacetylene Equipment

Major Parts. □ A portable oxyacetylene rig includes a cart, cylinders, valves, regulators, gauges, hoses, and torch assemblies, figures 22-3 and 22-4. A *torch* is an assembly that mixes gases and discharges them to support a controllable flame. A gas *cylinder* is a long round tank with extremely thick walls built to hold gases under great pressure. *Valves* and *regula-*

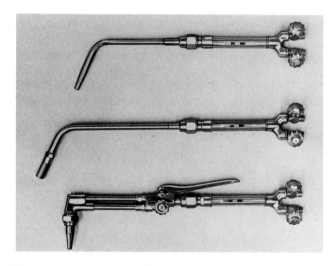

Figure 22-4. Torches designed for brazing and welding (top), heating (center), and cutting (bottom) (*courtesy of Victor Equipment Company*)

tors are devices that control or regulate the flow of the gas. A *gauge* measures and indicates the pressure in the hose, tank, or manifold, figure 22-5. *Hoses* are flexible lines that carry the gases. They are rubber reinforced with nylon or other material to withstand high pressures and heavy use.

The gas flow and control components of a portable oxyacetylene rig are:

Acetylene Side	*Oxygen Side*
• acetylene cylinder	• oxygen cylinder
• cylinder valve	• cylinder valve
• red cylinder pressure gauge	• green cylinder pressure gauge
• regulator	• regulator
• hose pressure gauge	• hose pressure gauge
• red hose	• green hose
• acetylene valve on torch	• oxygen valve on torch

• torch mixing chamber
• torch tip

Setting Up. □ When setting up oxyacetylene equipment, one must be aware of certain design features. The valve on most acetylene cylinders is protected with a high collar. The valve is turned on and off with a handle. On cylinders with removeable handles, care must be taken to leave the handle in place when turned on in case the gas must be turned off in an emergency. Acetylene equipment is color coded red. Another important distinction is that all acetylene couplings have left-hand threads and notched nuts.

Oxygen cylinders generally have valves that are not protected during use. However, they have heavy caps

Figure 22-3. Portable oxyacetylene rig without torch

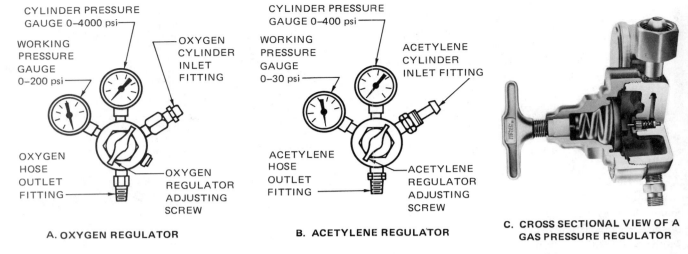

Figure 22–5. Oxygen and acetylene regulators and gauges *(courtesy of Victor Equipment Company)*

that screw in place over the valves to protect them when not in use. Oxygen equipment is color coded green. All oxygen couplings have right-hand threads.

Before attaching regulators to cylinders, *crack the cylinder* by turning the gas on and off quickly to blow any dust from the opening.

> **Caution:**
>
> **When attaching regulators, hoses, gauges, or torch connections, the connectors *must* be threaded in the correct direction. The red color coding and opposite threads for acetylene are means of preventing the accidental mixing of acetylene and oxygen, except in the torch itself.**

All connections are gastight when screwed on properly. Some parts such as regulators have metal-to-metal fittings. These draw together snugly and must be gastight. Some brass fittings used for connecting hoses are also metal-to-metal types. Gauge fittings have tapered threads. These are wrapped with teflon tape to ensure a gastight fit when screwed into regulators or pipe fittings.

> **Caution:**
>
> **Oxygen and acetylene valves on torches seal with light finger pressure. Overtightening will damage the seats of the valves. A *seat* is the point where the moveable part of a valve seals off the gas.**

Turning On the Acetylene and Oxygen. □ After all components are assembled, the following sequence is used to turn on the gases.

> **Caution:**
>
> **The area must be properly ventilated. Wear protective goggles and clothing.**

Procedure

1. Close the acetylene valve on the torch.
2. Close the oxygen valve on the torch.
3. Turn the acetylene regulator handle counterclockwise until no spring tension is felt.
4. Turn the oxygen regulator handle counterclockwise until no spring tension is felt.
5. Open the oxygen cylinder valve slowly until the pressure gauge responds. Then open the valve all the way.

> **Caution:**
>
> **Do not stand in front of the gauges when gas is being turned on.**

6. Open the acetylene cylinder valve slowly half a turn.

7. Open the oxygen torch valve an eighth of a turn. Turn the oxygen regulator handle clockwise until the pressure gauge reads 10 psi (pounds per square inch) pressure (the desired pressure will vary with the specific torch equipment being used). Close the oxygen valve on the torch.

8. Open the acetylene torch valve an eighth of a turn. Turn the acetylene regulator handle clockwise until the pressure gauge reads 5 psi (the desired pressure will vary with the equipment being used). Close the acetylene valve on the torch.

Steps 7 and 8 purge the lines and set the regulators at safe starting pressures. To *purge the lines* is to remove undesirable gases. The unit is now pressurized and ready for use.

Testing for Leaks. □ A leak test should be performed when equipment is first set up, when cylinders are changed, or if the odor of acetylene is present when the unit is not in use. To test for leaks, put a small amount of water in a small jar or can. Add a sliver of nondetergent hand soap. Use a 1-inch paint brush to produce a soapy lather; use the brush to apply the soap solution gently around each fitting and point where gas may escape. If a leak exists, it will cause bubbles in the solution.

If a leak is found, tighten the fitting. If this does not stop the leak, turn off both gases. Disassemble the joint and correct the problem. Turn on the gases again and recheck the entire system with the soap solution.

Lighting and Adjusting Torches. □ Once the system is pressurized, attention can be given to selecting torch parts. This is followed by lighting and adjusting the torch according to the work to be done.

Two types of torches are available: welding torches, figure 22–6, and cutting torches, figure 22–7. Both include a body or handle with hose connections and valves to control the oxygen and the fuel. Welding tips with mixing chambers are screwed onto the handle. Tips of different sizes are available. A cutting assembly with another set of valves may also be attached to the handle.

To light and adjust a welding or cutting torch, use the following procedure.

Procedure

1. Put on leather gloves and goggles with a No. 5 shaded lens, figure 22–8.
2. Open the acetylene valve an eighth of a turn.
3. Use a spark lighter to ignite the torch, figure 22–9.

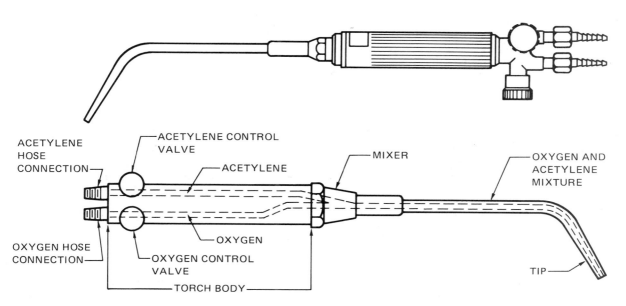

Figure 22–6. Exterior and interior views of a welding torch

ACETYLENE HOSE CONNECTION

ACETYLENE CONTROL VALVE

ACETYLENE

MIXER

OXYGEN AND ACETYLENE MIXTURE

OXYGEN HOSE CONNECTION

OXYGEN CONTROL VALVE

OXYGEN

TORCH BODY

TIP

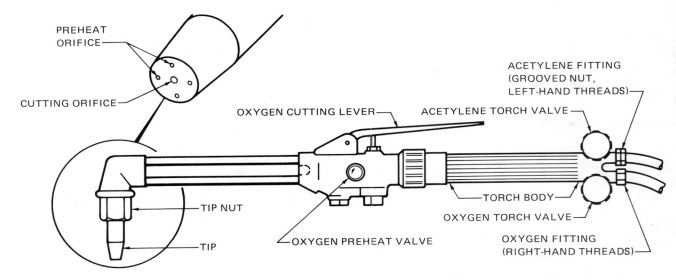

PREHEAT ORIFICE

CUTTING ORIFICE

ACETYLENE FITTING (GROOVED NUT, LEFT-HAND THREADS)

OXYGEN CUTTING LEVER

ACETYLENE TORCH VALVE

TIP NUT

TIP

OXYGEN PREHEAT VALVE

TORCH BODY

OXYGEN TORCH VALVE

OXYGEN FITTING (RIGHT-HAND THREADS)

Figure 22–7. An oxyacetylene cutting torch

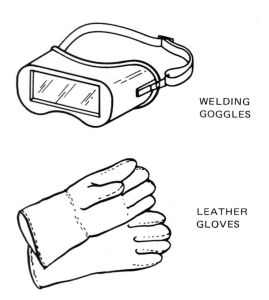

WELDING GOGGLES

LEATHER GLOVES

Figure 22–8. Before lighting a gas torch, the operator must put on the proper protective clothing.

Caution:

Do not point the torch directly into the lighter.

---

4. Open the acetylene valve slowly until the flame is ¼ inch off the tip of the torch. Increase or decrease the regulator pressure until the flame just touches the tip. At this point, the acetylene valve on the torch is open several turns. This will be

the correct pressure for the tip being used. The flame is called a carbonizing flame. A *carbonizing flame* is one with an excess of acetylene, figure 22–10. It is cooler than other types of flames.

5. Turn the oxygen valve on slowly and watch the inner flame shorten. Continue to add oxygen until the long inner flame just fits the cone. This is a *neutral flame,* or one with a correct balance of acetylene and oxygen. A neutral flame is correct for heating, cutting, and welding.

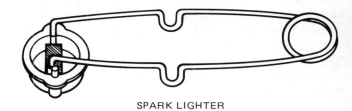

SPARK LIGHTER

CORRECT POSITION OF THE SPARK LIGHTER

Figure 22–9. For safe lighting of a gas torch, use a spark lighter.

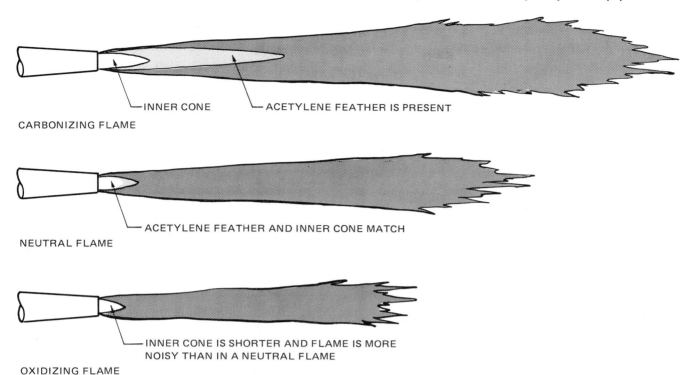

INNER CONE — ACETYLENE FEATHER IS PRESENT

CARBONIZING FLAME

ACETYLENE FEATHER AND INNER CONE MATCH

NEUTRAL FLAME

INNER CONE IS SHORTER AND FLAME IS MORE NOISY THAN IN A NEUTRAL FLAME

OXIDIZING FLAME

Figure 22–10. Carbonizing, neutral, and oxidizing flames. The neutral flame is correct for heating, cutting, and most welding jobs.

6. If additional oxygen is added, the cone becomes sharp and the flame noisy. This is an *oxidizing flame,* or one with an excess of oxygen. It is the hottest type of flame, but not recommended except for special applications. Additional instructions for lighting and adjusting cutting torches are provided in unit 23.

If the torch cannot be adjusted to produce a neutral flame, clean the tip, figure 22–11. *Tip cleaners* are rods with rough edges designed to remove soot, dirt, or metal residue from the hole in a tip. Use a tip cleaner that is equal to or smaller in size than the hole in the tip. If molten metal is fused to the tip, remove it with fine emery cloth.

Shutting Off Torches. □ Torches become a fire hazard if the proper procedure is not followed when shutting them off. Improper procedures may cause

Caution:

Use gentle pressure on the torch valves as they have soft internal parts that may be damaged.

excessive popping of the torch, soot from unburned gases, or carbon deposits in the tip. To shut off a torch correctly, first close the acetylene valve on the torch then close the oxygen valve.

Bleeding Lines. □ When the torch is not in use it is important to close every point where gas may escape. Gas should also be removed from all lines and equipment. This practice is called *bleeding the lines.* To bleed the lines use the following procedure.

Procedure

1. Turn off the acetylene at the cylinder.
2. Turn off the oxygen at the cylinder.
3. Open the acetylene valve at the torch until both regulators return to zero. Close the acetylene valve at the torch.
4. Open the oxygen valve at the torch until both regulators return to zero. Close the oxygen valve at the torch.
5. Turn both regulator handles counterclockwise until they are easy to turn (there is no pressure on the diaphragm).
6. Coil the hoses over the cart handles, special hose hangers, or the large part of the cylinders.

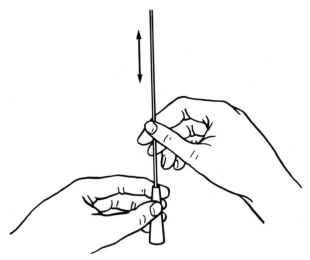

A. CLEANING A TIP WITH A STANDARD TIP CLEANER

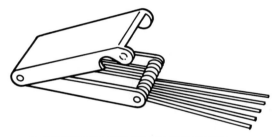

B. STANDARD SET OF TIP CLEANERS

Figure 22–11. Use a tip cleaner that is equal to or smaller in size than the hole in the tip.

Caution:

Hanging hoses on the regulators may break the regulators or cause a leak.

---

7. If portable, store the rig in a suitable place.

□ STUDENT ACTIVITIES

1. Define all the italicized terms in this unit.
2. Attach regulators to oxygen and acetylene cylinders.
3. Turn on the oxygen and acetylene.
4. Test the apparatus for gas leaks.
5. Clean the tip of the torch.
6. Light the torch and adjust the gas flow to achieve a neutral flame.

7. Shut off the torch and bleed the lines.
8. Store the apparatus properly.
9. Using a gas forge, do the following metal staple exercise. (The staple is pictured in figure 19–3.)

METAL STAPLE EXERCISE

1. Cut a piece of ⅜-inch round rod 6 inches long.
2. Heat and work one end to form a round tip.
3. Heat and shape the other end to form a square tip.
4. Smooth each end with a grinder or file.
5. Heat and bend the rod in a U-shape with a 1-inch width between legs of equal length.

□ SELF-EVALUATION

A. Multiple Choice. Select the best answer.
1. Gas can be compressed with a
   a. cylinder
   b. lever
   c. pump
   d. valve
2. Acetylene may be dangerous because it is
   a. compressed
   b. flammable
   c. explosive
   d. all of these
3. Which gas is not a fuel used for torches?
   a. acetylene
   b. oxygen
   c. propane
   d. none of these
4. The two main controls for managing the flame on a gas forge are the gas control and the
   a. air control
   b. firebox
   c. igniter
   d. lid handle
5. A properly adjusted forge flame has
   a. black tips
   b. a carbonizing flame
   c. no sound
   d. sharp tails
6. Oxygen and acetylene hoses can stand pressure because they are
   a. color coded
   b. extra thick
   c. made of steel
   d. reinforced

7. Oxygen hoses and related equipment are color coded
   a. green
   b. ivory
   c. orange
   d. red
8. Acetylene hoses and related equipment are color coded
   a. green
   b. ivory
   c. orange
   d. red
9. The acetylene pressure to light a torch should be
   a. 5 psi or less
   b. 15 psi
   c. 25 psi
   d. 50 psi
10. Gas leaks are checked with
    a. compressed air
    b. flame
    c. soapy water
    d. teflon

B. Matching. Match the terms in column I with those in column II.

Column I	Column II
1. weld	a. pressure
2. gas	b. heats metal efficiently
3. compress	c. cylinder
4. gas forge	d. balanced
5. igniter	e. tape
6. oxyacetylene outfit	f. expands without limit
7. tank	g. starts a gas forge
8. neutral flame	h. join by fusion
9. goggles	i. rig
10. teflon	j. No. 5

C. Completion. Fill in the blanks with the word or words that make the following statements correct.
1. Gas cylinders must be securely _____.
2. Never use gas-burning equipment without an approved _____ _____ in the area.
3. When no apparatus is attached, a _____ should be screwed onto oxygen cylinders.
4. The presence of acetylene and propane is detected by _____.
5. The term oxyacetylene comes from _____ and _____.
6. The valve on acetylene cylinders is protected by a _____.
7. Connectors for acetylene hoses have _____ threads.
8. Connectors for oxygen hoses have _____ threads.
9. Removing gas from oxyacetylene equipment is known as _____ .
10. Hanging hoses on regulators may result in _____.

Unit 23
Cutting with Oxyfuels

Objective

To use oxygen and selected fuels to cut steel with a flame torch.

Competencies to Be Developed

After studying this unit, you will be able to:
- Write the names and characteristics of common fuels used for cutting.
- State and apply recommended safety practices for using oxyfuels.
- Select appropriate pressures for using oxygen and common fuel gases for cutting.
- Cut steel with oxyfuels.
- Pierce steel with oxyfuels.

Terms to Know

oxyfuel	base metal
oxyfuel cutting	backfire
kerf	flashback
slag	slag box
brazing	pierce
brass	

Materials List

Oxyacetylene cutting rig and welding area
Welding safety equipment and protective clothing
Two pieces of steel, ¼" to ½" × 2" × 12"
Plate steel, ¼" to ½" × 6" × 12"
Black pipe up to 3-inch diameter, 1 to 2 feet long
Soapstone

Combustible gases combined with oxygen have been used since the 1800s for welding metals. Over the years, scientists have experimented to find the gases that work best. Unfortunately, no one gas is the safest and best for all jobs. A study of gas heating, cutting, and welding procedures is an important part of modern agricultural mechanics.

☐ OXYFUEL PROCESSES

The term *oxyfuel* refers to the combination of pure oxygen and a combustible fuel gas to produce a flame.

Oxyfuels are used for welding, brazing, cutting, and heating metals.

General Principles

Oxygen and Fuel Gases. ☐ These gases are stored under pressure in tanks or cylinders. They are released as individual gases through carefully designed valves, regulators, and hoses. The gases are mixed as they flow through torch assemblies. They burn as the mixture is discharged through carefully engineered tips.

The gas flame from an oxyfuel burns with intense heat. The temperature may range from 5000°F to 6000°F. These temperatures are hot enough to melt most metals and permit cutting and fusion welding of steel.

Cutting, Brazing, and Heating. ☐ These functions differ from welding. *Oxyfuel cutting* is a process in which steel is heated to the point where it burns and is removed to leave a thin slit called a *kerf.*

To burn, hot steel must combine with oxygen. *Slag* is a product formed during the process. Slag is a good insulator. It hinders the cutting process because it forms a layer between the burning steel and the torch flame.

Cutting torches are designed to send a forceful stream of oxygen into preheated cherry red steel. This oxygen supports the combustion. The force of the oxygen stream drives the slag out of the area and permits the heat from the torch to keep the steel burning. As a cutting torch moves forward, the oxygen stream pushes slag out to form the kerf. Torches can be adjusted to cut kerfs that are nearly as straight and clean as the cut produced by a saw.

Heating. ☐ In heating, the temperature may be raised enough to soften metal for bending or shaping. No melting takes place. The job may call for heat in a very narrow band to aid in making a sharp bend in a specific place. The heat may also be spread equally over a large area to reduce stress from welding. Different tips for oxyfuel torches are available for these purposes.

Brazing. ☐ *Brazing* is the process of bonding with metals and alloys that melt at or above 840°F. *Brass* is a mixture of copper and zinc. It is an example of an alloy used for brazing. The process of brazing is similar to soldering. When brazing, the *base metal* (the main piece of metal) is heated until the brass melts, flows, and bonds to it. The base metal is not melted, and the metals do not mix during brazing.

Fusion Welding. □ Steel may be joined by fusion welding. *Fusion welding* is joining metal by melting it together. An oxyacetylene welding torch may be used to heat two pieces of steel until the metal from each runs together to form a joint. When properly done, such joints are as strong as the base metal itself.

□ CHARACTERISTICS OF OXYFUELS

Oxygen

Oxygen is not a fuel and it will not burn. However, it combines with other substances and causes them to burn. According to the fire triangle, fuel plus heat plus oxygen equals fire. Oxygen is necessary if other gases are to be used in torches, figure 23–1.

Oxygen must be 99.5 percent pure to support the combustion of iron. When iron is hot enough, it burns in oxygen the way wood or paper burns in air. This principle makes cutting with a torch possible. To burn in the presence of oxygen, iron must be heated to between 1600°F and 1800°F. This is about 800° below the melting temperature of iron.

Acetylene

The fuel most suitable for welding is acetylene. This gas produces a cleaner weld than most other fuel gases. It also produces a more controllable flame. However, acetylene gas is unstable and, therefore, it is hazardous. Acetylene must be handled very carefully.

Acetylene is more expensive than other oxyfuels. Therefore, it is used in welding only when its burning characteristics are needed. In such cases, the safety hazards must be tolerated. For high-volume cutting and heating, cheaper fuels are generally preferred, figures 23–2 and 23–3. However, since acetylene performs most functions well, it is the choice of fuel for many agricultural mechanics shops.

Caution:

Acetylene must not be used at pressures greater than 15 psi. The lower the pressure, the more stable the gas is and the safer it is to use. Acetylene containers must not be subjected to electrical shock, rough handling, or excessive heat. Do not use acetylene in copper lines or in lines with any grease or oil residues.

Propane and Natural Gas

Propane gas is available in cylinders and tanks. Natural gas is piped into buildings in urban and suburban areas through public utility lines. Both gases are used extensively for general heating and may be used for torch heating and cutting, but not welding. For torch heating and cutting, however, both propane and natural gas consume large volumes of cylinder oxygen. This fact may offset other price advantages offered by these gases.

FUEL 1 CUBIC FOOT	TOTAL O$_2$ REQUIRED	O$_2$ SUPPLIED THROUGH TORCH*	% TOTAL THROUGH TORCH	CU FT O$_2$ PER LB OF FUEL
Acetylene— 5589°F/1470 Btu	2.5	1.3	50.0	18.9
MAPP® gas— 5301°F/2406 Btu	4.0	2.5	62.5	22.1
Natural gas— 4600°F/1000 Btu	2.0	1.9	95.0	44.9
Propane— 4579°F/2498 Btu	5.0	4.3	85.0	37.2
Propylene— 5193°F/2371 Btu	4.5	3.5	77.0	31.0

* Balance of total oxygen demand is entrained in the fuel-gas flame from the atmosphere.
Source: Airco Welding Products.

Figure 23–1. Cubic feet of oxygen needed per cubic foot of fuel burned

FUEL	NEUTRAL FLAME TEMP °F	PRIMARY FLAME BTU/FT³	SECONDARY FLAME BTU/FT³	TOTAL HEAT BTU/FT³
Acetylene	5589	507	963	1470
MAPP® gas	5301	517	1889	2406
Natural gas	4600	11	989	1000
Propane	4579	255	2243	2498
Propylene	5193	438	1962	2371

Source: Airco Welding Products.
Key: MAPP® gas—A multipurpose industrial fuel gas consisting of a mixture of methylacetylene and propadiene.
BTU/ft³—British thermal units per cubic foot of fuel. A BTU is a small unit of heat.

Figure 23–2. Heating values of major industrial fuel gases when burned with pure oxygen

APPLICATION	ACETYLENE	MAPP® GAS	PROPYLENE
Cutting			
Under ⅜ in thick	100	95	90
⅝ in to 5 in thick	95	100	95
Over 5 in thick	80	100	95
Cutting dirty or scaled surfaces	100	95	80
Cutting low-alloy specialty steels	100	90	80
Piercing	100	100	85
Welding	100	70	0
Braze welding	100	90	70
Brazing	100	100	90

Source: Airco Welding Products.

Figure 23–3. Average performance ratings of selected oxyfuel flames

MAPP® Gas

MAPP® gas is a formulated mixture of methylacetylene and propadiene gases. (MAPP® is a trade name of Airco, Inc.) The gas is reported to have many of the advantages of acetylene but to be more stable and so safer. Its high temperature flame is suitable for brazing, cutting, heating, and metallizing.

☐ CUTTING STEEL WITH OXYFUELS

Review the previous unit on using gas-burning equipment.

Caution:

Use every precaution in handling oxygen and fuel gases, setting up apparatus, and performing cutting and heating tasks.

The Flame

A description of a neutral flame and the procedure for obtaining it are given in unit 22. However, it is more difficult to determine when a flame is neutral on a cutting torch than on a welding torch. A cutting torch must be adjusted to obtain a neutral flame both with and without the oxygen jet, figure 23–4.

To obtain a neutral flame with a cutting torch the following procedure is recommended.

Procedure

1. Wear appropriate protective clothing and goggles with a No. 5 shaded lens.
2. Check the area and remove all fire hazards.
3. Set up the oxyacetylene equipment in a safe manner and check for leaks.
4. Set the pressure gauges at the recommended pressure for the tip being used. These will vary with the fuel and tip, figure 23–5.
5. Light the torch using acetylene only.
6. Increase the acetylene pressure until the flame

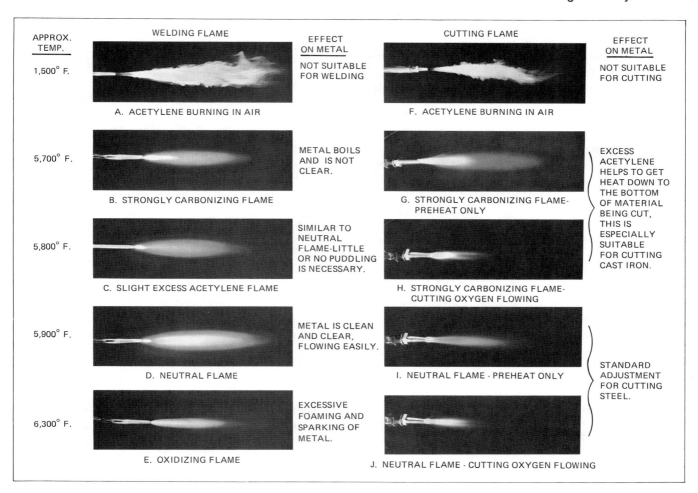

Figure 23–4. Oxyacetylene flame adjustments (*courtesy of Airco Welding Products, adapted*)

CUTTING GUIDE			
METAL THICKNESS (in.)	**AIRCO CUTTING TIP SIZE**	**PRESSURE PSI**	
		OXYGEN	**ACETYLENE**
⅛	00	30	1½
¼	0	30	3
⅜	1	30	3
½	1	40	3
¾	2	40	3
1	2	50	3
1½	3	45	3
2	4	50	3
3	5	45	4
4	5	60	4
5	6	50	5
6	6	55	5

Figure 23–5. Oxygen and acetylene pressures vary according to the size of the cutting tip (*courtesy of Airco Welding Products*).

just starts to leave the tip when the acetylene torch valve is opened several turns.

7. Close the acetylene torch valve until the flame touches the tip.

8. With the oxygen preheat valve closed, open the oxygen (O_2) torch valve several turns.

9. Slowly open the O_2 preheat valve until the acetylene feathers just match the inner cones of the preheat flames.

10. Press the oxygen cutting lever and observe whether the flames stay neutral.

11. If the flames do not stay neutral, make slight adjustments in the oxygen preheat valve and/ or acetylene torch valve until the flame remains neutral both with and without the oxygen lever pressed.

12. If a neutral flame cannot be obtained, clean the tip. Use fine emery cloth on the outside and the appropriate tip cleaners for the inside of the holes.

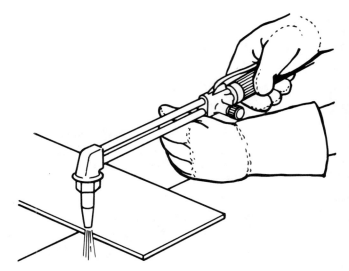

Figure 23–6. One gloved hand controls the tip clearance and the other controls the oxygen lever.

Caution:

Touching the tip against the work, overheating, incorrect torch adjustment, a loose tip, a dirty tip, or damaged valves may all cause backfire. *Backfire* **is a loud snap or popping noise which generally blows out the flame. The cause must be corrected before relighting.**

Caution:

Backfire sometimes causes a flashback. A *flashback* **is burning inside the torch that causes a squealing or hissing noise. When this occurs, quickly turn off the torch oxygen valve and then the torch acetylene valve. If fire is suspected in the hoses, rush to close the acetylene valve and then the oxygen valve at the tanks. After a flashback, only an experienced operator can determine if the torch is safe to relight.**

Cutting the Steel

After obtaining a neutral torch flame, position the work for cutting. This is done by placing the metal

to be cut over a slag box. A *slag box* is a metal container of water or sand placed to catch hot slag and metal from the cutting process. The work should be weighted or clamped so that it will not slip from the work area.

For general cutting, mark the line of the cut with soapstone. Place one gloved hand on the metal near the torch head. This permits control of the tip clearance. The other hand controls the handle and oxygen lever, figure 23–6.

To start the cut, hold the flame over the corner and edge of the metal, figure 23–7. Hold the torch at a slight angle away from the edge. The cones of the preheat flames should not quite touch the metal. Hold the torch steady until the edge of the metal turns cherry red. Press the oxygen lever and move the torch across the metal at a steady rate. Maintain the flame cones about ⅛ of an inch from the metal.

Different techniques may be used to control speed and clearance. When making short cuts, it is helpful to slide the torch over the gloved hand positioned near the torch head. In this case, one hand controls the clearance and the other controls the oxygen lever and movement across the metal.

When making cuts longer than about 2 inches, both hands grip the torch and slide over the metal. Special devices such as wheeled trolleys may be used to maintain the correct clearance, figure 23–8. A rod with a sliding center point can be attached to a torch to aid in cutting perfect circles, figure 23–9. A piece of steel angle or similar device can be used to guide the torch when making straight cuts, figure 23–10. A smoother cut will be obtained by placing the guide metal on the waste piece since most of the slag will collect on the piece with the guide metal.

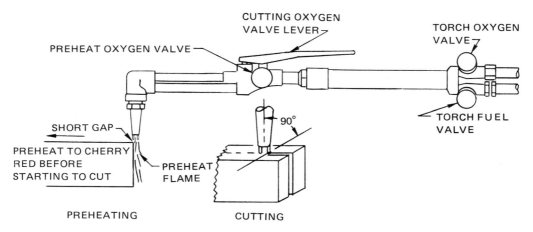

CUTTING OXYGEN
VALVE LEVER

PREHEAT OXYGEN VALVE

TORCH OXYGEN VALVE

TORCH FUEL VALVE

90°

SHORT GAP

PREHEAT TO CHERRY RED BEFORE STARTING TO CUT

PREHEAT FLAME

PREHEATING

CUTTING

Figure 23–7. Torch positions when preheating and cutting (*courtesy of Airco Welding Products*)

Improving the Cut

An examination of the metal along a cut reveals how the cut was made. Marks left by the flame provide clues to the preheat procedure, speed, and pressure.

Preheat. □ If the preheat flame is too hot or the torch travels too slowly, the surface melts before the metal is heated through. This leaves a melted or rounded appearance along the top, figure 23–11. The tip may be raised slightly to reduce the preheat. Increasing the speed slightly may also correct the problem.

Clearance. □ Clearance is the distance from the torch tip to the metal. Generally the clearance is cor-

rect when the tips of the primary flames are almost level with the surface of the metal.

Speed. □ Moving the torch too fast across the metal results in an incomplete cut and rough edges. Incomplete cuts generally occur at the bottom and end of the cut, figure 23–12. On the other hand, moving too slowly results in a melted top edge and leaves gouges where the cutting stream has wandered, figure 23–13.

Pressure. □ If the oxygen pressure is too high, the result may be a dish shape in the kerf near the top. On the other hand, if the pressure is too low, the cut may not be complete at the bottom.

Figure 23–8. Wheeled trolleys may be attached to cutting tips to maintain the proper distance between the tip and the metal for long cuts (*courtesy of Victor Equipment Company*).

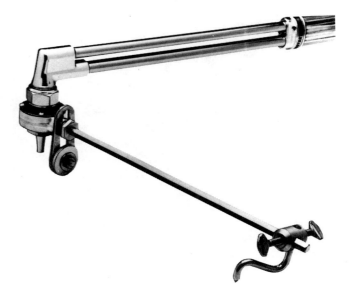

Figure 23–9. A rod with a sliding center point attached to a torch aids in cutting perfect circles (*courtesy of Victor Equipment Company*).

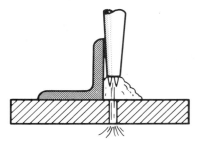
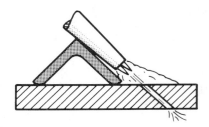

Figure 23–10. Steel angle used to guide the torch when cutting straight lines (*courtesy of L. Jeffus and H. Johnson,* Welding Principles and Applications, *Delmar Publishers Inc., 1984*).

Figure 23–11. Excessive preheating and/or traveling too slowly across the metal result in a melted top edge (*courtesy of L. Jeffus and H. Johnson,* Welding Principles and Applications, *Delmar Publishers Inc., 1984*).

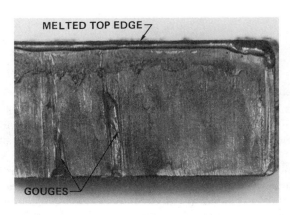

Figure 23–13. Moving the torch too slowly across the metal results in a melted top edge and gouges along the cut (*courtesy of L. Jeffus and H. Johnson,* Welding Principles and Applications, *Delmar Publishers Inc., 1984*).

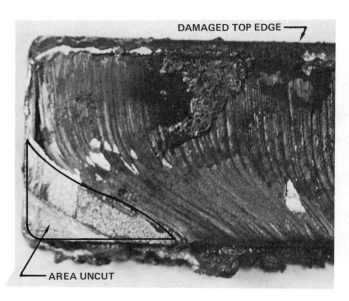

Figure 23–12. When the torch travels too fast across the metal, rough edges, and uncut metal result (*courtesy of L. Jeffus and H. Johnson,* Welding Principles and Applications, *Delmar Publishers Inc., 1984*).

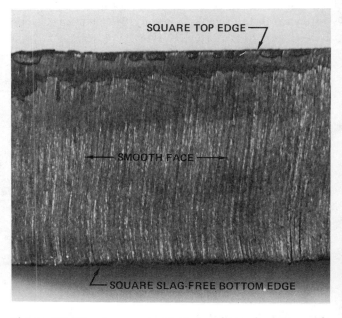

Figure 23–14. A correct cut is straight and square with a smooth face. Drag lines bend backward slightly at the bottom (*courtesy of L. Jeffus and H. Johnson,* Welding Principles and Applications, *Delmar Publishers Inc., 1984*).

Figure 23–15. During a good cut, there is a steady stream of sparks from the bottom of the kerf (*courtesy of L. Jeffus and H. Johnson,* Welding Principles and Applications, *Delmar Publishers Inc., 1984*).

The Correct Cut. □ A correct cut is straight and square with a smooth face. Drag lines bend backward slightly at the bottom. If preheat, clearance, speed, and pressure are all correct, then cutting will be fast, clean, and accurate, figure 23–14. Such a combination requires a steady hand and plenty of practice. While a good cut is being made there is a smooth, even

sound and a steady stream of sparks from the bottom of the kerf, figure 23–15.

Piercing Steel

A flame cutting torch can also be used to pierce steel. To *pierce* means to make a hole. The operator must take special care to avoid being burned by molten slag and metal during the procedure.

Caution:

Use full face, head, shoulder, and body protection when piercing steel.

To pierce steel, the cutting torch is held above the mark at the normal preheat distance. When the spot becomes cherry red, raise the torch half an inch or more to reduce the hazard of molten metal; then slowly press the oxygen lever.

Move the tip sideways and into a circular motion until the hole breaks through, figure 23–16. Enlarge the hole by cutting around the edges. If an inside cut is desired, proceed with the cut from inside the hole.

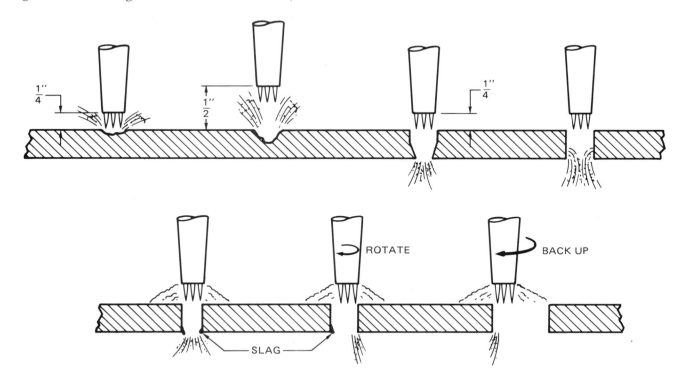

Figure 23–16 Using an oxyacetylene torch to pierce and enlarge a hole. After the flame breaks through the metal, the torch is rotated to enlarge the hole or backed up slightly before proceeding with the cut (*adapted from L. Jeffus and H. Johnson,* Welding Principles and Applications, *Delmar Publishers Inc., 1984*).

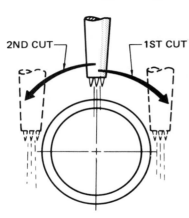

Figure 23–17. To cut pipe up to 3 inches in diameter, cut across the top, then rotate the pipe and cut the remaining section (*courtesy of L. Jeffus and H. Johnson,* Welding Principles and Applications, *Delmar Publishers Inc., 1984*).

Cutting Pipe

Cutting pipe is much like piercing and cutting thin plate. To cut pipe up to 3 inches in diameter, first pierce a hole in the top of the pipe, then cut a kerf to the left side, followed by one to the right side, figure 23–17. Rotate the pipe and repeat the process to cut the underside.

To cut large pipe, the torch is held at a right angle to the pipe. It is then moved around the pipe to make

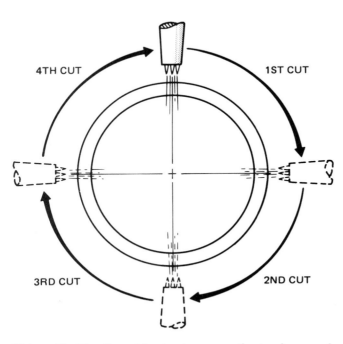

Figure 23–18. To cut large pipe move the torch around the pipe, or hold the torch in position and rotate the pipe (*courtesy of L. Jeffus and H. Johnson,* Welding Principles and Applications, *Delmar Publishers Inc., 1984*).

the cut. An alternate method is to rotate the pipe in steps or continuously, figure 23–18.

□ STUDENT ACTIVITIES

1. Define the terms italicized in this unit.
2. Light a cutting torch and adjust it to a neutral flame.
3. Obtain a piece of steel ¼" × 2" × 12" or longer. Square six lines across the metal at ½-inch intervals. Use soapstone to mark the lines. Practice cutting and then inspect each cut. Discuss your progress with your instructor.
4. Obtain a piece of scrap plate steel. Draw a straight line about ½ inch in from an edge. Practice making a long straight cut.
5. Obtain a piece of scrap steel ¼" × 2" × 12". Practice piercing holes in the strip.
6. Practice cutting off rings about ½ inch wide from the end of black pipe.

Caution:

Do not use galvanized pipe. The melted zinc coating gives off poisonous fumes.

□ SELF-EVALUATION

A. Multiple Choice. Select the best answer.
   1. Combustible gases were first used for welding
     a. in the American space exploration program
     b. in the 1800s
     c. in the early 1900s
     d. during World War II
   2. The temperatures from gas flames using oxyfuels are
     a. 150°–200°F
     b. 400°–500°F
     c. 1000°–2000°F
     d. 5000°–6000°F
   3. The result of burning iron in the presence of pure oxygen is
     a. brass
     b. propane
     c. slag
     d. weld
   4. When cutting steel, the oxygen stream
     a. aids in keeping the steel hot
     b. drives out slag
     c. supports combustion
     d. all of these

5. The fuel with the best qualities for welding and cutting is
   a. acetylene
   b. MAPP®
   c. natural gas
   d. propane

6. A mixture of gases with excellent qualities for cutting is
   a. acetylene
   b. MAPP®
   c. natural gas
   d. propane

7. A cutting torch must be adjusted so that it is neutral when
   a. cutting
   b. preheating
   c. the oxygen lever is down
   d. all of these

8. Correct oxygen and fuel pressure will vary with
   a. the tip
   b. the job
   c. the fuel
   d. all of these

9. When a correct torch cut is in progress there will be a
   a. smooth, even sound
   b. spray of sparks
   c. slightly dished kerf
   d. all of these

10. When piercing, the clearance is increased after preheating to
    a. increase the force of the oxygen stream
    b. introduce more air into the process
    c. provide time for heat to move through the metal
    d. reduce the hazard from molten metal

B. Matching. Match the terms in column I with those in column II.

Column I	Column II
1. fusion	a. poisonous fumes
2. brazing	b. results in fire
3. propane	c. maximum safe pressure is 15 psi
4. oxygen	d. cone and feather are one
5. fire triangle	e. pop or snap at tip
6. acetylene	f. bonding
7. neutral flame	g. oxyfuel not suitable for welding
8. backfire	h. supports combustion
9. flashback	i. fire inside the torch
10. hot galvanizing	j. melting metals together

C. Completion. Fill in the blanks with the word or words that will make the following statements correct.

1. When cutting, slag should be caught in a _____.

2. When cutting, the edge of the metal should be preheated until it is _____.

3. When cutting, the flame cones should be about _____ from the metal.

4. When using proper equipment with a clean tip, most cutting problems can be traced to incorrect _____, _____, _____, or _____.

5. Cutting pipe is much like cutting _____.

Unit 24
Brazing and Welding with Oxyacetylene

Objective

To braze and weld safely with oxyacetylene equipment.

Competencies to Be Developed

After studying this unit, you will be able to:
- Explain the nature and uses of braze welding.
- Prepare metal for welding.
- Select fluxes for welding.
- Identify joints commonly used in welding.
- Braze and braze weld butt, lap, and fillet joints.
- Fuse weld mild steel with and without filler rod.

Terms to Know

soldering	tinning
brazing	puddle
capillary action	bead
braze welding	butt weld
filler rod	tacking
fusion welding	fillet weld
oxide	root
oxidation	solidify
impurity	back-stepping
flux	burn-through
play the flame	

Materials List

Oxyacetylene rig, welding table, and protective clothing
Welding torch with No. 1–4 tips
Spool of 50-50 solder
Rosin flux
Brazing flux
Brazing rod, about $\frac{1}{16}$ inch in diameter
Welding rod, about $\frac{1}{16}$ inch in diameter
Steel plate (not galvanized), 16 gauge × 6″ × 16″
Steel plate (not galvanized), $\frac{1}{8}$-inch × 6″ × 10″
Copper pipe or tubing, $\frac{3}{8}$-inch or $\frac{1}{2}$-inch, or sheet copper
Power wire wheel or grinder
Fine emery cloth

☐ BASICS OF SOLDERING, BRAZING, AND FUSION WELDING

The term welding may be defined as uniting metal parts by heating or compression. Early American blacksmiths welded wagon rims and other parts by heating and hammering. Parts were heated to a certain color and a joint was formed by pounding the parts together using a hammer and anvil.

Soldering, brazing, and braze welding have much in common. All three processes are done without melting the base metal. All three use a metal or alloy as the bonding agent.

Soldering

The term *soldering* means bonding with metals and alloys that melt at temperatures *below* 840°F. The most common soldering material is 50–50 solder, consisting of 50 percent tin and 50 percent lead. Other alloys are available for special applications, figure 24–1. Tin-lead soldering is generally done with electric irons, soldering guns, and propane torches. Common applications of soldering are to join electrical wires, sweat copper pipes, and fasten thin tin-plated steel. The basic procedure for soldering is covered in unit 13, "Fastening Metal."

Brazing

The term *brazing* means bonding with metals and alloys that melt at or above 840°F where capillary action occurs, figure 24–2. The term also refers to joining parts that are fitted extremely well. Only a very thin layer (.025 inch or less) of alloy is needed to fill the void between parts, figure 24–3. This small spacing allows the alloy to be drawn into the joint by capillary action. *Capillary action* is the rising of the surface

ALLOY	USE ON
Tin-Lead	Copper and copper alloys Mild steel Galvanized metal
Tin-Antimony	Copper and copper alloys Mild steel
Cadmium-Silver	High strength for copper and copper alloys Mild steel Stainless steel
Cadmium-Zinc	Aluminum and aluminum alloys

Figure 24–1. Alloys used for soldering (*courtesy of L. Jeffus and H. Johnson, Welding Principles and Applications, Delmar Publishers Inc., 1984*)

BASE METAL	BRAZING FILLER METAL
Aluminum	BAlSi, aluminum silicon
Carbon Steel	BCuZn, brass (copper-zinc) BCu, copper alloy BAg, silver alloy
Alloy Steel	BAg, silver alloy BNi, nickel alloy
Stainless Steel	BAg, silver alloy BAu, gold base alloy BNi, nickel alloy
Cast Iron	BCuZn, brass (copper-zinc)
Galvanized Iron	BCuZn, brass (copper-zinc)
Nickel	BAu, gold base alloy BAg, silver alloy BNi, nickel alloy
Nickel-copper Alloy	BNi, nickel alloy BAg, silver alloy BCuZn, brass (copper-zinc)
Copper	BCuZn, brass (copper-zinc) BAg, silver alloy BCuP, copper-phosphorus
Silicon Bronze	BCuZn, brass (copper-zinc) BAg, silver alloy BCuP, copper-phosphorus
Tungsten	BCuP, copper-phosphorus

KEY: B = Brazing   Ag = Silver
      Al = Aluminum   Ni = Nickel
      Si = Silicon   Au = Gold
      Cu = Copper   P = Phosphorus
      Zn = Zinc

Figure 24–2. Base metals and the common brazing filler metals used for each (*courtesy of L. Jeffus and H. Johnson, Welding Principles and Applications, Delmar Publishers Inc., 1984*)

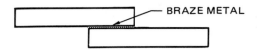

Figure 24–3. A very thin layer of braze metal joins the two pieces of base metal (*courtesy of L. Jeffus and H. Johnson*, Welding Principles and Applications, *Delmar Publishers Inc., 1984*).

Figure 24–5. Braze welding bonds pieces together and fills the voids between pieces with the braze metal (*courtesy of L. Jeffus and H. Johnson*, Welding Principles and Applications, *Delmar Publishers Inc., 1984*).

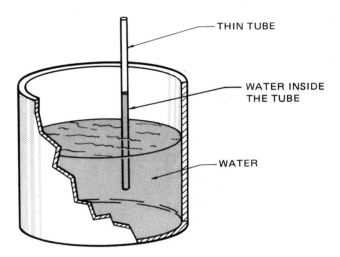

Figure 24–4. Capillary action pulls water up a thin tube (*courtesy of L. Jeffus and H. Johnson*, Welding, Principles and Applications, *Delmar Publishers Inc., 1984*).

of a liquid that is in contact with a solid. Capillary action may be observed by placing a thin tube in a container of water, figure 24–4.

The temperature of properly prepared metal must first be raised to the melting point of the brazing alloy to be used. The brazing alloy is then added by means of a long, thin metal rod called a *filler* rod. The most popular filler rods for brazing and braze welding are copper-zinc alloys. The alloy is drawn between the parts and spreads throughout the spaces that are narrow enough for capillary action to work. This principle makes brazing an attractive method for repairing broken castings.

Braze Welding

Braze welding refers to bonding with alloys that melt at or above 840°F where capillary action does *not* occur. Here the alloy is bonded to each part, and the void between or around the part is filled with the melted alloy, figure 24–5. Since many applications require brazing as well as braze welding, the term brazing is frequently used for both operations.

Advantages of Soldering and Brazing

The advantages of soldering and brazing are that they (1) are low temperature processes, (2) permit easy disassembly, (3) allow different metals to be joined, (4) can be done at high speed, (5) do little damage to parts, (6) permit easy realignment of parts, and (7) permit parts of various thicknesses to be joined.

Fusion Welding

The term *fusion welding* means joining parts by melting them together. The base metals are melted and mixed together to form a continuous piece of the same kind of metal. Fusion welding can be done with or without a filler rod. A sawed cross section of a skillfully made fusion weld will show such complete fusion that it looks like a single piece of metal.

Fusion welding is done extensively on steel. The strength of steel is such that only fusion welding is acceptable for most work. Oxyfuels are used only for welding on very thin steel. Most other fusion welding of steel is done using an electric arc welder as the source of heat.

Preparing Metals for Welding

Metals must be clean before soldering, brazing, or welding can take place. This means that oxides and other impurities must be removed from the metal. An *oxide* is the product resulting from oxidation of metal. *Oxidation* means combining with oxygen. Rust on iron and steel is an oxide. The dull coatings that form on brass and copper are also oxides.

Metal may be cleaned for soldering or welding by using a wire brush, wire wheel, grinder, sander, steel wool, file, or other mechanical process.

After cleaning metal by mechanical means, chemicals are used to remove any impurities that remain. An *impurity* is any product other than the base metal.

The term *flux* is given to any chemical used to clean metal. Fluxes remove tarnish and corrosion caused by oxidation. They also prevent corrosion from developing. Additionally, they promote wetting or move-

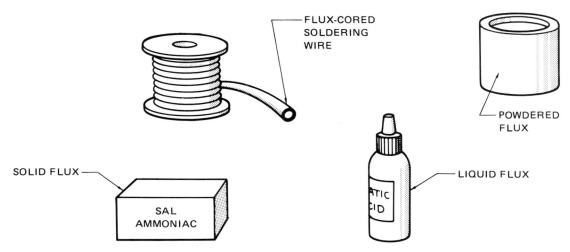

Figure 24–6. Fluxes come in many forms (*adapted from L. Jeffus and H. Johnson,* Welding Principles and Applications, *Delmar Publishers Inc., 1984*).

ment of molten alloys on metal. Finally, fluxes aid in capillary action which moves molten alloys into very small cracks. Most fluxes work best when hot. If the metal gets too hot, however, the flux will burn and must be cleaned off before proceeding. Fluxes are available as solids, pastes, powders, liquids, sheets, rings, and washers, figure 24–6. They may also be in the hollow core or on the surface of filler rods.

Using Flux. □ In order to control the heat needed to manage the fluxing action, the following procedure is recommended.

Procedure

1. Obtain a piece of copper pipe, tubing, or sheet. The copper should be clean but not shiny.
2. Set up, light, and adjust a small torch.
3. Put a pea-sized spot of rosin flux on the copper.
4. Heat the flux and surrounding area by playing the flame on the rosin. To *play the flame* means to alternately move a flame into and out of an area to achieve temperature control.
5. Observe the point when the copper suddenly becomes shiny under the flux. Remove the heat. The shiny surface indicates that the flux has reacted with the oxide and removed it, figure 24–7.
6. Heat an area near the shiny spot. The flux should flow to the newly heated area and clean it too.
7. Observe the effect of too much heat by continuing to heat the clean areas. Notice that too much heat

will burn the flux and the area will turn dark or black.
8. Remove the burnt flux with steel wool or fine sandpaper.
9. Reapply the flux and reheat. Note that the fluxing action can be repeated only after mechanical removal of the burnt flux. It is important to recognize fluxing action. It is necessary to learn how to control the fluxing process. Otherwise, soldering or brazing fillers will not bond to the metal.

Tinning. □ Tinning is the key to success in soldering and brazing. The term *tinning* means to bond filler material to a base metal. The term probably originated from the tinlike appearance of solder when it is applied to a base metal. Before two pieces of metal can

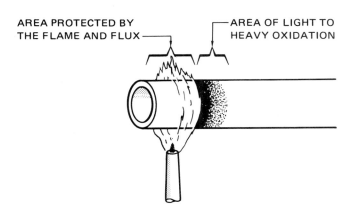

Figure 24–7. Fluxing copper pipe (*courtesy of L. Jeffus and H. Johnson,* Welding Principles and Applications, *Delmar Publishers Inc., 1984*)

be joined by soldering, brazing, or braze welding, both pieces must be tinned. Most unsuccessful attempts at soldering fail at this point. Learn to tin before attempting to solder, braze, or braze weld.

> **Caution:**
>
> **Never braze or weld galvanized steel without special training and special precautions. Galvanized coatings give off toxic fumes when melted or heated above soldering temperatures.**

The following procedure is suggested for tinning.

Procedure

A. Tinning with Solder
1. Obtain a piece of copper pipe or sheet.
2. Clean the metal with fine emery cloth.
3. Obtain a spool of tin-lead solder, paste flux, and a damp cloth.
4. Light and adjust an oxyfuel torch with a small tip and neutral flame. **Note:** A propane torch without oxygen may also be used.
5. Heat the copper until the fluxing action starts.
6. Touch the end of the solder to the hot metal. Feed the solder down to the metal as it flows across the fluxed area.
7. Use the torch to heat an enlarged area and attract flux and solder across the metal.
8. Play the torch over the tinned area to get a thin, smooth, and shiny surface.
9. To improve the appearance and to remove excess solder, wipe the solder-covered area quickly with a damp cloth. Note that the damp cloth removes excess solder and leaves a thin coating of solder bonded to the copper.

B. Tinning with Braze Filler
1. Obtain a piece of nonrusted steel plate approximately ⅛" × 2" × 6".
2. Clean the plate carefully with fine emery cloth.
3. Obtain a brazing filler rod and can of brazing flux.
4. Light and adjust an oxyfuel torch using a small tip and neutral flame.
5. Place the plate on firebricks.

6. Heat the center of the plate until it is dull red.

> **Caution:**
>
> **Do not permit the steel to melt.**

7. Heat the end of a braze filler rod by placing it in the flame as the base metal is heating. **Note:** The operator controls the torch with one hand. The other hand holds the filler rod like a pencil, about 12 inches from the end.
8. With the tip of the filler rod heated, dip it into the can of flux, figure 24–8.
9. Place the flux-covered end of the rod into the flame and touch the hot base metal.
10. When the flux reaches the correct temperature, fluxing action occurs (the chemical flows across the hot metal and removes impurities).
11. Once fluxing occurs, watch for the end of the filler rod to melt. Then play the flame on the filler metal, flux, and plate, causing the filler to flow as you create the correct temperature on the base metal.
12. Play the heat over the entire area until the filler material flows into a thin layer of bonded metal.

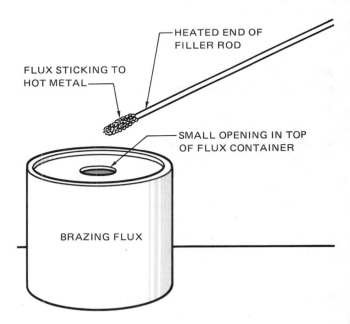

Figure 24–8. The heated end of a filler rod can be used to transfer flux from the container to the base metal.

Caution:

Do not overheat the filler material and base metal or the metals will turn dark from oxidation and will require mechanical cleaning again.

If difficulty is encountered with the process, clean the metal completely and start again. The critical task is to keep the filler rod heated very nearly to its melting point. Then, when it is needed for fluxing or filling, the operator can easily control these processes. Quick melting of the filler rod is achieved by touching the base metal with the rod while playing the torch on the rod.

Controlling Heat. □ Controlling heat is the key to successful soldering and brazing. Metals are excellent heat conductors. Therefore, heat applied in one place soon moves to surrounding areas. The larger the piece of metal, the faster the heat gets away from the heated area. This must be kept in mind when joining two pieces of metal. If the torch directs an equal amount of heat onto two pieces of different size, the smaller piece will get hot faster. When this occurs, the flux will burn on the small piece before the large piece is up to fluxing temperature.

The operator must manipulate the torch to heat pieces of different size equally. The following procedure is helpful.

Procedure

1. Obtain one piece of steel ⅛″ × 2″ × 2″ and one 16 gauge × 2″ × 2″.
2. Place the two pieces end to end on a firebrick.
3. Light and adjust the torch.
4. Play heat on the joint and in a circular area covering both pieces until both are dull red.

The flame must be directed to the thick piece more than to the thin one. Practice with pieces of various sizes until you can make both pieces turn dull red at the same time. This skill is essential for soldering, brazing, and welding.

□ BRAZING AND BRAZE WELDING PROCEDURES

Brazing and braze welding are generally done in agricultural mechanics using an oxyfuel torch. Acetylene is perhaps the most widely used fuel. However, acetylene has the disadvantage of producing a very hot inner flame and a relatively cold outer flame. For brazing and braze welding, a hot outer flame is useful for heating large areas uniformly. Such fuel gases as MAPP®, propane, butane, and natural gas have the advantage of providing more uniform heating.

Heating of large areas is necessary when entire assemblies, such as pipe fittings or large castings, are being bonded. A good way to control heat when brazing large assemblies is to first heat the basic assembly until it is near the melting point of the braze filler. Then it is fairly easy to add the additional heat required to melt the braze metal and fill the voids in specific locations.

Torch Tips. □ Tips used for oxyfuel welding have a single hole. They are identified by a number which designates the size of the hole. Tip numbers range from 000 (the smallest) to 10 or larger. A small tip is used for thin metal. The tip size must be larger as the thickness and mass of the metal increases. The torch manual should provide proper guidelines for tip selection.

Torch Adjustment. □ A good general rule for torch adjustment is that most soldering, brazing, and welding is done with a neutral flame. Only specialized applications require slightly carburizing or slightly oxidizing flames.

Brazing

Since brazing works by capillary action, it requires that parts fit together very closely. Some common

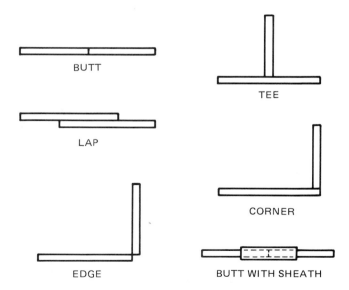

Figure 24–9. Some common types of joints used to fasten metal

types of joints used to fasten metal are shown in figure 24–9. The following procedure is suggested for brazing flat material in a lap joint.

Procedure

1. Obtain three perfectly flat and clean pieces of nongalvanized steel plate ⅛″ × 2″ × 2″.
2. Clean a 1-inch strip along one edge of each of two pieces.
3. Place these two pieces on firebricks so that clean, ½-inch strips overlap one another. Use the third piece to support the top piece, figure 24–10. If possible, clamp the assembly or hold the top piece in place with a brick.
4. Obtain brazing rod and flux.
5. Light and adjust the torch.
6. Apply heat to the clean joint until both pieces are dull red.
7. Add flux to the joint by way of the heated filler rod.
8. Continue to raise the temperature of the assembly and move the filler rod in and out of the flame.

9. When both pieces are dull red, touch the metal with the rod and feed filler into the joint.

    Note: Bend a loop into the cold end of the brazing rod before using.
10. Play heat more on the top piece and back from the edge to draw the filler into the joint.
11. Let the assembly cool.
12. Turn the assembly over and examine it for completeness of brazing. Filler metal should be present on the back edge of the joint.

Cast iron assemblies of irregular shapes may be brazed. Appropriate fluxes must be used to correct the presence of free graphite in some iron. Additionally, a large torch is needed to heat large assemblies. Brazing may be preferred over fusion welding of cast iron since there is less likelihood of heat distortion and breaking of the casting.

Braze Welding

When braze welding, many of the principles of brazing apply. The major difference is that a perfect fit is not needed. Still, good fits generally result in stronger joints.

Braze welding is useful on metal that is 16 gauge or thinner. Thicker metal can be braze welded, but

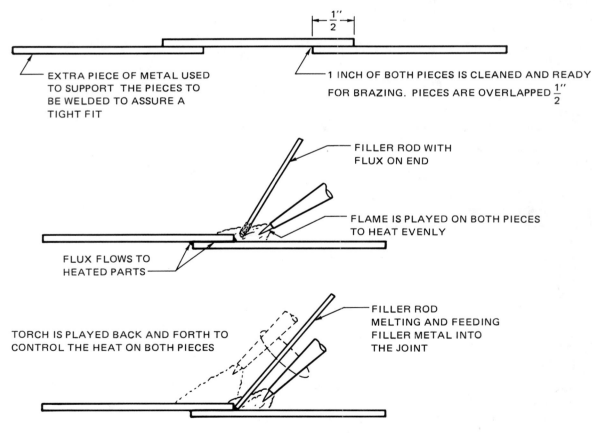

EXTRA PIECE OF METAL USED TO SUPPORT THE PIECES TO BE WELDED TO ASSURE A TIGHT FIT

1 INCH OF BOTH PIECES IS CLEANED AND READY FOR BRAZING. PIECES ARE OVERLAPPED $\frac{1''}{2}$

FILLER ROD WITH FLUX ON END

FLAME IS PLAYED ON BOTH PIECES TO HEAT EVENLY

FLUX FLOWS TO HEATED PARTS

TORCH IS PLAYED BACK AND FORTH TO CONTROL THE HEAT ON BOTH PIECES

FILLER ROD MELTING AND FEEDING FILLER METAL INTO THE JOINT

Figure 24–10. Brazing a lap joint

electric arc welding is faster and easier and generally the preferred process for thicker metals.

Running a Bead. □ Most welding requires the ability to run a bead. A *bead* is a continuous and uniform line of filler metal. To run a bead with a braze filler rod, the following procedure is suggested.

Procedure

1. Obtain a clean piece of ⅛″ × 2″ × 6″ steel.
2. Clean both sides with fine emery cloth.
3. Light and adjust the torch.
4. Position the metal on a flat firebrick. The bead will be run across the metal ½ inch in from one end.
5. Heat the end of a filler rod and a spot on one edge of the plate about ½ inch from the end.
6. Dip the hot rod into flux and touch the fluxed rod to the heated spot.
7. When the rod starts to melt, move it in and out of the flame to add filler metal as needed.
8. The bead is formed by depositing filler material along a fluxed path across the metal.
9. Run additional beads across the metal at 1-inch intervals until you can run complete, straight, and even beads.

Note: When running a bead, it helps to move the torch in a circular pattern. The inner cone is kept about ⅛ inch from the molten puddle. A *puddle* is a small pool of liquid metal. Heat is controlled by the clearance of the cone, the angle of the flame, and the movement of the rod into and out of the puddle. The filler rod must be dipped into the flux frequently. This provides flux to cover the new bead as it forms.

Braze Welding Butt Joints. □ A *butt weld* is essentially a bead laid between two pieces of metal set edge to edge. This may be end to end or side to side. When braze welding a butt joint, it is important that the filler metal is down between the pieces and bonds the edges at all points. To achieve this, the two pieces are placed with a slight gap between them. To braze weld a butt joint, the following procedure is suggested.

Procedure

1. Cut two pieces of nongalvanized steel 16 gauge × 2″ × 2″.
2. Clean the edges and faces with fine emery cloth.
3. Place them edge to edge on a flat firebrick with a gap between them of about 1/32 inch.

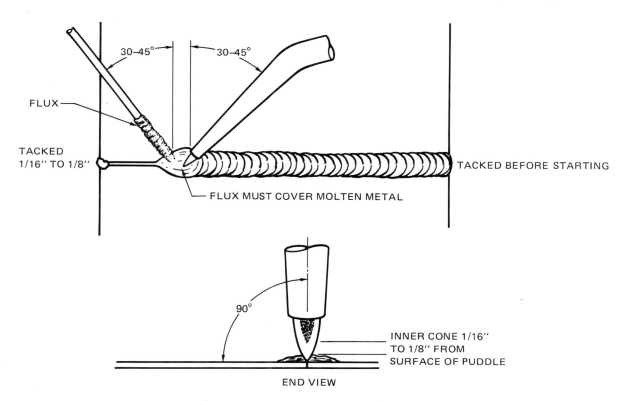

Figure 24-11. Braze welding a butt joint

4. Weight the two pieces with firebricks to help hold them in position.

5. Heat a spot at one end of the joint. Flux the area and deposit a spot of filler material.

6. Repeat step 5 at the other end of the joint. These small welds are called tacks. The process of making a small weld to hold two pieces together temporarily is called *tacking*.

7. Lay a bead over the entire length of the joint, figure 24–11.

Braze Welding Fillets. □ A *fillet weld* is a weld placed in a joint created by a 90° angle. The joint may be in the form of an L or a T.

When making a fillet weld the torch must be played on the flat and vertical pieces with care. If both pieces are equal in thickness and mass, the heat is applied equally. However, if one piece is thicker, that piece must receive more heat, figures 24–12 and 24–13.

Special care must be taken to ensure that sufficient heat reaches the root of the weld. The *root* is the deepest point in a weld. In this case, it is the place

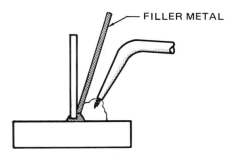

Figure 24–12. Torch and rod positions required to balance heat when the base is thicker than the vertical piece (*courtesy of L. Jeffus and H. Johnson*, Welding Principles and Applications, *Delmar Publishers, Inc., 1984*)

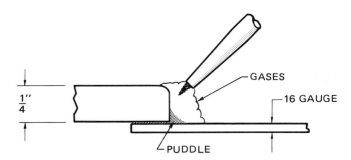

Figure 24–13. Torch position for directing heat to a vertical surface with a greater mass than the base (*courtesy of L. Jeffus and H. Johnson*, Welding Principles and Applications, *Delmar Publishers Inc., 1984*)

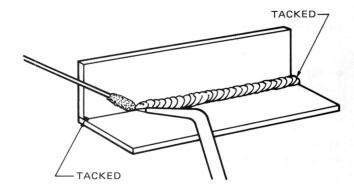

Figure 24–14. Making a fillet weld on pieces of equal thickness

where the two pieces melt to form an angle. A braze welded fillet should have the filler material under the vertical piece and a bead along the 90° corner, figure 24–14.

□ FUSION WELDING WITH OXYACETYLENE

Fusion welding of steel with oxyacetylene is generally limited to thin metal or small jobs where portability of equipment is a factor. Electric arc welders are now so versatile that they account for most fusion welding done in agricultural mechanics. Still, it is useful to be able to do small jobs and weld thin metal with oxyacetylene.

Factors Affecting the Weld

The Tip. □ Selection of the torch tip will depend upon the width and penetration of the bead desired. The larger the tip, the more heat the flame produces. The heat from a given tip is controlled by the distance the cone is held from the puddle and the angle of the tip to the work. The greatest melting ability is obtained when the angle is 90°. The ideal angle is 45°. The ideal cone distance is ⅛ to ¼ inch from the puddle.

The Rod. □ The filler rod for fusion welding should be made of the same material as that being welded. Remember, fusion welding is a melting and mixing of base metals. The filler rod adds volume so the joint will be filled and slightly reinforced, or strengthened. The larger the rod, the greater is its tendency to cool the molten puddle, increase the size of the bead, and reduce penetration.

Caution:

Always bend one end of the filler rod to form a loop. The loop identifies the unheated end and decreases chances of those in the area being hit by the sharp end or burned by the hot end.

Controlling the Process

The gases in the air will contaminate, or pollute, an unprotected welding puddle. Fortunately, the presence of the torch flame prevents air from getting to the molten puddle. However, this makes it necessary to keep the torch on or nearly on the puddle at all times. To complete a weld properly, the torch is lifted slowly so the puddle can cool and solidify without contamination. To *solidify* is to turn from a liquid to a solid.

During the welding process, some components of the base metal burn and produce sparks. Changes in welding temperature can be detected by observing the amount of sparks given off. Hence, the alert welder can detect when a burn-through is about to happen and reduce the heat to prevent it. *Burn-through* is the process whereby the cool side of metal becomes molten and a hole opens in the surface due to excessive heat from the opposite side. A burn-through is generally preceded by an increase in spark activity.

Pushing the Puddle

The first step in oxyacetylene welding is to control the puddle. This is called puddling. To create a bead by puddling without a filler rod is called pushing the puddle and is done by the following procedure.

Procedure

1. Obtain a clean piece of 16 gauge × 2″ × 6″ steel.
2. Light and adjust a torch with a number 2 or 3 tip, using 3 psi acetylene pressure and a neutral flame.
3. Place the strip of steel on a firebrick, and weight it with another firebrick.
4. Start at one end of the metal with the torch tip turned toward the plate and a gap of ⅛ inch from cone to metal.
5. When the metal starts to melt, move the torch in a small circular pattern down the plate to create a bead.

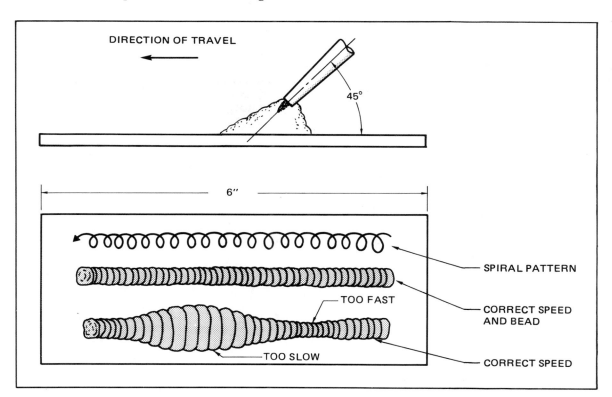

Figure 24–15. The torch speed affects the size of the bead (*adapted from L. Jeffus and H. Johnson, Welding Principles and Applications, Delmar Publishers Inc., 1984*).

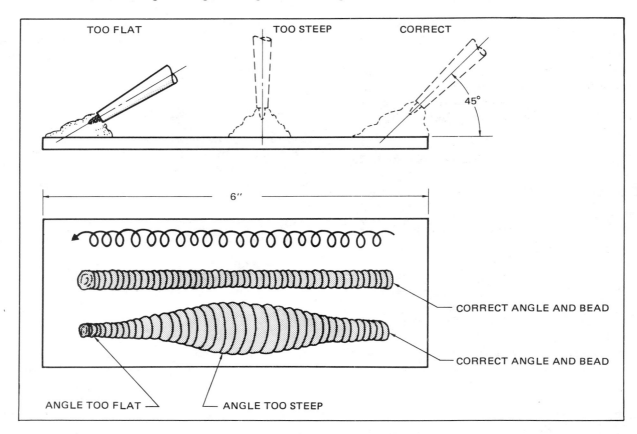

Figure 24–16. The torch angle also affects the shape of the bead (*adapted from L. Jeffus and H. Johnson*, Welding Principles and Applications, *Delmar Publishers Inc., 1984*).

6. Lift the tip slowly at the end of the bead. Make additional practice beads until the metal is covered.

The puddle should move ahead as the torch moves down the plate. A bead will form as the flame passes. If the size of the pool changes, speed up or slow down, figure 24–15. The size of the pool should be kept constant.

The ideal torch angle is 45°. As the angle of the torch increases up to 90°, the bead gets wider and wider. If the angle is too flat, such as 30°, the bead will be too narrow, figure 24–16.

Making a Corner Weld without Filler Rod

A corner weld is one made on edges laid together to form a corner, figure 24–17. To practice making a corner weld, use the following procedure.

Procedure

1. Obtain two pieces of clean 16 gauge × 1½″ × 6″ steel.

2. Place the pieces on their edges on a firebrick to form a tentlike shape with a 60° angle.

3. Tack weld the joint at both ends and two places in between.

4. Make a weld bead along the entire corner.

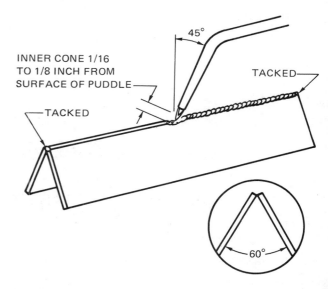

Figure 24–17. Making a corner weld without a filler rod

5. Evaluate your speed and torch angle by comparing your results with figure 24–16.
6. Open the joint, examine the penetration, and discuss the results with your instructor.
7. Grind off the rough edges and save the metal for the next exercise.

Laying a Bead with Filler Rod

The movements for laying a bead with a filler rod in fusion welding are similar to those used in brazing. Practice laying a bead with a filler rod in a butt weld as follows.

Procedure

1. Use the two pieces of 16 gauge × 1½″ × 6″ metal left from the previous exercise.
2. Place the two pieces edge to edge on a firebrick to form a 6-inch-long butt joint.
3. Adjust the pieces until there is a ¹⁄₁₆-inch gap between them. Weight them with firebricks.
4. Tack weld at both ends of the joint and at two places in between.
5. Point the torch toward one end of the joint at an angle of 45°.
6. Establish a molten puddle with a ¹⁄₁₆- to ⅛-inch cone clearance.
7. Add filler metal to the front edge of the puddle.
8. Rotate the flame and move the rod in and out to create an even bead, figure 24–18.
9. Complete the bead by slowly raising the torch while filling the puddle as you withdraw the torch.
10. Discuss the resulting weld with your instructor.

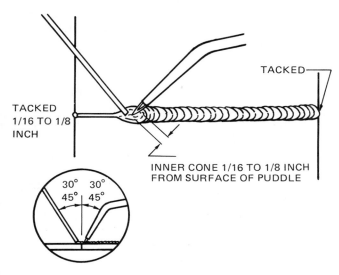

Figure 24–18. Laying a bead to form a butt weld

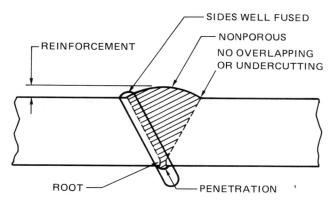

Figure 24–19. Characteristics of a proper fusion weld

The Finished Weld

The completed weld should be thoroughly fused to the base metal throughout the joint, figure 24–19. The weld metal should penetrate to the root of the joint with a small amount extending below the surface to assure a full section weld. At the face of the weld, there is usually a buildup known as "reinforcement." This reinforcement should be slight, blending smoothly with the base metal surfaces. It is especially important to avoid undercutting or overlapping at the juncture of the weld and base metal.

Bending or distortion of the metal can be reduced by back-stepping. *Back-stepping* is making short welds in the backward direction as the process progresses in the forward direction, figure 24–20. Repeat the previous exercise using the back-step method.

Making a Fillet Weld

A fillet weld is made by laying a bead in a 90° angle formed by two pieces of metal. To make a fillet weld, the following procedure is recommended.

Procedure

1. Obtain two pieces of clean 16 gauge × 1½″ × 6″ steel.
2. Place one piece flat on firebricks. Set the other piece on its edge on the first piece to form a 90° angle.
3. Prop the assembly with firebricks.
4. Tack weld both ends and the middle.
5. Lay a bead along the entire joint, figure 24–21.

Other joints can be welded using the procedures outlined in this unit. The student is encouraged to learn well all basic procedures presented.

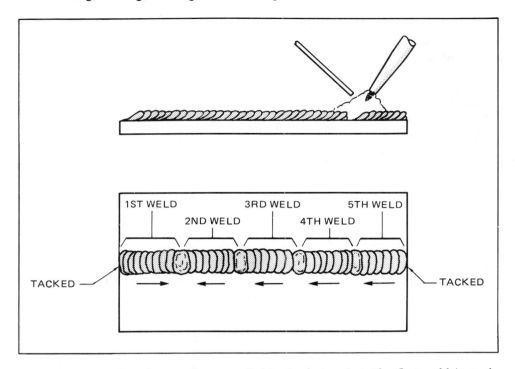

Figure 24–20. Distortion can be controlled by back-stepping. The first weld is made from left to right, but all subsequent welds are made from right to left (*adapted from L. Jeffus and H. Johnson,* Welding Principles and Applications, *Delmar Publishers Inc., 1984*).

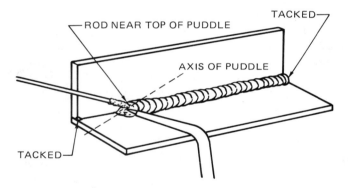

Figure 24–21. Making a fillet weld

□ STUDENT ACTIVITIES

1. Define the new terms italicized in this unit.
2. Using a torch and flux, heat a piece of copper and practice fluxing copper without overheating the flux.
3. Using a torch, flux, 50-50 solder, and copper, tin the copper surface.
4. Using an oxyfuel torch, brazing filler rod, flux, and ⅛″ × 2″ × 6″ steel plate, tin the plate without burning the flux.
5. Prepare an oxyfuel torch, a piece of ⅛″ × 2″ × 2″ steel and a piece of 16 gauge × 2″ × 2″ steel. Place the pieces side by side and heat both at the same time to an even red color.
6. Prepare a torch, flux, brazing filler rod, and ⅛″ × 2″ × 6″ steel plate. Practice laying braze weld beads across the plate at ½-inch intervals.
7. Prepare a torch, flux, brazing filler rod, and three pieces of ⅛″ × 2″ × 2″ steel plate. Use one plate for support and braze a ½-inch lap joint between the other two.
8. Prepare a torch, flux, brazing filler rod, and two pieces of 16 gauge × 2″ × 2″ steel. Butt weld the two pieces together.
9. Prepare a torch, flux, brazing filler rod, and two pieces of 16 gauge × 1″ × 2″ steel. Lay one piece flat on a firebrick. Bend the other piece slightly and place it on its side on top of the first piece. Braze weld the two pieces to form a fillet weld.
10. Prepare a torch, welding flux, and one piece of 16 gauge × 1½″ × 6″ steel. "Push the puddle" to form three beads across the steel without filler rod.
11. Prepare a torch, welding flux, and two pieces of 16 gauge × 1½″ × 6″ steel. Prop the two pieces edge to edge to form a 60° angle. Make a corner weld over the ridge of the assembly without a filler rod.

12. Prepare a torch, flux, $\frac{1}{16}$ inch welding filler rod, and two pieces of 16 gauge × 1½″ × 6″ metal. Use the filler rod to butt weld the two pieces together.

13. Prepare a torch, flux, $\frac{1}{16}$ inch welding filler rod, and two pieces of 16 gauge × 1½″ × 6″ steel. Join the two pieces to form a T using a fillet weld.

14. Prepare a torch, flux, $\frac{1}{16}$ inch welding filler rod, and two pieces of 16 gauge × 1½″ × 6″ steel. Weld a lap joint.

□ **SELF-EVALUATION**

A. Multiple Choice. Select the best answer.
1. Welding is
   a. uniting
   b. heating
   c. fusion
   d. all of these
2. Brazing is much like
   a. soldering
   b. painting
   c. fusion welding
   d. arc welding
3. The most popular soldering alloy is
   a. aluminum silicon
   b. copper-zinc
   c. silver
   d. tin-lead
4. The most popular brazing alloy is
   a. aluminum
   b. copper-zinc
   c. silver
   d. tin-lead
5. Rust is a form of
   a. dirt
   b. galvanize
   c. oxidation
   d. weathered paint
6. Metal may be cleaned before brazing with a
   a. brush
   b. flux
   c. emery cloth
   d. all of these
7. Two pieces lying flat and end to end may be joined by a
   a. butt weld
   b. corner weld
   c. fillet weld
   d. lap weld
8. Bonding solder or braze material to a piece of metal is called
   a. brazing
   b. fusion
   c. soldering
   d. tinning
9. The best distance between cone and puddle when fusion welding is
   a. ½ inch
   b. ¼ inch
   c. ⅛ inch
   d. $\frac{1}{32}$ inch
10. The best torch angle for flat welding is
   a. 30°
   b. 45°
   c. 70°
   d. 90°

B. Matching. Match the terms in column I with those in column II.

Column I	Column II
1. soldering	a. joins electrical wires
2. brazing process	b. fastens temporarily
3. fusion welding	c. used before fluxing
4. bead	d. capillary action
5. tacking	e. size 000
6. flux	f. size 5
7. small tip	g. mixes metals
8. large tip	h. line of filler metal
9. emery cloth	i. toxic when heated
10. galvanize	j. removes oxides

C. Brief Answers. Briefly answer the following questions.
1. What are three advantages of brazing over fusion welding?
2. What is the difference between brazing and braze welding?
3. Why is back-stepping used?

Unit 25
Selecting and Using Arc Welding Equipment

Objective

To select electric arc welders, equipment, and materials needed for welding in agricultural mechanics.

Competencies to Be Developed

After studying this unit, you will be able to:
- Describe the shielded metal arc welding process.
- Distinguish types of electric welding machines.
- Select suitable supplies and equipment for shielded metal arc welding.
- Recognize color and numerical code markings on electrodes.
- Select electrodes for use in agricultural mechanics.

Terms to Know

arc welder	transformer
arc	alternating current (AC)
weldor	60-cycle current
electrodes	generator
shielded metal	direct current (DC)
arc welding	straight polarity (SP)
(SMAW)	reverse polarity
arc welding	electrode holder
stick welding	ground clamp
duty cycle	chipping hammer
ampere (A)(amp)	National Electrical
conductor	Manufacturers
volt (V)	Association (NEMA)
voltage	end marking
watt (W)	spot marking

group marking	tensile strength
American Welding	carbon arc torch
Society (AWS)	

Materials List

A variety of welding machines
A fully equipped welding area
Electrodes for identification

Heat for arc welding is obtained by using electricity. In this process electric current flows from a transformer connected to lines from a power plant. Another source of electricity for welding is electric motors or gas engines driving special generators. Regardless of the source of the electrical energy, a machine that produces current for welding is known as an *arc welder*. An *arc* is the discharge of electricity through an air space. A person who welds is known as a *weldor*.

☐ ARC WELDING PROCESS AND PRINCIPLES

Shielded Metal Arc Welding

This unit focuses on a welding process that uses flux-coated metal welding rods called *electrodes*. The process is called *shielded metal arc welding*, abbreviated SMAW. Some weldors call the process *arc welding* or *stick welding*. The term "shielded" refers to the gaseous cloud formed around the weld by the burning flux. The gases help the metal electrode burn evenly

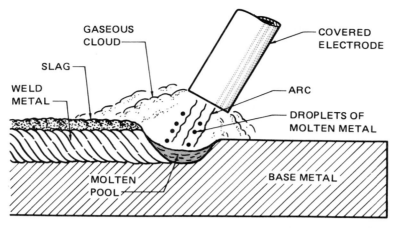

Figure 25–1. Components of the shielded metal arc welding process (*courtesy of L. Jeffus and H. Johnson,* Welding Principles and Applications, *Delmar Publishers Inc., 1984*)

as it mixes with the base metal, figure 25–1. The flux also removes impurities from the base metal. The flux and impurities float to the top of the weld to form a layer called *slag.*

Advantages of Arc Welding. □ Arc welding is used extensively in agricultural mechanics. Some of the reasons for its popularity are as follows:

• Electricity is relatively inexpensive as a source of heat for welding.
• Electric welders suitable for farm welding are relatively inexpensive.
• Welders are available that work on ordinary 230-volt household or farmstead wiring.

• Engine-driven portable welders are available.
• Arc welding is fast and reliable.
• Agricultural students and workers can become good arc weldors quickly.
• An arc welder can be used for heating, brazing, and hardsurfacing as well as welding.

Temperature. □ The arc from a welder has a temperature of about 9000°F. The exact temperature varies with the length of the arc, size of the electrode, and amperage setting. Typical amperage settings for welders used in agricultural mechanics range from about 20 amps to 225 amps. Upper limits for industrial and commercial welders are higher, figure 25–2.

MACHINE MODEL	TYPE (60 Hertz)	NEMA RATING CURRENT (amperes)	ARC VOLTS	DUTY CYCLE	OUTPUT CURRENT RANGE (amperes)	TYPE	REQUIRED POWER SYSTEM	BULLETIN NUMBER
AC-225-S	K-1170	225	25	20%	40-225	AC	1 phase	E320
AC-250	K-1051	250	30	30%	35-300	AC	1 phase	E330
AC/DC-250	K-1053	250	30	30%	40-250#	AC & DC	1 phase	E330
TM-300	K-1103	300	32	60%	30-450	AC	1 phase	E340
TM-400	K-1105	400	36	60%	40-600	AC	1 phase	E340
TM-500	K-1108	500	40	60%	50-750	AC	1 phase	E340
TM-300/300	K-1104	300	32	60%	45-375#	AC & DC	1 phase	E340
TM-400/400	K-1107	400	36	60%	60-500#	AC & DC	1 phase	E340
TM-500/500	K-1110	500	40	60%	75-625#	AC & DC	1 phase	E340
TM-650/650	K-1126	650	44	60%	75-750#	AC & DC	1 phase	E340
R3R-300	K-1284	300	32	60%	45-375	DC	3 phase	E351
R3R-400	K-1285	400	36	60%	60-500	DC	3 phase	E351
R3R-500	K-1286	500	40	60%	75-625	DC	3 phase	E351

DC ranges, AC ranges same as straight AC models.
Source: Lincoln Electric Company.

Figure 25–2. The range of welding machines available allows the consumer to select one with the performance characteristics needed for a variety of jobs.

Duty Cycle. □ Welding machines get hot from use. The design of the machine determines how long it can operate. The *duty cycle* is the percentage of time that a welder can operate without overheating. Stated another way, it is the number of minutes out of 60 that a welder can operate at full capacity. A 20 percent duty cycle welder should weld only 12 minutes out of every hour, or 20 percent of 60 minutes. Welders are available with duty cycles ranging from 20 percent to 100 percent.

Duty cycle ratings are based on a midrange setting. The duty cycle is shorter if the welder is used at higher settings. Similarly, it is longer when the welder operates at lower settings.

Electricity for Welding

Three terms are basic to understanding how electricity is used for welding. They are amperes, volts, and watts. An *ampere* (A) or *amp* is a measure of the rate of flow of current in a conductor. A *conductor* is any material that permits current to move through it. Amperes of electricity flowing through a wire can be compared to gallons per minute of water flowing through a pipe.

A *volt* (V) or *voltage* is a measure of electrical pressure. Volts in a wire can be compared to pounds per square inch of water pressure in a pipe.

A *watt* (W) is a measure of energy available or work that can be done. For example, a 100-watt bulb gives out a specific amount of light. To figure the number of watts consumed, multiply volts by amperes: W = VA. This is called the West Virginia (W.VA.) formula to make it easier to remember. If any two values are known, the third can be calculated, figure 25–3.

Welding machines put out a high amperage and relatively low voltage. It is the low voltage that permits the welder to be a relatively safe machine while putting out so much energy. It requires about 60 volts

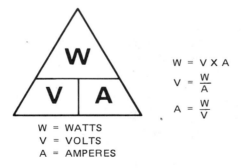

$$W = V \times A$$
$$V = \frac{W}{A}$$
$$A = \frac{W}{V}$$

W = WATTS
V = VOLTS
A = AMPERES

Figure 25–3. The third electrical value can be calculated when two out of three values are known (*courtesy of L. Jeffus and H. Johnson,* Welding Principles and Applications, *Delmar Publishers Inc., 1984*).

to push current through the human body. At low to moderate settings, some welders put out less than 60 volts. However, reasonable caution to prevent electrical shock must be exercised at all times.

Alternating and Direct Current. □ Some welders are transformers that receive current directly from utility power lines. *Transformers* convert high voltage and low amperage to low voltage and high amperage. These are called alternating current welders. *Alternating current* (AC) is current that reverses its direction of flow frequently. In the United States, power plants produce electricity that reverses its direction of flow 60 times per second. This electricity is referred to as *60-cycle current.*

Other welders are really generators, driven by an electric motor or a gas or diesel engine. A *generator* is a machine that produces direct current. *Direct current* (DC) flows in one direction only in accordance with how the welder is set. *Straight (negative) polarity* (SP) is the term given to DC current flowing in one direction, while *reverse (positive) polarity* (RP) is direct current flowing in the opposite direction. Most DC welders have a switch to operate at either straight or reverse polarity as needed for any given electrode or job. Otherwise, polarity can be switched by reversing the welding cables at the machine.

□ WELDING EQUIPMENT

AC Welders

AC welders rated at 180 and 225 amperes are popular for farm use, figure 25–4. Such welders can be purchased for just over $100 and are adequate for most farm welding jobs. They do not draw excessive current, so that electricity bills are usually not greatly affected by them. Large farming operations may benefit from larger and more versatile welders for specialized welding.

DC Welders

Some DC welders are driven by electric motors. Such welders must be located near an electrical outlet. Other welders are driven by gasoline or diesel engines. These units may be transported to the job by truck or trailer. Some machines are both DC and AC, figure 25–5. These machines can be used as welders and as portable generators to operate lights and equipment, figure 25–6. The control panel on such machines includes controls to operate the engine as well as the welder, figure 25–7.

Figure 25–4. AC welders rated at 180 and 225 ampere are popular for farm use (*courtesy of Lincoln Electric Company*).

Figure 25–5. Diesel engine-driven AC/DC welder and generator (*courtesy of Miller Electric Manufacturing Company of Appleton, WI*).

Figure 25–6. Some welders are portable and can act as generators to produce electricity for lights and power tools (*courtesy of Miller Electric Manufacturing Company of Appleton, WI*).

Welding Cables

The diameter of electrical wire is stated in terms of gauge. The lower the gauge number, the larger the size of the wire. Wire sizes range from 18 for lighting circuits in automobiles to 0 or lower for large service cables. The gauge used for electrical wiring in homes is generally 14 for lights and 12 for heavier loads.

Figure 25–7. Control panel of an engine-driven welder/ generator (*courtesy of Miller Electric Manufacturing Company of Appleton, WI*).

LENGTH OF CABLE		COPPER WELDING LEAD SIZES								
AMPERES		**100**	**150**	**200**	**250**	**300**	**350**	**400**	**450**	**500**
ft	m									
50	15	2	2	2	2	1	1/0	1/0	2/0	2/0
75	23	2	2	1	1/0	2/0	2/0	3/0	3/0	4/0
100	30	2	1	1/0	2/0	3/0	4/0	4/0		
125	38	2	1/0	2/0	3/0	4/0				
150	46	1	2/0	3/0	4/0					
175	53	1/0	3/0	4/0						
200	61	1/0	3/0	4/0						
250	76	2/0	4/0							
300	91	3/0								
350	107	3/0								
400	122	4/0								

LENGTH OF CABLE		ALUMINUM WELDING LEAD SIZES								
AMPERES		**100**	**150**	**200**	**250**	**300**	**350**	**400**	**450**	**500**
ft	m									
50	15	2	2	1/0	2/0	2/0	3/0	4/0		
75	23	2	1/0	2/0	3/0	4/0				
100	30	1/0	2/0	4/0						
125	38	2/0	3/0							
150	46	2/0	3/0							
175	53	3/0								
200	61	4/0								
225	69	4/0								

Figure 25–8. Copper and aluminum welding lead sizes (gauges) for various current loads and cable lengths (*courtesy of L. Jeffus and H. Johnson,* Welding Principles and Applications, *Delmar Publishers Inc., 1984*)

A welder puts out high amperage, so it needs large electrical cables. Cables carry current from the welder to the electrode, across the arc, and back to the welder. The cable that comes with a welder as original equipment is the correct diameter for its length. If longer cables are used, they may also need to be of larger diameter, figure 25–8. If the cables are too small they will not deliver as much current as the welder is designed to provide. Under these conditions, the force of the arc is not as intense as the machine is capable of delivering.

Caution:

Careful attention should be given to cable care. Always protect cables from forceful impact and contact with sharp objects that may damage their coverings. These coverings insulate the conductors and prevent dangerous shocks and electrical shorts.

Electrode Holders and Ground Clamps

A welder has two cables: one ends in an electrode holder and the other ends in a ground clamp. An *electrode holder* is a spring-loaded device with insulated handles used to grip a welding electrode, figure 25–9. Welding cables have a copper or aluminum core. This core is clamped or soldered to the metal part of the electrode holder. An insulated sleeve slides over the connection to protect the operator from electric shock. All other parts of the holder are insulated except the jaws.

A *ground clamp* is a spring-loaded clamp attached to an electrical cable, figure 25–10. It is not insulated and is not a shock hazard. It carries current between the welding table or project and to the welder.

Welding Table and Booth

A booth and metal table for welding are convenient and efficient, figure 25–11. The proper table permits the operator to stand or sit comfortably.

The booth must have curtains to protect other work-

Figure 25–9. Electrode holder (*courtesy of Lincoln Electric Company*)

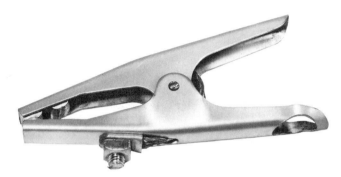

Figure 25–10. Ground clamp (*courtesy of Lincoln Electric Company*)

Figure 25–12. Curtains must cover the openings of a welding booth to protect workers in the area from light flashes.

ers from blinding light flashes, figure 25–12. The booth should also have panels for the storage of welding tools and equipment. Exhaust equipment must be provided to remove the fumes resulting from welding. The entire area must be fire-resistant and free of flammable materials.

Caution:

It is essential to have a fire blanket and appropriate fire extinguishers in the welding area.

Welding is often done on the floor of a shop or outdoors. In such situations, portable curtains must be set up to protect others in the area from welding flashes, figure 25–13. It is important that all combustible materials be removed before welding in such areas.

Equipment for Cleaning Welds

Welding areas should be equipped with a large grinder with a wire wheel. The grinder is used extensively

Figure 25–11. A welding booth and welding table are convenient and efficient.

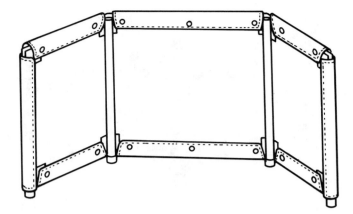

Figure 25–13. Portable curtains for use around welding done on the shop floor or outdoors

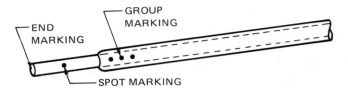

Figure 25–15. Locations on an electrode used in the color coding system of the National Electrical Manufacturers Association (NEMA)

in preparing metal for welding and the wire wheel is useful for cleaning beads after welding.

Tongs or pliers of various types are useful for handling hot metal. Hand wire brushes and chipping hammers are recommended for removing slag. A *chipping hammer* is a steel hammer with a sharp edge and/or point.

☐ ELECTRODES

Arc welding technology has developed to a high level. A wide variety of electrodes is available to enable a weldor to do many different jobs, including welding in all positions.

Welding positions include flat, horizontal, vertical up, vertical down, and overhead. (These are described in unit 26.) Welding in agricultural mechanics may call for any of these positions. Some electrodes are suitable for all-position welding. Others are only suitable for one or two positions. The parts of an electrode are shown in figure 25–14.

NEMA Color Coding

The *National Electrical Manufacturers Association* (*NEMA*) has developed a system to permit manufacturers to mark their electrodes by color codes. The markings are placed on three areas of an electrode: (1) the

exposed end of the metal rod, (2) the exposed surface of the metal rod, and (3) the flux near the exposed rod. The color markings are given specific names. Color on the end is called an *end marking;* color on the bare surface is called a *spot marking;* and colors on the flux make up the *group marking,* figure 25–15. Thus electrodes are identified by the colors of their end, spot, and group markings. Many manufacturers no longer use the NEMA system. Most manufacturers simply stamp the AWS classification number on each electrode.

AWS Numerical Coding

The *American Welding Society* (*AWS*) is an organization of individuals and agencies that support education in welding processes. The society has developed a numerical system for coding electrodes. The code condenses a lot of information into a four- or five-digit number for mild steel electrodes. The number is preceded by the letter "E" to indicate that the number describes an electrode.

The first two digits (or the first three if it is a five-digit number) refer to the tensile strength of one square inch of weld. *Tensile strength* refers to the amount of tension or pull the weld can withstand. The number represents thousands of pounds of tensile strength per square inch. For example, E60 stands for 60,000 pounds per square inch (psi) of tensile strength.

The two right-hand digits in the AWS number refer to the type of welding the electrode is capable of doing. The third digit (or fourth if it is a five-digit number) refers to the welding positions for which the electrode is suited. The fourth digit (or fifth if a five-digit number) refers to the depth of penetration and/or welding current, figures 25–16 and 25–17. For complete identification it is more useful to read the last two digits together as defined in tables available from the American Welding Society, figure 25–18.

Electrode Selection

All-purpose, all-position, AC/DC mild steel electrodes are the choice for most shops. Both the E6013 and the E6011 electrode fit this description.

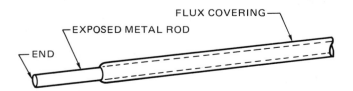

Figure 25–14. Parts of an arc welding electrode

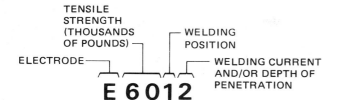

Figure 25–16. Components of a typical AWS classification number for electrodes. A five-digit AWS number is used if three digits are required to express the tensile strength. (*adapted from L. Jeffus and H. Johnson,* Welding Principles and Applications, *Delmar Publishers Inc., 1984*)

E __ __ 1 __	Usable in all directions
E __ __ 2 __	Usable in flat and horizontal positions only
E __ __ 4 __	Usable for vertical down only

THE SECOND DIGIT FROM THE RIGHT INDICATES WELDING POSITION

E __ __ __ 0	DC reverse polarity only
E __ __ __ 1	AC and DC reverse polarity
E __ __ __ 2	AC and DC straight polarity
E __ __ __ 3	AC and DC
E __ __ __ 4	AC and DC
E __ __ __ 5	DC reverse polarity
E __ __ __ 6	AC and DC reverse polarity
E __ __ __ 8	AC and DC reverse polarity

THE RIGHT-HAND DIGIT INDICATES TYPE OF WELDING

Figure 25–17. The last two digits in the AWS electrode numbering system specify the position of welding and the type of welding current.

The E6013 electrode is popular for beginning weldors. It is relatively inexpensive and easy to use. It deposits metal quickly and leaves an attractive bead. The slag is easy to remove. However, the E6013 electrode has shallow penetration. Therefore, it is recommended primarily for thin metal, ⅛ inch or less.

The E6011 electrode is perhaps the most commonly used in agricultural mechanics. It is not easy for the

ELECTRODE CLASSIFICATION

AWS CLASSIFICATION	TYPE OF COVERING	CAPABLE OF PRODUCING SATISFACTORY WELDS IN POSITION SHOWN[a]	TYPE OF CURRENT[b]
		E60 series electrodes	
E6010	High cellulose sodium	F, V, OH, H	DCEP
E6011	High cellulose potassium	F, V, OH, H	AC or DCEP
E6012	High titania sodium	F, V, OH, H	AC or DCEN
E6013	High titania potassium	F, V, OH, H	AC or DC, either polarity
E6020 ⎫	High iron oxide	H-fillets	AC or DCEN
E6022[c] ⎭	High iron oxide	F	AC or DC, either polarity
E6027	High iron oxide, iron powder	H-fillets, F	AC or DCEN
		E70 series electrodes	
E7014	Iron powder, titania	F, V, OH, H	AC or DC, either polarity
E7015	Low hydrogen sodium	F, V, OH, H	DCEP
E7016	Low hydrogen potassium	F, V, OH, H	AC or DCEP
E7018	Low hydrogen potassium, iron powder	F, V, OH, H	AC or DCEP
E7024	Iron powder, titania	H-fillets, F	AC or DC, either polarity
E7027	High iron oxide, iron powder	H-fillets, F	AC or DCEN
E7028	Low hydrogen potassium, iron powder	H-fillets, F	AC or DCEP
E7048	Low hydrogen potassium, iron powder	F, OH, H, V-down	AC or DCEP

a. The abbreviations, F, V, V-down, OH, H, and H-fillets indicate the welding positions as follows:

    F = Flat
    H = Horizontal
  H-fillets = Horizontal fillets
  V-down = Vertical down
    V = Vertical ⎱ ⎰ For electrodes ³⁄₁₆ in. (4.8 mm) and under, except ⁵⁄₃₂ in. (4.0 mm) and under for classifications
  OH = Overhead ⎰ ⎱ E7014, E7015, E7016, and E7018.

b. The term DCEP refers to direct current, electrode positive (DC reverse polarity). The term DCEN refers to direct current, electrode negative (DC straight polarity).

c. Electrodes of the E6022 classification are for single-pass welds.

Source: Reproduced from AWS A5.1-81, American Welding Society, with permission.

Figure 25–18. Specifications for covered carbon steel arc welding electrodes

beginner to make an attractive weld using the E6011. However, the deep penetration it provides generally results in a stronger weld than can be obtained with the E6013. Commercial welding shops can choose from many types and suppliers of electrodes, figure 25–19.

The experienced weldor needs at least a pound or more of additional electrodes in the shop. These include special electrodes for welding cast iron and stainless steel. Such electrodes are handy for specialized repair jobs.

Other useful electrodes include the hardsurfacing types. One type deposits material that resists soil abra-

sion. Layers deposited on parts such as plowshares, landsides, cultivator shovels, and bulldozer blades extend the life of the part. Another type of hardsurfacing electrode leaves a deposit on metal chains, drags, or bars to resist metal-to-metal wear.

Some shops have arc welders but do not have gas cutting and welding torches. Such shops may benefit from special arc welding electrodes designed for cutting and/or brazing. Additionally, the carbon arc torch may be a useful addition to the arc welder for heating. The *carbon arc torch* is a device which holds two carbon sticks and produces a flame from the energy of an electric welder.

See AWS A5.1-81, Specifications for Carbon
Steel Covered Arc Welding Electrodes

MANUFACTURERS	AWS CLASSIFICATION	E6010	E6011	E6012	E6013	E6020	E6022	E6027
Aga de Mexico, S.A.		AGA C10 C12	AGA C11	AGA R12	AGA R10 R11	—	—	—
Airco Welding Products		AIRCO 6010	AIRCO 6010 6011C 6011L0C	AIRCO 6012 6012C	AIRCO 6013 6013C 6013D	AIRCO 6020	—	EASY ARC 6027
Air Products and Chemicals, Inc.		AP6010W	AP6011W	AP6012W	AP6013W	—	—	—
A-L Welding Products, Inc.		AL6010	AL6011	AL6012	AL6013	—	—	—
Alloy Rods, Allegheny International, Inc.		AP100 SW610	SW14	SW612	SW15	—	—	—
Applied Power, Inc.		—	No. 130 Red-Rod	—	No. 140 Production Rod	—	—	—
Arcweld Products Limited		Easyarc 10	Arcweld 230 Easyarc 11	Arcweld 387	Arcweld 90 Satinarc	—	—	—
Bohler Bros. of America, Inc.		Fox CEL	—	—	Fox OHV Fox ETI	Fox UMZ	—	—
Brite-Weld		Brite-Weld E6010	Brite-Weld E6011	Brite-Weld E6012	Brite-Weld E6013	—	—	Brite-Weld E6027
Canadian Liquid Air Ltd.		LA6010	LA6011P	LA6012P	LA6013 LA6013P	—	—	—
Canadian Rockweld Ltd.		R60	R61	R62	R63 R63A	R620	—	R627
C-E Power Systems, Combustion Engineering, Inc.		—	—	—	—	—	—	—
Century Mfg. Co.		—	331 324	—	331 313	—	—	—
Champion Hobart, S.A. de C.V.		6010	6011	6012 Ducto P60	6013 Versa-T	—	—	—
CONARCO, Alambres y Soldaduras, S.A.		CONARCO 10 CONARCO 10P	CONARCO 11	CONARCO 12 CONARCO 12D	CONARCO 13A CONARCO 13	—	—	—
Cronatron Welding Systems, Inc.		—	Cronatron 6011	—	Cronatron 6013	—	—	—
Electromanufacturas, S.A.		West Arco XL-610M ZIP 10T	West Arco ACP611	West Arco FP-612 ZIP 12	West Arco SW-613M SUPER SW-613 SW-10	—	—	West Arco ZIP 27
ESAB		OK 22.45	OK 22.65	—	OK 46.00 OK 43.32 OK 50.10 OK 50.40	—	—	—

Figure 25–19. Carbon steel covered arc welding electrodes (*reproduced from AWS A5.0–83, American Welding Society, with permission*)

MANUFACTURERS	AWS CLASSIFICATION	E6010	E6011	E6012	E6013	E6020	E6022	E6027
Eureka Welding Alloys		Eureka 6010	Eureka 6011	Eureka 6012	Eureka 6013	—	—	Eureka 6027
Eutectic Corporation		—	—	—	DynaTrode 666	—	—	—
Hobart Brothers Company		Hobart 10 60AP	Hobart 335A	Hobart 12 12A	Hobart 413 447A		Hobart 1139	Hobart 27H
International Welding Products, Inc.		INWELD 6010	INWELD 6011	INWELD 6012	INWELD 6013	INWELD 6020	—	INWELD 6027
Kobe Steel, Ltd.		KOBE 6010	KOBE 6011	KOBE TB-62 RB-26D	ZERODE 44 RB-26 B-33	—	—	ZERODE 27 AUTOCON 27, B-27
Latamer Company, Inc.		Latco E6010	Latco E6011	Latco E6012	Latco E6013	Latco E6020	—	Latco E6027
The Lincoln Electric Company		Fleetweld 5P	Fleetweld 35 35LS 180	Fleetweld 7	Fleetweld 37 57	—	—	Jetweld 2
Liquid Carbonic, Inc.		RD 704D RD 610P	RD 504D RD 611P	RD 604	RD 613	—	—	—
Murex Welding Products		SPEEDEX 610	Type A 611C 611LV	Type N 13 GENEX M	Type U U13	F.H.P.	—	SPEEDEX 27
Oerlikon, Inc.		OERLIKON E6010	—	OERLIKON E6012	OERLIKON E6013	OERLIKON E6020	—	—
Sodel		—	SODEL 11	—	SODEL 31	—	—	—
Techalloy Maryland, Inc. Reid-Avery Div.		RACO 6010	RACO 6011	RACO 6012	RACO 6013	RACO 6020	—	RACO 6027
Teledyne Canada HARFAC		6010	6011	6012	6013	—	—	—
Teledyne McKay		6010	6011	6012	6013	—	—	—
Thyssen Draht AG		Thyssen Cel 70	—	Thyssen Blau Thyssen A5 Thyssen Grun	Thyssen Grun T SH Blau SH Gelb B SH Gelb T Sh Gelb S SH Gelb R SH Grun TB SH Lila R Union 6013	SH Gelb	SH Tiefbrand	—
WASAWELD		WASAWELD E6010	WASAWELD 6011	WASAWELD E6012	WASAWELD E6013	WASAWELD E6020	—	WASAWELD E6027
Weld Mold Company		—	—	—	—	—	—	—
Weldwire Co., Inc.		WELDWIRE 6010	WELDWIRE 6011	WELDWIRE 6012	WELDWIRE 6013	WELDWIRE 6020	WELDWIRE 6022	WELDWIRE 6027
Westinghouse Electric Corporation		ZIP 10 XL 610A	ACP ZIP-11R	ZIP 12	SW	DH 620	—	ZIP 27 ZIP 27M

MANUFACTURERS	AWS CLASSIFICATION	E7014	E7015	E7016	E7016-1	E7018	E7018-1	E7024
Aga de Mexico, S.A.		—	—	—	—	AGA B10	—	AGA RH10
Airco Welding Products		EASY ARC 7014	—	7016 7016M	—	EASY ARC 7018C 7018MR CODE ARC 7018MR	—	EASY ARC 7024 7024D
Air Products and Chemicals, Inc.		AP7014W	—	AP7016W	—	AP7018W	—	AP7024W
A-L Welding Products, Inc.		AL7014	—	—	—	AL7018 Nuclearc 7018	—	AL7024
Alloy Rods, Alleghany International, Inc.		SW15IP	—	70LA-2	Atom Arc 7016-1	Atom Arc 7018 SW-47	Atom Arc 7018-1	7024
Applied Power, Inc.		No. 146 Hy-Pro-Rod	—	—	—	No. 7018 Marq-Rod	—	—

Figure 25–19. Carbon steel covered arc welding electrodes (Cont'd.)

MANUFACTURERS	AWS CLASSIFICATION	E7014	E7015	E7016	E7016-1	E7018	E7018-1	E7024
Arcweld Products Limited		Easyarc 14	—	Arcweld 312	—	Easyarc 328 Easyarc 7018MR	—	Easyarc 12
Bohler Brothers of America, Inc.		—	—	Fox EV47	—	Fox EV50	—	Fox HL 180 Ti
Brite-Weld		Brite-Weld 7014	—	—	—	Brite-Weld E7018	—	Brite-Weld E7024
Canadian Liquid Air, Ltd.		LA7014	—	—	—	Super Arc 18 LA7018, AA7018, LA7018B	—	LA7024 LA24HD
Canadian Rockweld, Ltd.		R74	—	Tensilarc 76	—	Hyloarc 78	—	R724
C-E Power Systems, Combustion Engineering, Inc.		—	—	—	—	CE7018	—	—
Century Mfg. Co.		331 363	—	—	—	331 327	—	—
Champion Hobart, S.A. de C.V.		·	—	724	—	7018	718	MULTI-T
CONARCO, Alambres y Soldaduras, S.A.		CONARCO 14	CONARCO 15	CONARCO 16	—	CONARCO 18	CONARCO 18-1	CONARCO 24
Cronatron Welding Systems, Inc.		—	—	—	—	Cronatron 7018	—	—
Electromanufactures, S.A.		West Arco ZIP 14	—	West Arco WIZ 16	—	West Arco WIZ 18	—	West Arco ZIP 24
ESAB		OK 46.16	—	OK 53.00 OK 53.05	OK 53.68	OK 48.00 OK 48.04 OK 48.15	OK 48.68 OK 55.0	OK Femax 33.65 OK Femax 33.80
Eureka Welding Alloys		Eureka 7014	—	—	—	Eureka 7018	—	Eureka 7024
Eutectic Corporation		—	—	—	—	EutecTrode 7018	—	—
Hobart Brothers Company		Hobart 14A	—	—	—	Hobart 718 718LMP	—	Hobart 24 24H
International Welding Products, Inc.		INWELD 7014	INWELD 7015	INWELD 7016	—	INWELD 7018	—	INWELD 7024
Kobe Steel, Ltd.		KOBE RB-14	—	KOBE LB-52 LB-52U LB-26Vu ZERODE-52	—	KOBE LB-52-18 LTB-52A	—	KOBE RB-24 ZERODE 50F FB-24
Latamer Company, Inc.		Latco E7014	Latco E7015	Latco E7016	—	Latco E7018	—	Latco E7024
The Lincoln Electric Company		Fleetweld 47	—	—	—	Jetweld LH70 LH 73 LH 78 LH 75	Jetweld LH75	Jetweld 1 3
Liquid Carbonic, Inc.		RD 714	—	—	—	RD 718	—	RD 724
Murex Welding Products		SPEEDEX U	—	HTS HTS-18 HTS-180	—	SPEEDEX HTS-MR HTS-M-MR-718	—	SPEEDEX 24 24D
Oerlikon, Inc.		—	—	—	OERLIKON Spezials Extra Tenacito WZ	OERLIKON E7018	OERLIKON E7018-1	OERLIKON Ferromatic Ferrocito R
Sodel		—	—	—	—	—	SODEL 328	SODEL 314
Techalloy Maryland, Inc. Reid-Avery Div.		RACO 7014	RACO 7015	RACO 7016	—	RACO 7018	RACO 7018-1	RACO 7024
Teledyne Canada HARFAC		7014	—	7016	—	—	7018-1	7024
Teledyne McKay		7014	—	7016	—	7018 XLM	7018-1 XLM	7024

Figure 25-19. Carbon steel covered arc welding electrodes (Cont'd.)

MANUFACTURERS	AWS CLASSIFICATION	E7014	E7015	E7016	E7016-1	E7018	E7018-1	E7024
Thyssen Draht AG		SH Multifer 130	Thyssen K50 SH Grun K45 (6015)	SH Kb F Thyssen K50R SH Grun K50W Thyssen Kb Spezial SH Grun K70W Thyssen K90S	—	SH Grun K70	Thyssen 120K	Thyssen Rot R160 Thyssen Rot R160S SH Multifer 180 Thyssen Rot AR160
WASAWELD		WASAWELD E7014	—	—	—	WASAWELD E7018	—	WASAWELD E7024
Weld Mold Company		—	—	—	—	WELD MOLD 7018	—	—
Weldwire Co., Inc.		WELDWIRE 7014	WELDWIRE 7015	WELDWIRE 7016	WELDWIRE 7016-1	WELDWIRE 7018	WELDWIRE 7018-1	WELDWIRE 7024
Westinghouse Electric Corporation		ZIP 14	—	LOH 2	—	WIZ 18	—	ZIP 24

MANUFACTURERS	AWS CLASSIFICATION	E7024-1	E7027	E7028	E7048
Aga de Mexico, S.A.		—	—	—	—
Airco Welding Products		—	—	EASY ARC 7028	—
Air Products and Chemicals, Inc.		—	—	—	—
A-L Welding Products, Inc.		—	—	—	—
Alloy Rods, Allegheny International, Inc.		7024-1	—	—	—
Applied Power, Inc.		—	—	—	—
Arcweld Products Limited		—	—	Super 28	—
Bohler Bros. of America, Inc.		—	—	Fox HL 180 Kb	—
Brite-Weld		—	—	—	—
Canadian Liquid Air, Ltd.		LA7024	—	LA7028 LA7028B	LA7048B
Canadian Rockweld, Ltd.		—	R727	Hyloarc 728	—
C-E Power Systems, Combustion Engineering, Inc.		—	—	—	—
Century Mfg. Co.		—	—	—	—
Champion Hobart, S.A. de C.V.		—	—	—	—
CONARCO, Alambras y Soldaduras, S.A.		—	—	CONARCO 28	CONARCO 48
Cronatron Welding Systems, Inc.		—	—	—	—
Electromanufacturas, S.A.		—	—	—	—
ESAB		—	—	OK 38.48 OK 38.65 OK 38.85 OK 38.95	OK 53.35
Eureka Welding Alloys		—	—	—	—
Eutectic Corporation		—	—	—	—
Hobart Brothers Company		—	—	Hobart 728	—
International Welding Products, Inc.		—	—	INWELD 7028	—
Kobe Steel, Ltd.		—	—	—	KOBE LB-26V LB-52V ZERODE 6V
Latamer Company, Inc.		—	—	Latco E7028	—
The Lincoln Electric Company		Jetweld 1	—	Jetweld LH3800	—
Liquid Carbonic, Inc.		—	—	RD 728	—

Figure 25–19. Carbon steel covered arc welding electrodes (Cont'd.)

MANUFACTURERS	AWS CLASSIFICATION	E7024-1	E7027	E7028	E7048
Murex Welding Products		—	—	SPEEDEX 28	—
Oerlikon, Inc.		—	—	OERLIKON E7028	—
Sodel		—	—	—	—
Techalloy Maryland, Inc. Reid-Avery Div.		—	—	RACO 7028	RACO 7048
Teledyne Canada HARFAC		—	—	—	—
Teledyne McKay		—	—	—	—
Thyssen Draht AG		—	SH Multifer 200	SH Multifer 150 K11 Thyssen Rot BR160	—
WASAWELD		—	—	—	—
Weld Mold Company		—	—	—	—
Weldwire Co., Inc.		WELDWIRE 7024-1	WELDWIRE 7027	WELDWIRE 7028	WELDWIRE 7048
Westinghouse Electric Corporation		—	—	—	—

Source: Reproduced from AWS A5.0-83, American Welding Society, with permission.

Figure 25–19. Carbon steel covered arc welding electrodes (Cont'd.)

□ **STUDENT ACTIVITIES**

1. Define the new terms italicized in this unit.
2. Examine the welders in your school agricultural mechanics shop. Classify them according to AC or DC type and determine if the DC welders are motor driven or engine driven.
3. Examine the electrodes used in your agricultural mechanics shop. Is there an AWS code stamped on the flux? If not, ask your instructor what the color markings mean.
4. Identify the tools and equipment in the welding area of your shop.

□ **SELF-EVALUATION**

A. Multiple Choice. Select the best answer.
1. Voltage is a measure of
   a. rate of current
   b. electrical pressure
   c. available energy
   d. all of these
2. A welder that gets its energy directly from a utility power plant is
   a. an alternator
   b. a generator
   c. a rectifier
   d. a transformer

3. When welding, the flux on an electrode
   a. forms a gas
   b. creates slag
   c. shields the weld
   d. all of these
4. Shielded metal arc welding is also called
   a. brazing
   b. electrode welding
   c. oxyacetylene welding
   d. stick welding
5. Suitable arc welders for farm use are
   a. fairly easy to use
   b. relatively inexpensive
   c. reliable
   d. all of these
6. The output of a welder is relatively
   a. low voltage and high amperage
   b. high voltage and low amperage
   c. high voltage and high amperage
   d. low voltage and low amperage
7. Types of polarity on DC welders include
   a. reversed
   b. straight
   c. both of these
   d. none of these
8. The proportion of time that a welder can operate without overheating is known as its
   a. AC/DC
   b. AWS classification
   c. duty cycle
   d. voltage drop

9. A chipping hammer is used to
   a. prepare edges for welding
   b. remove scale from steel
   c. remove slag
   d. temper beads
10. An organization of people and agencies interested in promoting welding is the
    a. AWS
    b. EPA
    c. SAE
    d. WPA

B. Completion. Fill in the blanks with the word or words that will make the following statements correct.

1. The color markings on an electrode will be
   _____, _____, or _____
   markings.
2. The E in an electrode code stands for
   _____.
3. In the American Welding Society electrode code, the first two (or three) numbers refer to _____; the next number refers to _____; and the final number refers to _____.
4. Three types of welding electrodes other than those used for welding mild steel are those used for _____, _____, and _____.

Unit 26
Arc Welding Mild Steel

Objective

To use arc welding equipment and procedures in cutting and welding mild steel.

Competencies to Be Developed

After studying this unit, you will be able to:
- Use safety equipment and protective clothing for arc welding.
- Strike an arc and run beads.
- Make butt and fillet welds.
- Make flat, horizontal, and vertical welds.
- Weld pipe.
- Pierce holes and cut with electrodes.

Terms to Know

pad
stringer bead
weaving
crater
pass
root pass
vertical down
vertical up
spot welding
spot welder

Materials List

Welding booth
Electric welder, 180 amperes or larger
Protective clothing: welding helmet, clear goggles, leather gloves, welding jacket, leather apron, and leather leggings
Chipping hammer
Pliers
Wire brush
Bucket with water
Fire extinguisher and fire blanket

Welding pads, steel, ¼″ × 2″ × 6″, ¼″ × 4″ × 6″, ⅛″ × 1″ × 6″
Grinder with a coarse abrasive wheel and a wire wheel
Soapstone markers
Electrodes, ⅛-inch E6011 and E6013

The term "arc welding" as used in this unit refers to shielded metal arc welding (SMAW). In this type of welding, an electric welder and flux-covered electrodes are used. Before starting to weld, the operator must be thoroughly familiar with shop safety and fire prevention.

□ SAFETY PROCEDURES

Fire Protection

The temperature of the electric arc (about 9000°F) creates a very real danger of burns as well as fires. Never-

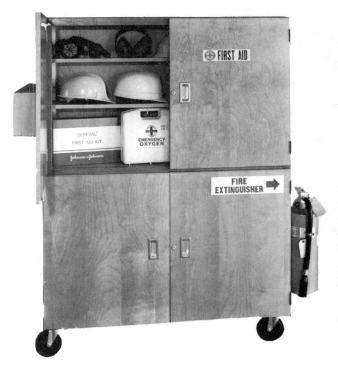

Figure 26–1. Fire extinguishers suitable for Class A, B, and C fires, safety equipment, and a first-aid kit should be within easy reach of every weldor.

theless, with reasonable care and use of the proper equipment, the welding area is no more dangerous than other areas of the shop.

The welding area should be equipped with metal benches. These serve both for fire protection and to electrically ground the work. Welding booths should be constructed of fireproof or fire-resistant materials, such as metal sheets or concrete block. Other materials such as pressed wood panels, plywood, and special fire-resistant canvas are also used.

Fire extinguishers suitable for Class A, B, and C fires, safety equipment, and a first aid kit should be within easy reach at all times, figure 26–1. A wool fire blanket is another important piece of fire control equipment, figure 26–2. If human hair or clothing catches on fire, the fire blanket is used to wrap the victim and smother the fire.

Buckets of water are frequently used to receive sparks and cool metal in the welding areas. Cooling of metal used for welding practice reduces the chances of personal burns. The water may also be useful in extinguishing accidental fires when electric shock is not a hazard. However, any water spilled on the floor of the work area must be removed because it will create an electric shock hazard.

All grease, oil, sawdust, paper, rags, and other flammable materials must be removed from areas where welding is done. Good housekeeping is a major factor in reducing fire hazards.

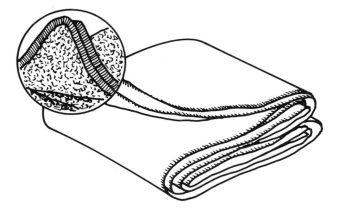

Figure 26–2. A wool fire blanket is the first line of defense if clothing catches on fire. The victim is wrapped in the blanket to smother the fire quickly.

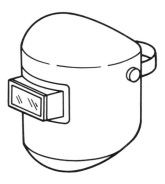

Figure 26–3. A welding helmet with a No. 10 shaded lens protects the eyes from damaging welding light rays

Personal Protection

Caution:

The human eye must not be exposed to direct light from a welding arc. The welding arc contains light rays that burn, even from a distance. Fortunately, it is easy to protect the eyes from these rays. Any material that blocks the light also stops the damaging rays. The weldor's face and eyes are protected by using a face shield with a dark viewing glass. The glass must be classified as a No. 10 shade, figure 26–3.

Figure 26–4. Some welding helmets have a flip-up shaded lens. When this lens is up, the weldor can see through the clear lens which backs up the shaded lens for nonwelding operations (*courtesy of J. I. Scott Company*).

The No. 10 shade lens is too dark to see through except when the welding arc is burning. Therefore, the operator must either wear safety glasses under

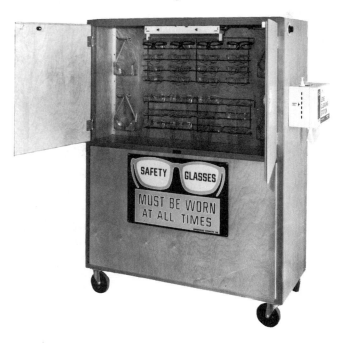

Figure 26–5. When using a welding helmet that does not have a flip-up shaded lens, the weldor should wear standard safety glasses or goggles under the helmet.

the helmet or use a helmet with a flip-up lens for processes such as chipping, figure 26–4. When the flip-up lens helmet is not used, clear safety glasses *must* be worn under the helmet. When not welding, the operator may exchange the helmet for another type of clear face and eye protective covering, figure 26–5. Prescription eye glasses are not approved safety glasses unless ordered as such.

> **Caution:**
>
> **Attempting to chip hot slag from welds without eye protection is likely to result in eye injury.**

Fire-resistant coveralls and high leather shoes are recommended as standard clothing when welding. In addition, leather gloves are needed to protect hands and wrists. All skin areas must be covered or "sunburning" will occur from the rays of the arc. For many welding operations, protective clothing such as a leather apron, sleeves, jacket, or pants may be needed to protect the body from sparks and hot metal, figure 26–6.

☐ SETTING UP

When preparing to weld, the appropriate electrode must be selected. For the beginning weldor this is likely to be an E6011 or E6013 electrode. As described in the preceding unit, the best single choice is probably the E6011. The E6013 is easier to use but is recommended primarily for thin metal.

The next step is to check the welding area to eliminate fire hazards. Gather all the necessary materials, including a piece of ¼″ × 4″ × 6″ plate steel for practice, wire brush, chipping hammer, pliers to hold hot metal, and bucket of water to cool the metal. The practice metal is called a *pad*. It may be thicker and larger than specified, if desired. To improve handling, an electrode may be bent into a U shape and the ends welded to the pad to provide an insulated handle.

The welding machine should have suitable welding cables. One cable will end in an electrode holder and the other in a ground clamp. The metal pad to be used for practice is clamped to the welding table using the ground clamp, a vise-grip, or a spring clamp. When an extra clamp is used, the ground clamp is attached to either the practice metal or the table.

The operator must wear a welding helmet or use a face shield with a No. 10 lens. It is recommended that tight-fitting coveralls, leather shoes, and leather gloves be worn. Pants must not have cuffs as these will catch burning steel in the form of sparks.

Once dressed, the operator is ready to set the welder. Most welders have a rotary switch or lever that moves a pointer to the desired ampere setting. On other welders there is a selection of sockets into which the cables can be plugged. The desired amperage setting is obtained by plugging the cables into the correct pair of sockets. For practice welding, electrodes ⅛ inch or ⁵⁄₃₂ inch in diameter are to be used. Appropriate welder settings range between 70 and 220 amperes, depending upon the welder and the electrode selected, figure 26–7. A recommended starting setting is 125 amperes for ⅛-inch E6011 or E6013 electrodes. The amperage must then be adjusted according to the individual machine.

☐ STRIKING THE ARC

Before striking an arc the metal should be placed on the welding table. It should be located so the operator can reach across the metal and weld in a comfortable position. The metal should be clean and free of grease,

Figure 26–6. Certain welding jobs require the operator to wear leather gloves and other protective leather clothing such as an apron, sleeves, jacket, and/or pants.

oil, or rust. The following procedure is suggested for striking an arc.

Procedure

1. Place a ⅛-inch E6011 electrode in a 90° position in the electrode holder.
2. Close all curtains in the welding booth.
3. Warn others that you are about to start welding by saying "Cover up!" This means they should place shielding in front of their eyes if they can see your work.
4. Turn on the welder.

5. Position the tip of the electrode about ¼ inch above the practice metal and lean it slightly in the direction you plan to move. Right-handed weldors move from left to right.
6. Drop the face shield or cover lens over your eyes.
7. Strike the arc by quickly lowering, touching, and lifting the electrode, figure 26–8. The action is similar to striking a match. The lift should only be about ⅛ inch.
   a. If you do not lift the electrode in time it will stick to the metal. If it does, use a quick whipping action to free it or release the electrode holder.

ELECTRODE SIZE	CLASSIFICATION					
	E6010	E6011	E6012	E6013	E7016	E7018
³⁄₃₂ in	40–80	50–70	40–90	40–85	75–105	70–110
⅛ in	70–130	85–125	75–130	70–125	100–150	90–165
⁵⁄₃₂ in	110–165	130–160	120–200	130–160	140–190	125–220

Figure 26–7. Welding amperage ranges for various electrode classifications (*courtesy of L. Jeffus and H. Johnson,* Welding Principles and Applications, *Delmar Publishers Inc., 1984*)

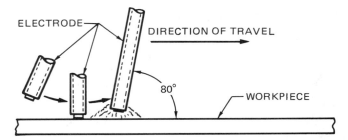

Figure 26–8. Striking the arc

Caution:

Do not turn the welder off with the electrode frozen in the circuit as the welder would be damaged. First, release the electrode holder from the electrode.

b. If the flux becomes cracked or broken on the end of the rod, lay the electrode aside for a more experienced welder to use. Obtain another electrode for your next attempt to strike an arc.

c. If the arc goes out, try again but keep the electrode closer to the metal after lifting it.

8. Feed the electrode very slowly as the weld metal burns away, keeping it about ⅛ inch from the metal. As you feed the electrode, move very slowly and evenly across the plate (generally from left to right for right-handed weldors).

9. Lift the electrode slowly to break the arc after traveling about 1 inch.
Note: As you practice this technique, strike the arc and hold a slightly longer arc for a second or two before lowering the rod and moving across the metal. This practice preheats the metal and results in a better bead.

10. Continue practicing steps 5 through 9 until you can start and stop the arc at will. When you stop the bead, lift the electrode slowly to permit the crater to fill. The *crater* is a low spot in metal where the force of a flame has pushed out molten metal.

11. Make 16 strikes with 1-inch beads. Compare your results with figure 26–9. Discuss your welds with your instructor.

The method described is called the scratch method of striking an arc. Some people find it easier simply to touch the metal and raise the electrode. This is known as the tapping method. Both methods are acceptable. Striking an arc and keeping it going takes a lot of practice. However, it is well worth the time

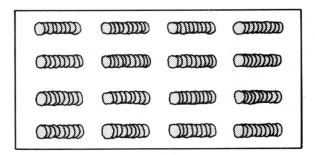

Figure 26–9. Practice striking and maintaining an arc until you can make a short bead of ½ to 1 inch long. The practice beads should be made in parallel lines across the metal (*adapted from L. Jeffus and H. Johnson,* Welding Principles and Applications, *Delmar Publishers Inc., 1984*).

and patience required, since it is the starting point for arc welding.

□ RUNNING BEADS

A bead is produced by handling the electrode in a way that results in a proper mix of filler and base metal. A bead is called a *stringer bead* if it is made without weaving. *Weaving* means moving the electrode back and forth sideways to create a bead that is wider than the electrode would normally make. After learning to strike and maintain an arc, the student should practice making stringer beads.

Making stringer beads helps the student learn to:

• start a bead at the desired location
• hold the electrode at about an 80° angle
• feed the electrode at an even rate
• move across the metal in a straight line
• move across the metal at an even rate of speed
• stop or terminate a weld at the desired spot and in an acceptable manner

Most of these skills must be developed at the same time. This means that attention to technique is important. It is helpful if someone watches the student practice who can make suggestions for correcting the student's technique.

Angle of the Electrode

Experienced weldors vary the angle of the electrode according to the electrode being used and the job. However, the beginner should lean the electrode slightly in the direction of travel. A 75° to 80° angle, 10° to 15° from the vertical (straight up) position, is suggested.

Arc Length

The arc length indicates the skill with which the electrode is fed as it burns away. The sound of the arc is a good guide. An arc that is the correct length sounds like bacon frying. If the arc is too long, a booing sound is heard. If the arc is too short, the electrode sticks to the metal.

Movement Across the Metal

The ability to weld in the intended direction is not easy. Looking through a dark lens, it is difficult to see anything except a white soapstone mark, a groove in the metal or another welded bead, and the molten puddle. However, the arc lights the surrounding area. With experience, it becomes possible to observe the welding process with ease.

Speed

To achieve a uniform rate of travel across the metal requires practice in arm movement. It requires the weldor to be in a comfortable position that permits such movement. The best way to control the movement is to watch the welded metal solidifying behind the puddle. When the electrode is moving at an even rate of speed, the weld metal forms evenly spaced semicircles behind the puddle.

Amperage Setting

The correct amperage can be obtained by observing the welding process and the weld that results. If the amperage is too low it will be difficult to strike the arc and keep it running. Low amperage results in a

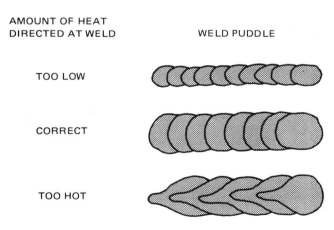

Figure 26–10. The amount of heat directed on the welding puddle affects the shape of the puddle (*courtesy of L. Jeffus and H. Johnson*, Welding Principles and Applications, *Delmar Publishers Inc., 1984*)

narrow, stringy bead. If the arc seems to struggle or start and stop, it is advisable to try a higher amperage. Increase the setting by 10 to 15 amperes and try again.

Learn to observe the shape of the puddle as it is forming. Long, narrow semicircles indicate that the heat is too low. Semicircles that are wider than they are long indicate that the correct heat is being used. Gourd-shaped marks indicate that the heat is too high, figure 26–10.

Other symptoms or indicators of excessive heat are

* the electrode covering turns brown
* the bead does not have clear markings
* the puddle burns through the plate
* it is difficult to manage the puddle
* the arc is very noisy

When these conditions exist, decrease the current setting by 10 to 15 amperes and try again.

Making a Practice Bead

To practice running beads it is best to make short beads about ¾ inch apart on the practice pad. This is far enough apart to evaluate the individual beads and also permit most efficient use of the metal.

It is advisable to use both hands to hold the electrode holder, or use the other hand to steady the wrist of the hand holding the electrode holder. It is important to wear proper protective clothing so that hot slag cannot reach the skin and a comfortable position can be maintained, figure 26–11. To practice running beads the following procedure is suggested.

Procedure

1. Clamp a ¼″ × 4″ × 6″ welding pad in a suitable position on the table.
2. Set the welder at 110 amperes and turn on the welder.
3. Insert a ⅛-inch E6011 electrode in the holder at 90°.
4. Cover up and remind others to do so too.
5. Strike the arc and quickly move the electrode to the starting point near the edge of the pad.
6. Watch the puddle, feed the electrode, move from left to right across the pad, observe the angle of the electrode, and concentrate on coordinating your movements.
7. At the end of the bead near the edge of the plate, lift the electrode gradually until the crater fills and the arc goes out.
8. Put on clear goggles or a clear face shield (if you are not wearing safety glasses or goggles

under your helmet) and remove the slag by striking the edges of the bead with a chipping hammer.

9. Compare your results with a welding chart or sample beads provided by the teacher, figure 26–12.

10. Repeat the process until you can strike the arc and run a good bead over and over again. Readjust the amperage setting if necessary. As the pad accumulates heat from welding, the amperage may need to be lowered to prevent overheating of the electrode and metal.

Weaving

After learning to run stringer beads, it is useful to practice running wider beads by weaving. Many patterns of weaving are used by experienced weldors to help control the heat in the puddle. These patterns are generally practiced as welding in different positions is learned. The beginning weldor should use a circular pattern, which provides good control of the bead width, figure 26–13.

□ WELDING JOINTS

The object of welding is generally to fasten metal pieces together. There are many types of joints, but most are variations of the butt joint and fillet joint. In a butt joint the pieces are placed end to end or edge to edge. In a fillet joint the two parts come together to form a 90° angle.

Butt Welds

Butt welds require the proper preparation. There must be a gap between the pieces to be welded of about the thickness of the electrode core. If the metal is thicker than the electrode core, it must be ground down. Grinding may be done on one or both sides of the metal. Grinding on both sides will produce stronger welds. The thickness of the unground metal should be no more than the diameter of the electrode core. For example, when welding with a ⅛-inch electrode, the top and/or bottom of the metal is ground at a 30° angle to leave a ⅛-inch thickness, figures 26–14 and 26–15.

When welding, one bead or layer of filler metal is called a *pass*, figure 26–16. The first pass made in a joint is called the *root pass*. The root pass is the most important pass in a weld.

When making a butt weld, the following procedure is suggested.

Figure 26–11. This student is wearing the proper protective clothing and is ready to weld in any position.

Procedure

1. Obtain two pieces of ¼″ × 2″ × 6″ flat steel.
2. Grind one edge of each piece so the matching surfaces are ⅛ inch thick.
3. Position the two pieces ⅛ inch apart.
4. Set the welder at 110 amperes and obtain ⅛-inch E6011 electrodes.
5. Wear protective clothing.
6. Tack weld each end of the joint with about ¼ inch of weld.
7. Straighten the joint so the two pieces are flat.
8. Strike the arc and keep a long arc as you move to the edge of the metal. Then shorten the arc to the correct length and proceed to weld. Rotate the electrode in a circular pattern to create an even bead about ¼ inch wide.
9. Chip the weld to remove all slag. Use a chipping hammer blade or point and a stiff wire brush to remove the slag.
10. Examine the weld for evenness and penetration. The answer should be "yes" to the following questions.
    a. Do the sides blend in evenly with the base metal?

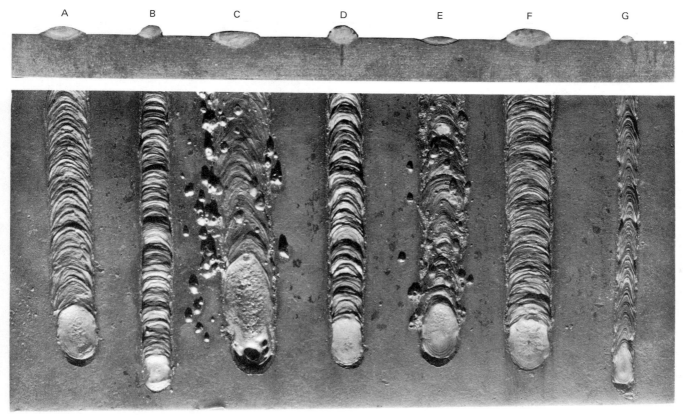

A. PROPER CURRENT,
   TRAVEL SPEED AND
   ARC LENGTH

B. CURRENT TOO LOW

C. CURRENT TOO HIGH

D. ARC LENGTH TOO SHORT

E. ARC LENGTH TOO LONG

F. TRAVEL SPEED TOO SLOW

G. TRAVEL SPEED TOO FAST

Figure 26–12. A careful study of welding beads will reveal the errors made while welding. In this way, the weldor can correct techniques and develop proficiency (*courtesy of Lincoln Electric Company*).

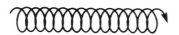

THIS WEAVE PATTERN RESULTS
IN A NARROW BEAD WITH
DEEP PENETRATION.

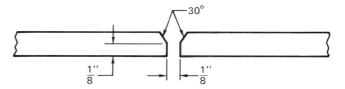

THIS WEAVE PATTERN RESULTS
IN A WIDE BEAD WITH
SHALLOW PENETRATION.

Figure 26–13. Weaving patterns are used to create wider beads and help control the heat when welding (*courtesy of L. Jeffus and H. Johnson,* Welding Principles and Applications, *Delmar Publishers Inc., 1984*).

b. Are all the semicircles of the bead evenly spaced?

c. Does the weld go all the way through to the bottom?

d. Does the weld start at the starting edge?

e. Does the weld fill the groove at the starting edge?

f. Does the weld go all the way to the finishing edge?

g. Does the weld fill the groove at the finishing edge?

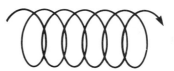

Figure 26–14. Before welding a butt joint, the metal pieces are ground and properly positioned. When welding with a ⅛-inch electrode, the top and/or bottom of the metal is ground at a 30° angle to leave a ⅛-inch thickness for the root weld.

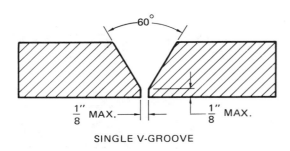

SINGLE V-GROOVE

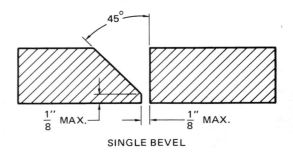

SINGLE BEVEL

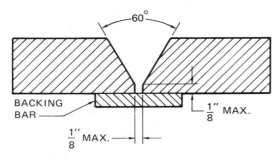

SINGLE V-GROOVE WITH BACKING

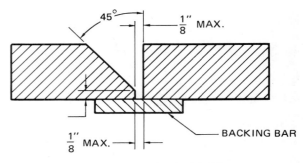

SINGLE BEVEL WITH BACKING

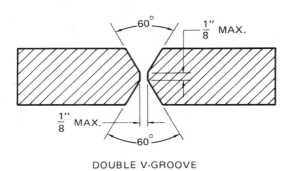

DOUBLE V-GROOVE

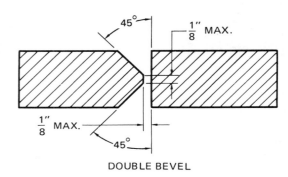

DOUBLE BEVEL

Figure 26–15. Typical butt joint preparations (*courtesy of L. Jeffus and H. Johnson,* Welding Principles and Applications, *Delmar Publishers Inc., 1984*)

11. If the bead does not fill the gap, run one additional wide bead over the first bead, figure 26–17. If thicker metal is used, two or more beads may be needed to complete the weld, figure 26–18.

Prepare another butt weld, correcting any problems observed in the first weld. To make an internal check, the procedure is as follows.

Procedure

1. Saw 1 inch off the end of the welded piece.
2. Examine the cross section of the weld. Voids that are open or filled with slag indicate weak parts of the weld, figure 26–19.

Caution:

All **slag must be removed between beads to assure a solid weld with no voids. Voids are caused when slag is trapped inside the weld.**

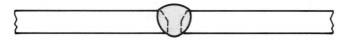

Figure 26–16. Butt weld using one heavy pass

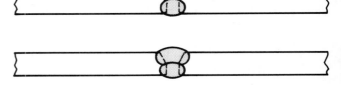

Figure 26–17. Butt weld using two passes

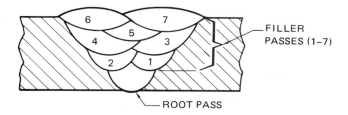

Figure 26–18. Butt weld using many passes (*courtesy of L. Jeffus and H. Johnson,* Welding Principles and Applications, *Delmar Publishers Inc., 1984*)

3. Saw off another inch-long section of the weld.
4. Clamp one side of the 1-inch section in a heavy vise with the welded section exposed.
5. Bend the welded section by driving the exposed piece with a heavy hammer. The welded section should be as tough and strong as the rest of the metal. If the welded section breaks, determine what steps are needed to improve the weld.

Fillet Welds

Most of the principles and procedures that apply to butt welds also apply to fillet welds. To practice making fillet welds, prepare the piece that will be vertical so that the weld metal will fuse both pieces completely. It is not necessary to grind ⅛-inch thick metal. However, thicker pieces need to have one or both edges beveled to permit welds to penetrate the entire thickness. A ⅛-inch electrode should penetrate about ⅛ inch deep. Therefore, metal that is thicker than ⅛ inch needs to be ground on one or both sides.

The recommended procedure for making a fillet weld is as follows:

Procedure

1. Obtain one piece of ⅛″ × 1″ × 6″ flat steel.
2. Lay the piece on its edge on the piece of metal used in the butt welding exercise described previously; or use a new piece of ⅛″ × 2″ × 6″ or ¼″ × 2″ × 6″ metal for the base piece.
3. Clamp the assembly to the table or steady the parts with firebricks.
4. Select a ⅛-inch E6011 electrode and set the welder at 110 amperes.
5. Tack weld both ends of the fillet joint.
6. Square the vertical piece so that it is at a 90° angle to the bottom piece.
7. Hold the electrode at a 45° angle to the table top and lean it in the direction of travel.
8. Run a bead from end to end along the joint, figure 26–20. Watch the puddle to be sure it penetrates both pieces equally. To create more penetration in one piece or the other, direct the end of the electrode more to that piece. Adjust the angle of the electrode as needed to obtain equal penetration. Adjust the amperes if needed. Check for undercut or overlap in the sides of bead, figure 26–21.
9. Run a bead on the other side of the vertical piece. Alternating passes from one side to the other reduces the problem of distortion.
10. Chip and examine the head.
11. Make additional passes following the steps shown in figure 26–22.

It is important to remember that heat rises in metal. The weldor must make allowances for this fact in certain types of welds. For example, when making a fillet weld to form a lap joint, the "J" weave is useful

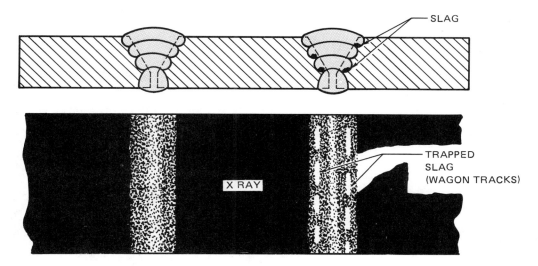

Figure 26–19. Slag trapped between passes will show on a cross-section of X-ray examination (*courtesy of L. Jeffus and H. Johnson,* Welding Principles and Applications, *Delmar Publishers Inc., 1984*).

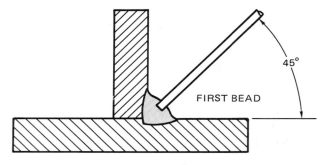

Figure 26–20. Making the first pass of a fillet weld

to help control heat. When welding a thin piece of metal on top of a thicker piece, the J pattern concentrates more heat on the thicker bottom piece, figure 26–23. This permits proper weld penetration without overheating the vertical piece.

☐ WELDING IN VARIOUS POSITIONS

Many welding jobs in agricultural mechanics require welding in a variety of positions. Fortunately, modern welding electrodes are available that make welding in all positions possible. The weldor should have reasonable skill in flat or down hand welding before attempting to weld in other positions. Even the ex-

perienced weldor will try to position materials so that they can be welded in a down hand manner.

Horizontal Welds

Welds made by moving horizontally across a vertical piece of metal are called *horizontal welds*. When making horizontal welds, the metal is placed so that it is comfortable to run a bead across the surface. The metal should be at eye level so welding from a sitting position is recommended. To make a horizontal bead, the following procedure is suggested.

Procedure

1. Obtain a ⅛″ or ¼″ × 4″ × 6″ welding pad.
2. Weld two small pieces on the edge of the pad so the pad can be set on edge and clamped securely to the table.
3. Put on helmet, gloves, arm and shoulder protection, and a leather apron.
4. Sit in front of the table.
5. Select a ⅛-inch E6011 electrode and set the welder at 110 amperes.
6. Hold the tip of the electrode near the left side of the pad. Lean it in the direction of travel at an angle of about 70° to the plate. Tilt the elec-

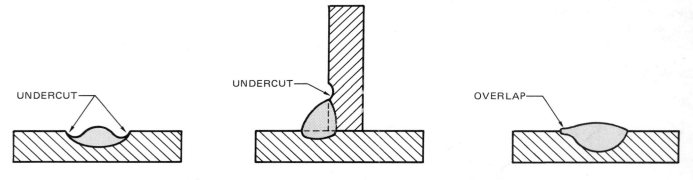

Figure 26–21. Undercut and overlap should be avoided in welding beads.

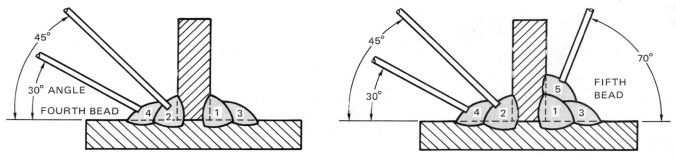

Figure 26–22. Alternating passes to control distortion

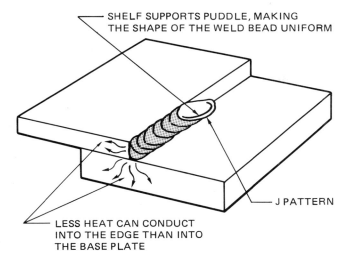

SHELF SUPPORTS PUDDLE, MAKING
THE SHAPE OF THE WELD BEAD UNIFORM

J PATTERN

LESS HEAT CAN CONDUCT
INTO THE EDGE THAN INTO
THE BASE PLATE

Figure 26–23. The "J" pattern allows heat to be concentrated on the thicker plate (*courtesy of L. Jeffus and H. Johnson,* Welding Principles and Applications, *Delmar Publishers Inc., 1984*).

trode so it points slightly upward at an 80° angle to the plate, figure 26–24.

7. Run a bead across the plate at a rate of speed that creates a proper bead without sagging, undercutting, or overlapping.

8. Adjust the amperes if needed and run additional beads to develop the skill.

Vertical Welds

Welds made by moving downward across a vertical piece of metal are called *vertical down* welds. Welds made by moving upward across the metal are called *vertical up* welds. Vertical down welds are easy to make. However, they are shallow in penetration and so should be made only on materials that are ⅛ inch or less in thickness.

To make a vertical down weld, set up a practice pad as described in the section on horizontal welds. A sitting position for welding is recommended. The electrode is held with the tip pointing upward at an 80° angle to the plate. Strike the arc at the top of the plate and let the force of the arc push the electrode downward across the plate at an even rate. Weaving is not recommended for this procedure. Adjust the amperes as needed to obtain an even bead.

Vertical up welding is rather difficult to manage, because heat builds up in the metal as the weld progresses up the metal. Special movements of the electrode are used to control the heat and permit individual parts of the puddle to cool. Unless this is done carefully, the weld metal will sag or drop out of the puddle. The shape of the puddle indicates the temperature of the surrounding base metal, figure 26–25.

A shorter arc, lower amperage, steeper electrode angle, and more rapid movement of the electrode may all be helpful in controlling the puddle. To make a vertical up weld on a flat area, the following procedure is recommended.

Procedure

1. Set up a practice pad as outlined under horizontal welding.

2. Start at the bottom of the pad and work upward.

3. Use a wide weaving pattern, moving the tip of the electrode from side to side. Some weldors create the bead by making a series of very short horizontal beads as they progress upward. Other weldors use a figure eight pattern. Others use a wide, crescent-shaped, back-and-forth motion.

4. If the bead gets too hot and is not controllable, stop. Chip and clean the weld. Then start again. Make corrections by reducing the amperage and using more sideways motion to permit parts of the puddle to cool more quickly. Make additional passes, if needed, after the root pass is cool and all slag chipped, figure 26–26.

To make a vertical up fillet weld, a T pattern is helpful. In this pattern, the electrode is directed first

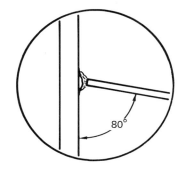

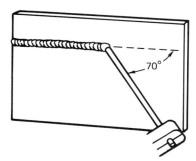

Figure 26–24.   Electrode angles for horizontal welding

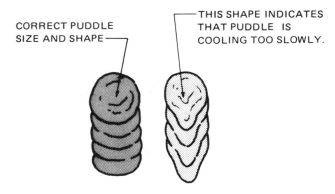

Figure 26–25. The shape of the puddle indicates the temperature of the surrounding base metal (*courtesy of L. Jeffus and H. Johnson,* Welding Principles and Applications, *Delmar Publishers Inc., 1984*).

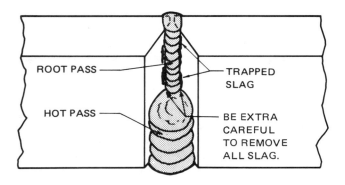

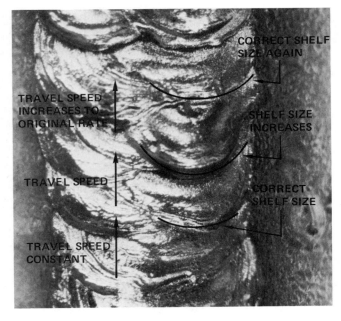

Figure 26–26. Making a vertical up weld on a flat area (*courtesy of L. Jeffus and H. Johnson,* Welding Principles and Applications, *Delmar Publishers Inc., 1984*).

at the root of the weld. The movement then is to one side, then to the other side, and finally back to the root to start the sequence again, figure 26–27.

Overhead Welding

Overhead welding is not a difficult procedure. However it can be dangerous without protective clothing. The metal is positioned above the weldor and the weldor assumes a sitting position, if possible. The electrode is held nearly straight up and down. Therefore, some weldors prefer to place the electrode so that it extends out the end of the holder. The holder is then gripped like a pencil.

When welding overhead, use an E6011 electrode, a normal amperage setting, electrode angle of 10–15 degrees from vertical and normal speed. Welds may be single pass or multiple pass according to the need, figure 26–28.

□ WELDING PIPE

Welding of pipe is done extensively in agricultural mechanics. Pipe may be used to make gates, trailers, wagon sides, athletic equipment, and other projects. Used pipe is a cheap source of construction material in many communities.

Pipe is made of thin metal. Therefore, pipe welding requires the use of electrodes which produce shallow penetration. The E6013 electrode is a good choice. It is a fast-filling electrode which makes it suitable when fits are not perfect.

Pipe is generally welded without grinding. As the electrode proceeds around a pipe, the rate of travel

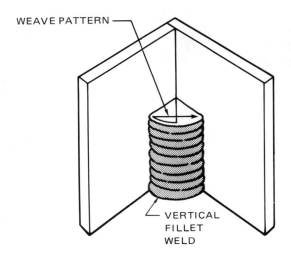

Figure 26–27. Making a vertical up fillet weld (*courtesy of L. Jeffus and H. Johnson,* Welding Principles and Applications, *Delmar Publishers Inc., 1984*)

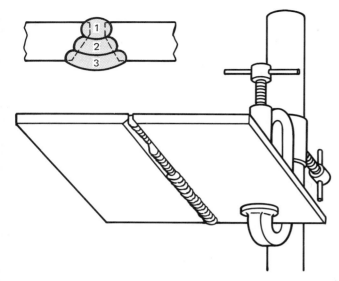

Figure 26–28. A welding pad showing multiple passes in the overhead position

must be faster at the electrode holder than at the electrode tip. It is important to maintain the same relationship of electrode-to-weld-surface as used for straight areas, figure 26–29. Welding of pipe ends is easier if the ends are first flattened. This permits the weldor to make straight passes.

Caution:

Do not weld galvanized pipe due to the hazard of poisonous fumes except under the supervision of an expert.

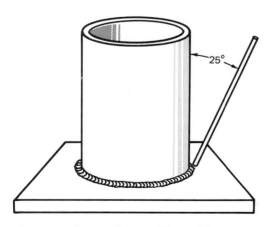

Figure 26–29. Making a fillet weld on pipe

□ PIERCING, CUTTING, AND SPOT WELDING

Piercing and cutting with arc welders is not as fast and clean as with an oxyacetylene torch. However, it is a useful procedure for those who have arc welders but not an oxyacetylene torch.

Piercing

To pierce a hole with an arc welder, the following procedure is recommended.

Procedure

1. Clamp a piece of ¼″ × 2″ × 6″ steel so that it projects 4 inches from the table top.
2. Place a bucket half filled with water or a sandbox under the metal.
3. Put on protective clothing, including leather shoes and a leather apron and/or leggings.
4. Determine the correct amperage for welding and increase this value by 50 percent. Set the amperage at the increased value.
5. Select a ⅛-inch E6011 electrode or a special cutting electrode.
6. Strike the arc above the metal and hold a long arc (about ¼ to ⅜ inch) until the metal is molten.
7. Rotate the tip of the rod to create an enlarged molten puddle.
8. Quickly push the electrode into the molten puddle to create a hole. If the rod does not go through, quickly withdraw the electrode or whip it loose if it sticks. Start again but make the molten puddle deeper before thrusting the electrode through the metal.
9. Keeping the arc burning, rotate the tip of the electrode around the hole to enlarge it as needed.
10. If a smooth hole is required, quickly place the hole in the metal over the hole in an anvil. Then drive a blacksmith's punch into the hole while the metal is still very hot.
11. Smooth the underside of the piece with a grinder as needed.
    Note: If additional amperage gives better results, use a higher setting. However, excessive amperage burns and overheats the rod and creates a heavier load on the machine.

Cutting

To cut with an arc welder, the following procedure is recommended.

Procedure

1. Use the same piece of metal as in the piercing exercise.
2. Use an E6011 electrode or a special cutting electrode.
3. Set the amperage 50 percent higher than for welding.
4. Strike the arc and direct a long arc at the edge of the metal until it becomes molten.
5. Use a quick up-and-down chopping movement to gouge out the molten metal and create a kerf across the metal.
6. Smooth the cut with a grinder, if needed.

Spot Welding

Electricity can be used to heat two pieces of metal in a small area between two electrodes. When the heat is sufficient to cause the metal to melt, welding occurs. The process is called *spot welding* and the device used to spot weld is called a *spot welder,* figure 26–30. A spot welder is used when sheet metal must be fastened in many places near the edges.

To operate a spot welder, two pieces of sheet metal are placed between the electrodes. The length of the electrode holders determines how far from the edge of the metal a spot weld can be made. The operator provides pressure to bring the electrodes into contact with the metal, and the machine provides the heat. The machine shuts off automatically when the weld is complete. The actual welding takes only a split second once it starts.

Figure 26–30. A spot welder

☐ STUDENT ACTIVITIES

1. In the welding area of the shop, examine all the protective clothing that is available.
2. Examine all fire protective equipment in the welding area. Report any irregularities to your instructor.
3. Ask your instructor to provide samples of weld showing good beads and poor beads due to improper welding.
4. Using a ¼″ × 4″ × 6″ welding pad, make 16 beads 1 inch long (four rows of four beads each).
5. Using a ¼″ × 4″ × 6″ welding pad, fill the pad with practice beads ¾ inch apart.
6. Using two pieces of ¼″ × 2″ × 6″ steel, make a butt weld 6 inches long with an even bead.
7. Make an internal check of the bead from activity 6 by examining a cross section of a 1-inch piece sawed from the assembly. Bend open the 1-inch weld and analyze the bead.
8. Weld a ⅛″ × 1″ × 6″ piece of steel to a ¼″ × 2″ × 6″ piece with a fillet weld on each side.
9. Run a straight 6-inch horizontal weld across a ⅛″ or ¼″ × 4″ × 6″ welding pad.
10. Make a 2-inch vertical down butt weld.
11. Make a 2-inch vertical up butt weld.
12. Make a 2-inch vertical up fillet weld.
13. Pierce a 1-inch hole in a ¼″ × 2″ × 6″ pad.
14. Cut three 1-inch strips from the pad used in activity 13, using an electric welder and electrode.

☐ SELF-EVALUATION

A. Multiple Choice. Select the best answer.
　1. The temperature of an electric welding arc is about
　　a. 400°F
　　b. 840°F
　　c. 1800°F
　　d. 9000°F
　2. Welding tables should be made of
　　a. concrete
　　b. masonite
　　c. metal
　　d. wood
　3. Fire extinguishers for welding areas should be suitable for
　　a. Class A fires
　　b. Class B fires
　　c. Class C fires
　　d. all of these
　4. Burning clothes on a human should be extinguished with

a. a fire blanket

b. a fire extinguisher

c. sand

d. any of these

5. Water in a welding area is useful for

a. receiving sparks from piercing

b. extinguishing fires

c. cooling metal

d. all of these

6. Injury to eyes can result from

a. chipping without goggles

b. viewing welding without shielding

c. welding with less than a No. 10 lens

d. all of these

7. If only one kind of electrode for all arc welding is to be purchased, the best choice is an

a. E6010

b. E6011

c. E6013

d. E7018

8. For most welding in agricultural mechanics the best electrode size is

a. ¹⁄₁₆ inch

b. ⅛ inch

c. ³⁄₁₆ inch

d. ¼ inch

9. Correct arc length is approximately

a. ⅛ inch

b. ¼ inch

c. ⅜ inch

d. ½ inch

10. When welding, the operator sees by

a. daylight

b. fluorescent light

c. light from the arc

d. all of these

11. The appearance and strength of a bead are influenced by

a. amps

b. angle

c. speed

d. all of these

12. The recommended position of the weldor for horizontal and vertical welding is

a. standing

b. sitting

c. the most comfortable one

d. laying flat

13. The recommended weave pattern for the beginning weldor doing down hand welding is

a. circular

b. figure eight

c. J

d. T

14. A second pass should never be done if

a. the first was a poor weld

b. the joint was ground

c. the slag has been removed

d. the slag has not been removed

15. In metal, the most rapid movement of heat is

a. down

b. equal in all directions

c. horizontal

d. up

B. Matching. Match the items in column I with those in column II.

Column I	Column II
1. lens for welding	a. protective clothing for weldors
2. lens for chipping	b. electrode for thin metal
3. leather	c. electrode for most welding
4. E6013	d. first pass
5. E6011	e. procedure results in a weld with shallow penetration
6. root pass	f. clear
7. vertical up	g. relatively difficult
8. vertical down	h. useful technique when welding metal of different thicknesses
9. J weave	i. E6013
10. electrode for pipe	j. No. 10

C. Completion. Fill in the blanks with the word or words that make the following statements correct.

1. The _____ _____ weld is the most difficult weld to make.

2. To make a hole with an electrode is called _____.

3. The best amperage setting for cutting is _____ percent higher than for welding.

4. The correct motion for cutting is _____.

Unit 27
Preparing Wood and Metal for Painting

Objective

To prepare wood and metal for painting.

Competencies to Be Developed

After studying this unit, you will be able to:
- Prepare unpainted wood for painting.
- Prepare previously painted wood surfaces for re-painting.
- Steam clean machinery.
- Prepare unpainted metal for painting.
- Remove rust and scale from metal surfaces.
- Feather chipped paint on metal surfaces.
- Mask tractors and other machinery for spray painting.
- Estimate materials for paint jobs.

Terms to Know

waterproof	sealer
paint	rot
pigment	warp
vehicle	air-dried lumber
paint film	old work
wood preservative	primer
creosote	caulk
pentachlorophenol	putty
zinc naphthenate	pliable
cuprinol	glazing compound
new wood	steam cleaner
new work	feather
seal	mask

Materials List

Paint can label
Varsol
Rags
Putty
Scraper and wire brush
Assortment of emery cloth and sandpaper
Silicon carbide paper, 200 and 400 grit
Window sash in need of repainting
Wooden project to be painted
Wooden project to be repainted
Piece of machinery that needs repainting
Portable power wire wheel
Steam cleaner

Wood, steel, and concrete are the primary materials used to construct agricultural buildings and equipment. Wood may rot and steel will rust if not protected from moisture. Concrete is not damaged by moisture in the air but may need special treatment to make it waterproof. *Waterproof* means that water cannot enter the material.

Most wood exposed to the weather must be painted or treated with a preservative. *Paint* is a substance consisting of pigment suspended in a liquid known as a vehicle. A *pigment* is a solid coloring substance. The liquid is called a *vehicle* because it carries the pigment.

Paint must be applied in thin layers so that some of the vehicle can evaporate. As the vehicle evaporates, the paint is said to be drying. After a paint dries, the material that is left is called a *paint film*. Wood and other materials must be properly prepared or the paint film will peel off rather than wear off. In such cases, the paint film first dries out excessively. Hairline

Figure 27–1. An advanced stage of paint peeling from siding on a building

Figure 27–2. Creosote and pentachlorophenol are two commonly used wood preservatives for home and farm applications.

cracks then start to appear. The film then breaks and starts to curl away from the wood. Finally, the paint film separates from the wood and leaves the surface exposed, figure 27–1.

□ PRESERVATIVES

Wood may be covered with a liquid which kills wood-rotting fungus and wood-eating insects such as termites. Such liquids are called *wood preservatives.* Creosote, pentachlorophenol, and zinc naphthenate are popular wood preservatives. *Creosote* is black and generally shows through if covered with paint. It also has a strong odor. Therefore, it is generally used for posts and poles in outside applications. *Pentachlorophenol* is sold in a nearly clear vehicle, does not bleed through, and makes an excellent undercoat for painting materials, figure 27–2. *Zinc naphthenate,* popularly known as *cuprinol,* is another nearly clear preservative used extensively for exterior surfaces such as shingles.

□ PREPARING WOOD FOR PAINTING

Wood that has never been painted or sealed is called *new wood.* Wood used in original construction is frequently referred to as *new work.* To *seal* wood means to apply a coating that fills or blocks the pores so that no material can pass through the surface. The coating is called a *sealer.*

A sealer prevents moisture from entering the wood and causing it to rot or warp. To *rot* means to decay or break down into other substances. To *warp* means to bend or twist out of shape. A sealer also keeps moisture and natural liquids inside the wood. This prevents the wood from drying out and cracking.

Preparing New Wood

New wood is easy to prepare for painting. See unit 11, "Finishing Wood," for procedures to prepare wood for clear or stained finishes. The procedures are the same to prepare wood for painting.

New wood must be dry when it is painted, that is, there should be no indication of moisture on the surface and it should have been air dried. *Air-dried lumber* means that the boards are separated by wooden strips when they are stacked and are then protected from rain and snow for six months or more. Air drying permits the excessive natural moisture to leave the wood. If the wood contains excess moisture, that moisture will tend to lift the paint as it tries to escape to the drier surrounding air.

The wood must be free of grease, wax, dirt, and other substances or, again, the paint will not adhere to the wood. Grease and oil are removed by wiping the wood with a cloth soaked in varsol. Wax is removed by washing the wood with a solution of household ammonia. Dirt from shoes and other sources is removed by using a stiff bristle brush.

Planing of rough lumber greatly reduces the amount of paint needed to cover the surfaces. No sanding is needed except to remove objectionable roughness or to remove grease, wax, and other residues.

Figure 27–3. A wire brush is useful for removing old paint.

Figure 27–4. To speed up the preparation of old work, a power drill can be used with a wheel and abrasive flaps, wire wheel, or sanding discs.

Bark must be removed before painting or planting posts and poles. Bark holds moisture and also serves as a home for disease organisms and insects which destroy wood. Posts may be dried by standing them on end around a tree trunk or other vertical support so air can flow around each one. Bark can then be removed with a draw knife or axe. After the bark is removed, the post must be left until all surfaces are thoroughly dry before painting is attempted.

Preparing Old Work

Wood that was previously painted is called *old work*. All loose paint must be removed from old work. This is done with a scraper, wire brush, and sandpaper, figure 27–3. The job is made easier with the use of abrasive flaps, wire wheels, and discs attached to a portable power drill, figure 27–4. If a smooth paint job is needed, the paint must be removed by a chemical paint remover or a torch and a scraper, or by complete sanding.

Paint manufacturers recommend the washing of previously painted surfaces before repainting. This can be done with a cloth and household detergent in water. The detergent removes grease, dirt, and weathered paint. A uniform, clean surface is left to which the new paint will adhere.

A final step in preparing old work is to fill all holes, caulk all cracks, and seal any problem spots. The holes and cracks must be sealed before filling or liquids in the filling material will be absorbed by the dry wood. This means that the filler will shrink and loosen, causing the paint to crack and peel around the filler.

A *primer* is a special paint used to seal bare wood. Primer should be applied to the wood with a brush so the material can be worked into holes, cracks, and crevices. A thin coat is recommended because the job of the primer is to penetrate the wood as it is applied. When the primer dries, the pores are sealed and only a thin film is left on the surface. Knots or other problem areas that seem to absorb too much of the primer should be sealed with a primer containing shellac.

After priming, all cracks must be filled and the joints between materials must be caulked. *Caulk* is a material that stretches, compresses, and rebounds as materials expand and contract. This characteristic permits it to bridge the air gap between two materials such as brick and wood, figure 27–5.

Holes must be filled and window glass secured with putty or glazing compound on outdoor work. Plaster-like fillers may be used for indoor surfaces. *Putty* and glazing compound are soft materials containing oils that help keep them pliable over a long period of time, figure 27–6. *Pliable* means it will give if pushed or pulled. Putty is workable and may be pushed into holes and formed into special shapes. An example is the triangular bead of putty used to seal glass in a window, figure 27–7. Priming followed by applying new putty or glazing compound are important steps in repainting window sashes.

Figure 27–5. Caulking between two different materials, such as brick and wood, permits them to expand and contract without breaking the paint film.

Scraping, brushing, washing, sanding, priming, caulking, and filling should prepare any old work for painting.

□ STEAM CLEANING MACHINERY

Tractors, engines, and machinery may be cleaned quickly and thoroughly with a steam cleaner. A *steam cleaner* is a portable machine that uses water, a pump, and a burner to produce steam, figure 27–8. Special chemicals in solution are added to the steam to dissolve grease and remove paint, dirt, manure, and other materials from metal, wood, and concrete.

A steam cleaner is powerful enough to strip paint from metal if the appropriate chemical solution and pressure are used. This combination of heat and chemicals can make the steam cleaner hazardous in the hands of careless operators.

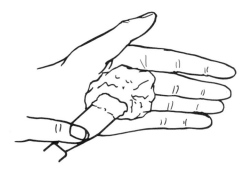

Figure 27–6. Putty and glazing compound are soft materials that contain oils to keep them pliable over long periods of time. Here the material is being worked in the hand to make it warm and pliable for application.

Figure 27–7. Window glass is sealed to wood or metal frames by applying a bead of putty and smoothing it to a triangular shape. Here paint is being applied to keep the putty and wood from drying out.

Caution:

Steam cleaners can cause serious burns to the operator or bystanders if not used carefully. The contols of the steam cleaner will vary with the manufacturer. Read the operator's manual before using a steam cleaner.

Figure 27–8. A steam cleaner removes grease and dirt from machinery, making it easier to repair or prepare the machinery for painting.

A general procedure for operating a steam cleaner is as follows.

Procedure

1. Cover all electrical parts on the machinery to be cleaned with plastic. Food bags, garbage bags, or vinyl film can be used. Wrap the parts with string or rubber bands to hold the plastic in place.
2. Place the machine at least fifteen feet from any building.
3. Plug the motor into a three-wire grounded electrical extension cord attached to a proper outlet.

> **Caution:**
>
> **Do not allow the plug to get wet. Place the plug up on the steam cleaner to keep it out of water that may accumulate in the area.**

4. Attach a garden hose to the water inlet of the cleaner and to a cold water faucet. Make all connections watertight.
5. Fill the fuel tank of the cleaner with kerosene or other recommended fuel.
6. Mix the cleaning chemical(s) in the solution tank according to the manufacturer's instructions.

> **Caution:**
>
> **Cleaning materials are powerful chemicals that may burn eyes and skin. Use the proper goggles and protective clothing.**

7. Position the steam cleaner so that you can observe all gauges and controls while operating the machine.

> **Caution:**
>
> **Have a helper to watch the gauges and manage the controls of the machine as you steam clean. The helper should wear the same protective clothing as the operator.**

8. Unwrap and lay out the steam hose.
9. Hold the steam handle with the nozzle pointed toward concrete or free air.

> **Caution:**
>
> **Never point the nozzle toward a person. Do not allow water or steam to contact electrical cords, wires, motors, outlets, or lights.**

10. Open the water faucet.
11. Turn on the switch to start the steamer motor.
12. Turn on the fuel valve.
13. Adjust the fuel valve to create the steam pressure recommended by the manufacturer.
14. When steam comes from the nozzle, turn on and adjust the solution valve to obtain the desired cleaning power.
15. Direct the steam to the high parts of the machine and move downward. The nozzle is held 8 to 12 inches from surfaces needing gentle cleaning and 2 to 4 inches from surfaces needing harsh cleaning.

Caution:

The nozzle should be positioned so steam and dirt are not forced into bearings or electrical components by the steam and water.

16. Resteam the entire machine lightly to wash off all loose dirt.

Shutting down and storing the steam cleaner properly is very important. If stored incorrectly the machine may freeze up, which will cause extensive damage. Some general recommendations for a safe shutdown are:

Procedure

1. Close the solution valve.
2. Turn off the fuel.
3. Hold the nozzle until cold water comes out.
4. Turn off the motor.
5. Unplug the extension cord at the building. Then unplug the steam cleaner.

Caution:

Since the surrounding area is wet, electric shock is a real hazard.

6. Disconnect the garden hose from the machine and hose down the work area.
7. If the machine will be exposed to freezing temperatures, straighten the steam hose. Use compressed air to push water from the inlet side through the coils and out through the steam nozzle on the hose.
8. Store the steam cleaner with all cords and hoses protected.
9. Remove all plastic coverings from the clean machine.
10. If the machine has an engine, start the engine and let it warm up to help the drying process.
11. Dry the machine quickly and thoroughly.

Steam cleaning may produce moisture in engine systems that will prevent starting. When this occurs, the trouble can usually be corrected by drying the spark plug wires and the inside of the distributor cap. A common hair dryer is effective for drying out electrical systems.

Caution:

Since steam cleaning may dissolve grease in bearings, these should be checked and repacked if needed.

□ PREPARING MACHINES FOR PAINTING

Machines used in shops, buildings, and fields all need to be painted to prevent rusting. The paint film on a machine is damaged by mechanical contact, normal wear, and weathering. Such damage results in chipping, cracking, peeling, flaking, and fading of paint.

New Metal

Painting must follow new construction or repairs such as welding which leave metal exposed. To prepare new metal, the following steps are suggested.

Procedure

1. Chip all welds and wire brush them thoroughly.
2. Wire brush or sand (using emery cloth) to remove all dirt or rust from the metal.
3. Steam clean or wipe down with varsol to remove grease.
4. Clean the metal with a commercial solvent designed to be a preparatory solvent.
5. Prime immediately.

Previously Painted Metal

Previously painted metal surfaces include castings, parts, and assemblies, and smooth surfaces such as guards, hoods, and fenders, where an automobile-type finish is used.

All surfaces must first be cleaned of dirt, mud, manure, and grease. Steam cleaning is the fastest and easiest method. If steaming is not possible, then high-pressure water cleaning machines may be used. Or, materials can be removed by hand with a solvent, hose, water, scrapers, and brushes.

Heavy grease deposits are removed by scraping, followed by treatment with varsol or another commercial grease solvent. Generally, high pressure water must then be used to remove the grease and solvent from the metal. Any grease or oil film left will prevent paint from holding to the metal.

Figure 27-9. When cleaning metal prior to painting, a power sander can remove rust, and smooth rough or pitted areas.

Figure 27-10. Cleaning metal in preparation for painting

Scaling paint and rust can be removed and rough or pitted areas can be smoothed by wire brushes, wire wheels, disc sanders, or emery paper. A combination of hand and power methods is commonly used, figures 27-9 and 27-10.

Smooth surfaces such as hoods and fenders need to be stripped of all paint if they are badly chipped. Otherwise, every chip will show in the new paint. Metal can be stripped using a paint remover or by sanding. If only a few chips exist, wet-type sandpaper can be used to feather the chipped areas. To *feather* means to sand so the chipped edge is tapered and no roughness can be felt between the painted and unpainted areas, figure 27-11. If a chip in a painted surface is feathered properly, the place where the chip starts and stops cannot be felt.

Feathering is done using the following procedure.

Procedure

1. Tear a piece of 200 grit silicon carbide waterproof paper into quarters.
2. Pour about one quart of water into a bucket or pan.
3. Dip the paper into the water.
4. Grip the paper between the thumb and index finger so the balls of the fingers serve as a backing pad for the paper.
5. Sand across the chipped area in all directions until a feathered edge is obtained. Dip the paper into water frequently to wash the pigment from the paper.

6. Finish with 400 or 600 grit paper to remove sanding marks and leave a smooth surface.
7. Wash the part thoroughly with water to remove all pigment.

Figure 27-11. Chipped areas on a painted surface are sanded smooth or "feathered" so that no roughness can be felt between the painted and sanded areas.

Caution:

Paint pigment remains suspended in water, so the part or the area must be thoroughly washed and rinsed with clean water to avoid leaving a paint residue after the water dries.

8. Dry the part thoroughly or rusting will start immediately.

□ MASKING FOR SPRAY PAINTING

A good-looking spray job requires careful masking of all glass, chrome, and other areas that should not be painted. To *mask* means to cover so paint will not touch. Masking may be done with tape or paper, figures 27–12 and 27–13.

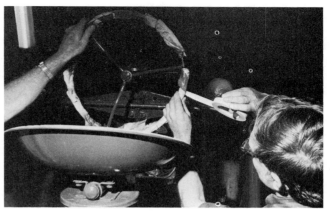

Figure 27–13. Applying masking tape over newspaper to provide a tight masking on the grip surfaces of a steering wheel.

Small items such as a light lens, distributor cap, and chrome trim are generally covered with masking tape. High-quality masking tape is required for good results. A sharp pocket knife is used to trim off the tape that extends beyond the part. For larger parts, paper is cut so that it is about ¼ inch short of covering the area. Then ½-inch or ¾-inch masking tape is used to finish covering the exposed area and tape the paper in place. Newspaper is frequently used for masking because of its availability and low cost. Very small areas such as grease fittings may be coated with a heavy film of general purpose grease. When the paint job is dry, the grease and paint are wiped from the fitting.

□ ESTIMATING PAINT JOBS

Flat Surfaces

For most types of paint, one gallon will cover 400 to 500 square feet or about 100 square feet per quart. On flat surfaces such as doors, it is easy to calculate the square footage. For example, if two shed doors cover an opening 12 by 20 feet, the square footage is determined by multiplying length (20 feet) by height (12 feet). The amount is 240 square feet. A quart of paint will cover 100 square feet, so 240 is divided by 100 to get 2.4, which is rounded to 3. Thus, three quarts of paint will do the job. However, paint generally is cheaper by the gallon. For example, a quart may cost $7 while a gallon costs $21. In that case, three quarts of paint in quart cans will cost as much as four quarts in a one-gallon can. A good rule to use is: When the estimate indicates that two quarts will not be enough for the job, then buy a gallon.

Figure 27–12. All surfaces that are not to be spray painted must be masked. The students are pointing to masking tape covering the lens of the light (left) and the snap couplings on the hydraulic hoses (right).

If the job called for two coats of paint, then the decision to buy a full gallon would be an easy one. If the gallon did not cover the entire area with a second coat, an additional quart of paint could be purchased.

When estimating paint for a whole building, the square footage for all four sides should be added up. The estimate for a shed 15 feet wide by 30 feet long with an average height of 11 feet is as follows:

Front = 11' × 30' = 330 square feet
Back = 11' × 30' = 330 square feet
End = 11' × 15' = 165 square feet
End = 11' × 15' = 165 square feet
Total 990 square feet

A gallon of paint is expected to cover 400 to 500 square feet. Dividing 990 square feet by 400 equals 2.5. If the building were very smooth two gallons might cover it. However, an extra quart or two may be needed to finish the job. If it is a long distance to the paint store, it would be wise to buy three gallons. This would eliminate the chance of running out of paint and avoid having to interrupt the painting once the job is started.

For buildings with shed-type roofs, the average height of the four sides is satisfactory for estimating purposes. Buildings with gable ends require an additional amount of paint to cover the areas in the triangles of the roof. The area (A) of a triangle is determined by multiplying the length of the base (b) by the height (h) of the triangle and dividing by 2. Thus, $A = \frac{1}{2} bh$.

Spray Painting Machinery

Machinery has many irregular surfaces. When spraying, a lot of paint is wasted. For instance, the cone of spray coming from the gun may be 12 inches wide and the part being sprayed may be only 3 inches wide. Therefore, only one fourth of the paint is being used. The rest is escaping into the air and settling on other surfaces.

Experience shows that a gallon of machinery enamel will paint a small farm tractor. Generally, a quart of paint is sufficient to paint the trim or apply a second color to a tractor. One quart of enamel thinner may be sufficient to thin five quarts of enamel paint, but it may be wise to purchase a gallon and have some available for cleanup purposes. A cheaper solvent such as varsol should be used first to clean the spray gun and equipment. Therefore, the estimate for a tractor paint job would be as follows:

Main color	1 gallon
Trim color	1 quart
Enamel thinner	1 gallon
Cleanup solvent	1 gallon

Estimates for other jobs can be made by comparing them with the example given. Some useful guidelines for purchasing paint are as follows:

- One quart of spray enamel will be sufficient for one piece of lawn or garden equipment.
- A gallon of enamel is needed for a small farm tractor.
- If the paint is custom mixed, be sure to buy enough paint for the job. A second batch may not match the original batch perfectly.
- If two quarts are not likely to do the job, buy a gallon.
- Estimate on the high side.
- To run out of paint during a spray paint job results in loss of time and money. Cleaning the spray equipment twice is time-consuming and requires extra solvent. A second trip for paint also costs time and money.
- When buying paint, purchase an extra amount with the understanding that unopened cans may be returned.

□ STUDENT ACTIVITIES

1. Define all italicized terms used in this unit.
2. Examine some buildings and machinery for signs of chipped paint and paint that has scaled or blistered.
3. Examine the label on a can of paint. Write down the ingredients listed as pigments. Write down the ingredients listed as the vehicle. Record the percentage of the total volume made up by each ingredient.
4. Examine a window with loose and missing putty. Note the cross-sectional view of the putty. Clean, prime, and apply new putty to the window. Repaint the window when the putty hardens.
5. Prepare a wooden project for painting.
6. Prepare a wooden project or building for repainting.
7. Steam clean a piece of machinery.
8. Wire brush and clean a piece of machinery for spray painting.
9. Sand a piece of metal with chipped paint. Feather the chipped area so that its presence cannot be felt with bare fingers.
10. Mask a machine for spray painting.
11. Estimate the amount of paint needed to apply

one coat to the sides and ends of a building with the following dimensions: L = 40 feet, W = 24 feet, H = 12 feet. Assume that there are no windows.

□ SELF-EVALUATION

A. Multiple Choice. Select the best answer.
1. The solid substance that gives color to paint is
   a. film
   b. pigment
   c. thinner
   d. vehicle
2. The vehicle in paint is a
   a. film
   b. liquid
   c. pigment
   d. solid
3. The material that helps steam to dissolve grease is
   a. electricity
   b. fuel
   c. pressure
   d. a chemical solution
4. The first step in starting a steam cleaner is to connect the
   a. electricity
   b. fuel
   c. solution
   d. water
5. The first step in shutting down a steam cleaner is to
   a. stop the motor
   b. shut off the fuel
   c. close the solution valve
   d. turn off the water
6. A hazard generally *not* associated with the steam cleaner is
   a. chemical damage
   b. electrical shock
   c. falling objects
   d. steam burns
7. Steam cleaning will not
   a. cause hard starting
   b. dry out bearings
   c. increase paint preparation time
   d. strip paint
8. Scaling paint is removed by
   a. brushes
   b. scrapers
   d. wire wheels
   d. all of these
9. A one-gallon can of paint generally costs about the same as
   a. two quart cans
   b. three quart cans
   c. five quart cans
   d. none of these
10. One gallon of paint is generally expected to cover
   a. 400–500 square feet
   b. 300–400 square feet
   c. 200–300 square feet
   d. 100–200 square feet

B. Matching. Match the terms in column I with those in column II.

Column I	Column II
1. new work	a. dry paint
2. old work	b. preservative
3. film	c. purpose of a primer
4. seal	d. correct by feathering
5. creosote	e. previously painted
6. pliable	f. silicon carbide
7. chipped paint	g. length × height
8. waterproof paper	h. masking
9. tape and paper	i. putty
10. square feet	j. never painted

C. Brief Answers. Briefly answer the following questions.
1. What is a reasonable estimate of the amount of paint thinner and cleanup solvent needed to paint a small farm tractor?
2. How many quarts of paint would be needed to give a 10-foot by 10-foot door one coat?

Unit 28
Selecting and Applying Painting Materials

Objective

To select and apply paint in agricultural settings.

Competencies to Be Developed

After studying this unit, you will be able to:
- Name major types of paint and paint components.
- Select paint for wood and metal.
- Apply paint with brushes.
- Spray paint with aerosol containers.
- Select and use spray painting equipment.
- Clean painting equipment.

Terms to Know

formulate	asphalt
chalking	undercoater
trim and shutter	compatible
paint	drop cloth
enamel	fingering
gloss	loading the brush
semigloss	paint roller
flat finish	nap
water-base	roller cover
oil-base	aerosol
latex	propellant
alkyd	spray gun
epoxy	air atomization
interior	regulator
exterior	extractor
hiding power	siphon system
lead	pressure feed system
titanium dioxide	CFM
iron	viscosimeter
zinc	viscosity
aluminum	sag
calcium	run
magnesium	tacky
silicon	

Materials List

Labels from oil-base and latex paints
Paint brushes
Paint thinner
Latex paint
Oil-base machinery enamel
Spray aerosols with enamel
Spray gun and equipment
Paint room or outdoor area for painting
Goggles, respirators, fire extinguishers, and other safety materials
Cloths, waste cans, paint storage cabinet, drop cloths, and other painting equipment
Projects to paint

Painting may be regarded as both a luxury and a necessity. Changing the colors of the surroundings just for fun is a luxury, adding enjoyment to life. However, most objects made of steel must be painted to prevent them from rusting. Similarly, wood that is exposed to high moisture levels must be painted or otherwise protected or it may rot. In these cases, painting is a necessity. Thus, there are two major reasons for painting: to improve appearance and to preserve.

The student is referred to unit 11, "Finishing Wood" for a discussion of thinners and solvents, procedures for painting with brushes, and brush cleaning and storage. Unit 27, "Preparing Wood and Metal for Painting," should also be reviewed before continuing with this unit.

☐ SELECTING PAINT

The chemistry of paint is complex. Therefore, the recommendations of the manufacturer must be relied upon heavily when choosing paints. However, some basic materials used in paints must be recognized in order to choose the right paint for the job and to buy paints that are a good value for the money.

Formulation

Paint is formulated for specific jobs. To *formulate* means to put together according to a formula. For example, paint for fences may be formulated so that a tiny bit of the film washes off each time it rains. This keeps the paint bright. The process is called *chalking.* In addition, fence paints must not contain any toxic chemicals, such as lead, that might poison livestock when they lick painted surfaces. Fence paint should not be used to paint windows on a brick house, because the dull pigment that washes off will run down over the brick and cause streaks. A special type of paint called *trim and shutter paint* is formulated to stay bright without chalking.

Types of Finishes

Paint with a gloss or semigloss finish is called *enamel. Gloss* means shiny. Paint with a slight shine is called *semigloss.* Most enamel is formulated to be used inside or outside. It is especially tough and useful on toys, boats, vehicles, tools, and machinery. It is also recommended for doors, trim, and baseboards in houses. Enamels are resistant to wear. Thus, they are useful on the floors of porches, patios, and other heavy-wear areas.

Some paints dry with a flat finish. *Flat finish* means dull or without shine. A flat finish paint is preferred for walls that are not generally touched or exposed to friction or moisture.

Vehicle

The vehicle is the liquid portion of the paint. The vehicle in most paints is either water or oil. Paints containing water are called *water-base* paints. Those containing oil are called *oil-base* paints.

The word *latex* on a paint label generally indicates a water-base paint. Most latex paints can be thinned with water, and brushes used with latex can be cleaned with water and detergent. The term *alkyd* indicates an oil-base paint. Such paint must be thinned with an oil, various petroleum products, or turpentine.

In recent years, epoxy paints have been added to the traditional line of paint materials. *Epoxy* is a synthetic material with special adhesive and wear-resistant qualities. Epoxy paints may be more difficult to apply, but their excellent durability makes them desirable for use in barns, milking parlors, public rest rooms, commercial kitchens, and other hard-use areas.

Interior Versus Exterior

The word *interior* on a label means the paint will not hold up if exposed to weather. The term *exterior* designates a paint that will withstand moisture and outside weather conditions. Both water-base and oil-base paints may be formulated for interior or exterior use.

Pigments

Pigments give color and hiding power to paint. *Hiding power* is the ability of a material to create color and mask out the presence of colors over which it is spread. Many materials are used for paint pigments. In the past, lead was used extensively. However *lead* is a metal that stays in the body once it is ingested.

> **Caution:**
>
> **Lead paint has been found to act as a slow poison to livestock who lick it. The same is true of children who put objects painted with lead paint in their mouths. Laws have been passed to stop future use of lead paint, but old paint containing lead remains a hazard.**

A high-quality pigment used in many paints is *titanium dioxide.* In addition, compounds of *iron, zinc,* and *aluminum* are associated with high-quality paints. Therefore, as a rule of thumb, the higher the proportion of these pigments in a paint, the better the quality.

Low-quality pigments provide little or no hiding power. Some low-quality pigments are *calcium, magnesium,* and *silicon* compounds. *Asphalt* in paints provides excellent black hiding power, but it wears away faster than the high-quality pigments.

Primers and Undercoaters

Primers and *undercoaters* are used to prepare surfaces for high-quality top coats. They stick to and seal surfaces. They must be compatible with the material being painted as well as with the top coats that follow. *Compatible* means they go together with no undesirable reactions. Primers and undercoaters will not stand weather, wear, or exposure. They must be covered with a proper top coat. It is generally wise to use the manufacturer's recommended primer or undercoat with any paint or finishing material.

Rust Resistance

Some paints consist of a vehicle and pigments that enable them to adhere well to rusted surfaces. Such paints work well if all loose rust is removed before the paint is applied. The primer and top coat must be compatible for best results.

□ USING BRUSHES

Brushes are useful for small paint jobs. They are especially adapted to use on irregular areas such as window sashes and building trim. The cost of equipment is low for both brush and roller painting. Therefore, both types of applicators are used extensively for home and farm projects. However, many commercial painters rely on the speed and professional results obtainable with modern spray painting equipment.

A new paint job can do almost as much to lift the spirits as a new purchase. However, a poorly done job may be ugly and messy, and may interfere with the proper operation of the object that is painted. For example, improperly painted windows may not open due to sticking. The recommended procedure for applying paint with brushes follows.

Procedure

1. Prepare the object or area as outlined in the previous unit.
2. Mask all parts that are not to be painted. Cover areas under surfaces to be painted with drop cloths. A *drop cloth* is any material used to protect floors, furniture, shrubbery, and other objects from paint spatter or droppings, figure 28–1.
3. Use masking tape to make a straight line where paint should stop, figure 28–2. When the tape

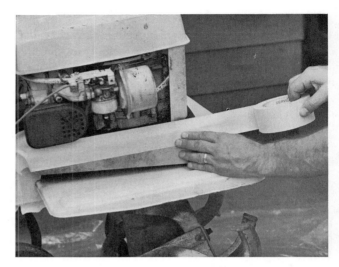

Figure 28–2. Use masking tape to make a straight line where paint should stop.

is removed, a straight paint line will be left, giving the job a professional appearance.

4. Mix paint thoroughly. Unopened oil-base paints should be shaken by machine. Once opened, they should be divided between two containers and stirred thoroughly, figure 28–3. Pouring the paint back and forth between containers is generally the final step in mixing. **Note:** Latex paints should not be shaken. They are mixed by stirring only.
5. Use a piece of screen or cheesecloth to strain paints containing lumpy particles, figure 28–4.
6. Use pure bristle brushes for oil paints and nylon or other artificial bristles for latex paints. Select

Figure 28–1. Most paint jobs require the use of drop cloths to protect shrubbery, walks, blacktop, floors, or furniture.

Figure 28–3. Oil-base paints should be divided between two containers, stirred thoroughly, and then poured back and forth between containers.

Figure 28–4.   Use a piece of screen or cheesecloth to strain paints containing lumpy particles.

Figure 28–6.   End all brush strokes in the freshly painted area.

a brush that has extra long and thick bristles and one that matches the job in size and shape—round, beveled, tapered, or flat.

7. Paint with the flat side of a flat brush. Painting with the edge causes *fingering* or dividing of the bristles, figure 28–5.

8. Dip about one third the length of the bristles into the paint. Touch the bristles to the side of the paint container to remove the excess paint

Figure 28–5.   Use the flat side of a brush when painting. Painting with the edge will cause the bristles to divide into fingerlike clusters.

and prevent dripping. This is called *loading the brush*.

9. Touch the loaded brush to the work in several places in a small area to deposit paint. Then smooth the paint by brushing back and forth to fill in the area. End all strokes in the paint rather than on unpainted surfaces, figure 28–6.

Caution:

Latex paints dry quickly and should not be brushed more than necessary to spread.

10. Paint adjacent sections before the paint dries. This reduces the marks that result from uneven blending of paint layers.

11. Stir the paint frequently.

12. If it is necessary to stop painting briefly, wrap the bristles in a plastic food bag to prevent the paint from drying in the brush.

13. At the end of the work day, clean latex paint brushes with warm water and detergent. Wrap them in a paper towel and place them in a warm place to dry. This is necessary to maintain brush shape and compactness. Oil-base paint brushes may be suspended in linseed oil or thinner until the next day.

14. When restarting the paint job, wipe excess thinner from the bristles with rags or paper towels. Proceed with painting.

15. When the painting is finished, clean oil paint from brushes using varsol or thinner. Wash the

brush thoroughly with soap and water to remove all paint and thinner residue. Clean bristles will squeak like clean hair. Wrap the brush in a paper towel to form the bristles in their proper shape. The bristles will then dry to provide a brush that is like new.

□ USING ROLLERS

Paint can be poured into a wide, shallow pan and applied with rollers more quickly and with less work than if brushes are used. A *paint roller* is a cylinder that turns on a handle. However, rollers tend to throw off tiny droplets of paint which spatter the surfaces below. Therefore, drop cloths are especially important when applying paint with a roller. Some suggestions for using rollers to apply paint are as follows.

- Use a 7-inch roller for narrow or uneven surfaces such as fence boards and siding. A 9-inch roller can be used for large, level surfaces such as plywood, wall sections, and ceilings.
- Use a roller with a short nap for smooth surfaces and one with a longer nap for rougher surfaces. *Nap* refers to the soft, woolly, threadlike surface of the roller cover. The *roller cover* is the hollow, fabric-covered cylinder that slides onto the roller handle.
- Line the paint pan with aluminum foil. The foil is discarded after painting, leaving the pan clean for the next job, figure 28–7.
- Use small rollers or sponges on bevels or grooves

Figure 28–8. Use small rollers or sponges on bevels or grooves in siding or on rounded surfaces such as rails and posts.

in siding or on rounded surfaces such as rails and posts, figure 28–8.
- Paint corners and edges first, then roll the larger areas. First apply paint in a diagonal or cross pattern, figure 28–9. Then roll it again in the direction of movement across the work. Finish the surface by rolling toward the painted area—not the unpainted area.
- Use a roller with a very long nap on chain-link and woven-wire fences, figure 28–10. Paint both sides of the fence and roll in several directions for complete coverage of all joints.
- Stir the paint frequently, especially when the pan is refilled.
- When work is stopped, temporarily or even for several days, store the roller in an airtight plastic bag or wrapper, figure 28–11.

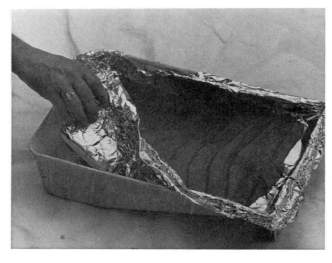

Figure 28–7. Cleanup time is saved by lining the paint tray with aluminum foil. After painting, the foil is removed and discarded.

Figure 28–9. Paint corners and edges first. Then roll the large areas by first applying paint in a cross pattern.

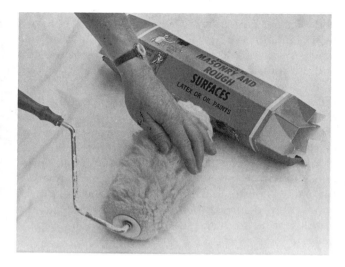

Figure 28–10.   Use a roller with a very long nap for painting wire fences.

- When the job is finished, discard low-cost roller covers. High-quality covers are generally made with a washable core and may be cleaned. However, the process requires large amounts of solvent. Therefore, cleaning may be practical only when using latex paints, which are removed with warm water and detergent. If all paint is not removed, the roller cover will not be usable again.

□ USING AEROSOLS

Small paint jobs can be done quickly at a low cost with aerosol spray cans. An *aerosol* can is a high-pressure container with a valve and spray nozzle, fig-

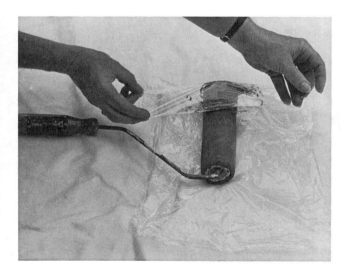

Figure 28–11.   A used roller may be stored for several days without cleaning by wrapping it in an airtight material such as plastic.

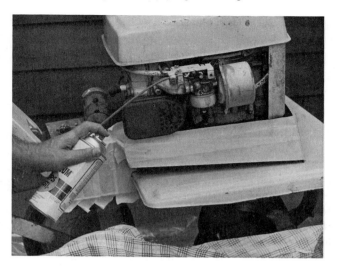

Figure 28–12.   Aerosol spray paint is especially useful on small jobs and on areas that are hard to paint with brushes.

ure 28–12. The container is partially filled with paint, which is put under pressure with compressed gas. The gas is called a *propellant,* since it pushes or propels the paint toward the object being painted.

When painting with aerosols, the following procedure is suggested.

Procedure

1. Degrease and sand the object to be painted.
2. Wash all dust and residue from the object.
3. Choose a clean, well-ventilated area in which to paint.
4. Cover or mask all adjacent areas.
5. Wipe areas to be painted with a tack rag and solvent.
6. Shake the can for several minutes.
7. Practice spraying on a piece of cardboard before starting on the project.
8. Spray with rapid, uniform strokes. Hold the can 10 to 12 inches from the object. Start the spray in the air and then move it across the surface to be painted.
9. Spray on a mist coat and let it become tacky or sticky. Gradually build a full cover by successive overlapped strokes to create an even paint film.

Caution:

Do not apply too much paint at one time on vertical surfaces or it will develop unsightly runs.

10. When finished, invert the can and press the nozzle until there is no more paint in the spray. This leaves the nozzle free of paint so that it can be used another time.

11. If additional coats are needed for good coverage, follow the directions on the can regarding sanding and drying time.

12. Allow to dry thoroughly, before heavy use. Paint may take 30 days to achieve maximum hardness.

A great selection of paints, varnishes, and other coatings is available in aerosol cans. Therefore, their use is encouraged for touch-up work on machinery and full paint jobs on small tools and equipment. Paint in aerosols is generally more expensive than an equal amount of paint in a regular can. However, the time saved and the attractive appearance obtained make the aerosol spray can the preferred method for many small jobs.

□ SPRAY PAINTING EQUIPMENT

Spray Guns

A *spray gun* is a device that releases paint in the form of a fine spray. Spray guns may be the air atomization type or they may be airless. *Air atomization* means that the paint is split into tiny droplets using compressed air, figure 28–13.

A typical air spray gun consists of a gun body, fluid control knob, pattern control knob, air valve, paint needle valve, packing nuts, fluid nozzle, air nozzle,

and trigger, figure 28–14. Air pressure is kept constant by a regulator. When the trigger is pulled a short distance, air flows through the passageways of the gun and out the air nozzle. The air flow creates a vacuum in the nozzle. When the trigger is pulled further, the fluid needle valve opens and paint rushes out the fluid nozzle. Holes are positioned in the air nozzle to discharge air at angles which force the paint into a fan or cone pattern, depending on the adjustment.

Airless spray guns are similar in design to air spray guns. However, no air is used in the gun. Instead, the paint is pumped through hoses under high pressure. The pressurized paint then rushes through the nozzle and is atomized as it is released into a spray pattern.

Caution:

Never point an airless gun at any part of your body or at anyone else. If the trigger is pulled, the paint is discharged at sufficient pressure to penetrate the skin. If this occurs, get medical attention immediately.

Fluid Systems

Air spray guns may be connected to paint supplies in one of three ways: (1) a siphon feed cup, (2) a pressure feed cup, or (3) a pressure feed tank. All three

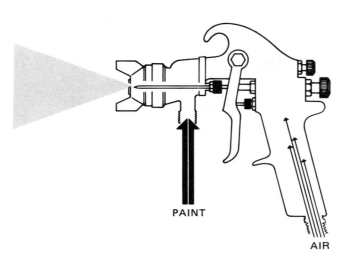

PAINT

AIR

Figure 28–13. An air spray gun uses compressed air to split paint into tiny droplets (*courtesy of Binks Manufacturing Co.*)

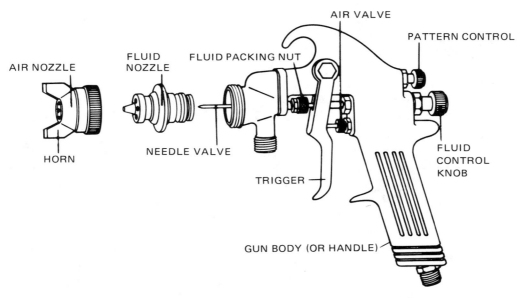

Figure 28–14. Parts of an air spray gun (*courtesy of Binks Manufacturing Co.*)

systems must include a regulator and extractor in the air line. The *regulator* keeps the air pressure at a set level. The *extractor* removes moisture and oil from the compressed air. Moisture and oil in the air interfere with the spray pattern and ruin the paint job.

The *siphon system*, figure 28–15, has a small vent hole in the cap of the cup. This hole must be kept open at all times to permit atmospheric pressure on the paint. Paint may cover the hole if the cup is tipped too far. As a result, there will be a vacuum in the cup and the gun will not work. The hole may be reopened by using a small piece of wire or a stiff broom straw. The siphon-type system works by creating a low pressure area in the nozzle where paint can move in by force of atmospheric pressure.

The *pressure feed cup* and *pressure feed tank systems* operate with the paint under air pressure at all times. The pressure pushes the paint through the system. These systems do not use a vent hole in the cap. The pressure feed cup holds only one quart of paint; the pressure feed tank may hold five gallons or more. The pressure tank system is good for large

jobs such as barns. The paint tank can remain on the ground while the operator is painting the sides or roof. The operator needs to carry only the gun and hoses.

Air Supply

Good spray painting systems permit fast and efficient painting. However, the initial cost is relatively high. For shop and farm use, a fairly large compressor is needed to operate a good spray gun. It is a good idea to buy the compressor, air hose, regulator, extractor, and related equipment from one source, figure 28–16. The supplier should provide assurance that the components work well together to provide an efficient system.

It is especially important that the compressor provide as many cubic feet per minute (*CFM*) of air as the gun requires. The air must also be delivered to the gun at the required pressure. To achieve this requires adequate sizing of the motor, compressor, electric wiring, air pipe, and hoses.

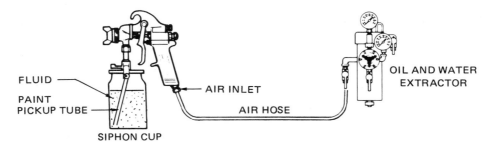

Figure 28–15. A siphon cup air spray system (*courtesy of Binks Manufacturing Co.*)

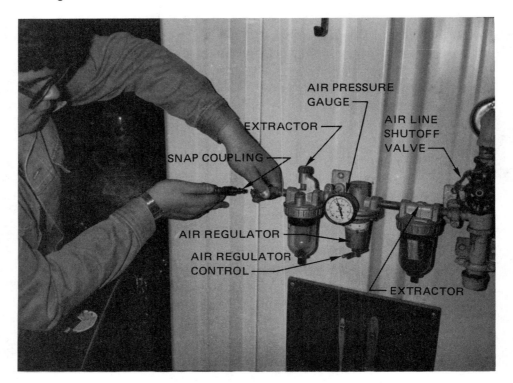

Figure 28–16. To prepare for spray painting, an air hose is attached between the gun and other parts of the spray system. This system includes a snap coupling, oil and water extractors, a pressure regulator with a gauge, a shutoff valve, an air line, tank, and compressor.

□ SPRAY PAINTING PROCEDURES

Preparing the Surface

Work should be prepared for painting as outlined in the previous unit. Expensive equipment and a skillful painter cannot make paint attractive and permanent if the surface has not been prepared properly. All surfaces must be clean of blistered paint, rust, dirt, oil, and wax. Glossy surfaces must be sanded to create a dull finish to which paint can stick.

Choosing a Location

Gun spraying is not recommended indoors unless a commercial spray booth is available. Paint spray is a fire and explosion hazard. In addition, the spray will settle on surfaces and ruin ventilation equipment. Spraying may be done outdoors if the temperature is 50° F or above and there is no wind.

Following Safety Rules

Spray painting can be hazardous. Some safety rules for spray painting are:

- Mix paints and solvents in well-ventilated areas.
- Spray paint outdoors or in special booths with recommended electrical fixtures and ventilation.

Caution:

All automobiles and other objects with good paint jobs should be at least 100 feet from the painting area. Otherwise, particles of paint will drift, settle on them, and ruin their appearance.

- Never smoke or have any source of fire in the vicinity.
- Wear a respirator or paint mask when painting.
- Use ladders that are in good condition. Do not use ladders with rungs or sides cracked or damaged.
- To avoid electrocution, be especially careful to keep aluminum ladders away from all electrical wires.

Preparing to Paint

Nothing is more frustrating than to have the project ready, spray equipment set up, paint in the gun, and then find that the gun will not spray paint. Such problems occur from the improper preparation of paint and equipment. Paint is prepared for spraying as follows.

Procedure

1. Thoroughly mix and stir the paint until no streaks can be seen. If more than one container of paint is to be used, mix them together to assure a uniform color.
2. Thin the paint at room temperature according to the manufacturer's instructions. Since all thinners and paints are not compatible, mix a small amount of paint and thinner to see if they form a smooth paint with uniform color.
3. To be safe, use a viscosimeter to confirm that the paint is thin enough. A *viscosimeter* is an instrument used to measure the rate of flow of a liquid. A liquid's tendancy to flow is called its *viscosity*.
4. Using a commercial paint strainer, strain the paint into a clean container. Strainers are available from paint and automotive parts stores.
5. Pour the thinned paint into a clean sprayer cup or tank.

Adjusting the Spray Gun

Fluid and air pressures should be set as low as possible, yet do the job. Pressure requirements vary with the design of the gun, the viscosity of the paint, and the temperature of the paint. All guns should be adjusted according to the manufacturer's instructions. However, some general recommendations for adjusting spray guns are as follows.

Procedure

1. Set up all equipment.
2. Put on safety glasses and a paint respirator.
3. Set up a sheet of cardboard to use as a practice surface.

4. Turn on the air and set the regulator at the level specified for the gun.
5. Open the pattern control and fluid control knobs.
6. Prepare to make a pass by pointing the gun off to the left side of the panel. Pull the trigger slowly until air is heard coming from the nozzle, if using an air-type gun. Then pull the trigger further until a faint paint fan is seen.
7. Position the gun with the nozzle 6 to 10 inches from the work and perpendicular to it. Slowly move the gun across the surface, maintaining the same distance from it, to create an even spray band, figure 28–17.
8. Readjust the pattern control to obtain a fan-shaped pattern.
9. Adjust the fluid control knob to permit the desired amount of paint to flow when the trigger is pulled all the way.
10. Adjust the distance from the object and the rate of travel until a good, even pattern is obtained. Each stroke should overlap the previous one by 50 percent.

Operating the Spray Gun

Spray guns must be held the correct distance from the work, figure 28–18. The gun stroke is made by moving the gun parallel to the work.

The closer the gun is held to the work, the faster the gun must be moved to prevent sags and runs. A *sag* occurs when a large area of paint shifts downward because it has been applied too thickly. A *run* is a narrow stream of paint flowing downward due to excessive buildup. On the other hand, holding the gun too far from the work causes dry spray and excessive spray dust. These result in a flat, dull, and rough finish.

When painting tractors and machinery, many irreg-

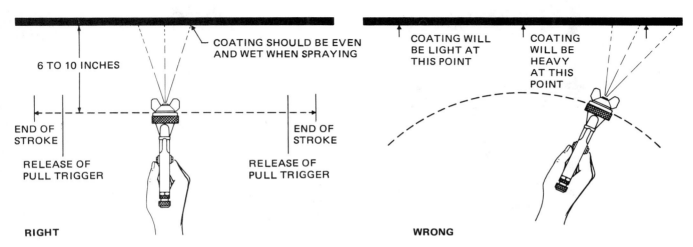

Figure 28–17. Moving the gun parallel to the surface assures that the coating is even (*courtesy of Binks Manufacturing Company*).

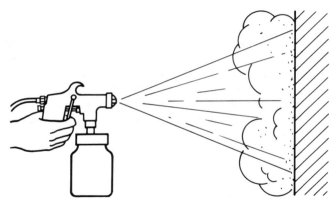

A. NOZZLE TOO FAR FROM THE SURFACE. DUSTING RESULTS IN A ROUGH PAINT FILM WITH A DULL FINISH.

B. NOZZLE IS TOO CLOSE. A THICK AND UNEVEN FILM RESULTS IN SAGS AND RUNS.

Figure 28–18. When the spray nozzle is held at an incorrect distance from the object being painted, an undesirable effect is obtained.

ular surfaces must be sprayed. Under these conditions the spray pattern should be narrow. This permits paint to reach parts that are farther than the ideal distance from the nozzle. Large smooth areas such as hoods and fenders require a broad fan. The shape of the fan is controlled by adjusting the pattern control knob.

Modern machinery enamels are designed to be sprayed in two or more layers at one time. To do this, spray on a thin layer and let it become tacky. *Tacky* means sticky. By the time a machine such as a tractor or baler has been given one coat, the paint applied at the beginning of the job will be tacky. Go back to the starting point and apply a heavy top coat. The paint layer should shine but not run or sag. When a run or sag occurs, the entire part or panel must be wiped clean with solvent and then repainted. Another method of correcting sags or runs is to allow the part to dry thoroughly, then sand the area and repaint.

Correcting Faulty Spray Patterns

Faulty spray patterns can be caused by incorrect pressure, clogged air passages, loose packing nuts, dry packings, thick paint, and unstrained paint.

After each use, a drop of light machine oil should be placed on each needle in a spray gun where it enters the packing nut. This procedure helps eliminate some problems with faulty patterns. Other recommendations for correcting faulty patterns are given in figure 28–19.

CAUSE
Dried material in side port "A" restricts passage of air through port on one side. Results in full pressure of air from clean side of port in a fan pattern in direction of clogged side.

REMEDY
Dissolve material in side port with thinner. Do not use metal devices to probe into air nozzle openings.

CAUSE
Dried material around the outside of the fluid nozzle tip at position "B" restricts the passage of atomizing air at one point through the center ring opening of the air nozzle. This faulty pattern can also be caused by loose air nozzle, or a bent fluid nozzle or needle tip.

REMEDY
If dried material is causing the trouble remove air nozzle and wipe off fluid tip, using rag wet with thinner. Tighten air nozzle. Replace fluid nozzle or needle if bent.

CAUSE
A split spray pattern (heavy on each end of a fan pattern and weak in the middle) is usually caused by: (1) atomizing air pressure, too high (2) attempting to get too wide a spray with thin material, (3) not enough material available.

REMEDY
(1) Reduce air pressure. (2) Open material control "D" to full position by turning to left. At the same time turn spray width adjustment "C" to right. This reduces width of spray but will correct split spray pattern.

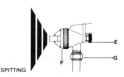

SPITTING

CAUSE
■ Air entering the fluid supply.
■ Dried packing or missing packing around the material needle valve which permits air to get into fluid passageway.
■ Dirt between the fluid nozzle seat and body or a loosely installed fluid nozzle.
■ A loose or defective swivel nut, siphon cup or material hose.

REMEDY
■ Be sure all fittings and connections are tight.
■ Back up knurled nut "E", place two drops of machine oil on packing, replace nut and finger tighten. In aggravated cases, replace packing.
■ Remove air and fluid nozzles "F" and clean back of fluid nozzle and nozzle seat in the gun body, using a rag wet with thinner. Replace and tighten fluid nozzle using wrench supplied with the gun. Replace air nozzle.
■ Tighten or replace swivel nut "G".

CAUSE
A fan spray pattern that is heavy in the middle, or a pattern that has an unatomized "salt-and-pepper" effect indicates that the atomizing air pressure is not sufficiently high, or there is too much material being fed to the gun.

REMEDY
Increase pressure from air supply. Correct air pressures as discussed elsewhere in this manual.

Figure 28–19. Faulty spray patterns and how to correct them (*courtesy of Binks Manufacturing Co.*)

Cleaning Spray Equipment

Cleaning spray guns and equipment requires a lot of solvent. Varsol is recommended for oil paints because it is low in cost. Another advantage of varsol is that the fire hazard is not as great as it is with other solvents.

To clean the gun, cup, and other equipment, the following procedure is suggested.

Procedure

1. Pour all unused paint back into the container and reseal it.
2. Wipe most of the paint from the interior of the paint can and gun parts.
3. Pour about 1 inch of varsol into the cup.
4. Attach the cup to the gun and shake the gun so the varsol can dissolve the paint residue.
5. Spray a piece of cardboard until the cup is empty. Cardboard absorbs the thinner.
6. Wipe all visible surfaces, both inside and outside of the gun and cup, with clean varsol on a cloth.
7. Repeat steps 3 to 6 one or two times using paint thinner until all traces of paint are absent from the thinner.
8. Remove the air nozzle and paint nozzle.
9. Use a cloth with thinner to wipe away any remaining evidence of paint.
10. Dry all parts and reassemble the gun.

Caution:

The spray gun will not work well the next time unless every trace of paint is removed. A good rule to follow is that you should not be able to tell the color of the paint used previously when examining the gun.

---

11. Clean all other equipment using a cloth and solvent.
12. Discard all paper, cloths, used strainers, and containers in an approved metal trash container.
13. Store all equipment properly.
14. Turn off the air pressure and/or compressor.

Spray painting can be very satisfying. However, care must be taken to prepare the work and paint carefully. Absolute cleanliness is required to avoid spray gun problems. Since paint and thinners are very flammable, there is always a danger of fire. Every precaution must be exercised to keep fire and electrical sparks away from the spray painting area.

☐ STUDENT ACTIVITIES

1. Define the italicized terms used in this unit.
2. Examine the paint on a wooden building, a wooden fence, and a piece of used farm machinery or garden equipment. What evidence is there to indicate a need for repainting?
3. Examine the labels on a can of oil-base paint and a can of latex paint. List the ingredients and the percentage of each. What is recommended as a thinner for each?
4. Prepare a wooden project for painting. Brush paint the project.
5. Prepare a metal project for painting. Spray paint the project using an aerosol spray container.
6. Prepare a piece of machinery for spray painting. Use a spray gun to paint the machine.
7. Use a pan and roller to paint a fence, room, or building.

☐ SELF-EVALUATION

A. Multiple Choice. Select the best answer.
   1. Painting is not a necessity when it
      a. is done for personal preference
      b. preserves wood
      c. prevents rot
      d. prevents rust
   2. An example of a low quality pigment is
      a. aluminum
      b. calcium
      c. titanium
      d. zinc
   3. Lumpy particles in paint should be
      a. ignored
      b. mashed up
      c. reduced by thinning
      d. removed by straining
   4. Perfectly straight lines are obtained when painting by use of
      a. a steady hand
      b. masking tape
      c. a ruler and pencil
      d. a yardstick
   5. Oil paints should not be applied with
      a. sponges
      b. nylon bristle brushes
      c. pure bristle brushes
      d. rollers
   6. Latex paints should not be
      a. brushed excessively
      b. applied with rollers
      c. used outdoors
      d. used on wood

7. An aerosol for painting is
   a. not pressurized
   b. relatively expensive
   c. difficult to use
   d. too expensive for small jobs
8. All spray guns are
   a. airless
   b. pressurized
   c. siphon-type
   d. none of these
9. Painting may be hazardous because of
   a. explosive materials
   b. flammable materials
   c. compressed air
   d. all of these
10. Painting hazards are greatly increased by
    a. excessive ventilation
    b. smoking
    c. spraying outdoors
    d. using a respirator

B. Matching. Match the terms in column I with those in column II.

Column I	Column II
1. formulate	a. dull
2. chalking	b. thin with water
3. gloss	c. first coat
4. flat	d. poisonous
5. latex	e. removes water
6. titanium dioxide	f. controls pressure
7. primer	g. put together
8. lead	h. safe, high-quality pigment
9. extractor	i. shiny
10. regulator	j. wash off

C. Completion. Fill in the blanks with the word or words that make the following statements correct.

1. Gun spray painting should not be done indoors unless a _____ is available.
2. Aluminum ladders are a special hazard around _____ .
3. Dry paint in a spray gun will _____ .
4. Spray guns should be lubricated with thin machine oil on each _____ where it enters the _____ .
5. The tendency of a liquid flow is known as its _____ .

Unit 29
Small Engine Maintenance

Objective

To maintain and perform minor repairs on small engines.

Competencies to Be Developed

After studying this unit, you will be able to:
* Practice appropriate safety precautions.
* Identify the major parts and systems of small engines.
* Describe the general operation of two- and four-cycle engines.
* Conduct recommended maintenance procedures on small engines.
* Solve minor small engine problems.
* Prepare small engines for storage.

Terms to Know

internal combustion engine	reed valve
piston	intake stroke
crankshaft	compression stroke
reciprocating	compression ratio
horizontal shaft engine	top dead center (TDC)
vertical shaft engine	bottom dead center (BDC)
stroke	fire
cycle	ignition
four-stroke cycle	power stroke
four-cycle engine	exhaust
two-stroke cycle	exhaust stroke
two-cycle engine	revolution
poppet valve	momentum
port	maintenance

troubleshooting	score
repair	torqued
adjust	speed indicator
contaminant	RPM
saturate	leaner
dual	richer
precleaner	labor
shroud	

Materials List

Protective goggles and coveralls
General shop hand tools and wrenches
Four-cycle engine in need of maintenance
Two-cycle engine in need of maintenance
Clean gasoline in an approved fuel can
Oil of recommended type and grade for the engine
Spark plug gauge
Rags and approved waste disposal cans
Varsol, kerosene, or approved parts cleaner.

Caution:

Do not use gasoline as a parts cleaner due to the fire hazard

Gasoline and diesel engines have become the backbone of the world in motion. They power automobiles, trucks, tractors, diesel electric trains, and ships at sea.

Small gasoline engines are used on boats, recreational vehicles, snowmobiles, and bikes. Small engines run chain saws, portable grass cutters, leaf blowers, and lawnmowers, common in most neighborhoods. Small engines power garden tractors, tillers, bale throwers, and elevators on farms. The ability to keep engines working is a valuable skill for most workers in agriculture.

□ SAFETY FIRST

There is some danger inherent in using all machines. Gasoline is an added hazard in gasoline engines. Suggestions for avoiding fire and injury around small engines are as follows:

- Wear safety glasses and leather shoes.
- Keep hands and feet a safe distance from all moving parts.
- Do not smoke around oil, solvents, or gasoline. Do not smoke near the openings of gasoline tanks and containers.
- Stop the engine before refueling.
- Do not spill gasoline on hot engine parts.
- Whenever possible, handle gasoline outdoors.
- Disconnect the spark plug cable when making any adjustments on machines to prevent the possibility of accidental engine starting.
- Keep all shields in place.
- Do not operate engines above the specified speed.
- Do not overload engines or force equipment beyond the designed capacity.

□ HOW AN ENGINE RUNS

An engine requires fuel, oxygen, and heat to run. It harnesses the energy of fuels such as gasoline, diesel fuel, propane gas, or kerosene. The fuel mixes with air, which provides the oxygen. The mixture is ignited by a spark, which provides the heat.

Internal combustion engines have been around for about a hundred years. An *internal combustion engine* is a device that burns fuel inside a cylinder to create a force that drives a *piston*, or cylindrical device within the cylinder. The moving piston pushes a connecting rod, which turns a *crankshaft*, figure 29–1. A *crankshaft* is a device which converts circular motion to linear motion, or vise versa. In this way, the *reciprocating*, or back-and-forth, motion of the piston is changed to circular motion.

All of an engine's parts and systems exist to convert the energy of burning fuel to circular motion. Once circular motion is obtained, a system of gears, sprockets, chains, belts, and pulleys is used to drive machines, figure 29–2.

Engine Classification

Engines can be classified by the normal position of the crankshaft. If the crankshaft is placed sideways in an engine, the engine is classified as a *horizontal shaft engine*, figure 29–3. If the crankshaft is in an

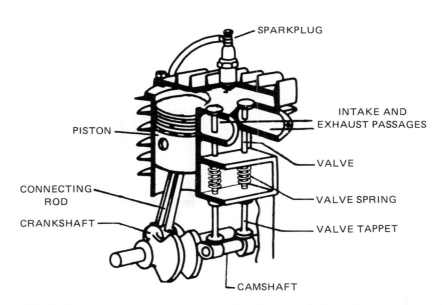

Figure 29–1. In an internal combusion engine, burning fuel pushes a piston, which is attached to a connecting rod. The connecting rod turns a crankshaft (*courtesy of The Pennsylvania State University, Department of Agricultural Education*).

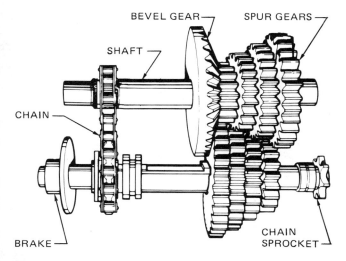

Figure 29–2. Once circular motion is obtained, a system of gears, sprockets, chains, belts, and pulleys is used to drive machines (*courtesy of Tecumseh Products Company, Engine and Gear Service Division*).

up-and-down position, the engine is known as a *vertical shaft engine*, figure 29–4. Engines designed to operate in all positions, such as those used on chain saws and weed trimmers, may be regarded as multiposition shaft engines.

Engines are also classified according to the number of strokes in a cycle. A *stroke* is the movement of the piston from top to bottom or from bottom to top.

The top of the stroke refers to the position of the piston when it is most distant from the crankshaft. A *cycle* is all the events that take place as an engine: (1) takes in air and fuel, (2) compresses the air-fuel mixture, (3) burns the mixture, and (4) expels the burned gases.

In a *four-stroke-cycle* engine, one event is accomplished during each stroke. This means that four strokes are required to complete a cycle. The name is shortened to *four-cycle engine*. In a *two-stroke-cycle* engine, all four functions are completed in just two strokes. It is called a *two-cycle engine* for short. Both four-cycle and two-cycle engines are available with either horizontal or vertical crankshaft.

Engine Parts

The major parts of engines are shown in the cross-section views in figures 29–3 and 29–4. On the four-cycle engine, figure 29–3, the major parts are the oil pump, connecting rod, piston, cylinder, fins, cylinder head, spark plug, exhaust valve, intake valve (not shown), intake port, blower fan, carburetor (not shown), gas tank (not shown), muffler (not shown), valve lifters, flywheel, camshaft drive gear, crankcase (also called the assembly), camshaft, and crankshaft.

The two-cycle engine has most of the same parts as the four-cycle engine. The main difference is in the valve systems. A four-cycle engine has two poppet

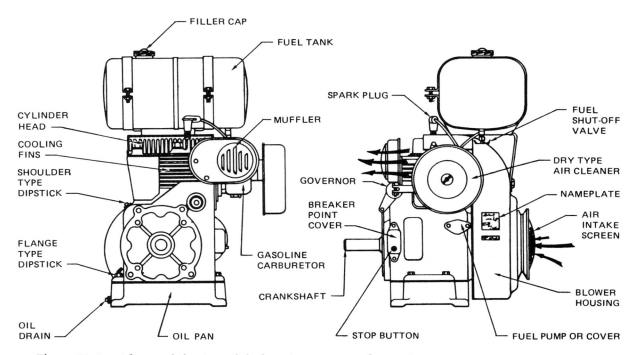

Figure 29–3. A four-cycle horizontal shaft engine (*courtesy of G. Stephenson,* Small Gasoline Engines, 4th ed., *Delmar Publishers Inc., 1984*)

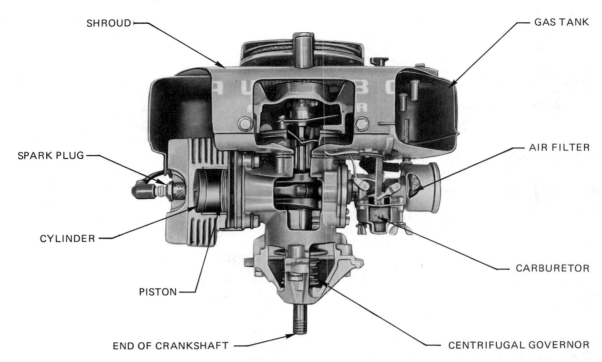

SHROUD

GAS TANK

SPARK PLUG

AIR FILTER

CYLINDER

CARBURETOR

PISTON

END OF CRANKSHAFT

CENTRIFUGAL GOVERNOR

Figure 29–4. A two-cycle vertical shaft engine (*courtesy of G. Stephenson*, Small Gasoline Engines, 4th ed., *Delmar Publishers Inc., 1984*)

valves per cylinder. A *poppet valve* controls the flow of air and gases by moving up and down. It is pushed open by a cam and closes by spring action. (A *cam* is a raised area on a shaft.) On the other hand, a two-cycle engine has one or more intake ports and exhaust ports per cylinder. A *port* is a special hole in the cylinder wall closed by the piston moving over the opening. A two-cycle engine also has a reed valve. A *reed valve* is a flat, flexible plate that lets air and fuel in but will not let the mixture out.

Operation of a Four-Cycle Engine

The four strokes of a four-cycle engine are shown in figure 29–5.

Intake Stroke. □ The intake stroke starts with the piston in its top position (most distant from the crankshaft). As the piston moves toward the crankshaft, it creates a vacuum in the cylinder. During this movement, the intake valve is open and a proper mixture of air and fuel from the carburetor rushes in (hence the term *intake stroke*). It is the function of the carburetor to supply a mixture of air and fuel in the correct proportion. The intake valve closes at the end of the stroke.

Compression Stroke. □ At the bottom of the intake stroke and with both valves closed, the crankshaft moves across the bottom of its swing. The movement of the engine parts pushes the piston upward and

squeezes or compresses the air-fuel mixture. This action gives rise to the term *compression stroke*. At the beginning of this stroke, the compression chamber and the cylinder are full of air and fuel. At the top of the compression stroke, the air and fuel mixture has been compressed into just the area in the head called the combustion chamber. The relationship between the volume of the cylinder plus combustion chamber at the beginning and end of this stroke is known as the *compression ratio*. Compression ratios for small gas engines are around 6 to 1. The ratio is written as 6:1.

Power Stroke. □ When the piston is at its highest point (greatest distance from the crankshaft), it is said to be at *top dead center* (*TDC*). When it is at its lowest point (closest to the crankshaft), it is said to be at *bottom dead center* (*BDC*). At or near TDC on the compression stroke, the spark plug fires. To *fire* means to make a spark jump across an air gap. This spark ignites the compressed air-fuel mixture. The process is called *ignition*. The burning fuel expands rapidly but evenly and drives the piston down. This stroke is thus known as the *power stroke*. It is the only stroke that receives any force from the fuel.

Exhaust Stroke. □ At or near the bottom of the power stroke, the exhaust valve opens. As the piston moves upward it pushes out the burned or *exhaust* gases. Hence the term *exhaust stroke*. After the ex-

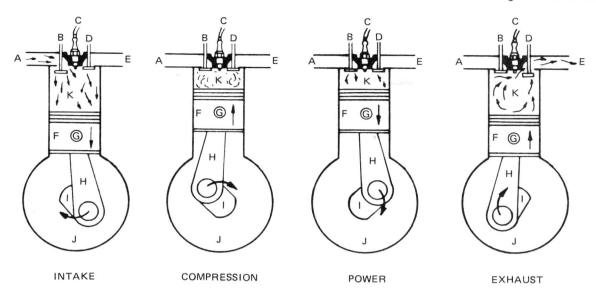

INTAKE	COMPRESSION
POWER	EXHAUST

KEY:
A = INTAKE PASSAGE
B = INTAKE VALVE
C = SPARKPLUG
D = EXHAUST VALVE

E = EXHAUST PASSAGE
F = PISTON
G = PISTON PIN
H = CONNECTING ROD

I = CRANKSHAFT
J = CRANKCASE
K = COMBUSTION
    CHAMBER

Figure 29–5. Operation of a four-stroke-cycle engine *(courtesy of The Pennsylvania State University, Department of Agricultural Education)*

haust stroke, the exhaust valve closes and the piston is at the top and ready for another intake stroke to start the cycle again.

Notice that the crankshaft made two complete turns or *revolutions* in the cycle. It is the *momentum* or turning force of the flywheel and other moving parts that carries the engine through the three nonpower strokes. In engines with more than one cylinder, one cylinder will be on the power stroke while the other cylinders are on their nonpower strokes.

Another important point about the four-cycle engine is the use of the crankcase as an oil reservoir. Oil is splashed on engine parts by a special dipper attached to the crankshaft. Further, oil is splashed by the crankshaft and connecting rod parts that pass through the oil as the engine runs. In addition, most four-cycle engines have an oil pump to pump oil to the valves and bearings. This lubrication is needed to minimize friction and keep the engine parts operating smoothly.

Operation of a Two-Cycle Engine

As stated before, the two-cycle engine completes the intake, compression, power, and exhaust stages in two strokes. The crankcase of the engine does not contain oil. Instead, it is airtight and contains a reed valve to admit the air-fuel mixture from the carburetor when the piston moves away from the crankshaft. The mixture is compressed when the piston returns toward the crankshaft. The crankcase then holds the mixture under pressure until the cylinder is ready to receive it. Oil is mixed with the gasoline to lubricate the moving parts of a two-cycle engine. Hence, there is no need for oil in the crankcase.

> **Caution:**
>
> **It is essential that the gasoline and the proper type of oil be mixed in exact proportions as recommended by the engine manufacturer. Since a two-cycle engine does not contain oil in the crankcase, it depends totally on the oil in the gasoline to lubricate all engine parts.**

The operation of a two-cycle engine is shown in figure 29–6. To begin a cycle, the piston moves upward and creates a vacuum in the crankcase of the engine.

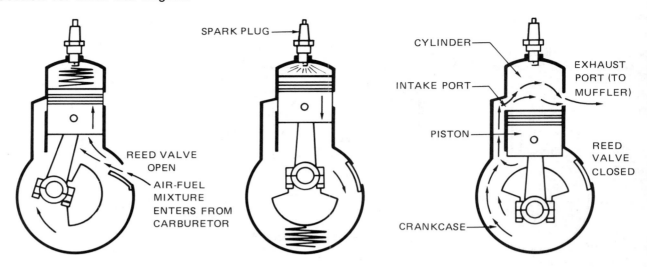

Figure 29–6. Operation of a two-stroke-cycle engine (*adapted from* Small Gasoline Engines Student Handbook, *The Pennsylvania State University, Department of Agricultural Education*)

The air-fuel mixture from the carburetor rushes into the crankcase to fill the vacuum. When the piston stops its upward movement, the reed valve closes by its own spring action. The piston then moves down and puts pressure on the mixture in the crankcase. Now the engine is ready to start a normal cycle.

Intake and Exhaust. □ As the piston nears the bottom of its stroke, it uncovers the intake and exhaust port(s), figure 29–6(C). Since the air-fuel mixture in the crankcase is under pressure, it rushes through a passage to the intake port and enters the cylinder. This incoming gaseous mixture pushes air or exhaust out of the cylinder. Therefore, intake and exhaust functions occur with very little movement of the piston.

Compression Stroke. □ The cylinder now is filled with an air-fuel mixture. The piston moves upwards, closes the intake and exhaust ports and compresses the air-fuel mixture trapped in the cylinder. At the same time, a new supply of air and fuel rushes into the crankcase, figure 29–6(A).

Power Stroke. □ At or near top dead center (TDC) the spark plug fires to ignite the mixture. The burning and expanding gases drive the piston downward

through the power stroke. This same downward movement puts pressure on the new air-fuel mixture in the crankcase, figure 29–6(B). Thus the engine completes its cycle of intake, compression, power, and exhaust with only two strokes of the piston.

□ SMALL ENGINE MAINTENANCE

An understanding of what happens in a cylinder is useful for engine maintenance and troubleshooting. *Maintenance* is doing the tasks that keep a machine in good condition. *Troubleshooting* is determining what is wrong with a machine so that the problems can be corrected. *Repairing* means replacing a faulty part or making it work correctly. *Adjusting* means setting a part or parts so the machine functions as designed.

Maintenance procedures for small engines are concerned with keeping the engine and everything that enters the engine clean. All oil, fuel, and air must be clean when they enter the engine or the engine will not run properly for very long.

Dirt or water in the fuel will make the engine miss or stall. Dirt or other contaminants in the oil will cause excessive wear of all parts. A *contaminant* is any material that does not belong in a substance. Dust in the air that enters the cylinder will scratch the cylinder walls. The engine will then lose compression and power.

Air Cleaners and Filters

An air cleaner is attached to the carburetor. It is designed to remove dirt and dust from the air before the air mixes with fuel. An engine may be ruined in just a few hours of operation if the air cleaner is not working properly. This is especially true under dusty conditions. Most small engine air filters should be cleaned after every 25 hours of operation or more frequently under dusty conditions.

Oil Foam Type. □ To service an oil foam air cleaner, the following general procedure is suggested. This procedure is also shown in figure 29–7.

Procedure

1. Remove the wing nut and cover.
2. Carefully remove the spongelike foam element from the base.

Caution:

Do not rip the foam.

3. Wash the foam in kerosene, varsol, or water with detergent.
4. Squeeze out the excess liquid.
5. Dry the foam by squeezing it with a dry cloth.
6. Saturate the foam with the motor oil recommended for the engine. (To *saturate* means to add a substance until the excess starts to run out.)
7. Squeeze the foam to remove the excess oil.
8. Carefully reassemble the air cleaner.

Dry Element Type. □ Dry element cleaners are efficient and easy to maintain. To clean a dry element, simply remove the metal cover and element, figure 29–8. Tap the bottom of the element on a flat surface to dislodge large particles of dirt and chaff and then reinstall the element.

Caution:

Use of compressed air to clean the element may result in damage to the element.

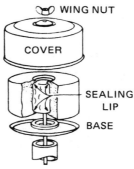

WING NUT

COVER

SEALING LIP

BASE

REMOVE THE WING NUT AND COVER; REMOVE FOAM ELEMENT FROM THE BASE

CUP

PUSH DOWN FOAM ELEMENT AND PULL OUT AIR CLEANER CUP

WASH THE FOAM IN KEROSENE, VARSOL, OR WATER WITH DETERGENT

CLOTH

SQUEEZE OUT EXCESS LIQUID AND THEN DRY THE FOAM BY SQUEEZING IT WITH A DRY CLOTH

SATURATE THE FOAM WITH THE RECOMMENDED MOTOR OIL

SQUEEZE OUT THE EXCESS OIL

CAREFULLY REASSEMBLE THE AIR CLEANER

Figure 29–7. Servicing an oil foam air cleaner (*adapted from* Engine Maintenance Guide, *Briggs and Stratton Corporation*)

Replace the element at the time interval recommended by the engine manufacturer. Replacement is recommended whenever the element becomes damaged or oily, or appears to be restricting air flow. Any element that has even a small hole or slight tear must be replaced.

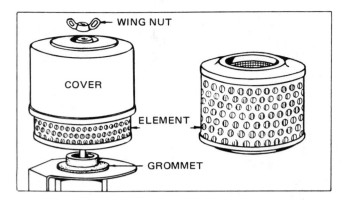

Figure 29–8. To service a dry element air cleaner, remove the metal cover and element. Tap the bottom of the element on a flat surface to dislodge the dirt (*courtesy of Briggs and Stratton Corporation*).

Caution:

When assembling the cleaner element be careful to center the element over its seat or grommet. Any crack or opening will permit dirt to bypass the filter.

Dual Element Type. ☐ A dual element air cleaner has both a dry element and an oil foam cover, figure 29–9. *Dual* means two. The same procedures described for the oil foam and dry element cleaners are used here. The oil foam cover keeps most dirt from entering the cleaner and so serves as a *precleaner*. The foam should be serviced every 25 hours. The dry element needs servicing less frequently. Some manu-

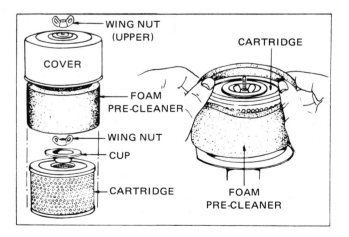

Figure 29–9. A dual element air cleaner has both a dry element and an oil foam cover (*courtesy of Briggs and Stratton Corporation*).

facturers recommend changing the dry element after about 100 hours of service.

Oil Bath Type. ☐ Oil bath cleaners are found on some older engines. They consist of a metal container with a metal mesh core. The following procedure is used to service this type of cleaner.

Procedure

1. Remove the nut and top cover.
2. Lift out the wire mesh core.
3. Lift off the metal container.
4. Pour the oil and dirt out of the container. Wash the container in kerosene or varsol. Wipe dry.
5. Add motor oil to the mark on the container.
6. Flush the wire mesh core in kerosene or varsol.
7. Shake excess solvent off the core.
8. Reassemble the unit.

Crankcase Oil

It is important to keep the crankcase of a four-cycle engine filled to the correct level with the recommended type of oil. The correct oil level for most engines is indicated by a mark on a dip stick or by a filler plug located so that oil will fill to the level of the plug, figure 29–10.

Oil Functions and Classifications. ☐ Oil has four major functions: (1) to lubricate, (2) to cool, (3) to seal, and (4) to clean by carrying contaminants away from the engine. Oils for internal combusion engines are classified according to their performance under various operating conditions. Performance ratings are then designated by code letters in the classification. Most manufacturers of small gas engines call for oil with a service classification of SF, SE, SD, or SC. The service classification is clearly marked on every container of motor oil.

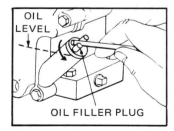

Figure 29–10. On some engines, oil fills to the level of the filler plug making oil level easy to check (*courtesy of Briggs and Stratton Corporation*).

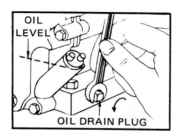

Figure 29-12. To change engine oil, remove the oil drain plug and drain the oil in the crankcase while the engine is warm. Replace and tighten the drain plug. Refill crankcase with new oil through the oil fill plug (*courtesy of Briggs and Stratton Corporation*).

Caution:

Using any oil other than the type recommended by the engine manufacturer may void the engine warranty and prove costly in other ways in the long run.

Oil Viscosity. □ Engines need lighter weight oil as temperatures drop. Oil thickens as it gets cold so that it cannot get between tight-fitting moving parts. Therefore, the oil must be changed as the seasons change in many areas. It is important to use oil with the correct viscosity grade as rated by the Society of Automotive Engineers (SAE), figure 29-11.

Oil Changes. □ After each 25 hours of engine use, the crankcase oil should be drained while the engine is warm, figure 29-12. Hot oil flows easily and carries the contaminants with it. The crankcase should then be refilled to the proper level using the recommended oil.

Additional attention may be needed if the engine has a reduction gear box. In this case, there is generally one hole for filling, one hole for checking the oil level, and one hole for draining the oil, figure 29-13. Each hole has a plug. Again, the manufacturer's

Caution:

Do not overfill the crankcase or the gear reduction box.

recommendations for the oil type and change interval should be followed. The oil level can be checked between oil changes by removing the oil level plug. If oil does not drip out when the plug is removed, oil is added until it does.

Cooling System

The cooling system on air-cooled engines consists of:

* fins on the flywheel, which serve as a fan to move air
* fins on engine parts to transfer heat away from the parts
* shrouds to route air over these fins. A *shroud* is a cover. It is important to keep grass, chaff, and dirt from blocking any of the areas under an engine shroud or between fins, figure 29-14. Such materials may be removed by brushing or using compressed air.

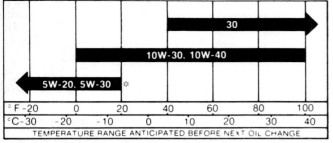

RECOMMENDED SAE VISCOSITY GRADES

*If not available, a synthetic oil may be used having — 5W-20, 5W-30 or 5W-40 viscosity.

Figure 29-11. Recommended viscosity of oil for use in the crankcase of four-cycle engines under various temperature conditions (*courtesy of Briggs and Stratton Corporation*)

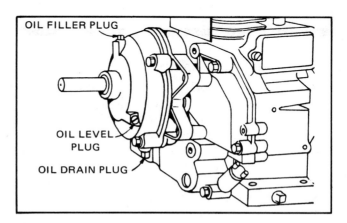

Figure 29-13. An engine equipped with a reduction gear box generally has one hole for filling, one hole for checking the oil level, and one hole for draining the oil (*courtesy of Briggs and Stratton Corporation*).

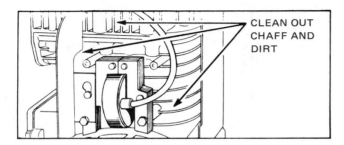

Figure 29–14. Keep grass, chaff, and dirt out of the fins and airways under the shroud (*courtesy of Briggs and Stratton Corporation*).

Spark Plug

After 100 hours of operation, the spark plug should be serviced or replaced. To service the plug, scrape off all deposits using a pocket knife, then soak the plug in a commercial solvent. Wipe the plug dry.

File the electrodes with a point file to restore their flat surfaces. Regap the plug to .030 inch, or as recommended by the manufacturer. Blow out all dirt particles with compressed air. Replace the plug and tighten until the gasket is partially flattened.

Caution:

Do not sandblast plugs used in small engines. Sand left by blasting *scores,* or scratches, the cylinder walls and ruins the engine. Use of sandblasted plugs will void the warranty on some engines. In view of the low cost of a new plug and the time and problems involved in cleaning a plug, it is generally better to replace the plug than to clean it.

Combustion Chamber

Carbon deposits build up on the tops of pistons, valves, and other surfaces in the combustion chamber. For this reason, it may be necessary to remove the cylinder head every 150 hours and scrape the carbon from all surfaces. An engine manual should be consulted for this work. Before scraping, make sure the piston is all the way up and the valves are closed. Use compressed air to blow away all carbon chips.

When reinstalling the head, clean all gasket surfaces with a smooth scraper. It is advisable to always replace the head gasket, figure 29–15. Head bolts should be tightened in the correct sequence, figure 29–16. All bolts on an engine should be *torqued* (tightened),

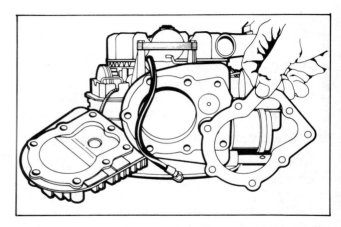

Figure 29–15. The head gasket must be replaced if it is damaged (*courtesy of Tecumseh Products Company, Engine and Gear Service Division*).

according to the manufacturer's specifications. In the absence of specific information, refer to a general torque specifications table such as the one shown in figure 29–17.

All manufacturers provide specific maintenance schedules for their engines. While there are many similarities, it is wise to follow specific manufacturer recommendations, if available, figure 29–18.

□ TROUBLESHOOTING

Small engines are generally reliable, and adjustments are not normally required. If an engine does not start or if it stops during use, the simplest things should be checked first.

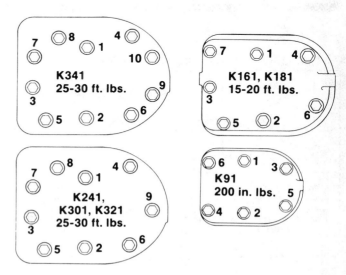

Figure 29–16. Recommended sequence for tightening head bolts (*courtesy of Kohler Co., Kohler Engines Division*)

Cast Iron or Steel

Size	Grade 2	Grade 5*	Grade 8
8-32	20 in. lb.	25 in. lb.	
10-24	32 in. lb.	40 in. lb.	
10-32	32 in. lb.	40 in. lb.	
1/4-20	70 in. lb.	115 in. lb.	165 in. lb.
1/4-28	85 in. lb.	140 in. lb.	200 in. lb.
5/16-18	150 in. lb.	250 in. lb.	350 in. lb.
5/16-24	165 in. lb.	270 in. lb.	30 ft. lb.
3/8-16	260 in. lb.	35 ft. lb.	50 ft. lb.
3/8-24	300 in. lb.	40 ft. lb.	60 ft. lb.
7/16-14	35 ft. lb.	55 ft. lb.	80 ft. lb.
7/16-20	45 ft. lb.	75 ft. lb.	105 ft. lb.
1/2-13	50 ft. lb.	80 ft. lb.	115 ft. lb.
1/2-20	70 ft. lb.	105 ft. lb.	165 ft. lb.
9/16-12	75 ft. lb.	125 ft. lb.	175 ft. lb.
9/16-18	100 ft. lb.	165 ft. lb.	230 ft. lb.
5/8-11	110 ft. lb.	180 ft. lb.	260 ft. lb.
5/8-18	140 ft. lb.	230 ft. lb.	330 ft. lb.
3/4-10	150 ft. lb.	245 ft. lb.	350 ft. lb.
3/4-16	200 ft. lb.	325 ft. lb.	470 ft. lb.

Aluminum

8-32	20 in. lb.	20 in. lb.	20 in. lb.
10-24	32 in. lb.	32 in. lb.	32 in. lb.
1/4-20	70 in. lb.	70 in. lb.	70 in. lb.
5/16-18	150 in. lb.	150 in. lb.	150 in. lb.

*Also Self Tapping Screws

Conversions

in. lbs. x .083 = ft. lbs.
ft. lbs. x 12 = in. lbs.
ft. lbs. x .1383 = kgm
ft. lbs. x 1.3558 = N m

Figure 29–17. Standard torque settings to be used when specific values are not specified (*courtesy of Kohler Co., Kohler Engines Division*).

Caution:

Do not adjust the carburetor or other parts until the problem is diagnosed correctly. Suggested steps for troubleshooting, or diagnosing gas engine problems follows.

Cold Engine Will Not Start

If the engine is cold and will not start, the problem may be that:

- no gas is in the tank—add gas.
- the fuel valve is shut off—turn on the fuel.
- the engine is flooded—leave it for about 10 minutes and then try again. Flooding refers to an excessive amount of gasoline in the cylinder. Such a condition may be caused by applying the choke too long. A more difficult cause to correct is the float needle valve sticking in the carburetor. If this is the case, the carburetor must be disassembled and corrections made.
- one or more control knobs are slipping on their shafts and the controls are not moving—tighten or replace the knobs or move the control by hand.

Hot Engine Will Not start

If the engine has been in use and will not restart, the problem may be that:

- no gas is in the tank—refill.
- the implement is clogged—clear the implement.
- the engine is overheated—leave it for 10 minutes and try again.

Starting Problems Persist

The next step in troubleshooting is to determine if the cylinder is getting fuel and spark. The following procedure is recommended.

Procedure

1. Remove and examine the spark plug. If the odor of gasoline is present on the plug or in the cylinder then the fuel system is probably working correctly. If the odor of gasoline is not present, trace the flow of fuel from the tank to the carburetor and look for fuel line blockages or restrictions.
2. To test for spark, attach the spark plug wire to the plug and lay the plug on its side on the engine head. Hold it down with a piece of rubber to avoid shock.
3. Crank the engine briskly and observe the spark at the gap. If there is no spark or if the spark is yellow, try a new plug. A good spark is blue in color.
4. If the engine still will not start or other problems are encountered, consult a troubleshooting guide, figure 29–19.

FREQUENCY	TYPE OF MAINTENANCE	SERVICE LOG	DATE SERVICED
DAILY	Check Oil Level Replenish Fuel Supply Clean Air Intake Screen	**DAILY Before Starting Engine**	
FIRST 5 Hours	On new or rebuilt engine, change oil after first 5 hours	[1] [2] [3] [4] [5] Change Oil	--------------
EVERY 25 Hours	Service Precleaner Change Oil	[6] ... [25]	--------------
EVERY 50 Hours	Clean Cooling Fins and External Surfaces Inspect Air Cleaner Element Check Oil Level in Reduction Gear Unit Check Belt Tension — Motor-Generator Set Clean Fuel Filter	[26] ... [50]	--------------
EVERY 100 Hours	Check Spark Plug Change Oil — Wet Type Clutch Check Battery Electrolyte Level	[51] ... [75] [76] ... [100]	--------------

ANNUALLY (or Every 500 Hours) These services require the attention of a trained mechanic and the use of special tools and equipment. Your Kohler engine service dealer has the facilities, training and genuine replacement parts necessary to properly perform these services.

| 500 Hours | ☐ Ignition Timing
☐ Breaker Points
☐ Valve / Tappet Clearance
☐ Starter Motor
☐ Alternator Connections
☐ Voltage Regulator
☐ Crankcase Breather
☐ Crankcase Vacuum
☐ Compression | **500 Hours*** | ☐ Cylinder Heads Serviced — Constant Load & Speed
*250 hours when leaded gasoline is used. |
| | | **1000 Hours**** | ☐ Cylinder Heads Serviced — Variable Load & Speed
**500 hours when leaded gasoline is used. |

ENGINE: Model _____ **Spec.** _____ **Serial No.** _____

Figure 29–18. Manufacturer's maintenance schedule for a single-cylinder engine. One block is crossed out for each hour of use (*courtesy of Kohler Co., Kohler Engines Division*).

Sheared Flywheel Keys

Many other problems can be diagnosed and corrected by the operator using common tools and a trouble-shooting guide. If an engine has stalled abruptly, such as when a rotary mower blade strikes a root, the flywheel key is probably sheared slightly or completely. In this case, the operator's manual should be used for guidance in replacing the key. If service

A. COMMON TROUBLES AND REMEDIES.
The following charts list the most common troubles experienced
with gasoline engines. Possible causes of trouble are given along
with probable remedy. Paragraph references direct the reader to the
correct portion of the handbook.

B. 4-CYCLE ENGINE TROUBLESHOOTING CHART

Cause	Remedy and Reference
ENGINE FAILS TO START OR STARTS WITH DIFFICULTY	
No fuel in tank	Fill tank with clean, fresh fuel.
Shut-off valve closed	Open valve.
Obstructed fuel line	Clean fuel screen and line. If necessary, remove and clean carburetor.
Tank cap vent obstructed	Open vent in fuel tank cap.
Water in fuel	Drain tank. Clean carburetor and fuel lines. Dry spark plug points. Fill tank with clean, fresh fuel.
Engine overchoked	Close fuel shut-off and pull starter until engine starts. Reopen fuel shut-off for normal fuel flow.
Improper carburetor adjustment	Adjust carburetor.
Loose or defective magneto wiring	Check magneto wiring for shorts or grounds; repair if necessary.
Faulty Magneto	Check timing, point gap, and if necessary, overhaul magneto.
Spark plug fouled	Clean and regap spark plug.
Spark plug porcelain cracked	Replace spark plug.
Poor Compression	Overhaul engine.
No spark at plug	Disconnect ignition cut off wire at the engine. Crank engine. If spark at spark plug, ignition switch, safety switch or interlock switch is inoperative. If no spark, check magneto. Check wires for poor connections, cuts or breaks.
Electric starter does not crank engine	See 12 volt starter troubleshooting chart.
ENGINE KNOCKS	
Carbon in combustion chamber	Remove cylinder head and clean carbon from head and piston.
Loose or worn connecting rod	Replace connecting rod.
Loose flywheel	Check flywheel key and keyway; replace parts if necessary. Tighten flywheel nut to proper torque.
Worn cylinder	Replace cylinder.
Improper magneto timing	Time magneto.

Figure 29–19. Troubleshooting guide (*courtesy of Tecumseh Products Company, Engine and Gear Service Division*)

4-CYCLE ENGINE TROUBLESHOOTING CHART (Cont.)

Cause	Remedy and Reference
ENGINE MISSES UNDER LOAD	
Spark plug fouled	Clean and regap spark plug.
Spark plug porcelain cracked	Replace spark plug.
Improper spark plug gap	Regap spark plug.
Pitted magneto breaker points	Replace pitted breaker points.
Magneto breaker arm sluggish	Clean and lubricate breaker point arm.
Faulty condenser (except on Tecumseh Magneto)	Check condenser on a tester; replace if defective.
Improper carburetor adjustment	Adjust carburetor.
Improper valve clearance	Adjust valve clearance to .010 cold.
Weak valve spring	Replace valve spring.
ENGINE LACKS POWER	
Choke partially closed	Open choke.
Improper carburetor adjustment	Adjust carburetor.
Magneto improperly timed	Time magneto.
Worn rings	Replace rings.
Lack of lubrication	Fill crankcase to the proper level.
Air cleaner fouled	Clean air cleaner.
Valves leaking	Grind valves and set to .010 cold.
ENGINE OVERHEATS	
Engine improperly timed	Time engine.
Carburetor improperly adjusted	Adjust carburetor.
Air flow obstructed	Remove any obstructions from air passages in shrouds.
Cooling fins clogged	Clean cooling fins.
Excessive load on engine	Check operation of associated equipment. Reduce excessive load.
Carbon in combustion chamber	Remove cylinder head and clean carbon from head and piston.
Lack of lubrication	Fill crankcase to proper level.

Figure 29–19. *(Cont'd.)*

4-CYCLE ENGINE TROUBLESHOOTING CHART (Cont.)

Cause	Remedy and Reference
ENGINE SURGES OR RUNS UNEVENLY	
Fuel tank cap vent hole clogged	Open vent hole.
Governor parts sticking or binding	Clean, and if necessary repair governor parts.
Carburetor throttle linkage or throttle shaft and/or butterfly binding or sticking	Clean, lubricate, or adjust linkage and deburr throttle shaft or butterfly.
Intermittent spark at spark plug	Disconnect ignition cut-off wire at the engine. Crank engine. If spark, check ignition switch, safety switch and interlock switch. If no spark, check magneto. Check wires for poor connections, cuts or breaks.
Improper carburetor adjustment	Adjust carburetor
Dirty carburetor	Clean carburetor.
ENGINE VIBRATES EXCESSIVELY	
Engine not securely mounted	Tighten loose mounting bolts.
Bent crankshaft	Replace crankshaft.
Associated equipment out of balance	Check associated equipment.
ENGINE USES EXCESSIVE AMOUNT OF OIL	
Engine speed too fast.	Using tachometer adjust engine RPM to spec.
Oil level too high.	To check level turn dipstick cap tightly into receptacle for accurate level reading.
Oil filler cap loose or gasket damaged causing spillage out of breather.	Replace ring gasket under cap and tighten cap securely.
Breather mechanism damaged or dirty causing leakage.	Replace breather assembly.
Drain hole in breather box clogged causing oil to spill out of breather.	Clean hole with wire to allow oil to return to crankcase.
Gaskets damaged or gasket surfaces nicked causing oil to leak out.	Clean and smooth gasket surfaces. Always use new gaskets.
Valve guides worn excessively thus passing oil into combustion chamber.	Ream valve guide oversize and install 1/32" oversize valve.
Cylinder wall worn or glazed, allowing oil to bypass rings into combustion chamber.	Bore hole, or deglaze cylinder as necessary.
Piston rings and grooves worn excessively.	Reinstall new rings and check land clearance and correct as necessary.
Piston fit undersized.	Measure and replace as necessary.
Piston oil control ring return holes clogged.	Remove oil control ring and clean return holes.
Oil passages obstructed.	Clean out all oil passages.

Figure 29–19. *(Cont'd.)*

4-CYCLE ENGINE TROUBLESHOOTING CHART (Cont.)

Cause	Remedy and Reference
OIL SEAL LEAKS	
Crankcase breather.	Clean or replace breather.
Old seal hardened and worn.	Replace seal.
Crankshaft seal contact surface is worn undersize causing seal to leak.	Check crankshaft size and replace if worn excessively.
Crankshaft bearing under seal is worn excessively, causing crankshaft to wobble in oil seal.	Check crankshaft bearings for wear and replace if necessary.
Seal outside seat in cylinder or side cover is damaged, allowing oil to seep around outer edge of seal.	Visually check seal receptacle for nicks and damage. Replace P.T.O. cylinder cover, or small cylinder cover on the magneto end if necessary.
New seal installed without correct seal driver and not seating squarely in cavity.	Replace with new seal, using proper tools and methods.
New seal damaged upon installation.	Use proper seal protector tools and methods for installing another new seal.
Bent crankshaft causing seal to leak.	Check crankshaft for straightness and replace if necessary.
Oil seal driven too far into cavity.	Remove seal and replace with new seal, using the correct driver tool and procedures.
BREATHER PASSING OIL	
Engine speed too fast.	Use tachometer to adjust correct RPM.
Loose oil fill cap or gasket damaged or missing.	Install new ring gasket under cap and tighten securely.
Oil level too high.	Check oil level - Turn dipstick cap tightly into receptacle for accurate level reading. DO NOT fill above full mark.
Breather mechanism damaged.	Replace reed plate assy.
Breather mechanism dirty.	Clean throughly in solvent. Use new gaskets when reinstalling unit.
Drain hole in breather box clogged.	Clean hole with wire to allow oil to return to crankcase.
Piston ring end gaps aligned.	Rotate end gaps so as to be staggered 90° apart.
Breather mechanism installed upside down.	Small oil drain holes must be down to drain oil from mechanism.
Breather mechanism loose or gaskets leaking.	Install new gaskets and tighten securely.
Damaged or worn oil seals on end of crankshaft.	Replace seals.
Rings not properly seated.	Check for worn, or out of round cylinder. Replace rings. Break in new rings with engine working under a varying load. Rings must be seated under high compression, or in other words, under varied load conditions.
Breather assembly not assembled correctly.	See section on Breather Assembly.
Cylinder cover gasket leaking.	Replace cover gasket.

Figure 29–19. *(Cont'd.)*

C. TROUBLESHOOTING CARBURETION.

POINTS TO CHECK FOR CARBURETOR MALFUNCTION

TROUBLE	CORRECTIONS
Carburetor out of adjustment	3-11-12-13-15-19
Engine will not start	1-2-3-4-5-6-8-11-12-14-15-22
Engine will not accelerate	2-3-11-12
Engine hunts (at idle or high speed)	3-4-8-9-10-11-12-14-20-21-24
Engine will not idle	4-8-9-11-12-13-14-19-20-21-22
Engine lacks power at high speed	2-3-6-8-11-12-19-20-22-23
Carburetor floods	4-7-17-20-21-22-23
Carburetor leaks	6-7-10-18
Engine overspeeds	8-9-11-14-15-18-19
Idle speed is excessive	8-9-13-14-15-18-19-22-23
Choke does not open fully	8-9-15
Engine starves for fuel at high speed (leans out)	1-3-4-6-11-15-17-20-23
Carburetor runs rich with main adjustment needle shut off	7-11-17-18-20-22-23
Performance unsatisfactory after being serviced.	1-2-3-4-5-6-7-8-9-10-11-15-16-17-18-19- 20-22-23

1. Open fuel-shut off valve at fuel tank–Fill tank with fuel.

2. Check ignition, spark plug and compression.

3. Clean air cleaner–Service as required.

4. Dirt or restriction in fuel system–Clean tank and fuel strainers, check for kinks or sharp bends.

5. Check for stale fuel or water in fuel–Fill with fresh fuel.

6. Examine fuel line and pick-up for sealing at fittings.

7. Check and clean atmospheric vent holes.

8. Examine throttle and choke shafts for binding or excessive play–Remove all dirt or paint, replace shaft.

9. Examine throttle and choke return springs for operation.

10. Examine idle and main mixture adjustment screws and "O" rings for cracks and damage.

11. Adjust main mixture adjustment screw; some models require finger-tight adjustment. Check to see that it is the correct screw.

12. Adjust idle mixture adjustment screw. Check to see that it is the correct screw.

13. Adjust idle speed screw.

14. Check for bent choke and throttle plates.

15. Adjust control cable or linkage, to assure full choke and carburetor control.

16. Clean carburetor after removing all non-metallic parts that are serviceable. Trace all passages.

17. Check inlet needle and seat for condition and proper installation.

18. Check sealing of welch plugs, cups, plugs and gaskets.

19. Adjust governor linkage.

SPECIFIC CARBURETOR CHECKS FOR FLOAT

20. Adjust float setting, if float type carburetor.

21. Check float shaft for wear and float for leaks or dents.

SPECIFIC CARBURETOR CHECKS FOR DIAPHRAGM

22. Check diaphragm for cracks or distortion and check nylon check ball for function.

23. Check sequence of gasket and diaphragm for the particular carburetor being repaired.

Figure 29–19. *(Cont'd.)*

D.TROUBLESHOOTING 12 VOLT STARTERS

PROBLEM	PROBABLE CAUSE	FIX
Does not function	Weak or dead battery	Check charge and/or replace battery.
	Corroded battery terminals and/or electrical connections	Clean terminals and/or connections.
	Brushes sticking	Free brushes. Replace worn brushes and those which have come in contact with grease and oil.
	Dirty or oily commutator	Clean and dress commutator.
	Armature binding or bent	Free armature and adjust end play, replace armature, or replace starter.
	Open or shorted armature	Replace armature.
	Shorted, open or grounded field coil	Repair or replace housing.
	Loose or faulty electrical connections	Correct.
	Load on engine	Disengage all drive apparartus and relieve all belt and chain tension.
	Electric starter cranks, but no spark at spark plug	Disconnect ignition cut-off wire at the engine. Crank engine. If spark at spark plug, ignition switch, safety switch is inoperative. If no spark, check magneto. Check wires for poor connections, cuts or breaks.
	Electric starter does not crank engine	Remove wire from starter. Use a jumper battery and cables and attach directly to starter. If starter cranks engine the starter is okay; check solenoid, starter switches, safety switches and interlock switch. Check wires for poor connections, cuts or breaks.
Low RPM	Unit controls engaged	Insure all unit controls are in neutral or disengaged.
	Worn bearings in cap assemblies	Clean bearings or replace cap assemblies.
	Bent armature	Replace armature.
	Binding armature	Free up armature. Adjust armature end play.
	Brushes not seated properly	Correct.
	Weak or annealed brush springs	Replace springs.
	Incorrect engine oil	Ensure the correct weight of oil is being used.
	Dirty armature commutator	Clean commutator.
	Shorted or open armature	Replace armature.
	Loose or faulty electrical connections in motor	Correct.

Figure 29–19. (Cont'd.)

TROUBLESHOOTING 12 VOLT STARTERS (Cont.)

PROBLEM	PROBABLE CAUSE	FIX
Motor stalls under load	Shorted or open armature	Replace armature.
	Shorted field coil	Correct, or replace housing assembly.
*Intermittent operation	Brushes binding in holders	Free up brushes. Replace worn brushes and those which have come in contact with grease and oil.
	Dirty or oily commutator	Clean and dress commutator.
	Loose or faulty electrical connections	Correct
	Open armature	Replace armature and interlock switch.
	Break in electrical circuit	Disconnect ignition cut-off wire at the engine. Crank engine, if spark, check ignition switch, safety switch and interlock switch. Check wires for poor connections, cuts or breaks.
Sluggish disengagement of the drive assembly pinion gear	Dirt and oil on assembly and armature shaft	Clean drive assembly and armature shaft and relubricate shaft splines.
	Bent armature	Replace armature.

E. TROUBLESHOOTING 120 VOLT STARTER SYSTEM

NOTE: The power supply for the rectifier assembly must be 120 volts A.C. with a grounding type outlet. The A.C. circuit cable should be no smaller than #14 and should be fused with a 30 ampere fuse or a 25 ampere Fustate or Fusetron.

TROUBLE	PROBABLE CAUSE	REMEDY
Low R.P.M.	Incorrect oil viscosity.	Install correct amount of oil.
	Binding armature.	Free up armature. Adjust armature end play.
	Brushes not seated properly.	Correct.
	Weak or annealed brush springs.	Replace springs.
	Dirty armature commutator.	Clean commutator.
	Shorted or open armature.	Replace armature.
	Low line voltage in A.C. circuit.	
Motor stalls under load.	Shorted or open armature.	Replace armature.
	Shorted field coil.	Correct.

Figure 29–19. (*Cont'd.*)

TROUBLE SHOOTING (Cont.)

TROUBLE	PROBABLE CAUSE	REMEDY
Intermittent operation.	Brushes binding in holders.	Free up brushes. Replace worn brushes (see each starter) and those that have come into contact with grease and oil.
	Dirty or oily commutator.	Clean and dress commutator. Replace any brush that has come into contact with grease or oil.
	Faulty electrical connections in motor or rectifier assembly.	Correct
	Open armature.	Replace armature.
Will not operate	Armature binding.	Free up armature. Adjust armature end play.
	Brushes sticking in brush holders.	Free up brushes. Replace worn brushes (see each starter) and those which have come into contact with grease or oil.
	Dirty or oily commutator.	Clean and dress commutator. Replace any brush which has come into contact with grease or oil.
	Faulty electrical conections.	Correct.
	Open field coil.	Correct.
	"Blown" fuse in A.C. circuit.	Replace fuse. Check starting motor to determine if overload was caused by faulty motor. Check rectifiers for serviceablity.
	Open or shorted armature.	Replace armature.
	2 rectifiers open in one heat sink.	Replace rectifier assembly
Sluggish disengagement of the drive assembly pinion gear	Dirt and oil on assembly and armature shaft.	Clean drive assembly and armature shaft and relubricate shaft splines.

Figure 29-19. *(Cont'd.)*

procedures are not covered in the operator's manual, it is advisable to consult a small engine technician or dealer.

Carburetor Adjustment

Small engine carburetors are adjusted correctly at the factory. They seldom get out of adjustment on their own. Unfortunately, carburetor settings are frequently changed by operators when engines do not start or when they run incorrectly due to other problems. Then when the real cause of the problem is corrected, the engine cannot perform well because the carburetor is out of adjustment. Typical carburetor adjustment points are shown in figure 29–20.

Idle Speed Adjustment. □ Most carburetors have an idle speed adjusting screw. The engine idle speed is increased when the screw is turned clockwise and decreased when the screw is turned counterclockwise. The engine idle speed should be set with a speed indicator. A *speed indicator* is a device used to measure the speed at which the crankshaft or other parts turn in revolutions per minute (*RPM* or r/min). It is important to keep the idle speed within the range specified by the manufacturer.

Air-Fuel Mixture. □ The air-fuel mixture must be carefully controlled if small engines are to run correctly. Most engines have one slotted head needle valve to adjust the air-fuel mixture at idle and one for intermediate and high speed operation. These valves are found on the carburetor and are called the idle mixture adjustment and the main mixture adjustment.

On most engines, these adjustment screws are turned clockwise to reduce the proportion of fuel and make the mixture leaner. A *leaner* misture has less fuel and more air. Turning the screw counterclock-

wise increases the proportion of fuel to air, creating a *richer* mixture.

Adjusting the Idle Mixture. □ The operator's manual should be consulted before any attempt is made to adjust a small engine. However, in the absence of a manual, the following steps are generally followed to make an idle mixture adjustment.

Procedure

1. Start the engine and run it until it is up to the recommended operating temperature.
2. Set the idle speed adjustment so the engine idles at the recommended speed.
3. Slowly turn the idle mixture adjustment screw clockwise until the engine slows down and starts to labor. To *labor* means to struggle or work hard to keep running.
4. Slowly turn the screw back counterclockwise until the engine runs smoothly and then decreases in speed.
5. Slowly turn the screw in the opposite direction to the point where the engine runs the most smoothly.
6. Reset the idle speed adjustment so the engine idles at the specified speed.

Adjusting the Main Mixture. □ When making a main mixture air-fuel adjustment, the following steps are commonly used.

Procedure

1. Warm the engine up to operating temperature.
2. Set the idle speed to the manufacturer's specification.

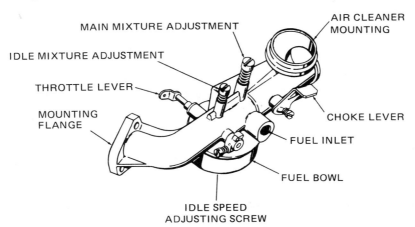

Figure 29–20. Typical carburetor adjustment points (*courtesy of Engine Service Association, Inc.*)

3. Be certain the idle mixture is correctly adjusted.
4. Set the throttle on the engine at its high-speed operating position.
5. Slowly turn the main mixture adjustment screw clockwise until the engine labors.
6. Slowly turn the screw counterclockwise until the engine runs fast and then begins to slow down.
7. Slowly turn the screw in the opposite direction until the engine runs smoothly.
8. Put the engine under load to check its performance.
9. If the engine does not accelerate or pull well, turn the screw counterclockwise slightly to make the mixture richer.

Caution:

When adjusting a carburetor needle valve, the screw should be moved only $\frac{1}{32}$ of a turn before observing the engine's performance. It takes a few seconds for the new valve setting to influence the performance of the engine.

☐ STORING SMALL ENGINES

Whenever a gas engine is to be stored, it is important to leave it in good condition. A little care when storing can save time when the engine is to be restarted.

When storing for a month or less, use the following procedure.

Procedure

1. Close the fuel shut-off valve and drain the carburetor if it has a drain plug or drain plunger.
2. Fill the fuel tank with gasoline of the correct grade. If it is a two-cycle engine, fill the tank with gasoline and oil mixed in the proper proportion.
3. Store the engine in a dry place.
4. If the engine must be stored outside, cover it with plastic or a tarpaulin.
5. If the engine is supplied with a battery, be sure that the battery is fully charged.

When storing an engine for longer than a month, the following procedure is recommended.

Procedure

1. Drain the fuel tank.
2. With the fuel shut-off valve open, run the engine until all fuel is burned.

---

3. For a four-cycle engine, drain the oil and refill with the correct oil for the weather when the engine will be put back into service.
4. Remove the spark plug. (For a two-cycle engine, turn the crankshaft until the piston covers the ports.) Pour 2 to 3 tablespoons of clean engine oil into the cylinder to prevent rust.
5. Reinstall the spark plug.
6. Turn the crankshaft slowly three or four times to distribute the oil.
7. Drain, clean, and reinstall the fuel filter bowl, if so equipped.
8. Remove the shrouds and guards so that the surface of the engine can be cleaned completely.

Caution:

When removing the shroud, be careful to avoid bending any carburetor or governor linkage. Also observe how the linkage is assembled so that you will be able to reassemble it correctly.

---

9. Clean the exterior.
10. Remove all debris from between the fins.
11. Check to see that all linkages move freely.
12. Loosen all belts.
13. Clean and lubricate all chains.
14. Lubricate all fittings.
15. Reinstall all shrouds and guards
16. Store the engine off the ground indoors. If it is stored outside, cover it to protect it from moisture.

☐ STUDENT ACTIVITIES

1. Define the italicized terms used in this unit.
2. Name the parts of a four-cycle engine.
3. Name the parts of a two-cycle engine.
4. Complete the table on page 357 by filling in the information requested for a four-cycle, horizontal shaft engine during one complete cycle.
5. Obtain a used small engine and its operator's manual. Perform the following service procedures:
   a. Use an air gun or brush and small screwdriver to clean the fins and air passageways.
   b. Service the air cleaner.
   c. Check the fuel tank for dirt, chaff, grass, or leaves. If dirty, drain the tank and remove all foreign matter. Refill with fresh gasoline after all service procedures are completed.

Stroke	Piston Movement	Intake Valve Position	Exhaust Valve Position	General Action/ Activity
Intake	Down	Open	Closed	Air-fuel mixture is drawn into the cylinder.
Compression				
Power				
Exhaust				

Caution:

Wear safety glasses and work in a well-ventilated area. Do not spill any gasoline. Wipe up oil spills immediately.

---

    d. Change the oil in the crankcase.
    e. Service or replace the spark plug.

Caution:

The plug must be gapped using a wire gauge to the manufacturer's specifications.

---

    f. If the engine has several hundred hours of use, ask your instructor if you should remove the head to clean the combustion chamber.
  6. Study a small engine troubleshooting guide. Place a mark beside the procedures you can perform to correct engine problems.
  7. Prepare a small engine for storage.

□ SELF-EVALUATION

A. Multiple Choice. Select the best answer.
  1. An engine takes in air to provide
    a. compression
    b. fuel
    c. lubrication
    d. oxygen

2. Internal combusion engines have been around for about
    a. 20 years
    b. 25 years
    c. 50 years
    d. 100 years

3. Reciprocal means
    a. circular
    b. expanding
    c. explosive
    d. back and forth

4. On the power stroke of a four-cycle engine the
    a. fuel is compressed
    b. exhaust valve is open
    c. intake valve is open
    d. spark plug fires

5. A dry element air cleaner is serviced by
    a. washing in varsol
    b. tapping on a level surface
    c. blowing with compressed air
    d. dipping in detergent

6. If specifications are not available, the gap of a four-cycle small engine spark plug should be set at
    a. .030 inch
    b. .020 inch
    c. .010 inch
    d. .005 inch

7. Sandblasting of spark plugs is not recommended because
    a. the practice may void the warranty
    b. the practice may damage the cylinder wall
    c. the savings over the cost of a new plug is not worth the risk
    d. all of these

8. The troubleshooting procedure starts with
    a. checking the simple things
    b. draining and replacing the fuel
    c. removing the shroud
    d. servicing the spark plug

9. Before storing an engine, 2 or 3 tablespoons of oil are placed in the cylinder to
    a. prevent rust
    b. dissolve carbon
    c. prevent freezing
    d. mix with gasoline

10. A good spark at the spark plug gap is
    a. blue
    b. orange
    c. red
    d. yellow

B. Matching. Match the terms in column I with those in column II.

Column I

1. reed
2. stroke
3. hole
4. 6:1
5. TDC
6. BDC
7. two-cycle engine
8. SF, SE, SD, SC
9. exhaust
10. viscosity

Column II

a. valve in two-cycle engines
b. top dead center
c. mix oil with gasoline
d. oils for four-cycle engines
e. tendency to flow
f. one piston movement
g. burned gases
h. port
i. bottom dead center
j. compression ratio

C. Completion. Fill in the blanks with the word or words that will make the following statements correct.

1. The spongelike foam in an oil foam air cleaner may be washed in _____, _____ , or _____ .

2. Crankcase oil should be changed after every _____ hours of operation in small, four-cycle engines.

3. Most small engines for home and farm use are cooled by _____ .

Section **11**
ELECTRICITY

Unit 30
Electrical Principles and Wiring Materials

Objective

To use principles of electricity and safety for planning simple wiring systems.

Competencies to Be Developed

After studying this unit, you will be able to:
- Describe some basic principles of electricity and magnetism.
- Use safety practices with electricity.
- Describe the relationships among volts, amperes, and watts.
- Select materials for electrical wiring.
- Design simple wiring systems.

Terms to Know

electricity	milliampere
filament	magnetism
fluorescent light	permanent magnet
resistance	pole
conductor	north pole
insulator	south pole
amperes	repel
volts	magnetic flux
watts	magnetic field
ohm	reverse polarity
Ohm's law	electromagnet
volt-ohm-milliampere	commutator
meter (VOM)	armature

field	kilo
generator	watthour
alternator	kilowatthour
turbine	branch circuit
circuit	fuse
open circuit	circuit breaker
short circuit	nonmetallic sheathed
grounding	cable
shock	armored cable
ground fault interrupter	conduit
(GFI)	strands
transformer	voltage drop
service drop	positive (hot) wire
service entrance panel	neutral wire
meter	

Materials List

Horseshoe magnet
Samples of wire, nonmetallic sheathed cable, armored cable, and conduit
Electric meter dial

Electricity is the major power source for stationary equipment in houses, farm buildings, and agribusinesses. It is the energy source most commonly used for driving machinery, lighting, heating, and cooling. Some knowledge of electricity is essential for the safe use of electrical equipment. Understanding how to wire simple circuits and make minor electrical repairs is also useful. Finally, it is important to maintain electrical circuits and equipment properly to ensure their long life and safe operation.

☐ PRINCIPLES OF ELECTRICITY

Heat and Light

Electricity is a form of energy that can produce light, heat, magnetism, and chemical changes. Light can be produced by heating a special metal element or *filament* in a vacuum tube called a bulb. The flow of electricity into the bulb must be carefully controlled so that the filament glows without burning out. In addition, electricity flowing through certain gases causes them to glow. A *fluorescent lamp* glows as a result of electricity flowing through a gas.

Heat is produced when electricity flows through metals with some difficulty. Any tendency of a material to prevent electricity flow is called *resistance*. If electricity flows easily, the metal is said to be a good *conductor*. Silver is an excellent conductor, copper is a very good conductor, and aluminum is a good conductor. Because silver is so expensive, copper is generally used in wiring systems; aluminum may be used when the price of copper is high. Aluminum is also used extensively in outside lines where heat buildup is not a problem. Applications include the high-voltage power lines that cross the countryside, as well as the overhead wires that stretch from building to building in the city or on the farm.

Heat in wires, switches, outlets, motors, and lights is not desirable. It wastes electrical energy, causes materials to deteriorate, heats up the surrounding areas, and may cause fires. Therefore, the proper design of wiring systems is important. In properly designed electrical systems, heat is kept within acceptable limits. In heating systems, the heat is produced by using elements with just the right amount of resistance. Heat from electricity is clean and easy to control.

A material that provides great resistance to the flow of electricity is called an *insulator*. Examples of good insulators include rubber, glass, vinyl, asbestos, and air. Anyone working with electricity must be aware that insulation is relative. For example, a very thin layer of rubber or vinyl will prevent electrical flow if the voltage is low, such as 12 volts for automobile lights. However, to prevent electricity at 30,000 volts in a spark plug wire from jumping to ground requires a much thicker layer of rubber.

Amperes, Volts, and Watts

Amperes, volts, and watts must be understood to design electrical circuits or install electrical materials safely. *Amperes* are a measure of the rate of flow of electricity in a conductor. *Volts* are a measure of electrical pressure. *Watts* are a measure of the amount of energy or work that can be done by amperes and volts.

The following relationships exist among amperes, volts, and watts:

$$\text{Watts} = \text{Volts} \times \text{Amperes}$$
$$\text{Volts} = \frac{\text{Watts}}{\text{Amperes}}$$
$$\text{Amperes} = \frac{\text{Watts}}{\text{Volts}}$$

For example, if a 200-watt light bulb operates at 120 volts, it will draw 1.67 amperes of electricity

$$A = \frac{W}{V} \text{ or } 200 \div 120 = 1.67$$

A 5-ampere motor running on 120 volts will consume 600 watts of electricity

$$W = V \times A \text{ or } 120 \times 5 = 600$$

Finally, a refrigerator motor rated at 3 amperes that consumes 360 watts of electricity should be plugged into a 120-volt circuit

$$V = \frac{W}{A} \text{ or } 360 \div 3 = 120$$

These relationships are known as the West Virginia formula: $W = VA$. By remembering the basic formula, any one of the three quantities can be determined if the other two are known.

Ohm's Law

A physicist named Georg Simon Ohm made a number of important discoveries about electric current. As a result, the unit used to measure a material's resistance to the flow of electrical current is known as the *ohm*. Ohm discovered that the flow of electricity through a conductor is directly proportional to the electrical or electromotive force that produces it. The relationship he discovered between electric current, electromotive force, and resistance is called *Ohm's Law*.

Electromotive force is measured in volts and is represented by a capital E in Ohm's Law. The rate of flow of electricity through a conductor is measured in amperes, represented by a capital I in Ohm's Law. The tendency of a material to prevent electrical flow, or resistance, is represented by a capital R.

$$E = \text{Volts} \quad I = \text{Amperes} \quad R = \text{Resistance}$$

Ohm's Law states that electromotive force is equal to amperes times resistance, or E = IR. From this basic formula, amperes (I) and resistance (R) can also be computed:

$$I = \frac{E}{R} \qquad R = \frac{E}{I}$$

In any situation, one of the values can be computed if the other two are known. This knowledge is useful when designing, testing, or using electrical equipment.

Volts, amperes, and resistance can be measured by meters. Voltage is measured with a voltmeter. Amperes are measured with an ammeter. Resistance in ohms is measured by an ohmmeter. A popular combination meter used in electronics is a *volt-ohm-milliampere meter,* commonly called a VOM meter. A *milliampere* is a thousandth of an ampere.

Magnetism and Electricity

Electricity flowing through a conductor results in magnetism. *Magnetism* is a force that attracts or repels iron or steel, figure 30–1. Magnetism can be created by exposing iron or steel to magnetic forces. If iron or steel holds its magnetism, it is said to be a *permanent magnet.* Permanent magnets may be used to create or generate electricity.

Magnets have two ends, called *poles.* These are designated as the north pole and the south pole. The *north pole* of one magnet *attracts* the south pole of another magnet but *repels,* or pushes away, the north pole of another magnet. In a similar manner, the *south pole* of one magnet will repel the south pole of another but attract the north pole. In summary, like poles repel and unlike poles attract, or opposites attract, figure 30–2.

Lines of magnetic force, often called *magnetic flux,* occur in patterns in the air between the two poles. The pattern is referred to as the *magnetic field,* figure 30–3.

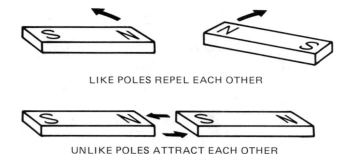

LIKE POLES REPEL EACH OTHER

UNLIKE POLES ATTRACT EACH OTHER

Figure 30–2. Like poles repel and unlike poles attract.

Electric Motors

The principle of magnetism is the basis upon which electric motors operate. A strong magnet can be made by wrapping many coils of insulated wire around an iron core and passing electrical current through the wire, figure 30–4. When the current stops flowing, the magnetic field ceases to exist, figure 30–5. If the wires are switched to opposite battery terminals, the poles of the magnet will reverse, but the magnetic field will remain, figure 30–6. However, reversing the direction of current flow *reverses the polarity* of the magnet. The unit is called an *electromagnet.*

A motor may be made by bending the electromagnet into the shape of a horseshoe and suspending a permanent magnet on a bearing between the poles. When current flows, the unlike poles of the electromagnet and the permanent magnet will attract each other, causing the permanent magnet to make a half turn, figure 30–7. If the wires are reversed at the battery, the direction of the current and the polarity of the magnet will be reversed, and the permanent magnet will make another half turn, figure 30–8. In a motor, the current is reversed by a part called a *commutator.* The commutator causes the motor to run continuously in the same direction. The rotating (turning)

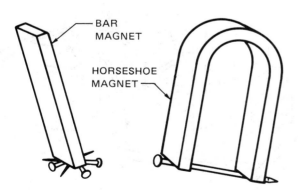

Figure 30–1. Magnetism is a force that attracts or repels iron or steel.

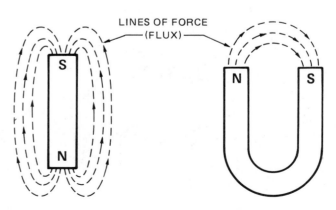

Figure 30–3. Lines of magnetic force occur in a pattern called the magnetic field.

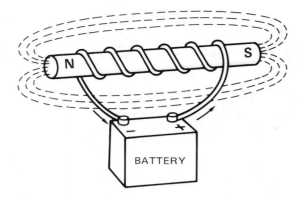

Figure 30–4. A strong magnet can be made by wrapping many coils of insulated wire around an iron core and passing electrical current through the wire.

magnet is called the *armature*, and the magnetic forces around the armature are called the *field*.

Devices that produce electricity by magnetism are called generators and alternators. A *generator* produces direct current which means the electricity flows in one direction. An *alternator* produces alternating current, which reverses in direction, or alternates, each time the armature turns. Generators and alternators must be powered by hand, gas or diesel engine, turbine, or an electric motor getting electricity from one of these sources. A *turbine* is a high-speed rotary engine driven by water, steam, or other gases.

Electric Motor Maintenance □ Electric motors are designed for long life with little maintenance. Some have permanently lubricated bearings. With these motors, the only maintenance is an occasional cleaning with compressed air, figure 30–9. Under extreme conditions the motor may accumulate greasy dirt on inside parts. In those cases, disassembly and cleaning with an electric motor cleaner may be necessary, figure 30–10.

Many motors have two oil cups or tubes with a wick to hold oil for the two bearings. The operator's

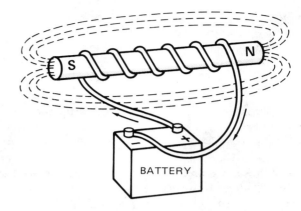

Figure 30–6. If the wires are switched to opposite battery terminals, the poles of the magnet reverse, but the magnetic field remains.

manual supplied by the manufacturer typically calls for about 10 drops of oil per year in these tubes. More oil than specified will cause dirty oil accumulations on interior parts. On the other hand, failure to use oil as specified by the manufacturer shortens the life of the motor due to excessive wear. Therefore, it is important to follow the instructions provided.

Circuits

A source of electricity plus two wires connected to a light, heater, or motor are known collectively as a *circuit*. A circuit may be likened to a circle. If the circle is broken so that current cannot flow through it, it is said to be an *open circuit*.

All circuits must include an object with resistance that falls within a certain range. Otherwise the electricity will flow around the circuit and back to its source too rapidly and blow fuses, burn wires, and drain batteries. Such a condition is called a *short circuit*.

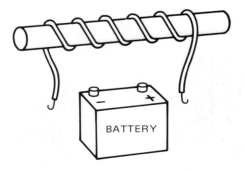

Figure 30–5. When the current stops flowing, the magnetic field ceases to exist.

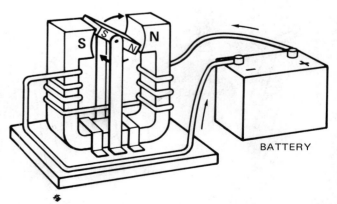

Figure 30–7. Unlike poles attract each other to move the permanent magnet a half turn.

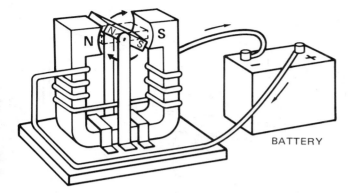

Figure 30–8. If the wires are reversed at the battery, the polarity of the electromagnet is reversed and the permanent magnet makes another half turn.

Figure 30–10. If motors are extremely dirty, they may need to be disassembled for cleaning.

Electricity can travel back to its source through the earth as well as through a wire. A circuit always includes a wire to carry the current back to its source. However, safety requires that an additional wire or conductor be provided in the system in case the current gets out of its circuit. Making this additional connection between a piece of equipment and the earth is called *grounding*. It is accomplished by driving a solid copper rod or a larger hollow steel pipe into soil that is always moist. The rod or pipe, called a ground *rod*, and the electrical equipment are connected by a conductor called a ground *wire*. Grounding of equipment is a standard safety practice. In this way electricity that accidentally gets out of its circuit is channeled to the earth through a ground wire rather than through the body of a human or animal.

□ ELECTRICAL SAFETY

There are two deadly hazards associated with electric current: shock and fire. *Shock* refers to the body's reaction to an electric current. Shock can interfere with a normal heart beat and result in the injury or death of the victim. Fire may result when electrical conductors overheat or when a spark is produced by an electrical current jumping an air space.

All people who work around electric current should observe safety practices at all times to avoid shock or fire. Some precautions advised by authorities are:

- Never disconnect or damage any safety device that is provided by the manufacturer.
- Do not touch electrical appliances, boxes, or wiring with wet hands or wet feet.
- Do not remove the long ground prong from three-prong 120-volt plugs.
- Use ground fault interrupters in kitchen, bathroom, laundry, and outdoor circuits, or wherever moisture may increase shock hazard. A *ground fault interrupter (GFI)* is a device that cuts off the electricity if even very tiny amounts of current leave the normal circuit. Ground fault interrupters should be installed in accordance with the instructions provided.
- Immediately discontinue the use of any extension cord that feels warm or smells like burning rubber.

Figure 30–9. Electric motors need occasional cleaning with compressed air to remove dust and dirt from interior surfaces.

- Do not place extension cords under carpeting.
- Install all electrical wiring according to the specifications of the National Electrical Code.
- Use only double insulated portable tools or tools with three-wire grounded cords.
- If a fuse is blown or circuit breaker is tripped, determine and correct the problem before inserting a new fuse or resetting the breaker.
- Fuses and circuit breakers are designed to prevent the circuit from being overloaded. Do not create a fire hazard by installing higher capacity fuses or breakers then the system is designed to handle.
- Do not leave heat-producing appliances such as irons, hair dryers, and soldering irons unattended.
- Place all heaters and lamps away from combustible materials.
- Keep the metal cases or cabinets of electrical appliances grounded at all times.
- Do not remove the back of a television set. There is danger of shock by 20,000 or 30,000 volts even with the unit unplugged.
- Keep electrical motors lubricated and free of grease and dirt.
- Keep appliances dry to reduce shock hazard and prevent rust.
- Do not use any switches, outlets, fixtures, or extension cords that are cracked or damaged in any way.
- Follow manufacturer's instructions for installation and use of all electrical equipment.

□ ELECTRICAL WIRING

An electrical system must meet several conditions to be satisfactory. It must:

- be safe
- be convenient
- be expandable
- look neat
- provide sufficient current

Service Entrance

Electrical power comes to the home or farm by overhead or underground wires. The power company provides a transformer, service drop, and appropriate wiring to an entrance head. An *entrance head* is a waterproof device used to attach exterior wires to interior wires of a building. The *transformer* converts high voltage from the power lines to 240 volts for home and farm installations. The *service drop* is an assembly of electrical wires, connectors, and fasteners used to transmit electricity from the transformer to the entrance head and on to the service entrance

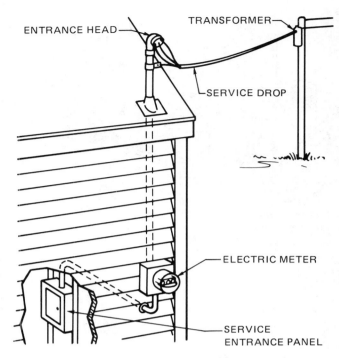

Figure 30–11. Electrical power is supplied to local homes and farms through a transformer, service drop, entrance head, meter, and service-entrance panel.

panel. The *service entrance panel* is a box with fuses or circuit breakers where electricity enters a building. Wires feed from the service drop through the insulated entrance head, through cable or pipe to a meter box. The electric meter plugs into the meter box. The power continues on to the service entrance panel for distribution to branch circuits, figure 30–11.

Meter

All electricity that passes through the system is measured by the *meter,* in kilowatthours, figure 30–12. *Kilo* means 1000. A *watthour* is the use of 1 watt for 1 hour. (A 100-watt bulb that burns for 1 hour will consume 100 watthours of electricity.) A *kilowatthour* is the use of 1000 watts for 1 hour.

The meter will convert any amount of electricity into kilowatthours. A power company representative reads the meter at regular intervals and the customer is billed according to the number of kilowatthours used. When reading the meter, the last number passed by each pointer is read. These four numbers constitute the meter reading. Note that the first and third dials turn counterclockwise, and the second and fourth dials turn clockwise. In figure 30–12, the dials indicate that 0019 kilowatthours of electricity have passed through the meter. The third dial appears to read 2, but since the fourth dial has not completed its revolution, the third dial cannot quite have reached 2.

Figure 30–12.   An electric meter

Branch Circuits

Most circuits begin from the service entrance panel. They are called *branch circuits* since they branch out into a variety of places and for a variety of purposes, figure 30–13. A branch circuit generally includes only one motor or a series of outlets or a series of lighting fixtures, figure 30–14. It is important to provide the correct size of wire and fuse or circuit breaker for the load on each circuit.

A *fuse* is a plug or cartridge containing a strip of metal that melts when more than a specified amount of current passes through it. A *circuit breaker* is a switch that trips and breaks the circuit when more than a specified amount of current passes through it. A circuit breaker can be reset after it trips. A fuse must be replaced if it blows.

Caution:

Fuses and circuit breakers are used to protect circuits from damage and fire. Never destroy this protection by installing fuses or breakers with a larger ampere rating than is recommended for the size of wire being protected.

When planning electrical wiring systems or reading blueprints, it is useful to know standard electrical symbols, figure 30–15. The symbols are used on drawings to indicate the location of outlets, receptacles, switches, and special appliances.

Types of Cable

Three systems of wiring are in general use for most home, farm, and commercial installations: (1) nonme-

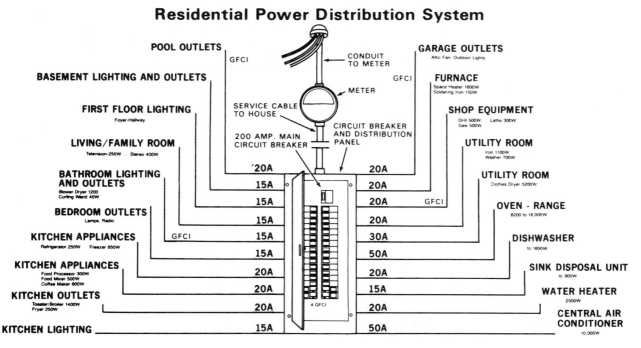

Residential Power Distribution System

Figure 30–13.   Branch circuits fan out from the service entrance panel (*courtesy American Association for Vocational Instructional Materials*).

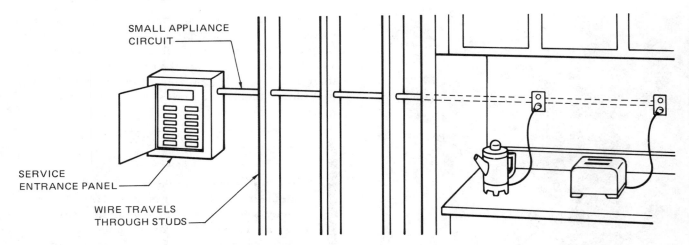

SMALL APPLIANCE
CIRCUIT

SERVICE
ENTRANCE PANEL

WIRE TRAVELS
THROUGH STUDS

Figure 30–14. A branch circuit generally includes a series of outlets, a series of lights, or one electric motor.

tallic sheathed cable, (2) armored cable, and (3) conduit.

Nonmetallic sheathed cable consists of copper or aluminum wires covered with paper, rubber, or vinyl for insulation and protection, figure 30–16. Certain types of nonmetallic sheathed cable are waterproof and suitable for burial in soil.

Armored cable consists of a flexible metal sheath with individual wires inside, figure 30–17. The wires are insulated with rubber or vinyl and wrapped in paper for protection. Armored cable provides good protection against mechanical damage. It is fairly easy to install in narrow and irregular building sections. However, its tendency to rust and develop short circuits makes it unsuitable for use in damp areas.

Conduit is tubing which contains individual insulated wires, figure 30–18. Such tubes are available in ½-inch, ¾-inch, 1-inch or larger diameters. Conduit may be rigid or bendable. The bendable type is known as electrical metallic tube (*EMT*). Special benders permit shaping EMT in bends of up to 90°. Couplings and connectors are used to assemble conduit and attach it to boxes. Waterproof plastic fittings provide waterproof seals where needed. Conduit systems are considered to provide the most protection for wiring and are required for most commercial jobs.

Wire Type and Size

Individual wires within cable or conduit may be aluminum or copper. Smaller sizes include No. 14 wire for 15-ampere circuits, No. 12 for 20-ampere circuits, and No. 10 for 30-ampere circuits, figure 30–19. These ampere ratings apply to copper wire. For aluminum, it is necessary to use one wire size larger than is required for copper wire. Wire size is designated by gauge. The lower the gauge number, the larger the wire size.

Small wires called *strands* are placed together to form bundles for wire sizes No. 8 and larger in order to improve flexibility and conductivity. Electricity is carried on the outer surfaces of wire. Thus stranded wire has more total surface area than solid wire of equal diameter and carries more current. Either solid wire or stranded wire may be placed in cables or conduit.

Voltage Drop

A common problem encountered in homes and on farms is voltage drop. *Voltage drop* refers to a loss of voltage as it travels along a wire.

Voltage drop causes lights to dim, heaters to put out less heat, and motors to put out less power and to overheat. The larger the wire, the less problem there is with voltage drop for a given amount of current. The longer the wire, however, the greater the problem.

Allowance for long wires must be made by using either lighter loads or larger wires. For example, to obtain 20 amperes of current with a standard voltage at the end of a circuit, the wire size must be increased according to the distance, figure 30–20. This is especially true when installing special circuits for motors.

Wire Identification

The type of outer covering, individual wire covering, cable construction, and number of wires all help determine where a cable can be used. The wire type is stamped on the outer surface of the wire or cable or cable covering. Some common types of wire and cable are:

- Type T—for use in dry locations
- Type TW—for use in dry or wet locations

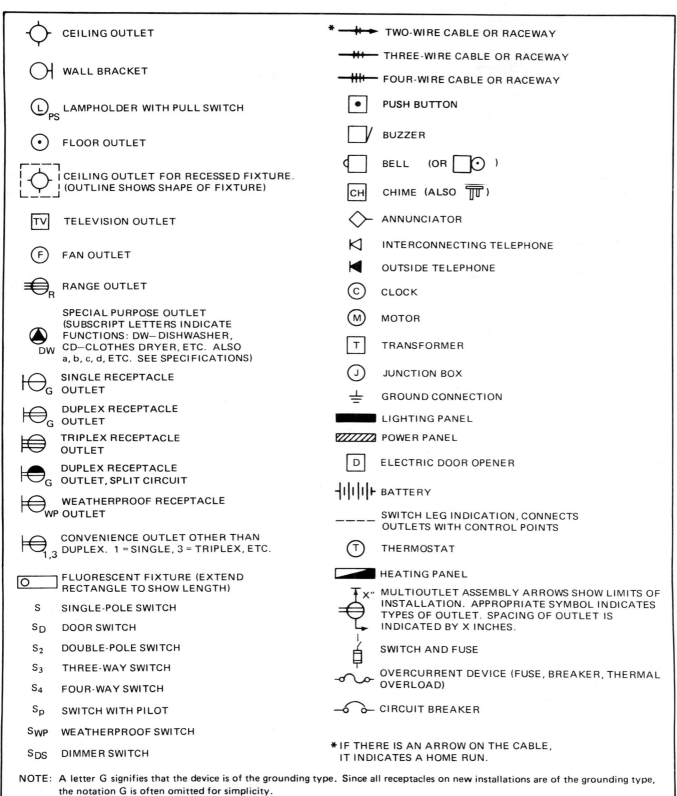

Figure 30–15. Standard symbols identify locations of outlets, switches, and equipment (*courtesy R. Mullin,* Electrical Wiring—Residential, *8th ed., Delmar Publishers Inc., 1984*).

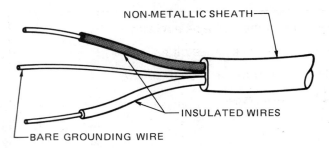

NON-METALLIC SHEATH

INSULATED WIRES

BARE GROUNDING WIRE

Figure 30–16. Nonmetallic sheathed cable

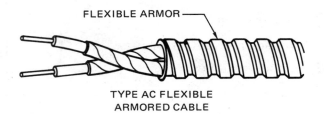

FLEXIBLE ARMOR

TYPE AC FLEXIBLE
ARMORED CABLE

Figure 30–17. Armored cable

- Type THHN—for use in dry locations with high temperature
- Type THW and THWN—for use in wet locations with high temperature
- Type XHHW—for high moisture and heat resistance
- Type UF—for direct burial in soil but not concrete

The National Electrical Code and local building codes must be consulted for specific guidance in wire selection.

Individual wires are color coded to help identify their function in the circuit. Black wires, red wires, and blue wires are *positive* or *hot wires,* which carry current to appliances. White wires are *neutral wires,* which carry current from the appliance back to the source. Green wires and bare wires are used to ground all metal boxes and appliances in the circuit.

Nonmetallic sheathed cable is generally stamped at regular intervals with a mark that identifies its type, wire size, and number of conductors. Typical markings and their meanings are as follows:

- 12–1—two strands of No. 12 wire, one black and one white
- 12–2 w/g—two strands of No. 12 wire plus a ground wire (w/g means "with ground"), one black, one white, and one green or bare
- 12–3—three strands of No. 12 wire, probably one black, one red, and one white
- 12–3 w/g—same as 12–3 cable with the addition of a green or bare wire for grounding the circuit

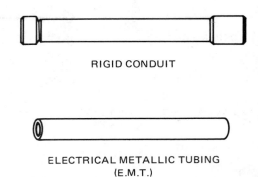

RIGID CONDUIT

ELECTRICAL METALLIC TUBING
(E.M.T.)

Figure 30–18. Conduit (*courtesy of American Association for Vocational Instructional Materials*)

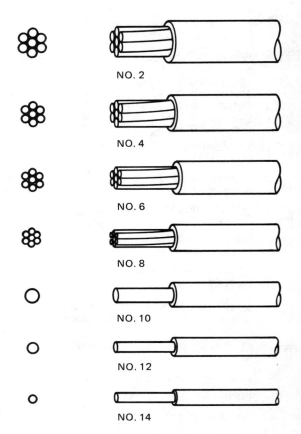

NO. 2

NO. 4

NO. 6

NO. 8

NO. 10

NO. 12

NO. 14

Figure 30–19. Common sizes of wire used in cables and conduit

The same system is used for other wire sizes. Letters such as T, TW, or THHN on the cable indicate its type.

□ STUDENT ACTIVITIES

1. Collect samples of cable and label them with the name, type of conductor, and type of insulation.

2. Use the formula W = VA to calculate the unknown value in the following problems.
   a. A 100-watt bulb in a 120-volt circuit draws _____ amperes.
   b. A 120-volt hair dryer is rated at 1200 watts. It will draw _____ amperes.
   c. A toaster draws 13 amperes at 120 volts. How many watts does it use?
   d. A circuit has three 100-watt lamp bulbs and a 1600-watt toaster. How many watts are used when all four appliances are in operation?
   e. Would a 15-ampere fuse carry the load described in item d? Why?
   f. A grain elevator motor has a 1-horsepower motor. A general rule of thumb is that 10 amperes at 120 volts equals 1 horsepower. How many amperes does this motor draw?
   g. A microwave oven draws 1400 watts at 120 v. on a 20-ampere circuit. It is 40 feet between the outlet and the service entrance panel. What size of copper wire is needed for the circuit? What size of aluminum wire?

3. Cover a horseshoe magnet with a piece of paper. Sprinkle iron filings over the paper. Observe the magnetic lines of force exhibited by the filings.

4. Read the electric meter at your home. Read it again two days later. How many kilowatthours of electricity did your family use in two days?

[handwritten note, top right:] If the motor was 308 feet from the power source what Size cond. would you use. Al + Cu

COPPER up to 200 Amperes, 115-120 Volts, Single Phase, Based on 2% Voltage Drop

Minimum Allowable Size of Conductor — Length of Run in Feet. Compare size shown below with size shown to left of double line. Use the larger size.

Load in Amps	Types R, T, TW	Types RH, RHW, THW	Bare & Covered Conductors	30	40	50	60	75	100	125	150	175	200	225	250	275	300	350	400	450	500	550	600	650	700
5	12	12	10	12	12	12	12	12	12	12	10	10	10	10	8	8	8	8	6	6	6	6	4	4	4
7	12	12	10	12	12	12	12	12	10	10	8	8	8	6	6	6	6	4	4	4	4	4	4	3
10	12	12	10	12	12	12	12	10	10	8	8	8	6	6	6	6	4	4	4	4	3	3	2	2	2
15	12	12	10	12	12	10	10	10	8	6	6	6	4	4	4	4	3	2	2	1	1	1	0	0
20	12	12	10	12	10	8	8	6	6	4	4	4	3	3	2	2	1	1	0	0	0	00	00	00
25	10	10	10	10	10	8	8	6	6	4	4	4	3	3	2	2	1	1	0	0	00	00	000	000	000
30	10	10	10	10	8	8	8	6	4	4	4	3	2	2	1	1	1	0	00	00	000	000	000	4/0	4/0
35	8	8	10	10	8	8	6	6	4	4	3	2	2	1	1	0	0	00	000	000	4/0	4/0	4/0	4/0	250
40	8	8	10	8	8	6	6	4	4	3	2	2	1	1	0	0	00	00	000	000	4/0	4/0	250	250	300
45	6	6	10	8	8	6	6	4	4	3	2	1	1	0	0	00	00	000	000	4/0	4/0	250	250	300	300
50	6	6	10	8	6	6	4	4	3	2	1	1	0	0	00	00	000	000	4/0	4/0	250	250	300	300	350
60	4	6	8	8	6	4	4	4	2	1	1	0	00	00	000	000	000	4/0	250	250	300	300	350	400	400
70	4	4	8	6	6	4	4	3	2	1	0	00	00	000	000	4/0	4/0	250	300	300	350	400	400	500	500
80	2	4	6	6	4	4	3	2	1	0	00	00	000	000	4/0	4/0	250	300	300	350	400	400	500	500	600
90	2	3	6	6	4	4	3	2	1	0	00	000	000	4/0	4/0	250	250	300	350	400	500	500	600	600	600
100	1	3	6	4	4	3	2	1	0	00	000	000	4/0	4/0	250	250	300	350	400	500	500	500	600	600	700
115	0	2	4	4	4	3	2	1	0	00	000	4/0	4/0	250	300	300	350	400	500	500	600	600	700	700	750
130	00	1	4	4	3	2	1	0	00	000	4/0	4/0	250	300	300	350	400	500	500	600	600	700	750	800	900
150	000	0	2	4	2	1	1	0	000	4/0	4/0	250	300	350	350	400	500	500	600	700	700	800	900	900	1M
175	4/0	00	2	3	2	1	0	00	000	4/0	250	300	350	400	400	500	500	600	700	750	800	900	1M		
200	250	000	1	2	1	0	00	000	4/0	250	300	350	400	500	500	500	600	700	750	900	1M				

ALUMINUM up to 200 Amperes, 115-120 Volts, Single Phase, Based on 2% Voltage Drop

Minimum Allowable Size of Conductor — Length of Run in Feet. Compare size shown below with size shown to left of double line. Use the larger size.

Load in Amps	Types R, T, TW	Types RH, RHW, THW	Bare & Covered Conductors Single Triplex	30	40	50	60	75	100	125	150	175	200	225	250	275	300	350	400	450	500	550	600	650	700
5	12	12	10	12	12	12	12	12	10	10	8	8	8	8	6	6	6	6	4	4	4	3	3	3
7	12	12	10	12	12	12	12	10	10	8	8	6	6	6	4	4	4	4	3	2	2	2	2	1
10	12	12	10	12	12	10	10	8	8	6	6	4	4	4	4	3	3	2	2	1	1	0	0	0
15	12	12	10	12	10	8	8	6	4	4	4	3	3	2	2	1	0	0	00	00	000	000
20	10	10	10	10	8	6	6	4	4	3	3	2	2	1	1	0	0	00	00	000	000	4/0	4/0	4/0
25	10	10	10	8	8	6	6	4	4	3	2	2	1	1	0	0	00	00	000	000	4/0	4/0	250	250	300
30	8	8	10	8	6	6	6	4	3	2	2	1	0	0	00	00	00	000	4/0	4/0	250	250	300	300	350
35	6	8	10	8	6	6	4	4	3	2	1	0	0	00	00	000	000	4/0	4/0	250	250	300	300	350	400
40	6	8	10	6	6	4	4	3	2	1	0	0	00	00	000	000	4/0	4/0	250	300	300	350	350	400	500
45	4	6	10	6	6	4	4	3	2	1	0	00	00	000	000	4/0	4/0	250	300	300	350	400	400	500	500
50	4	6	8	6	4	4	3	2	1	0	00	00	000	000	4/0	4/0	250	300	300	350	400	400	500	500	500
60	2	4	6	6	4	3	3	2	0	00	00	000	4/0	4/0	250	250	300	350	350	400	500	500	500	600	700
70	2	2	6	4	4	3	2	1	0	00	000	4/0	4/0	250	300	300	350	400	500	500	600	600	700	700	750
80	1	2	6	4	3	2	1	0	00	000	4/0	4/0	250	300	300	350	500	500	600	600	700	750	800	900
90	0	2	4	4	3	2	1	0	00	000	4/0	250	300	300	350	400	400	500	600	600	700	750	800	900	1M
100	0	1	4	3	2	1	0	00	000	4/0	250	300	300	350	400	400	500	600	600	700	750	900	900	1M
115	00	0	2	3	1	1	0	00	000	4/0	300	300	350	400	500	500	600	600	700	700	800	900	1M
130	000	00	2	2	1	0	00	000	4/0	250	300	350	400	500	500	600	600	700	800	900	1M
150	4/0	000	1	2	0	00	00	000	250	300	350	400	500	500	600	600	700	800	900	1M
175	300	4/0	0	1	0	00	000	4/0	300	350	400	500	600	600	700	750	800	900
200	350	250	00	0	00	000	4/0	250	300	400	500	600	600	700	750	900	900

Figure 30–20. Size of copper wire needed to carry given loads with 2 percent or less voltage drop (*courtesy of National Food and Energy Council, Inc., Columbia, MO*) continued on page 370.

COPPER up to 400 Amperes, *230-240 Volts,* Single Phase, Based on 2% Voltage Drop

Minimum Allowable Size of Conductor — **Length of Run in Feet**
Compare size shown below with size shown to left of double line. Use the larger size.

Load in Amps	Types R, T, TW	Types RH, RHW, THW	Overhead in Air* Bare & Covered Conductors	50	60	75	100	125	150	175	200	225	250	275	300	350	400	450	500	550	600	650	700	750	800
5	12	12	10	12	12	12	12	12	12	12	12	12	12	12	10	10	10	10	8	8	8	8	8	6	6
7	12	12	10	12	12	12	12	12	12	12	12	10	10	10	10	8	8	8	8	6	6	6	6	6	6
10	12	12	10	12	12	12	12	10	10	10	10	10	8	8	8	6	6	6	6	4	4	4	4	4	4
15	12	12	10	12	12	12	10	10	10	8	8	8	6	6	6	6	4	4	4	4	3	3	3	3	2
20	12	12	10	12	12	10	10	8	8	8	6	6	6	6	4	4	4	4	3	3	2	2	2	1	1
25	10	10	10	12	10	10	8	8	6	6	6	6	4	4	4	4	3	3	2	2	1	1	1	0	0
30	10	10	10	10	10	10	8	6	6	6	4	4	4	4	3	2	2	1	1	1	0	0	0	0	00
35	8	8	10	10	10	8	8	6	6	4	4	4	4	3	3	2	2	1	1	0	0	0	00	00	00
40	8	8	10	10	8	8	6	6	4	4	4	4	3	3	2	1	1	0	0	0	00	00	000	000	000
45	6	8	10	10	8	8	6	6	4	4	4	3	3	2	2	1	1	0	0	00	00	00	000	000	000
50	6	6	10	8	8	6	6	4	4	4	3	3	2	2	1	1	0	0	00	00	000	000	000	4/0	4/0
60	4	6	8	8	8	6	4	4	4	3	2	2	1	1	1	0	00	00	000	000	000	4/0	4/0	4/0	250
70	4	4	8	8	6	6	4	4	4	3	2	2	1	1	0	0	00	000	000	4/0	4/0	250	250	300	300
80	2	4	6	6	6	4	4	3	2	2	1	1	0	0	00	00	000	000	4/0	4/0	250	250	300	300	300
90	2	3	6	6	6	4	4	3	2	2	1	1	0	0	00	000	000	4/0	4/0	250	250	300	300	350	350
100	1	3	6	6	4	4	3	2	1	1	0	0	00	00	000	000	4/0	4/0	250	250	300	300	350	350	400
115	0	2	4	6	4	4	3	2	1	0	0	00	000	000	000	4/0	4/0	250	300	300	350	350	400	400	500
130	00	1	4	4	4	3	2	1	0	0	00	000	000	4/0	4/0	250	300	300	350	400	400	500	500	500	600
150	000	0	2	4	4	3	1	0	0	00	000	000	4/0	4/0	4/0	250	300	350	400	400	500	500	500	600	600
175	4/0	00	2	4	3	2	1	0	00	000	000	4/0	4/0	250	250	300	350	400	500	500	500	600	600	600	700
200	250	000	1	3	2	1	0	00	000	000	4/0	4/0	250	250	300	350	400	500	500	600	600	700	700	700	750
225	300	4/0	0	3	2	1	0	00	000	4/0	4/0	250	300	300	350	400	500	500	600	600	700	700	750	800	900
250	350	250	00	2	1	0	00	000	4/0	4/0	250	300	300	350	350	400	500	600	600	700	700	750	800	900	1M
275	400	300	00	2	1	0	00	000	4/0	250	250	300	350	350	400	500	500	600	600	700	700	800	900	900	1M
300	500	350	000	1	1	0	000	4/0	4/0	250	300	350	350	400	500	500	600	700	700	800	900	900	1M		
325	600	400	4/0	1	0	00	000	4/0	250	300	300	350	400	500	500	600	600	700	750	900	900	1M			
350	600	500	4/0	1	0	00	000	4/0	250	300	350	400	400	500	500	600	700	750	800	900	1M				
375	700	500	250	0	0	00	4/0	250	300	350	400	500	500	500	600	700	750	900	1M						
400	750	600	250	0	00	000	4/0	250	300	350	400	500	500	500	600	700	750	900	1M						

Conductors in overhead spans must be at least No. 10 for spans up to 50 feet and No. 8 for longer spans. See NEC, Sec. 225-6 (a).
See Note 3 of NEC "Notes to Tables 310-16 through 310-19".

ALUMINUM up to 400 Amperes, *230-240 Volts,* Single Phase, Based on 2% Voltage Drop

Minimum Allowable Size of Conductor — **Length of Run in Feet**
Compare size shown below with size shown to left of double line. Use the larger size.

Load in Amps	Types R, T, TW	Types RH, RHW, THW	Overhead in Air* Bare & Covered — Single	Triplex	50	60	75	100	125	150	175	200	225	250	275	300	350	400	450	500	550	600	650	700	750	800
5	12	12	10		12	12	12	12	12	12	12	12	10	10	10	10	8	8	8	8	6	6	6	6	4	4
7	12	12	10		12	12	12	12	12	10	10	10	8	8	8	8	6	6	6	6	4	4	4	4	4	4
10	12	12	10		12	12	12	10	10	8	8	8	8	6	6	6	4	4	4	4	3	3	3	2	2	2
15	12	12	10		12	12	10	8	8	8	6	6	6	4	4	4	3	3	2	2	1	1	1	1	0	0
20	10	10	10		10	10	8	8	6	6	6	4	4	4	3	3	2	2	1	1	0	0	0	0	00	00
25	10	10	10		10	8	8	6	6	4	4	4	3	3	2	2	1	1	0	0	00	00	00	000	000	000
30	8	8	10		8	8	8	6	4	4	4	3	2	2	1	1	0	00	00	00	000	000	000	4/0	4/0	4/0
35	6	8	10		8	8	6	4	4	3	3	2	2	1	1	0	0	00	00	000	000	000	4/0	4/0	4/0	4/0
40	6	8	10		8	6	6	4	4	3	2	2	1	1	0	0	00	00	000	000	000	4/0	4/0	4/0	250	250
45	4	6	10		8	6	6	4	4	3	2	2	1	1	0	0	00	00	000	000	4/0	4/0	4/0	250	250	300
50	4	6	8		6	6	4	4	3	2	2	1	1	0	0	00	00	000	000	4/0	4/0	250	250	300	300	300
60	2	4	6	6	6	6	4	3	2	2	1	0	0	00	00	00	000	4/0	4/0	250	250	300	300	350	350	350
70	2	2(a)	6	4	6	4	4	3	2	1	1	0	0	00	00	000	000	4/0	4/0	250	300	300	350	350	400	400
80	1	2(a)	6	4	4	4	3	2	1	0	0	00	00	000	000	4/0	4/0	250	300	300	350	350	400	500	500	500
90	0	2(a)	4	2	4	4	3	2	1	0	00	00	000	000	4/0	250	300	300	350	400	400	500	500	500	600	600
100	0	1(a)	4	2	4	3	2	1	0	00	00	000	000	4/0	4/0	250	300	300	350	400	400	500	500	600	600	600
115	00	0(a)	2	1	4	3	2	1	0	00	000	000	4/0	4/0	250	300	300	350	400	500	500	600	600	600	700	700
130	000	00(a)	2	0	3	2	1	0	00	000	000	4/0	250	250	300	300	350	400	500	500	600	600	700	700	750	800
150	4/0	000(a)	1	00	2	2	1	00	000	000	4/0	250	300	300	350	400	500	500	600	600	700	750	800	900	900	1M
175	300	4/0(a)	0	000	2	1	0	00	000	4/0	250	300	300	350	400	400	500	600	600	700	750	800	900	900	1M	
200	350	250	00	4/0	1	0	00	000	4/0	250	300	300	350	400	400	500	600	600	700	750	900	900	1M	1M		
225	400	300	000		1	0	00	000	4/0	250	300	350	400	500	500	500	600	700	750	900	1M	1M				
250	500	350	000		0	00	000	4/0	250	300	350	400	500	500	600	700	750	900	1M							
275	600	500	4/0		0	00	000	4/0	250	300	400	400	500	500	600	600	750	900	1M							
300	700	500	250		00	00	000	250	300	350	400	500	500	600	600	700	800	900	1M							
325	800	600	300		00	000	4/0	250	300	400	500	500	600	600	700	750	900	1M								
350	900	700	300		00	000	4/0	300	350	400	500	500	600	600	700	750	800	900								
375	1M	700	350		000	000	4/0	300	350	500	500	600	700	700	800	900	1M									
400		900	350		000	4/0	250	300	400	500	500	600	600	700	750	900	900									

Conductors in overhead spans must be at least No. 10 for spans up to 50 feet and No. 8 for longer spans. See NEC, Sec. 225-6 (a).
See Note 3 of NEC "Notes to Tables 310-16 through 310-19".

Figure 30–20 *(Cont'd.)*

5. Clean an electric motor. Oil it according to instructions in the operator's manual.

☐ SELF-EVALUATION

A. Multiple Choice. Select the best answer.

1. The major power source for stationary equipment in houses, farm buildings, and agribusinesses is
   a. diesel fuel
   b. electricity
   c. gasoline
   d. steam

2. Electricity produces
   a. chemical changes
   b. heat and light
   c. magnetism
   d. all of these

3. A device that produces direct current by means of magnetism is
   a. a generator
   b. an alternator
   c. a turbine
   d. a pole

4. A device used to protect circuits that can be reset is a
   a. three-way switch
   b. ground fault interruptor
   c. fuse
   d. circuit breaker

5. A device in circuits to protect against human shock is a
   a. three-way switch
   b. ground fault interruptor
   c. fuse
   d. circuit breaker

6. Electricity is distributed to branch circuits by
   a. an electric meter
   b. an entrance head
   c. a service drop
   d. a service entrance panel

7. Tubes used to carry wires are called
   a. armored cable
   b. conduit
   c. nonmetallic sheathed cable
   d. pipe

8. A suitable wire for high temperature, high moisture locations is
   a. Type T
   b. Type THHN
   c. Type THW
   d. WVA

9. A cable consisting of No. 12 wire, one black, one red, one white, and a ground wire will be stamped
   a. 12–2
   b. 12–3
   c. 12–3 w/g
   d. 12–3 BRW

10. In order to run copper wire in a building 100 feet to a 10-ampere, 120-volt motor, and hold the voltage drop to 2 percent, the size of wire should be
    a. No. 12
    b. No. 10
    c. No. 8
    d. No. 6

11. The first job in maintaining electric motors is to
    a. install them in series
    b. keep them clean
    c. change plugs frequently
    d. avoid frequent use

12. Motors that need lubricating
    a. require only small amounts of oil at one time
    b. must be oiled at least every three months
    c. have grease gun fittings
    d. generally need to be oiled in many places

B. Matching. Match the terms in column I with those in column II.

Column I	Column II
1. conductor	a. great resistance
2. fluorescent	b. electrical pressure
3. filament	c. electrical energy or work
4. insulator	d. north and south poles
5. amperes	e. circle
6. volts	f. rotating part of motor
7. watts	g. rate of electrical flow
8. magnet	h. glowing element
9. armature	i. glowing gas
10. circuit	j. carries electricity

C. Completion. Fill in the blanks with the word or words that make the following statements correct.

1. Two deadly hazards of electricity are _____ and _____ .

2. Electric meters measure the amount of electricity used and express it in _____ .

3. Electrical circuits in a building are called _____ circuits.

4. Three types of cable used for wiring are _____ , _____ , and _____ .

5. Strands of wire are gathered together in bundles to increase the _____ of the wire and therefore its conductivity.

Unit 31
Installing Branch Circuits

Objective

To run wires and safely install boxes, switches, outlets, and fixtures.

Competencies to Be Developed

After studying this unit, you will be able to:
- Select electrical boxes, outlets, and switches.
- Install and replace switches, outlets, and fixtures.
- Install, extend, and modify branch circuits.
- Test electric circuits.

Terms to Know

new work	knockout
fixture	ground (gee) clip
wire nut	continuity tester
receptacle	continuity
switch	three-way switch
duplex receptacle	switching (traveler)
single-pole switch	wires

Materials List

Hand tools suitable for electrical wiring
Cable ripper
Wire stripper
Test light
14-2 cable with ground
14-3 cable with ground
Two switch boxes
One octagon box
One single-pole switch
Two three-way switches
One porcelain lamp holder
One duplex receptacle
Two switch covers
One outlet cover
Wire nuts
Ground clips
Electrical tape

A knowledge of electrical principles and wiring materials is important. The rules for the proper and safe installation of electrical devices and circuits given in the National Electrical Code must be strictly observed at all times.

Cable and conduit are run through the floors, ceilings, and partitions of new buildings during construction. Such installations are called *new work.* Electrical codes may permit wiring to be placed in view in old work installations. This is due to the difficulty of concealing wires after a building is completed. In surface installations, the wiring must be protected from mechanical damage.

Utility buildings and other farm structures generally do not have finished interior walls. Such structures are relatively easy to wire at any time. However, the National Electrical Code contains special requirements for such wiring.

☐ WIRING BOXES

All wiring systems require the use of fuses or circuit breakers, protected wires, and boxes. The boxes are metal or plastic, figure 31–1. Boxes have several important functions:

- They hold the cable or conduit so stress cannot be placed on the wire connections.
- They are nailed, screwed, or clamped to the building to support switches, outlets, or fixtures. A *fixture* is a base or housing for light bulbs, fan motors, and other electrical devices.
- They contain all electrical connections made outside of fixtures.

Electrical boxes are available in rectangular or octagonal shapes and in various depths. Some switch and outlet boxes have removable sides so the boxes can be installed in gangs (series). Such boxes will hold multiple switches or outlets.

Large steel boxes are used for service entrances. They contain the main fuses or circuit breakers. They also contain fuses or circuit breakers for one or more branch circuits.

Certain basic rules apply to the wiring of all boxes. The National Electrical Code and local building codes must be checked for specific regulations. Some basic requirements are as follows:

- The box must be fastened securely to the building.
- The cable or conduit must be clamped securely to the box.
- Cables running from box to box must run through the interior of the building's walls, floors, and ceilings, be secured by staples or clamps near each box, and secured as needed to prevent the cable from being accidentally caught and pulled.
- The box must be grounded if it is metal. The ground wire from the cable is attached to the box by a screw or by a grounding clip, figure 31-2.
- Wires in boxes must be connected to each other by soldering or by insulated solderless connectors called *wire nuts*. Bare wire areas are not permitted except on ground wires.
- Ground wires must be held together by a special metal clamp or a solderless connector.
- Wires must be attached to terminals of switches and receptacles by tightening no more than one wire under one screw. A *receptacle* is a device for receiving electric plugs. A *switch* is a device used to stop the flow of electricity. If the receptacle or switch is equipped with special clamps, one wire may be inserted into each clamp provided.
- Positive or hot wires (black, red, or blue) must always be attached to yellow screws. Neutral wires (white) must always be attached to white screws. Ground wires (bare or green) are attached to green or uncolored screws.
- When a white wire must be used as a positive or hot wire, the insulation showing in the box should be painted black or marked with black tape.

The most common type of receptacle is the *duplex* (double) *receptacle* wired so that both outlets are on the same circuit, figure 31-3. The current comes in on a black wire, flows from one screw through a metal strap to the other screw, and, when so wired, continues on to the next electrical box. The two receptacles in the duplex may be wired to two different circuits by breaking the metal strap between the two screws. This is called splitting the receptacle.

A standard switch box will safely hold four wires only. To determine the number of wires in a box only the positive and neutral wires are counted, not the ground wires. In other words, one incoming cable with a black wire, a white wire, and a ground wire is counted as two wires. If an identical cable takes the current on to the next box, the total of four wires has been reached.

□ WIRING A SWITCH AND LIGHT FIXTURE

As stated, a switch is a device used to stop the flow of electricity. A *single-pole* switch is a switch designed to be the only switch in a circuit. The cable carrying current from the service entrance panel may come

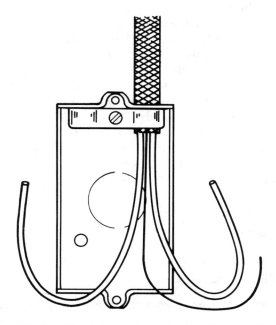

Figure 31-1. All wiring systems require the use of boxes. (*courtesy of University of Illinois, Vocational Agriculture Service*).

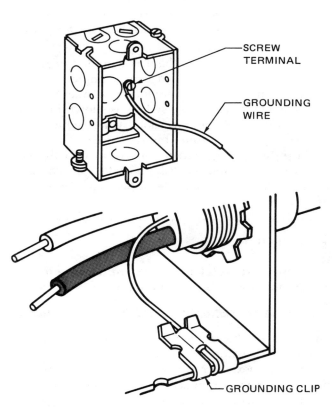

Figure 31-2. Ground wires are attached to metal boxes by means of a screw or grounding clip.

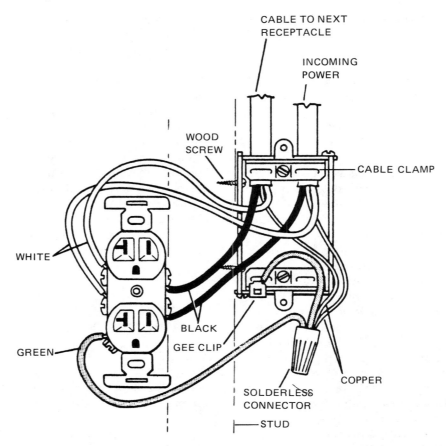

Figure 31–3. Typical duplex receptacle (*courtesy of The Pennsylvania State University, Department of Agricultural Education*).

to an outlet box where a light is mounted or to one where a switch is mounted. For example, a power cable coming to a box where a light is mounted, and controlled by a single-pole switch, is shown in figure 31–4. To wire this circuit, cable with a white wire, a black wire, and a ground wire is used.

The cable is prepared for insertion into the box by slitting 6 to 8 inches of the outside cable covering with a knife or special cable ripper, figure 31–5.

Caution:

Care must be taken to prevent damage to the insulation on the individual wires.

The wires are then separated from the jacket and the excess jacket material is cut off.

Electrical boxes are provided with a *knockout*, or partially punched impression, which can be punched out and removed. The cable is pushed through this hole. Some boxes have cable clamps provided with

the box, while others require the addition of a cable clamp, figure 31–4. The cable is inserted until $\frac{1}{16}$ inch of the cable jacket extends beyond the clamp. The clamp is then tightened and cable is run between the switch box and the light box. If there is more than one light in the circuit, a third cable is installed to carry current from one light box to the next. All cables must be clamped securely in the boxes and stapled or clamped to the structure within 12 inches of each box and at intervals of 4½ feet or less.

To install the switch, a knife or wire stripper is used to remove about ¾ inch of insulation from the ends of individual wires, figure 31–6.

Caution:

Care must be taken not to nick the wire during this procedure.

A round loop is then made in the end of each wire. Each loop is wrapped around a yellow screw in the

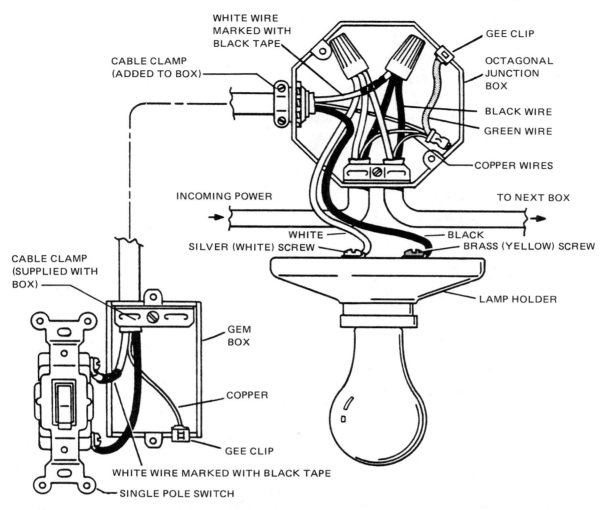

Figure 31–4. A light fixture controlled by a single-pole switch (*courtesy of The Pennsylvania State University, Department of Agricultural Education*)

direction the screw turns when being tightened. The screws are then tightened securely.

Switches are placed in hot wires only—never in neutral wires. Thus a piece of black tape is placed on the white wire to mark it as a hot wire. The ground wire is attached to the metal box with a *ground clip*, also called a *gee clip*. The switch is then screwed in place and a switch cover installed.

Wiring the lamp holder box is more difficult. However, the work progresses in a logical sequence according to the following steps.

Procedure

1. Strip the ends of all wires leaving ⅝ inch of wire exposed. The wires are now ready for connection with wire nuts.

2. Strip ⅝ inch from both ends of an 8-inch length of green wire and ground it to the box with a ground clip or screw.

3. Hold the four ground (green) wires so they form a bundle. Twist a wire nut onto the bundle and tighten it as much as possible with the hands. For large bundles, use a wire nut handle or pliers to tighten the wire nut.

4. Mark the white wire coming from the switch with black tape. This is now regarded as a black wire.

5. Attach the black wire from the switch to the yellow terminal of the fixture or lamp holder.

6. Cut an 8-inch piece of white wire and strip both ends. Attach one end to the white screw of the lamp holder.

7. Use a wire nut to connect the loose ends of the three white wires.

8. Use another wire nut to connect the ends of the

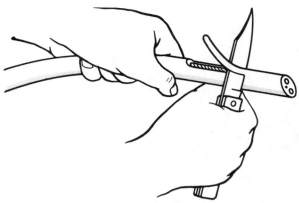

A. SPLITTING CABLE WITH KNIFE

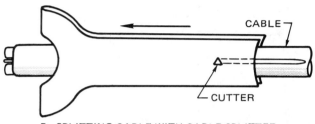

CABLE

CUTTER

B. SPLITTING CABLE WITH CABLE SPLITTER

Figure 31–5. Seven inches of outer covering of nonmetallic sheathed cable is slit by a knife or cable splitter. The individual wires are then pulled out of the jacket and the loose jacket is cut off and discarded.

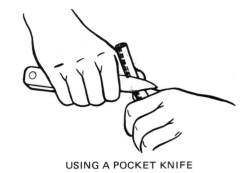

USING A POCKET KNIFE

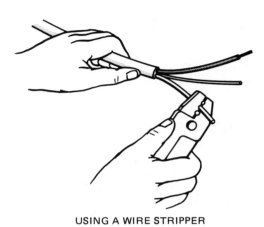

USING A WIRE STRIPPER

Figure 31–6. When preparing the ends of wires for screws, care must be taken to remove ¾-inch of insulation without nicking the wire. Remove only ⅝-inch for wire nuts.

three remaining black wires. Remember, the white wire with the black tape is treated as a black, or hot wire.

9. Check each connection for tightness by holding the wire nut and pulling hard on each of the wires. If one or more wires is loose, the entire bundle must be retightened.

The flow of electricity in this circuit proceeds from the service entrance box to the light box on the black wire of the incoming cable. It flows to the switch by way of the white wire marked with black tape. If the switch is on, it flows through the switch and through the black wire to the lampholder. Current flows through the bulb and back to the service entrance box by way of white (neutral) wires.

□ TESTING A CIRCUIT

After all lights and switches are wired, a circuit should be tested before turning on the circuit breaker or inserting the fuse. Use a continuity tester to be sure the circuit is not open. A *continuity tester* is a device used to determine if electricity can flow between two points. *Continuity* means connectedness. A circuit is open if there is a break or poor connection anywhere in it. The circuit should also be checked for shorts, or places where current can get from the black wires

to the ground wires or boxes or can bypass the bulb and flow directly to the neutral wires. The final step is to test all boxes to ensure that they are properly grounded.

The following procedure may be helpful in testing a circuit.

Procedure

1. Place switches in the on position.
2. Remove all bulbs.
3. Test the circuit by connecting to the wires before the cable is wired into the fuse block or circuit breaker. Using an ohmmeter, first touch one lead to the black wire and the other to the white wire of the cable. There should be no reading. This indicates no continuity. That is, there are no points where electricity can flow from the black wires to the white wires.
4. Install a good light bulb in the last fixture of the circuit. Repeat the test in step 3. There should be a reading. This indicates that current can flow through all black wires, through the bulb, and back through the white wires.
5. Test to see if each metal box in the circuit is

grounded. This is done by touching one test lead to the box and the other to the ground wire. There should be a reading.

When all cover plates and fixtures have been installed, the cable can be wired into the fuse block or circuit breaker.

Caution:

Do not add electricity to a circuit until a qualified electrician has checked it.

□ WIRING LIGHTS WITH THREE-WAY SWITCHES

A *three-way switch* is a switch that, when used in a pair, permits a light or receptacle to be controlled from two different locations, figure 31–7. The wiring of a circuit with three-way switches will vary according to the location of the incoming power cable, the light, and the two switches. However, certain basic rules apply regardless of the location of components.

A three-way switch has three terminals. The current feeds into a common terminal identified by its dark screw. The current flows out through either one of the remaining two terminals identified by light-colored screws. The light-colored screw that carries the current is determined by the position of the switch. The two wires attached to the light-colored screws are called the *switching wires* or *traveler wires*. Either wire can carry the current to the second switch. A wire running from the common terminal of the second switch is attached to the yellow screw on the light fixture.

Basic rules for wiring three-way switches are as follows.

Procedure

1. Connect the white wire of the incoming power cable to the silver terminal of the light fixture.
2. Connect the black wire of the incoming power cable to the dark screw (common terminal) of the switch.
3. Connect a black wire from the dark screw of the other switch to the yellow terminal of the light fixture.
4. Run a pair of traveler wires between the light-colored terminals of the two switches.

Three-wire cables are useful when wiring circuits controlled by three-way switches, figures 31–8 to 31–10. Such cables typically have a black, red, white, and ground wire. The white wire may be used as a hot wire if it is connected to a switch or to a black wire and is clearly marked with black tape.

Black and red wires are generally used as the traveler wires in a three-way switch circuit. If metal boxes are used, the ground wire grounds the boxes.

Wiring Lights Using Four-Way Switches

A four-way switch is a switch connected in the pair of traveler wires between two three-way switches, figure 31–11. Any number of four-way switches can be added to a three-way switch circuit to accommodate the total number of switch locations needed. The four-way switch permits the electricity to be changed from one traveler to the other depending on the position of the switch lever, figure 31–12.

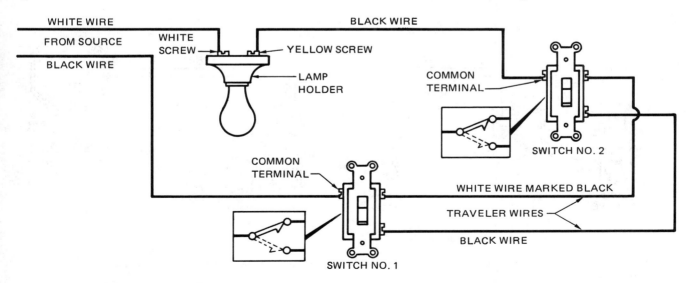

Figure 31–7. A three-way switch circuit

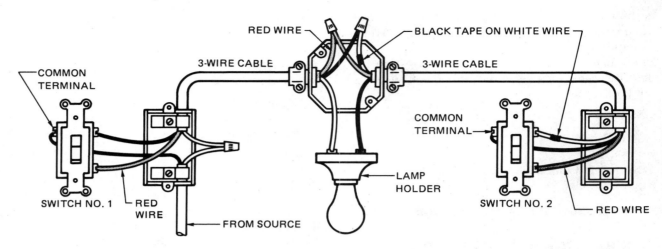

Figure 31-8. Wiring for a three-way circuit with current coming to one switch and operating a light located between two three-way switches.

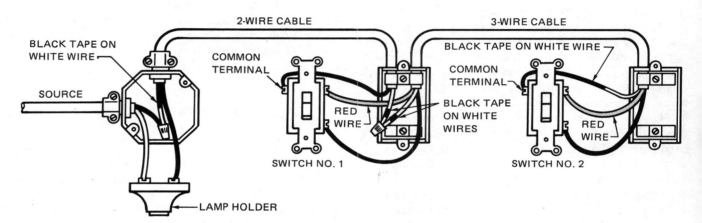

Figure 31-9. Wiring for three-way switches with current coming in at the light.

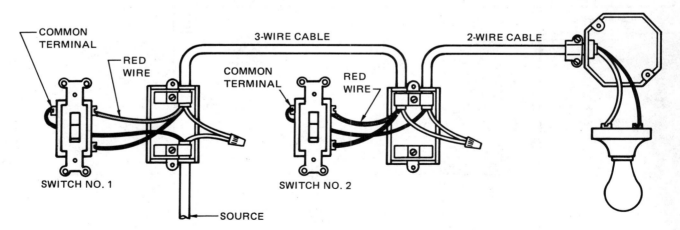

Figure 31-10. Wiring for three-way switches with the light at the end of the run and current coming in to one switch.

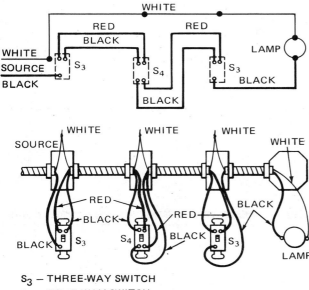

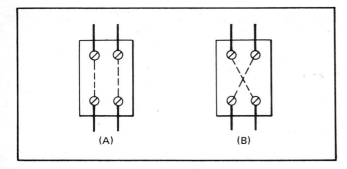

S₃ — THREE-WAY SWITCH
S₄ — FOUR-WAY SWITCH

Figure 31–11. A circuit using a four-way switch. Four-way switches (S₄) can only be installed in the traveler wires between two three-way switches (S₃) (*courtesy of R. Mullin, Electrical Wiring—Residential, 8th ed., Delmar Publishers Inc., 1984*).

Figure 31–12. Two positions of a four-way switch (*courtesy of R. Mullin, Electrical Wiring—Residential, 8th ed., Delmar Publishers Inc., 1984*).

□ STUDENT ACTIVITIES

1. Wire a circuit with an outlet.
2. Wire a circuit with several outlets.
3. Wire a circuit with a light and single-pole switch.
4. Wire a circuit with a light and two three-way switches.

□ SELF-EVALUATION

A. Multiple Choice. Select the best answer.
 1. All electrical connections in a circuit are made
    a. in boxes or fixtures
    b. by screws
    c. with solder
    d. with tape
 2. All metal electrical boxes must
    a. be grounded
    b. be securely fastened
    c. secure the cable or conduit
    d. all of these
 3. Neutral wires are attached to screws colored
    a. white
    b. green
    c. silver
    d. yellow
 4. The device that receives electrical plugs is a
    a. box
    b. cap
    c. circuit breaker
    d. receptacle
 5. White wires used as positive wires must be
    a. connected to black wires
    b. connected to fixtures
    c. stripped of all insulation
    d. taped or painted black
 6. A properly wired circuit will be
    a. grounded
    b. open
    c. shorted
    d. all of these
 7. In three-way switch circuits electricity passes from one switch to the other by
    a. traveler wires
    b. neutral wires
    c. common terminals
    d. none of these
 8. Three-way switch circuits usually include
    a. two-wire cables
    b. three-wire cables
    c. grounded boxes
    d. all of these

B. Matching. Match the terms in column I with those in column II.

Column I	Column II
1. old work	a. used in place of black wire
2. duplex	
3. solderless connector	b. always white
4. switch	c. traveler
5. positive or hot	d. fuse
6. red wire	e. extensions to existing systems
7. neutral wire	
8. circuit protector	f. receptacle
9. electrical contact	g. continuity
10. switching wire	h. stops current
	i. black wire
	j. wire nut

Section 12
PLUMBING

Unit 32
Plumbing

Objective

To identify plumbing materials and perform basic plumbing procedures.

Competencies to Be Developed

After studying this unit, you will be able to:
* Identify plumbing tools.
* Identify and select pipe.
* Identify common pipe fittings.
* Assemble pipe.
* Maintain water systems.

Terms to Know

plumbing	double extra heavy
pipe	K
fitting	L
tubing	M
bench yoke vise	compression fitting
chain vise	plastic
chain wrench	polyethlene (PE)
tubing cutter	polyvinyl chloride (PVC)
ratchet die stock	acrylonitrite-butadiene-
flaring block	styrene (ABS)
flaring tool	cast iron soil pipe
pipe stand	oakum
nominal	neoprene
ID	adaptor
OD	bushing
black pipe	reducer
galvanized pipe	elbow (ell)
standard pipe	tee
extra heavy	Y

coupling	valve
union	O ring
plug	packing
cap	diaphragm
nipple	faucet washer
pipe dope	aerator
teflon tape	float valve assembly
faucet	flush valve

Materials List

Common pipe fittings:
  a. bushing
  b. 90° elbow
  c. 45° elbow
  d. coupling
  e. reducer
  f. cap
  g. street elbow
  h. plug
  i. tee
  j. Y
  k. union
  l. floor flange
  m. pipe nipples
Examples of steel, copper, and plastic pipe
All materials in the bill of materials, figure 32–32

The term *plumbing* means installing and repairing water pipes and fixtures, including pipes for handling waste water and sewerage. In agricultural mechanics, all work done with pipe and pipe fittings is considered to be part of plumbing. *Pipe* refers to rigid tubelike material. A *fitting* is a part used to connect pieces of pipe or to connect other objects to pipe. Tubing is also used in plumbing installations. *Tubing* generally refers to pipe that is flexible enough to bend.

□ TOOLS FOR PLUMBING

Most of the tools used for plumbing have been discussed in previous units. The following tools are commonly used when working with pipe and tubing.

Tools useful for holding pipe and fittings are the *bench yoke vise* and the *chain vise*, figure 32–1. The most useful tools for turning pipe and fittings are the *chain wrench* and pipe wrench, figure 32–2. The hacksaw, file, pipe cutter, reamer, and *tubing cutter* are useful for cutting and smoothing pipe and tubing, figure 32–3. Taps and dies are used for cutting threads in pipe. A *rachet die stock* is used to turn the die.

The ends of soft copper tubing may be flared to fit bell-shaped fittings, figure 32–4. Copper tubing is cut to length with a tubing cutter. The pressure and rolling action of the cutter wheel create a burr on the inside of the tubing at the cut. This burr is removed using the burr remover on the tubing cutter. The flaring tool and *flaring block* are used to hold the tubing and to form a flared end on the tubing, figure 32–5. This end is held against bell-type fittings by a flare nut to create a gas or liquid tight seal.

A tripod-type pipe stand with a vise makes an excellent work area for cutting, reaming, and threading pipe, figure 32–6.

□ IDENTIFYING, SELECTING, AND CONNECTING PIPE

The *nominal*, or identifying, size of pipe is generally based on the inside diameter. Inside diameter is abbreviated *ID* and outside diameter, *OD*. Common sizes of pipe for home and farm applications (ID in inches) are ¼, ⅜, ½, ¾, 1, 1¼, 1½, 2, 2½, 3, 4, and 6.

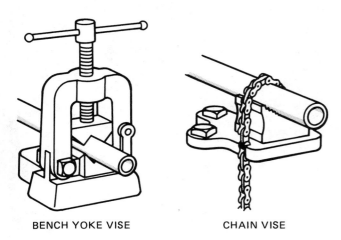

BENCH YOKE VISE   **CHAIN VISE**

Figure 32–1.   The bench yoke vise and chain vise are used for holding pipe and pipe fittings.

FOR POSITION ONLY

CHAIN WRENCH

PIPE WRENCH

Figure 32–2.   Chain wrenches and pipe wrenches are useful tools for turning pipe and pipe fittings (*pipe wrench courtesy of Klein Tools, Inc.; chain wrench courtesy of Mac Tools, Inc.*)

Steel Pipe

Steel pipe may be purchased as black pipe or galvanized pipe. *Black pipe* is painted black and has little resistance to rusting. Galvanized pipe is coated with zinc inside and outside. It resists rust for many years.

> **Caution:**
>
> **The zinc coating on galvanized pipe gives off toxic fumes when welded or heated with a torch. Special precautions must be taken when using galvanized steel pipe for construction.**

Steel pipe of each nominal size has a specified wall thickness and is classified as *standard pipe*. Special pipe may be ordered with thicker walls. This pipe is classified as *extra heavy* or *double extra heavy* pipe. It is generally used for construction rather than plumbing.

The outside diameter of standard, extra heavy, and double extra heavy pipe is the same. The additional wall thickness of the extra heavy and double extra heavy pipe results in reduced inside diameters, figure 32–7. Because the outside diameter is consistent, standard pipe fittings can be used with all three types of pipe.

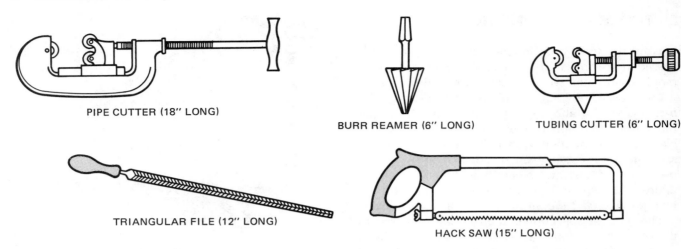

PIPE CUTTER (18" LONG)

BURR REAMER (6" LONG)

TUBING CUTTER (6" LONG)

TRIANGULAR FILE (12" LONG)

HACK SAW (15" LONG)

Figure 32–3. Tools used for cutting and smoothing pipe and tubing

Copper Tubing and Pipe

Pipe made of copper is frequently referred to as tubing. It may be purchased in the soft, annealed form or the rigid form. For convenience, this unit refers to the soft form as tubing and the rigid form as pipe. The tubing can be bent around irregular parts of build-ings. This makes it fast and easy to install. However, tubing does not have the neat appearance of rigid pipe. Therefore, it is used primarily in hidden spaces or where appearance is of little concern.

Copper is frequently preferred to steel for water lines because it resists corrosion, is easy to handle, and has been known to withstand freezing without breaking. Some disadvantages of copper are its high initial cost, high degree of expansion, and the bad taste created if the water is acid. Slightly acid water also reacts with copper to leave green stains in sinks and tubs.

Copper pipe and tubing are available in three types based on wall thickness, figure 32–8. The types are K, L, and M. Type *K* has the thickest wall, type *L* has medium wall thickness, and type *M* has the thinnest wall. Standard copper fittings are used for all

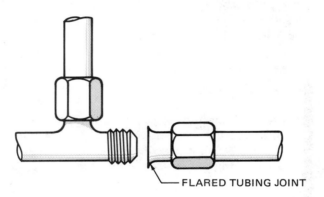

FLARED TUBING JOINT

Figure 32–4. The ends of soft copper tubing may be flared to fit bell-shaped fittings.

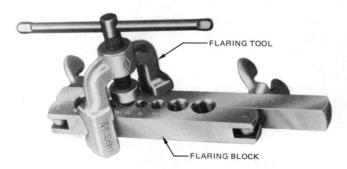

FLARING TOOL

FLARING BLOCK

Figure 32–5. A flaring tool is used to flare the ends of soft copper tubing (*courtesy of Mac Tools, Inc.*)

Figure 32–6. A pipe stand equipped with a chain vise and storage shelf

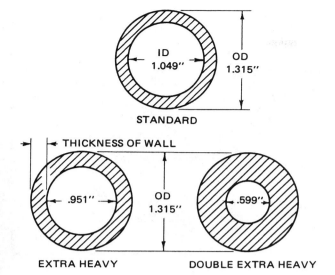

STANDARD

THICKNESS OF WALL

EXTRA HEAVY   **DOUBLE EXTRA HEAVY**

Figure 32–7. Standard, extra heavy, and double extra heavy steel pipe all have the same outside diameter for a given size. This permits pipe threading with the same die and the use of the same sized pipe fittings for all three types (*courtesy of H. Slater and L. Smith,* Basic Plumbing, *Delmar Publishers Inc., 1979*).

three types. Generally, type L is specified by plumbing codes as the minimum acceptable size for use in buildings. Type K is generally required when the pipes are to be buried.

Flexible copper tubing may be joined by flaring the ends of the tubing and connecting them with bell-type fittings. Very small tubing can also be joined with compression fittings. A *compression fitting* grips the pipe by compressing a special collar with a threaded nut. This procedure does not require the use of a flaring tool or soldering torch. A brief review of sweating or soldering copper tubing is shown in figure 32–9.

Plastic Pipe

In addition to steel and copper pipe, plastic pipe is used extensively in water systems. Plastic is the easiest type of pipe to install. The term *plastic* is used here to include all pipe made from synthetic materials. Such materials have long been used for cold water lines. Some plumbing codes now permit the use of certain plastic pipe materials for hot water lines. Most plastic pipe is fastened together by the simple procedure shown in figure 32–10. The steps of the procedure are given on page 386.

STANDARD WATER TUBE SIZE	ACTUAL OUTSIDE DIAMETER	NOMINAL WALL THICKNESS		
		TYPE K	TYPE L	TYPE M
INCHES	INCHES	INCHES	INCHES	INCHES
⅜	.500	.049	.035	.025
½	.625	.049	.040	.028
⅝	.750	.049	.042	. . . .
¾	.875	.065	.045	.032
1	1.125	.065	.050	.035
1¼	1.375	.065	.055	.042
1½	1.625	.072	.060	.049
2	2.125	.083	.070	.058
2½	2.625	.095	.080	.065
3	3.125	.109	.090	.072
3½	3.625	.120	.100	.083
4	4.125	.134	.110	.095
5	5.125	.160	.125	.109
6	6.125	.192	.140	.122
8	8.125	.271	.200	.170
10	10.125	.338	.250	.212
12	12.125	.405	.280	.254

Note: Type M is not made in ⅝″ size.

Figure 32–8. Sizes and weights of copper tubing and pipe (*courtesy of H. Slater and L. Smith,* Basic Plumbing, *Delmar Publishers Inc., 1979*)

CLEAN THE OUTSIDE OF COPPER TUBE WITH FINE STEEL WOOL OR EMERY CLOTH TO A BRIGHT FINISH.

1

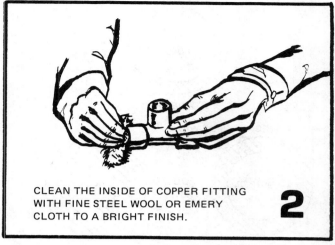

CLEAN THE INSIDE OF COPPER FITTING WITH FINE STEEL WOOL OR EMERY CLOTH TO A BRIGHT FINISH.

2

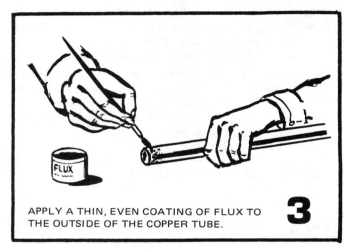

APPLY A THIN, EVEN COATING OF FLUX TO THE OUTSIDE OF THE COPPER TUBE.

3

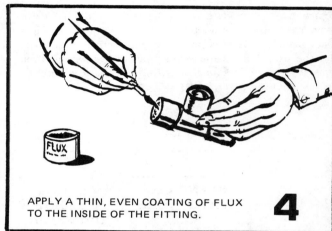

APPLY A THIN, EVEN COATING OF FLUX TO THE INSIDE OF THE FITTING.

4

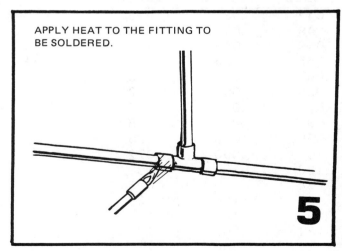

APPLY HEAT TO THE FITTING TO BE SOLDERED.

5

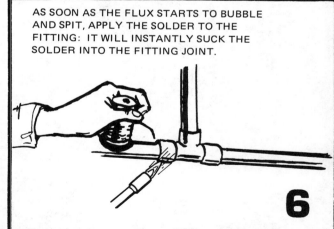

AS SOON AS THE FLUX STARTS TO BUBBLE AND SPIT, APPLY THE SOLDER TO THE FITTING: IT WILL INSTANTLY SUCK THE SOLDER INTO THE FITTING JOINT.

6

Figure 32–9. Six simple steps for soldering copper tubing (*courtesy of Step by Step Guide Book Co., 1981*)

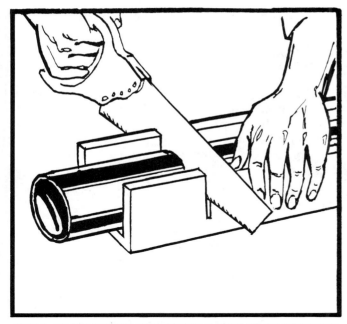

STEP 1 CUT PIPE WITH FINE-TOOTH HANDSAW OR HACKSAW.

STEP 2 WITH A PIECE OF CLOTH, CLEAN THE INSIDE OF THE FITTING AND THE OUTSIDE OF THE PIPE.

STEP 3 SWAB CONTACT AREAS WITH A SPECIAL CLEANER. APPLY A FULL, EVEN COATING OF CEMENT TO THE INSIDE OF THE FITTING AND TO THE OUTSIDE OF THE PIPE.

STEP 4 SLIP FITTING OVER PIPE, GIVING IT A QUARTER OF A TURN; HOLD FOR A FEW SECONDS UNTIL GLUE SETS.

Figure 32–10. Steps in cementing plastic pipe (*courtesy of Step by Step Guide Book Co., 1981*)

Procedure

1. Use a fine-tooth saw to cut the pipe so the end is square.
2. Remove the rough or ragged edges using a knife or special cutter.
3. Clean the inside of the fitting and the outside of the pipe with a cloth and then use a special plastic cleaner.
4. Swab plastic cement on the outside of the pipe and in the fitting where the parts will join.
5. Press the fitting onto the pipe and give it a quarter of a turn. The fitting will be difficult to remove after a few seconds. It will set permanently shortly thereafter.

Plastic pipe is made from several different materials. *Polyethlene (PE)* has been in use for several decades. It is used extensively for cold water lines. PE pipe is usually black, somewhat flexible, and is available in rolls. It is popular for direct burial, pump installations, and surface water lines, figure 32–11. PE pipe is assembled by pushing the pipe over the grooved section of the fittings. The pipe is secured to the fitting using a rustproof stainless steel clamp, figure 32–12. There is no need for cement or pipe sealing compounds. Plastic pipe is easily removed by sawing with a hacksaw.

Polyvinyl chloride (PVC) and chlorinated polyvinyl chloride (CPVC) are rigid and generally white or gray in appearance. PVC pipe is used for cold water lines and CPVC is approved by some plumbing codes for hot water lines. Figure 32–13 shows plastic pipe being used for the vent and drain system for a tub, basin, and toilet. PVC pipe is especially easy to install. It is

light in weight, may be cut with any saw, and is assembled using a liquid cement.

Acrylonitrite-butadiene-styrene (ABS) is used for sewerage and underground applications. Tests seem to indicate that it will last almost indefinitely. Manufacturers reportedly guarantee the material for 50 years. It is assembled by cementing the pipe to the fittings. The light weight of the pipe and the ease of assembly make plastic pipe especially attractive for sewerage installations where large diameters are required.

Cast Iron Soil Pipe

Cast iron soil pipe has thick walls and is used mostly for sewerage systems. Popular sizes range from 2 to 6 inches in diameter. Cast iron pipe is available in 5- and 10-foot lengths and in two weights—service and extra heavy. It can be purchased with straight ends, with an enlarged area called a hub on one or both ends, or with a bell-shaped end, figure 32–14.

Cast iron systems are joined by one of several methods. The oldest method is to place the straight or hub end into the bell end and pack the space with oakum. *Oakum* is a ropelike material that makes a watertight seal when it is compacted. After the oakum is added, the joint is sealed with molten lead, figure 32–15.

Two newer methods of connecting cast iron pipe which permit greater speed and ease in assembling joints have become popular. Both methods use a neoprene gasket to make a watertight seal. *Neoprene* is

Figure 32–11. Polyethylene pipe is excellent for cold water lines either in underground or surface areas.

Figure 32–12. Polyethylene pipe is attached to special grooved pipe fittings with stainless steel bands.

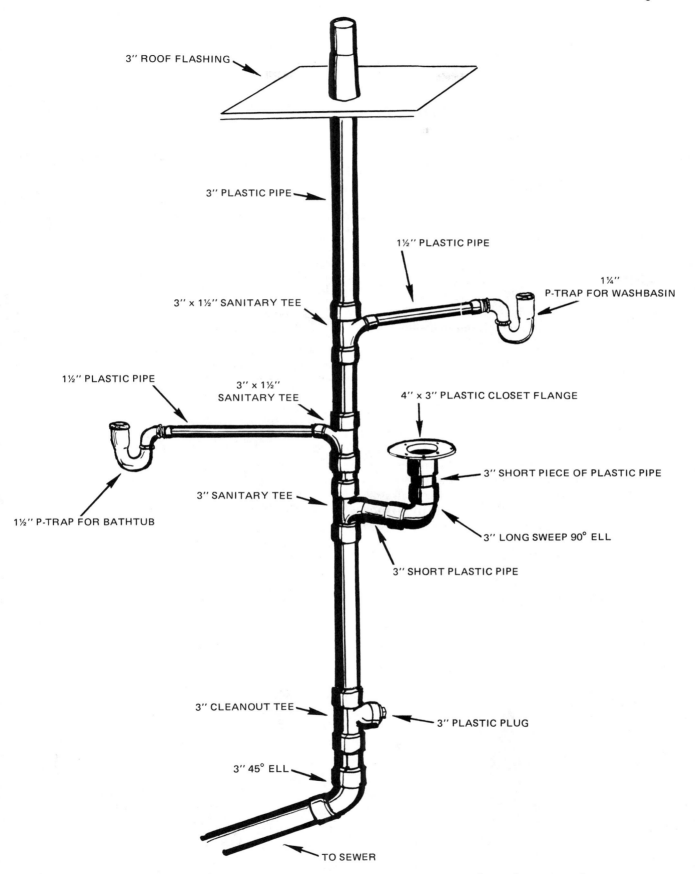

3″ ROOF FLASHING

3″ PLASTIC PIPE

1½″ PLASTIC PIPE

1¼″ P-TRAP FOR WASHBASIN

3″ × 1½″ SANITARY TEE

1½″ PLASTIC PIPE

3″ × 1½″ SANITARY TEE

4″ × 3″ PLASTIC CLOSET FLANGE

3″ SHORT PIECE OF PLASTIC PIPE

1½″ P-TRAP FOR BATHTUB

3″ SANITARY TEE

3″ LONG SWEEP 90° ELL

3″ SHORT PLASTIC PIPE

3″ CLEANOUT TEE

3″ PLASTIC PLUG

3″ 45° ELL

TO SEWER

Figure 32–13. Plastic vent and drain system for tub, basin, and toilet (*courtesy of Step by Step Guide Book Co., 1981*)

WEIGHT OF SOIL PIPE

SERVICE

SIZE	LB PER FT	SIZE	LB PER FT
2"	4	6"	15
3"	6	7"	20
4"	9	8"	25
5"	12		

EXTRA HEAVY

SIZE	LB PER FT	SIZE	LB PER FT
2"	5	8"	30
3"	9	10"	43
4"	12	12"	54
5"	15	15"	75
6"	19		

Figure 32–14. Cast iron soil pipe is available in a variety of diameters and two weights. It may be purchased with straight ends, with a hub on one or both ends, or with a bell-shaped end (*courtesy of H. Slater and L. Smith,* Basic Plumbing, *Delmar Publishers Inc., 1979*).

a tough, rubberlike material that is resistant to moisture and rotting.

One method uses pipe with one bell-shaped end and one straight end. The neoprene gasket is mounted in the bell, and the inside of the gasket is lubricated, figure 32–16. The straight end of the pipe or fitting is then slipped into the gasket. The resulting joint is watertight and flexible.

The other method uses pipe and fittings with straight ends. The straight ends are slipped into a lubricated neoprene gasket. Then a rustproof stainless

steel band is tightened over the joint, figure 32–17. This method has the advantage of requiring less space than the other methods because there are no bells on the pipe and the bands are thin.

Cast iron is heavy and hard to handle. Since copper and plastic are lightweight, require less space, and are easy to assemble, they are used extensively for indoor sewer lines. However, cast iron is frequently preferred for buried sewer lines. It resists rust well, is durable, and if properly installed is quite resistant to crushing.

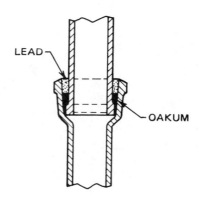

Figure 32–15. Cast iron pipe joint made with oakum and lead (*courtesy of H. Slater and L. Smith,* Basic Plumbing, *Delmar Publishers Inc., 1979*)

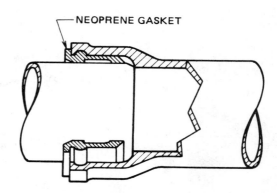

Figure 32–16. Cast iron pipe joint made with a neoprene gasket in a bell end (*courtesy of H. Slater and L. Smith,* Basic Plumbing, *Delmar Publishers Inc., 1979*)

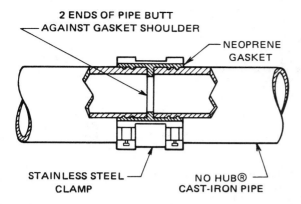

2 ENDS OF PIPE BUTT
AGAINST GASKET SHOULDER

NEOPRENE
GASKET

STAINLESS STEEL
CLAMP

NO HUB®
CAST-IRON PIPE

Figure 32–17. Cast iron pipe joint made with a neoprene gasket and a stainless steel clamp on straight ends (*courtesy of H. Slater and L. Smith,* Basic Plumbing, *Delmar Publishers Inc., 1979*)

Selecting Pipe Material

Each kind of pipe has advantages and disadvantages. When making repairs it is generally easier to repair systems with the kind of pipe used in the original installation. When planning a new system, however, it is important to consider the relative merits of each kind of pipe, figure 32–18.

□ IDENTIFYING PIPE FITTINGS

Pipe fittings are used to connect pipe. It is important to learn the names of basic pipe fittings in order to plan plumbing systems, order plumbing materials, and communicate with fellow workers about the job. The names of the fittings are the same for all kinds of pipe. Figure 32–19 shows some of these fittings.

Fittings used to connect pipes of different types (such as copper tubing and steel pipe) are called *adaptors. Bushings* and *reducers* are used to connect pipes of different sizes (such as ½-inch steel pipe to ¾-inch steel pipe). Pipes coming together from different directions may be connected by an elbow, tee, or Y. The *elbow,* or *ell,* is used where a single line changes direction, the *tee* and *Y* are used where a line splits in two directions. A *coupling* is used to connect two pieces of similar pipe. A *union* also connects two pieces of similar pipe. An advantage of the union is that it permits opening the line without cutting the pipe. A pipe fitting may be closed with a *plug;* a pipe end may be closed with a *cap.*

Short pieces of steel pipe are hard to thread on the job. Therefore, they are manufactured and sold in various lengths with both ends threaded. A short piece of pipe is called a *nipple.* Whenever fittings are threaded onto steel pipe it is necessary to use *pipe*

dope or *teflon tape* to make a watertight seal. Teflon tape is also recommended for sealing threads where gas will be used in the lines.

□ MAINTAINING WATER SYSTEMS

Properly installed water systems are generally quite durable. However, certain types of water react with steel and copper and may shorten the life of pipe, fittings, faucets, valves, and pumps unless the proper water conditioners are used. A *faucet* is a device that controls the flow of water from a pipe or container. A *valve* is a device that controls the flow of water in a pipe.

A green stain beneath the faucet in a tub or sink generally indicates a reaction between the water and copper pipes. Reddish water coming from a faucet generally means a rusting condition caused by a reaction between water and steel pipes or tanks.

The green stain or red water problems can generally be corrected by a water treatment system. Such systems are expensive but may be worth the investment. Money spent for water treatment may be offset by savings on repairs to the plumbing system.

Repairing Faucets

Normal maintenance of water systems includes occasionally tightening packing material or changing faucet washers or O rings to stop faucets from dripping. An *O ring* is a piece of rubber shaped like an *O* which fits into a groove on a shaft. It prevents liquids from passing between a rotating shaft like a faucet stem and the surrounding hole. *Packing* is a soft, slippery, and wear-resistant material. It, too, is used to seal the space between a moving shaft and the surrounding area. A *faucet washer* is a rubberlike part that creates a seal when pressed against a metal seat. A *seat* is a nonmoving part that is designed to seal when a moving part is pressed against it.

Toilet tanks and automatic livestock waterers have shut-off valves that need periodic service. Typically the problem is a failure to shut off completely. The replacement of O rings, diaphragms, or rubber seats generally corrects the problem. A *diaphragm* is a flat piece of rubber that creates a seal between a part moving a short distance up and down, and a nonmoving part.

The basic parts of valves and single faucets are the handle, stem, packing nut, packing, bonnet, washer, washer screw, seat, and body, figure 32–20. Mixing faucets that control the flow of hot and cold water with one handle have a different design, figure 32–21.

FACTORS TO CONSIDER	GALVANIZED STEEL (3 OZ. COATING MIN.)	COPPER		PLASTIC
		TYPE K (HEAVY DUTY)	TYPE L (STANDARD)	
Underground soil corrosion—Probable life expectancy (1)	30 plus yrs. under most soil conditions. (If no corrosion inside pipe, life could extend to 100 yrs. or more.)	40–100 yrs. under most conditions	30–80 yrs. under most conditions	Experience indicates durability is satisfactory under most soil conditions.
	Waterlogged soils under most conditions—12–16 years. May be less than 10 yrs. in very high acid soils	14–20 yrs. in high sulfide conditions. May be less than 10 yrs. in cinders.	12–14 yrs. in high sulfide conditions. May be less than 10 yrs. in cinders.	
Resistance to corrosion inside pipe	Will corrode in acid, alkaline and hard waters or with electrolytic action. (2)	Normally very resistant. May penetrate rapidly in water containing free carbon dioxide.		Very resistant
Resistance to deposits forming inside pipe	Will accumulate lime deposits from hard water. (2)	Subject to lime scale and encrustation from suspended materials.		Resistant, but occasional deposits will form. (3)
Effect of freezing	Bursts if frozen solidly	Will stand mild freezes.		PE—will stand some freezing. PVC—will stand mild freezes.
Safe working pressures (lbs. per sq. in.)	Adequate for pressures developed by small water systems	Adequate for pressures developed by small water systems.		Working pressures at 73°F. PE 80 to 160 PVC 180 to 600
Resistance to puncturing and rodents	Highly resistant to both	Resistant to both		PE—Very limited resistance to puncture and rodents PVC—resistant
Effect of sunlight	No effect	No effect		PE—Weakens with prolonged exposure— PVC—Highly Resistant
Effect on water flavor	Little effect	Very acid water dissolves enough copper to cause off flavor.		Little effect (4)
Lengths available	21 ft. lengths	Soft temper: 60-ft.—100-ft. coils up to 1″ diameter 60-ft. coils above 1″ diameter Hard temper: 12- and 20-ft. lengths		PE usually in 100-ft. coils, or longer PVC usually in 20-ft. lengths
Comparative Weight (Approx. lbs. per foot) Inside diameter ½ ¾ 1 1¼ 1½ 2 2½	.85 lb. 1.13 1.60 2.27 2.72 3.65 5.79	.34 lb. .64 .84 1.04 1.36 2.06 2.93	.29 lb. .46 .66 .88 1.14 1.75 2.48	.06 lb. .09 .14 .24 .30

Figure 32–18. Relative merits of different kinds of piping materials (*courtesy of American Association for Vocational Instructional Materials*)

FACTORS TO CONSIDER	GALVANIZED STEEL (3 OZ. COATING MIN.)	COPPER		PLASTIC	
		TYPE K (HEAVY DUTY)	TYPE L (STANDARD)		
Ease of bending	Difficult to bend except for slight bends over long lengths.	Soft temper bends readily, will collapse on short bends. Hard temper difficult to bend except for slight bends over long lengths.		**PE** Bends readily, will collapse on short bends.	**PVC** Rigid Bends on long radius
Conductor of electricity	Yes	Yes		No	
Comparative cost index Pipe size (in.)				**PE**	**PVC**
½	10	21	17	3	3
¾	13	39	28	5	4
1	18	54	42	9	5
1¼	24	66	59	14	7
1½	28	87	74	18	10
2	39	130	110	27	15

(1) Derived from studies reported by Dennison, Irving A. and Romanoff, Melvin, "Soil-corrosion Studies, 1946 and 1948: Copper Alloys, Lead and Zinc," Research paper RP2077, Vol. 44, March 1950 and "Corrosion of Galvanized Steel in Soils," Research paper 2366, Vol. 49, No. 5, 1952. National Bureau of Standards, U.S. Dept. of Commerce.

(2) It is possible to greatly reduce corrosion and prevent lime scale in steel pipe by adding a phosphate material. It coats the inside of pipes, as well as the lining of all connected equipment. Prevents further lime scale and greatly reduces corrosion.

(3) Jones, Elmer E., Jr., "New Concepts in Farmstead Water System Design," Am. Society of Agricultural Eng., paper No. 67-216, 1967.

(4) Tiedeman, Walter D., "Studies on Plastic Pipe for Potable Water Supplies," Journal American Water Works Association, Vol. 46, No. 8, Aug. 1954.

Figure 32–18. (*Cont'd.*)

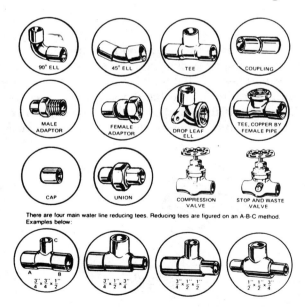

Figure 32–19. Names of pipe fittings (*courtesy of Step by Step Guide Book Co., 1981*)

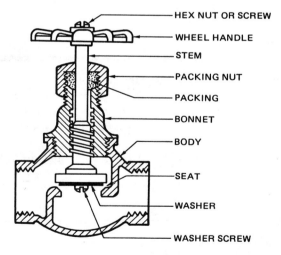

Figure 32–20. Parts of typical water valves and single faucets

Aerators mix air with the water as it comes from the faucet to create a smooth stream. Aerators have a screen to catch particles of rust or dirt. This screen and other parts of the aerator must be removed and cleaned occasionally to work properly.

Stopping Leaks Around Faucet Stems. □ To repair a faucet that leaks around the stem the following procedure is suggested.

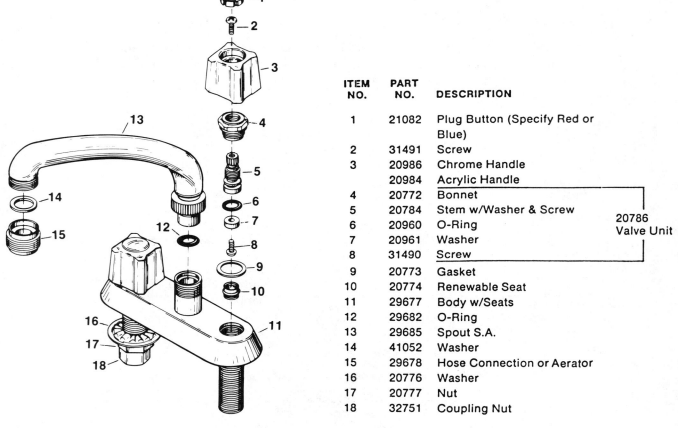

ITEM NO.	PART NO.	DESCRIPTION
1	21082	Plug Button (Specify Red or Blue)
2	31491	Screw
3	20986	Chrome Handle
	20984	Acrylic Handle
4	20772	Bonnet
5	20784	Stem w/Washer & Screw
6	20960	O-Ring
7	20961	Washer
8	31490	Screw
9	20773	Gasket
10	20774	Renewable Seat
11	29677	Body w/Seats
12	29682	O-Ring
13	29685	Spout S.A.
14	41052	Washer
15	29678	Hose Connection or Aerator
16	20776	Washer
17	20777	Nut
18	32751	Coupling Nut

20786 Valve Unit (items 4–8)

Figure 32–21. Parts of a typical mixing faucet used on kitchen or laundry sinks (*courtesy of Kohler Co.*)

Procedure

1. Tighten the bonnet or packing nut.
2. If the stem still leaks, unscrew the bonnet or packing nut and the stem, figures 32–22 and 32–23.
3. Replace the packing or O ring and reassemble, figure 32–24.
4. Tighten the packing nut or bonnet just enough to prevent leaking.

Caution:

Overtightening will make the faucet hard to turn on and off.

Faucet spouts that swing may leak at the base. To correct this problem simply tighten the nut that attaches the spout to the faucet body. If necessary, replace the O ring or packing on the spout.

Loose handles are tightened by prying off or unscrewing the plug in the handle and tightening the handle screw. If the handle is broken or slips, a replacement handle can generally be purchased. Most

Figure 32–22. Older type of faucet with the stem removed

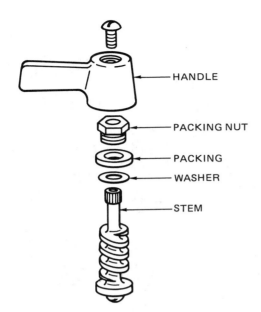

Figure 32–23. The handle must be removed to get to the packing nut on this type of faucet. The stem may be removed after the packing nut is unscrewed.

faucet parts can be purchased from a hardware store or a plumber.

Stopping Faucets from Hammering or Dripping. □ A common problem with faucets and valves is a hammering noise when the water is running slowly. This may be caused by worn threads on the stem or a loose washer, figure 32–25. If the problem is worn threads, the stem or entire faucet may need replacing. The problem may be reduced or corrected by tightening the packing nut.

The more likely cause of a hammering faucet is a loose washer. This is corrected by removing the stem and tightening the washer. It is generally advisable to replace the washer while the faucet is open. When replacing washers, be sure to use one of the correct size, shape, and style, figure 32–26. The size is determined by the diameter. When tightening the screw,

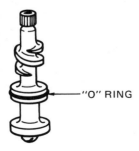

Figure 32–24. Some faucet stems have an O ring instead of packing.

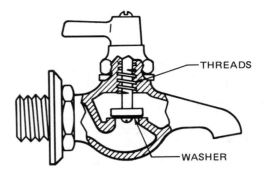

Figure 32–25. Worn threads or a loose washer may cause a faucet to make a hammering sound when the water is running slowly.

draw it down snugly but not so tightly that the washer is forced out of shape, figure 32–27.

Replacing Seats. □ The seats of faucets may become eroded by chemical reaction with the water. This may happen within weeks if the faucet is permitted to drip. The seats in many faucets can be replaced using an allen wrench, figure 32–28.

Repairing Concealed Faucets. □ Bathtub faucets are generally inside the wall with only the handles in view. To repair the faucet, the handle must be removed first. A large, deep socket is then used to unscrew the bonnet, figure 32–29. The bonnet and stem are removed as an assembly. The washer or seat is then serviced as described previously.

Repairing Float Valve Assemblies

Float valve assemblies are used to control the water flow to flush toilets. Another important use is in livestock watering tanks with automatic filling devices. Float assemblies consist of a body with a diaphragm or washer valve and a float that is lifted by the water as the tank fills, figure 32–30.

The action of the rising float puts pressure on the valve when the water in the tank reaches a specified level. This causes the valve to shut off the flow of incoming water. When the water level drops in the tank, the float drops, opens the valve, and permits water to enter until the tank refills.

Repairs to float valve assemblies may involve tightening the nut or replacing a gasket to stop leaks at the point where the assembly passes through the bot-

Figure 32–26. It is important to use the correct size and shape of washer.

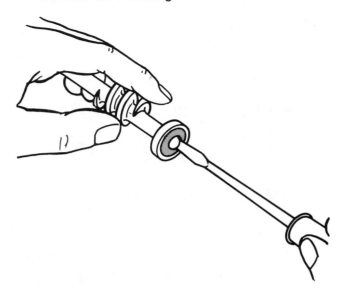

Figure 32–27. Faucet washers should be tightened snugly but not so tightly that they are forced out of shape.

tom of the tank; or a washer or diaphragm in the valve may need to be replaced to stop a flooding condition caused by the valve not shutting off completely. Some systems use a hollow ball for a float. If the ball develops a leak, it will fill with water and no longer float. In this case, the float and its rod are unscrewed from the assembly and replaced.

Repairing Flush Valves

Flush valves on toilets control the water used in the flushing process, figure 32–31. The valves frequently fail to seat properly and permit water to flow continuously from the tank to the toilet bowl. This causes the filler valve to remain open and water to flow in continuously until an adjustment is made. Such a condition, if not corrected, can cause a well to be pumped dry or a septic system to be flooded.

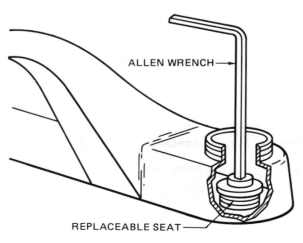

Figure 32–28. The seats in many faucets can be replaced.

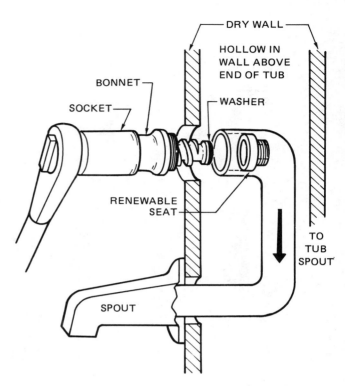

Figure 32–29. Bathtub faucets require a large, deep socket wrench to remove the bonnet.

In most cases the problem is caused by the chain or linkage getting twisted or caught on another part. This can be corrected by simply flipping the trip level up and down. If this does not permit the valve to reseal, then the flapper or ball seal and other parts should be cleaned of all rust, slime, or other deposits. If this does not correct the problem, replace the flapper and other parts as required.

□ STUDENT ACTIVITIES

1. Sketch and label each pipe fitting illustrated in the unit.
2. Repair a faucet that leaks at the stem.
3. Repair a faucet that makes a hammering noise when the water is running slowly.
4. Repair a dripping faucet.
5. Repair the float mechanism in a water tank.
6. Do the plumbing exercise in pipe fitting on page 396.

Caution:

This exercise involves the use of chemicals and extreme heat. Therefore follow directions carefully.

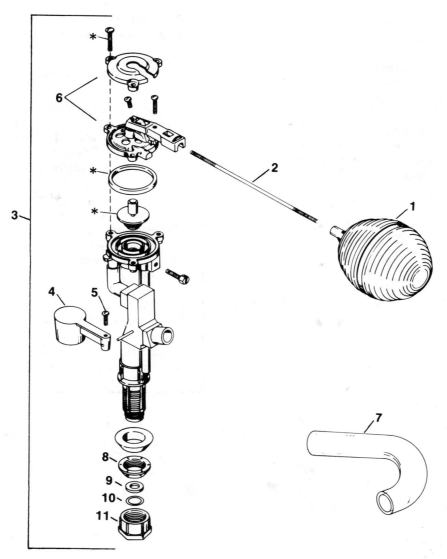

ITEM NO.	PART NO.	DESCRIPTION
1	51162	Float Ball
2	42386	Float Rod
3	42483	Float Valve Assembly
4	42381	Float
5	31490	Screw
6	42482	Cap
7	42389	Rimfeed Hose
8	42396	Nut
9	42398	Gasket
10	42399	Washer
11	42397	Nut
*		REPLACEMENTS PARTS AVAILABLE IN REPAIR KIT

Figure 32–30. Parts of a typical float valve assembly used in toilet tanks and livestock watering tanks (*courtesy of Kohler Co.*)

PLUMBING EXERCISE IN PIPE FITTING

(Courtesy of the Department of Agricultural and Extension Education, The Pennsylvania State University)

Purpose: The purpose of the pipe fitting exercise is to obtain experience in assembling a watertight unit using different types of plumbing materials. Skills to learn include the following:

- Using various types of adapters and fittings
- Cutting pipe materials to length
- Reaming and cleaning pipe ends
- Threading and making galvanized connections
- Making cemented plastic connections
- Soldering copper connections
- Making polyethylene connections using clamps

BILL OF MATERIALS

1 ½-inch hose adapter (F × F)
1 ½-inch sillcock (F)
1 ½-inch × 4″ nipple (galv.)
1 ½-inch × 6″ nipple (galv.)
1 ½-inch × ½″ × ½″ tee (galv.)
1 ½-inch coupling (galv.)
1 ½-inch elbow (galv.)
2 ½-inch adapters (M × PVC)
1 foot ½-inch PVC pipe
2 ½-inch adapters (M × insert)
2 ½-inch stainless steel pipe clamps
1 foot ½-inch polyethylene pipe
1 ½-inch adapter (F × c)
1 ½-inch adapting elbow (F × c)
1 ½-inch tee (c × c × c)
1 ½-inch union (c × c)
1 ½-inch adapter (c × M)
15 inches ½-inch type L copper pipe
Other materials include appropriate pipe joint compound, solvent cement, flux and solder, and the necessary tools for performing the job.

Key: F–female threads
     M–male threads
     c–copper
     galv.–galvanized

Procedure

1. Measure and cut pipe to length as specified in the bill of materials.
2. Ream and clean ends of pipe.
3. Thread galvanized pipe.
4. Apply pipe joint compound to threads.
5. Assemble galvanized pipe and hose adapter as shown, figure 32–32.
6. Working with one connection at a time, apply cement to PVC pipe and its fitting, then assemble quickly using slight turning motion.
7. Assemble polyethylene pipe using insert connectors and clamps.
8. Apply soldering flux to copper pipe and fittings.
9. Assemble and solder copper joints.
10. Cool copper parts before proceeding.
11. Assemble unit as shown, attaching the threaded sillcock. Make use of the union on the copper pipe for assembly purposes.
12. Attach garden hose and test for leaks.

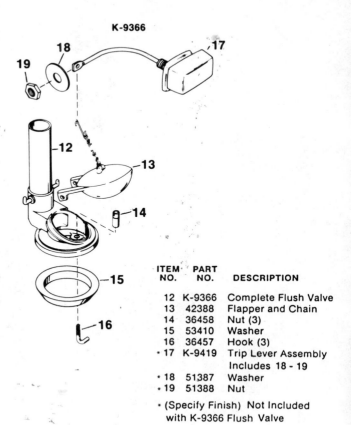

K-9366

ITEM NO.	PART NO.	DESCRIPTION
12	K-9366	Complete Flush Valve
13	42388	Flapper and Chain
14	36458	Nut (3)
15	53410	Washer
16	36457	Hook (3)
* 17	K-9419	Trip Lever Assembly Includes 18 - 19
* 18	51387	Washer
* 19	51388	Nut

* (Specify Finish) Not Included with K-9366 Flush Valve

Figure 32–31. Parts of a flush valve used in toilet tanks (*courtesy of Kohler Co.*)

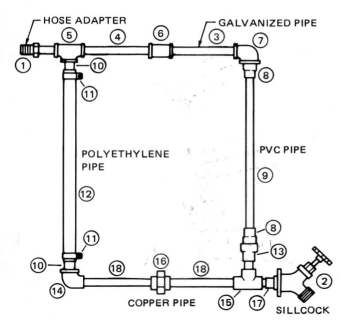

HOSE ADAPTER

GALVANIZED PIPE

POLYETHYLENE PIPE

PVC PIPE

COPPER PIPE

SILLCOCK

Figure 32–32. Plumbing exercise in pipe fitting (*courtesy of The Pennsylvania State University, Department of Agricultural and Extension Education*)

□ **SELF-EVALUATION**

A. Multiple Choice. Select the best answer.
  1. Extra heavy steel pipe has
     a. thicker walls than standard pipe
     b. smaller ID than standard pipe
     c. the same OD as standard pipe
     d. all of these
  2. Galvanized pipe is coated with
     a. zinc
     b. paint
     c. oil
     d. galvanoleum
  3. Galvanized pipe may be hazardous when
     a. used with acid water
     b. mixed with black pipe in the system
     c. exposed to soil
     d. cut with a torch or welded
  4. Pipe that is available in both flexible and rigid form is
     a. steel
     b. PVC
     c. copper
     d. none of these
  5. The heaviest copper pipe is type
     a. K
     b. M
     c. L
     d. Z

  6. A kind of flexible pipe that is preferred for long runs of outdoor cold water lines is
     a. copper
     b. PE
     c. PVC
     d. steel
  7. Green stain in a tub indicates
     a. poor quality finish on the tub
     b. reaction of water and steel
     c. reaction of water and copper
     d. reaction of water and plastic
  8. Faucets may drip or leak due to faulty
     a. washers
     b. packing
     c. O rings
     d. any of these
  9. A common problem with toilet tanks and automatic livestock waterers is
     a. incomplete shut-off
     b. green stain
     c. dripping
     d. all of these
  10. The easiest water or sewerage system to install uses pipe made of
     a. cast iron
     b. copper
     c. plastic
     d. steel

B. Matching. Match the terms in column I with those in column II.

Column I	Column II
1. steel pipe	a. nominal pipe size
2. copper pipe	b. changes direction of single line
3. cast iron pipe	c. connect several pipes
4. plastic pipe	d. oakum and lead joints
5. ID	e. connect similar pipe in line
6. elbows	f. short length of pipe
7. tee and Y	g. used to close pipe and fittings
8. cap and plug	h. connected by threading
9. coupling and union	i. connected by cementing
10. nipple	j. connected by soldering

C. Completion. Fill in the blanks with the word or words that make the following statements correct.
  1. A faucet that leaks around the stem may be corrected by
     a. _____
     b. _____
  2. Water reacting with steel pipe may be _____ in color.
  3. Two problems that may cause a faucet to make a hammering noise when water runs slowly are:

a. _____

b. _____

4. Three parts of a faucet are: a. _____

    b. _____ c. _____

5. The water level in a flush tank or automatic livestock waterer is controlled by a _____

6. The valve that stops water entering a flush tank or automatic livestock waterer is shut off by the action of a _____

Section 13
CONCRETE AND MASONRY

Unit 33
Concrete and Masonry

Objective

To mix and place concrete and use masonry materials.

Competencies to Be Developed

After studying this unit, you will be able to:
• Identify tools used for concrete work.
• Select ingredients for mixing concrete.
• Make a workable masonry mix.
• Prepare forms for concreting.
• Pour concrete.
• Finish concrete.
• Calculate concrete and block for a job.
• Lay masonry block.

Terms to Know

masonry	footer form (footing)
portland cement	construction joint
concrete	control joint
sand	reinforced concrete
mortar	air-entrained concrete
finishing lime	screeding
fine aggregate	bull float
gravel	broom finish
coarse aggregate	curing
silt	masonry units
clay	laying block
washed sand	mortar bed
cement paste	core
ratio	hollow core block
mix	ears
workable mix	stretcher block

sash block	frost line
jamb block	corner pole
course	

Materials List

Portland cement	Push broom
Sand	Oil for forms
Gravel	Level, 4 foot
Mortar mix	Mason's trowel
Wall reinforcement	Corner poles
Lumber for forms	Wheelbarrow
Stakes	Mortar pan
Bull float	Striking tool
Hand float	Rakes
Edging tool	Shovels
Grooving tool	Quart jar with lid

Concrete and masonry construction is basic to most farm and agribusiness facilities. *Masonry* refers to anything constructed of brick, stone, tile, or concrete units set or held in place with portland cement. *Portland cement* is a dry powder made by burning limestone and clay, then grinding and mixing to an even consistency.

Concrete and mortar are two construction materials with portland cement as a component. *Concrete* is a mixture of stone aggregates, sand, portland cement, and water that hardens as it dries. *Sand* consists of small particles of stone. *Mortar* is a mixture of sand, portland cement, water, and finishing lime. *Finishing lime* is a powder made by grinding and treating limestone. Like concrete, mortar is a mixture of materials mixed with water which hardens as it dries.

Nearly all buildings have concrete and masonry components somewhere in their construction. The ability to use masonry materials is therefore essential to many maintenance and repair operations.

399

□ MASONRY CONSTRUCTION

Masonry construction has been used for thousands of years. Its basic materials come from the earth and are plentiful in most areas. The pyramids of Egypt, the colosseum of ancient Rome, the temples of the Mayan Indians, and the skyscrapers of modern America all demonstrate the use of masonry construction. Masonry construction has most of the advantages of concrete. However, the hardness, durability, and moisture-resistant characteristics of the masonry units and mortar will determine the characteristics of the masonry structure.

□ CONCRETE CONSTRUCTION

Concrete has many advantages for farm use and in other agricultural enterprises. Concrete is:

- fireproof
- insect and rodent proof
- decay resistant
- highly storm resistant
- wear resistant
- waterproof
- strong
- attractive
- easy to make on the job without expensive equipment
- available locally
- low in original and maintenance costs
- sanitary and easy to keep clean
- easily broken up and used as fill material when the structure becomes obsolete

The strength and durability of concrete depend upon several factors: (1) the strength of the stone particles, (2) the proportion of stone particles by size, (3) the type of portland cement, (4) the purity of the water, (5) the uniformity of the mixture, and (6) the procedures used in placing, finishing, and curing the concrete.

Many types of stone found in fields, streams of water, and solid rock formations are suitable for making concrete. Occasionally the correct type and mixture of stone particles may be found in stream beds or gravel deposits. Usually, however, stone particles must be made by crushing stone and grading it to size. The correct proportion of small particles, called sand or *fine aggregate*, and large particles, called *gravel* or *coarse aggregate*, can then be mixed to meet the requirements of specific jobs.

To make good concrete, aggregates of various sizes should fit together to form a fairly solid mass, figure

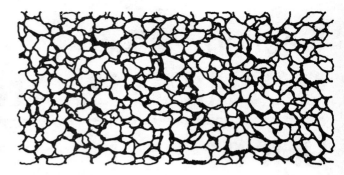

Figure 33–1.   In good concrete, aggregates of various sizes fit together to form a fairly solid mass.

33–1. The stone particles must be clean and free of clay, silt, chaff, or any other material. *Silt* is composed of intermediate sized soil particles. *Clay* consists of the smallest group of soil particles. Both silt and clay particles are too small for use as aggregates in concrete and decrease the quality of concrete if present.

Sand sold commercially for use in concrete generally contains little clay or silt. If the clay or silt content is above an acceptable level when the material comes from the quarry, the sand is flushed with water to remove the clay or silt. The material is then referred to as *washed sand*.

Testing Sand for Silt or Clay Content

Sand from streambeds or other farm areas may be suitable for making concrete. However, when sand from such places is being considered, a test for clay and silt must be made. The procedure is as follows.

Procedure

1. Fill a one-quart glass jar to a depth of 2 inches with the sand to be tested, figure 33–2.
2. Add water until the jar is three-fourths full.
3. Screw on a lid and shake the mixture vigorously for one minute to mix all particles with the water.
4. Shake the jar sideways several times to level the sand.
5. Place the jar where it will not be disturbed for 1 hour for a silt test or 12 hours for a clay and silt test.
6. After 1 hour, measure the thickness of the silt layer on top of the sand.
7. If the layer is more than ⅛ inch thick, the sand is not suitable for use in concrete unless the silt is removed by washing.
8. If the layer is not ⅛ inch thick in 1 hour, let the

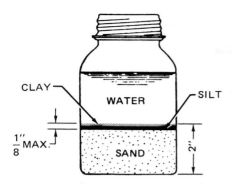

Figure 33–2. This simple test will determine the silt content of material that is being considered for use in concrete (*courtesy of G. White,* Concrete Technology, *Delmar Publishers Inc., 1977*).

mixture stand for 12 hours. Then remeasure the layer(s) that have settled on the sand.

9. If the silt plus clay layer exceeds ⅛ inch, wash the sand before using it in concrete.

□ MIXING CONCRETE

Cement paste is made by mixing portland cement and clean water in precise proportions. The reaction between the cement and the water is a chemical one, and the strength of the concrete is directly affected by the proportions of the materials. As a result, precision in mixing portland cement and water is as important as precision in mixing ingredients when baking a cake.

Caution:

A mistake in measuring will ruin the product.

All sand has some water attached to its particles. This water must be estimated and an allowance made for it when deciding how much water to use in the concrete mix. Sand is described as "damp," "wet," or "very wet." The more moisture there is in the sand, the less water is added from other sources when mixing concrete or mortar. To obtain a more durable concrete to withstand severe weather conditions and/or traffic, add less water, figure 33–3.

The ratio of cement to fine aggregate (sand) and coarse aggregate (gravel) is important. However, the exact proportion will vary with the makeup of each aggregate, figure 33–4. A correct mixture will assure that: (1) each particle of sand and gravel is covered with cement paste, and (2) each particle is bound to others when the cement paste dries and hardens.

The proportion of cement to fine and coarse aggregate is known as the *ratio.* The ratio is expressed as a three-digit number called a *mix.* For example, a 1–2–3 mix has one part cement, two parts fine aggregate, and three parts coarse aggregate. Once the mix for a job has been determined, a table can be used to help estimate the amount of material needed for a cubic yard of concrete, figure 33–5. A rule of thumb is that the resulting concrete will be about two-thirds the combined volume of the cement and aggregate used in the mix.

It should be noted that portland cement is sold in

INTENDED USE OF CONCRETE	FOR MAXIMUM AGGREGATE SIZE OF:	WATER (GALLONS) ADDED TO 1 CU FT OF CEMENT IF SAND IS:			SUGGESTED MIXTURE FOR 1 CU FT TRIAL BATCHES*		
		DAMP	WET (AVERAGE SAND)	VERY WET	CEMENT CU FT	FINE CU FT	COARSE CU FT
Mild Exposure	1½ in.	6¼	5½	4¾	1	3	4
Normal Exposure	1 in	5½	5	4¼	1	2¼	3
Severe Exposure	1 in	4½	4	3½	1	2	2¼

* Mix proportions will vary slightly depending on the gradation of aggregate sizes.
Note: Batches may be made by using 100 pounds (1 cwt) of each ingredient where 1 cubic foot is indicated.

Figure 33–3. Ratio of ingredients for mixing concrete (*courtesy of G. White,* Concrete Technology, *Delmar Publishers Inc., 1977*)

FOR MAXIMUM AGGREGATE SIZE OF:	SUGGESTED MIXTURE FOR 1 CWT TRIAL BATCHES*		
	CEMENT CWT** OR CU FT	AGGREGATES	
		FINE CU FT	COARSE CU FT
¾ in	1	2	2¼
1 in	1	2¼	3
(preferred mix) 1½ in	1	2½	3½
(Optional mix) 1½ in	1	3	4

* Mix proportions will vary slightly depending on the gradation of the aggregates. A 10-percent allowance for normal wastage is included in the fine and coarse aggregate values.
** 1 cwt of cement equals about 1 cubic foot.

Figure 33–4. Materials needed to make trial batches of concrete with separated fine and coarse aggregates (*courtesy of G. White,* Concrete Technology, *Delmar Publishers Inc., 1977*).

bags containing 94 pounds, or exactly 1 cubic foot, of cement. A 1–2–3 mix would consist of 1 cubic foot of cement (one bag), 2 cubic feet of sand, and 3 cubic feet of coarse aggregate. For the purpose of mixing, one bag of cement may be considered as 100 pounds or 1 hundredweight (1 cwt). This is especially useful when the aggregates being used are measured by weight rather than volume. Proportions may be based on either weight or volume.

The term *workable mix* refers to the consistency of the wet concrete after the various ingredients have been mixed together. A workable mix has the following characteristics:

- All aggregates are clean.
- All portland cement is mixed with water to form a cement paste. No dry powder is present.
- Every particle of aggregate is covered with cement paste.
- Aggregates are distributed evenly throughout the mix.
- No more than the recommended amount of water is added.
- No lumps are present.
- The mixture has uniform color and consistency.
- The mixture can be mixed, moved, and placed with a shovel or spade.

MIX RECOMMENDED FOR MAXIMUM AGGREGATE SIZE OF:	MATERIALS NEEDED TO MAKE ONE CU YD OF CONCRETE				
	CEMENT CWT OR CU FT	AGGREGATES			
		FINE		COARSE	
		CU FT	LB	CU FT	LB
¾ in 1–2–2¼	7	17	1550	19.5	1950
1 in 1–2¼–3	5.5	15.5	1400	21	2100
(preferred mix) 1½ in 1–2½–3½	5.25	16.5	1500	23	2300
(optional mix) 1½ in 1–3–4	4.5	16.5	1500	22	2200

Note: 1 cwt of cement or aggregate equals approximately 1 cubic foot.

Figure 33–5. Materials needed to make one cubic yard of placed concrete (*courtesy of G. White,* Concrete Technology, *Delmar Publishers Inc., 1977*)

☐ ESTIMATING MATERIALS FOR A JOB

It is important to know how to estimate the amount of concrete needed for a job. If the estimate is inaccurate, the job will cost more than is necessary and ingredients may be wasted. If insufficient material is available, the job must be stopped and then started again at a considerable increase in cost and time. The following example will show how to determine the amounts of materials needed for a concrete floor 4 inches thick, 14 feet wide, and 24 feet long.

Procedure

1. Determine the cubic yards of concrete needed by multiplying the thickness of the floor by its width and length in feet: Cu Ft = T′ × W′ × L′. In this case, thickness equals 4 inches which must be converted to ⅓ foot. Therefore, ⅓′ × 14′ × 24′ = 112 cubic foot. There are 27 cubic feet in 1 cubic yard. Thus 112 cubic feet ÷ 27 = 4.15 cubic yards.
2. Select the ratio of the mix needed for the job. In this example, a 1:2¼:3 ratio is selected.
3. Determine the amount of each ingredient needed to make one cubic yard of concrete. According to figure 33–5, 1 cubic yard of 1-2¼-3 mix requires: 5.5 cwt (550 pounds) of cement, 1400 pounds of fine aggregate, and 2100 pounds of coarse aggregate.
4. Multiply each of those figures by the number of cubic yards in the project. In this case, multiply each ingredient by 4.15:

   4.15 × 550 = 2282.5 pounds of cement
   4.15 × 1400 = 5810 pounds of fine aggregate
   4.15 × 2100 = 8715 pounds of coarse aggregate

Note: If the concrete is to be purchased in ready-mixed form, the estimate is complete once it is determined how many cubic yards are needed. In this case, 4.15 plus a little extra in case the estimate is too low might come to 4.5 cubic yards. If the base is not leveled carefully, 5 cubic yards could be ordered and any excess used in another location.

☐ PREPARING FORMS FOR CONCRETE

A *form* is a metal or wooden structure that confines the concrete to the desired shape or form until it hardens. Preparing forms for concrete can be as simple as nailing a board in place or as complex as construct-

ing forms for the side of a highway bridge. On farms, around homes, and in landscape projects, form construction is usually required for walks, floors, feedlots, steps, ornamental projects, and building footers. A *footer* or *footing,* is the concrete base that provides a solid, level foundation for brick, stone, or block walls. Footers, walks, and concrete slabs are generally constructed with 2″ × 4″ lumber nailed to 2″ × 4″ stakes driven into the soil. Forms for walls are generally constructed by nailing ¾-inch tongue and groove lumber or plywood to 2″ × 4″ materials, figure 33–6.

When constructing forms, the following points may be helpful:

- Use soft, clean, straight lumber.
- Sharpen stakes evenly so they can be driven in straight.
- Place stakes about 30 inches apart along the outside of forms for 4-inch-thick concrete. Place the stakes closer when the concrete is more than 4 inches thick.
- Use a transit or level to adjust the height of forms for the desired slope or "fall" of the slab.
- Drive nails through the form and into, but not through, the stakes.
- Be sure the stakes do not stick up above the top of the forms. If they do, saw them off so they are level with or tapered down from the form.

Caution:

Do not saw into the top of the form.

- Construct the inside surfaces of the forms to create the desired shape of the finished concrete. For example, if you want a rounded extension on the front of a concrete step, the shape of the extension is determined by the inside of the form.
- Brush used motor oil on wood surfaces that will be touched by concrete. This is a low-cost product that will prevent the wood from sticking to the concrete and permits easy removal of the forms.

Making Joints

Construction joints are needed when slabs of concrete larger than about 10 feet by 10 feet are poured. A *construction joint* is a place where one pouring of concrete stops and another starts. A *control joint* is a planned break which permits concrete to expand

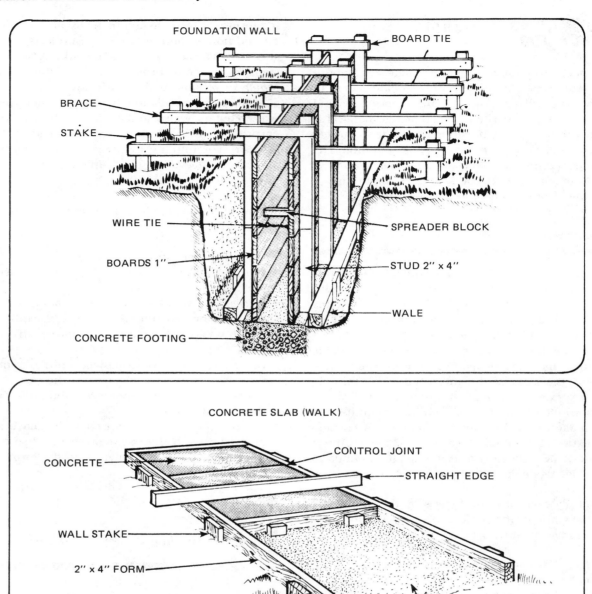

Figure 33–6. Proper form construction (*courtesy of Texas A & M University, Vocational Instructional Services*)

and contract without cracking. Control joints are placed at equally spaced intervals in the slab.

Control joints are made by placing a wood or galvanized metal "key" on the form, figure 33–7. A beveled 1″ × 2″ wood strip is adequate for most slabs. A saw kerf in the key prevents breakage of the concrete when moisture expands the wood strip. The groove left in the concrete after the form is removed provides an indentation into which the next section locks. Before the second section is poured, the first section is coated

with a curing compound. This prevents the second section from sticking to the first.

Another popular type of control joint is made by placing asphalt material between sections to absorb the expansion and contraction of the two slabs. This system, however, does not prevent one slab from rising or falling during freezing and thawing, which may leave an uneven surface where the two slabs come together.

Sidewalks need shallow grooves or joints across the

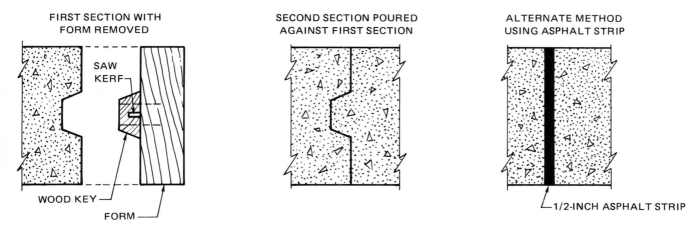

FIRST SECTION WITH
FORM REMOVED

SAW
KERF

WOOD KEY

FORM

SECOND SECTION POURED
AGAINST FIRST SECTION

ALTERNATE METHOD
USING ASPHALT STRIP

1/2-INCH ASPHALT STRIP

Figure 33–7. Control joints (*courtesy of G. White,* Concrete Technology, *Delmar Publishers Inc., 1977*)

surface at 3-foot intervals to control cracking due to expansion. When joints are present the concrete will crack under the joint if stress occurs. A crack under the joint will not create an unsightly appearance.

□ REINFORCING CONCRETE

Concrete slabs may be greatly strengthened by using steel reinforcing rods or wire mesh, figure 33–8. Such concrete is referred to as *reinforced concrete.* Any concrete slab that will carry vehicles such as tractors, loaders, or machinery must be reinforced. Otherwise, money is wasted on extra thicknesses of concrete that are still likely to crack under stress. Concrete posts and other thin concrete structures must be reinforced with steel to withstand normal stresses.

Concrete reinforcing rods are made of steel and have a rough, pebbly surface. This surface prevents them from slipping when embedded in concrete. They are classified according to diameter and are available in diameters from ¼ to 1 inch and over. They may be purchased in 20-, 40-, or 60-foot lengths.

Two or more bars may be joined with wire or by welding to make longer units. These should be wired so that they lap one another by at least a foot. A rule of thumb is to lap bars by 24 times the diameter. Therefore, a ½-inch bar should be lapped 12 inches and a ¾-inch bar lapped 18 inches. Rods should be placed at least ¾ inch from all surfaces of the concrete. For most applications, 1½ inches or more is preferred.

Rods may be placed in concrete slabs in a cross-sectional pattern. Such reinforcement is stronger if the bars are wired together where they cross one another. Reinforcing bars should be free of rust, dirt, oil, or other materials that will reduce adhesion by the concrete.

Concrete slabs are generally reinforced with wire mesh, or fabric. Wire fabric is steel wire welded to form a cross-sectional pattern. It is generally available in a 6-by-6-inch pattern and consists of No. 6, 8, or 10 gauge wire. Both wire mesh and reinforcing rods are more effective if they are not rusted.

Wire fabric is first rolled out on a flat surface to eliminate curves in the material. Then it is laid inside the concrete forms. Multiple pieces are wired together to form a continuous piece. Where two or more widths are needed they should overlap each other

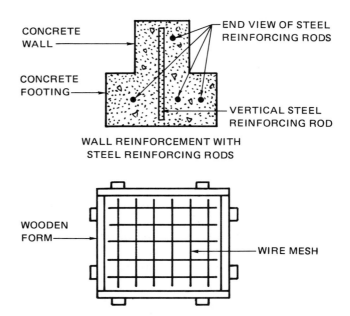

CONCRETE
WALL

CONCRETE
FOOTING

END VIEW OF STEEL
REINFORCING RODS

VERTICAL STEEL
REINFORCING ROD

WALL REINFORCEMENT WITH
STEEL REINFORCING RODS

WOODEN
FORM

WIRE MESH

SLAB REINFORCEMENT WITH WIRE MESH

Figure 33–8. Concrete is strengthened greatly by the addition of steel rods or wire mesh.

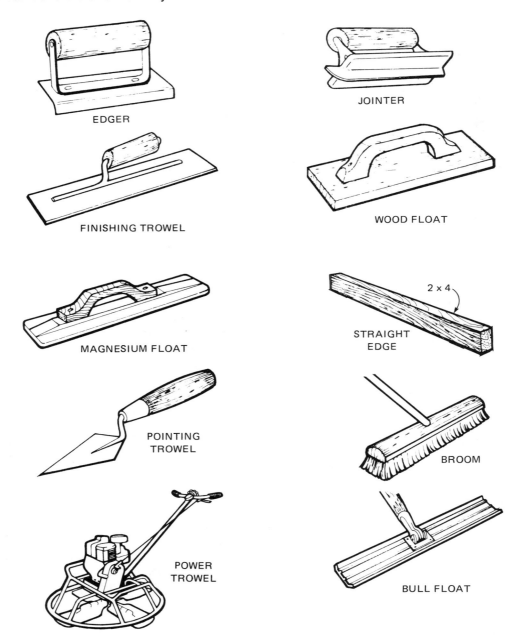

Figure 33–9. Tools for placing and finishing concrete (*courtesy of Texas A & M University, Vocational Instructional Services*)

by at least one and a half squares. The pieces should be tied securely at regular intervals.

The wire fabric should be supported by stones to keep it off the base material. As the concrete is being poured it is important to pull the wire up until it is approximately centered in the slab. This ensures maximum reinforcement.

Finally, ready-mixed concrete may be ordered with tiny air bubbles trapped throughout the mixture. Such concrete is known as *air-entrained concrete*. It is stronger and more resistant to acid, salt, and frost than regular concrete.

□ POURING, FINISHING, AND CURING CONCRETE

When pouring concrete, common tools such as shovels, spades, and rakes are needed to move and spread the material. Other tools are needed to finish the concrete, figure 33–9. Since concrete is so heavy, it is best to move it downhill with a chute and across level surfaces with a wheelbarrow. This permits the concrete to be pushed, pulled, or lowered rather than lifted.

Concrete starts to harden about 15 minutes after

it is mixed. Thus, it is very important to have all materials on hand before mixing starts.

Before pouring concrete, the stone base should be wetted down with water so that water does not move from the concrete into the dry base. Loss of water from the mixture will weaken the concrete.

Screeding, Floating, and Finishing

After spreading concrete, a straight 2″ × 4″ or 2″ × 6″ plank is used to strike off the excess concrete and create a smooth and level surface. This process is called *screeding*, figure 33–10. When the screeding is finished, a smooth board attached to a handle is pushed and pulled across the surface. This process is called floating, and the tool is called a *bull float.* The process brings the fine aggregate and cement paste to the surface for a smooth finish. Small areas may be finished with wooden or magnesium floats. Hand floats, edging tools, jointing tools, and trowels are used to finish the edges and/or place grooves or joints in the concrete, figure 33–11. All of these procedures must be done before the concrete begins to set.

After the concrete begins to set, a coarse bristled broom may be pulled across it to create a rough surface and improve footing or traction. This is called a *broom finish.* A very rough, pebbly finish may be obtained by washing away some of the cement paste, leaving the aggregate exposed, figure 33–12. A smooth finish is created by troweling after the concrete starts to harden.

When an extremely smooth surface is desirable, the concrete is rubbed with a metal trowel. Generally the trowel is dipped in water and worked over the semi-hardened surface. This action brings the cement paste to the surface. Excessive troweling will leave the surface weak and easily damaged by frost and chemicals.

Curing

Concrete must dry slowly or it will crack, crumble, and break up long before its intended lifetime has passed. The proper drying of concrete is called *curing.* To cure concrete, it must be protected from drying air, excessive heat, and freezing temperatures for several days after it is poured.

Concrete is kept moist by covering it with plastic or canvas to prevent the water from evaporating. The surface may also be covered with straw, sawdust, or other insulating material and sprinkled with water occasionally to keep it moist. Insulating materials help protect the concrete from air that is too hot or too cold as well as preventing moisture from escaping. Wooden forms help to protect concrete from drying out. Therefore, forms and other materials used for curing concrete should be left in place for up to a week until curing is complete.

□ LAYING MASONRY UNITS

Blocks made from concrete, cinders, or other aggregates are used extensively for agricultural buildings. Such blocks are called *masonry units.* They are referred to as "block" in both singular and plural. Masonry units are held together with mortar. The process of mixing mortar, applying it to block, and placing the block to create walls is called *laying block.*

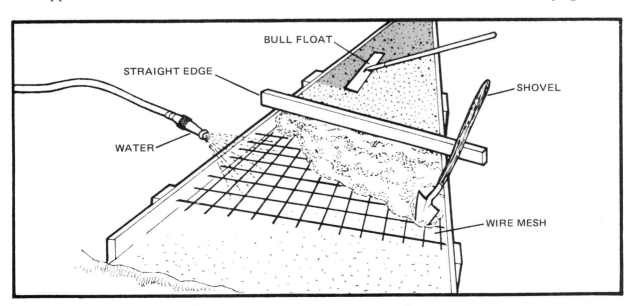

Figure 33–10. Pouring, screeding, and floating concrete (*courtesy of Texas A & M University, Vocational Instructional Services*)

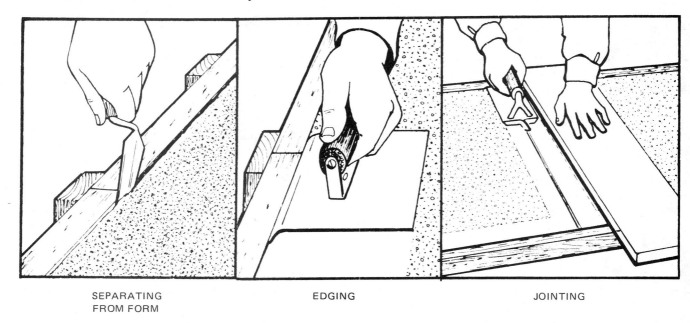

SEPARATING FROM FORM   EDGING   JOINTING

Figure 33–11. Separating, edging, and jointing procedures (*courtesy of Texas A & M University, Vocational Instructional Services*)

Types of Block

Concrete and cinder block are available in a variety of sizes and shapes. Standard block are 15⅝ inches long and 7⅝ inches high. When laid with ⅜-inch mortar joints they cover an area 16 inches long and 8 inches high. Standard block are available in 4-, 6-, 8-, 10-, and 12-inch widths. The 4-, 8-, and 12-inch sizes are used most frequently.

Block are wider when viewed from the top than when viewed from the bottom. This is because the block is poured into forms that are tapered slightly to permit the block to be removed easily. Block are usually laid with the thicker part up to provide a larger area for the mortar bed. A layer of mortar is called a *mortar bed.*

Block are generally manufactured with hollow spaces called *cores.* They are available with two or three cores per block. Such block are referred to as *hollow core block.* Some block, called stretcher block, have slight extensions on the ends called *ears.* When laid end to end, the ears create a core. *Stretcher block* are used in straight wall sections. Corner block have one flat end to create attractive walls at corners. Block with special grooves can be laid to receive window (*sash block*) and door (*jamb block*) parts so the openings are attractive and secure, figure 33–13.

Estimating Block

As stated before, standard block are 8 inches or ⅔ foot high and 16 inches or 1⅓ foot long when they are layed with a ⅜-inch mortar joint. One foot, or 12 inches, is ¾ the length of one block. Therefore, when estimating the number of block needed for a job, the length of the wall in feet can simply be multiplied by ¾. This gives the number of block needed

TYPE OF FINISH	USE	OBTAINED BY
Gritty, nonslip surface	Walks, barn floors, drive-ways, and ramps	Using wood float, stiff broom, or burlap covering
Smooth, trow-eled finish	Feed troughs, certain type of floors	Using steel trowel
Pea grain or ex-posed aggregate	Porches and patios	Washing with water hose before curing

Figure 33–12. Concrete can be finished in a number of ways (*courtesy of Texas A & M University, Vocational Instructional Services*).

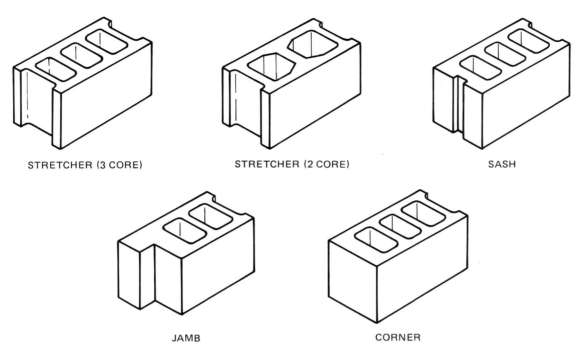

Figure 33–13. Stretcher block is used in straight wall sections; sash block hold windows; jamb block hold doors; and corner block are used wherever the ends of block are in view.

for one row, which is called a *course*. Similarly, 1 foot is 1⅜ or ¾ of the height of a block. Therefore, the height of the wall can be multiplied by ¾ to determine the number of courses needed. The number of block per course is then multiplied by the number of courses to obtain the number of block needed for the wall.

For example, to estimate the number of block needed to build a wall 12 feet long and 8 feet high, the procedure is as follows:

12' (length in feet) × ¾ = 9 block per course
8' (height in feet) × 3⁄2 = 12 courses
9 (block per course) × 12 (courses) =
108 block needed for the wall

When estimating block for buildings, this procedure is used for each of the four walls. The estimate is then decreased by the area of the doors and windows. **Note:** When planning a building it is important that the dimensions between corners, windows, and doors utilize full or half length block. Otherwise, large numbers of block must be cut, a procedure that is expensive and time consuming and that results in an unattractive appearance. Figure 33–14 has no cut block.

Constructing Footers

A footer or footing is a continuous slab of concrete which provides a solid, level foundation for block and brick walls. Footers should be at least as deep as the wall is wide, figure 33–15. A wall that is 8 inches wide rests on a footer that is at least 8 inches deep. The width of the footer should be at least twice that of the wall (or more where there is poor soil support). Hence, the 8-inch wall would rest on a footer 16 inches wide. In localities where temperatures drop to 32°F or lower, footers must be placed below the frost line to prevent damage by freezing in cold weather. The *frost line* is the maximum depth that the soil freezes in cold weather.

When starting a building or wall, a trench is dug by back hoe or shovel. The bottom of the trench should be the width of the footer and as deep as the intended bottom of the footer. Stakes are driven into the bottom of the trench so their tops are level with the top of the intended footer. Concrete is then poured and leveled to the height of the stakes. The stakes are pulled out and the holes filled after the concrete is leveled but is still workable.

Mixing Mortar

Before block can be laid, the mortar must be mixed to a working consistency. Premixed mortar can be purchased that requires only the addition of water. Mortar can also be made at the job site by mixing various combinations of portland cement, finishing lime, sand, and water. The different combinations yield different characteristics, such as greater strength, higher waterproof level, and greater ease of handling. Unfortunately, no combination has all the ideal qualities.

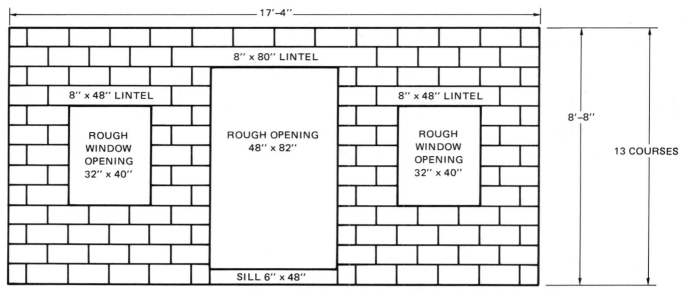

NOTES: 1. DOOR OPENING WILL HAVE 2" JAMBS
ON BOTH SIDES AND TOP.
2. DOOR WILL BE A STANDARD HEIGHT
(44" × 6'8").
3. DOOR SILL WILL ACCOMMODATE A
CONCRETE FLOOR INSIDE.
4. WINDOW OPENINGS WILL ACCOMMODATE
STANDARD SIZE WINDOWS.
5. SPACE ABOVE THE TOP OF THE DOOR WILL
ACCOMMODATE APPROPRIATE TRIM UNDER
THE EAVES OF THE ROOF.
6. REINFORCED CONCRETE LINTELS CARRY
THE WEIGHT OVER DOOR AND WINDOW
OPENINGS.
7. THERE ARE NO CUT BLOCK IN THE WALL.
THEREFORE, THE WALL CAN BE BUILT WITH
A MINIMUM OF COST AND LABOR.

Figure 33–14.   A building front laid out using only whole and half block.

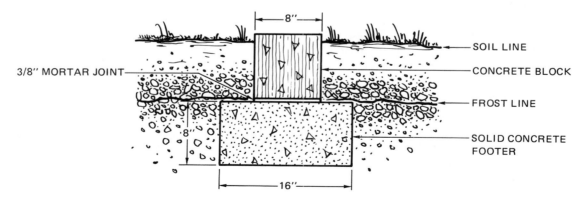

Figure 33–15.   The footer should be as thick as the wall resting on it is wide. The footer should also be at least twice as wide as the wall. Concrete footers are placed below the frost line to provide a solid base for masonry walls.

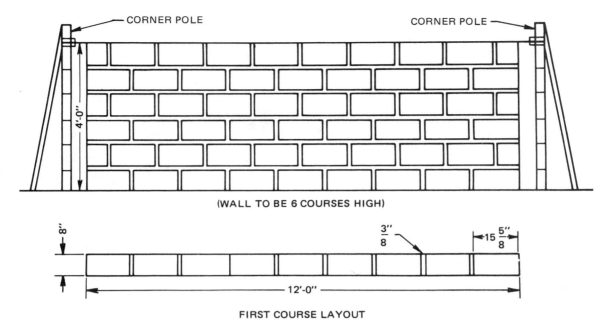

FIRST COURSE LAYOUT

Figure 33–16. Placement of corner poles for constructing a block wall (*courtesy of R. T. Kreh,* Masonry Skills, *Delmar Publishers Inc., 1982*)

Small batches of mortar may be mixed with a hoe in a mortar pan or wheelbarrow. Jobs requiring more than a wheelbarrow or two of mortar can be mixed more thoroughly and efficiently in a motor- or engine-driven mixer.

Block must be laid with mortar of the correct wetness. If the mortar is too stiff it will not bond tightly to the block. If it is too thin, it will be squeezed out of the joint by the weight of the block. In such a case, the joint will be less than ⅜ inch thick and the block must be relaid.

Laying Block

To ensure that block are laid straight and plumb, usually a corner pole is set at each corner of the building, figure 33–16. A *corner pole* is a straight piece of wood or metal held plumb by diagonal supports. It is used to support the lines to which the wall is built. This process is described further in the procedure that follows.

An alternate method is to lay block corners using a level to guide the construction. Since the level has bubble vials placed in different positions, it can be used to plumb the block as well as to level them. The corners are built, then the areas between them are filled with courses of stretcher block using a line to keep the courses straight, figure 33–17.

The procedure for laying block is illustrated in figure 33–18 and also listed on page 414.

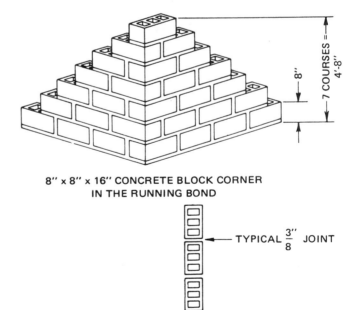

8" x 8" x 16" CONCRETE BLOCK CORNER IN THE RUNNING BOND

PLAN VIEW OF FIRST COURSE LAYOUT

Figure 33–17. Construction of corners for block walls (*courtesy of R. T. Kreh,* Masonry Skills, *Delmar Publishers Inc., 1982*)

SPREAD THE MORTAR BED
FOR THE FIRST COURSE.

LEVEL THE BLOCK.

APPLY MORTAR TO THE EARS OF THE
STRETCHER BLOCK. NOTE THE DOWN-
WARD WIPING OF THE TROWEL.

CHECK THE FIRST
COURSE TO BE SURE
IT IS LEVEL.

CHECK THE FIRST COURSE
TO BE SURE IT IS PLUMB.

CHECK THE FIRST COURSE TO SEE THAT THE
BLOCKS FORM AN EVEN, STRAIGHT LINE.

Figure 33–18. Procedure for laying block

APPLY THE MORTAR BED
FOR THE SECOND COURSE.

LAY REINFORCEMENT WIRE
IN THE MORTAR BED.

REMOVE EXCESS MORTAR.

LAY BLOCK TO THE LINE TO ENSURE THE COURSES
ARE STRAIGHT. THE LINE IS POSITIONED TO BE LEVEL
ALONG THE TOP EDGE OF THE BLOCK.

CUT BLOCK BY MAKING A GROOVE WITH
MODERATE BLOWS OF A MASON'S HAMMER.

FOLLOW BY A SHARP RAP
ON THE EDGE OF THE WEB.

Figure 33–18. (Cont'd.)

MEASURE A CORNER TO BE
SURE THAT EACH COURSE
IS 8 INCHES HIGH.

FINISH A FLUSH OR SMOOTH
JOINT WITH A PIECE OF
BROKEN BLOCK.

STRIKE A JOINT WITH A
SLED RUNNER JOINTER.

Figure 33-18. *(Cont'd.)*

Procedure

1. Spread a layer of mortar, called a mortar bed, on the footer.
2. Position the block on the mortar bed so that its outside corner is exactly where the outside corner of the wall should be. Level the block by first placing the level across the block and then lengthwise along the block.
3. Turn several stretcher block on end and apply mortar to the ears with a wiping or swiping stroke of the trowel.
4. Lay several stretcher block in place by working away from the end or corner block.
5. Use the end of the trowel handle to tap the block until each block is plumb, level, and the course is straight.
6. Apply a mortar bed on top of the first course in preparation for the second course.
7. If extra strength is needed in the wall, install reinforcement in the mortar bed.
8. As the block laying progresses, cut off excess mortar with the trowel.
9. Use a line to keep the courses straight. The line is positioned to be level along the top of the block.
10. When a block must be cut, use a mason's hammer and make multiple strikes along the line of cut; then make one sharp strike on the web.
11. Check the height to be sure each new course is an additional 8 inches high.
12. After the mortar dries and hardens slightly, finish the joint by rubbing it with a broken piece of block.

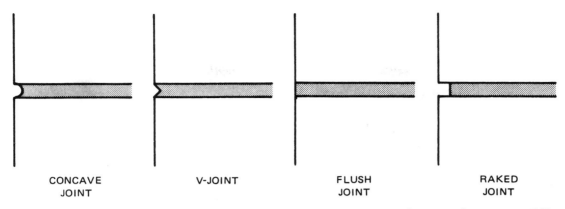

CONCAVE   V-JOINT   FLUSH   RAKED
JOINT   JOINT   JOINT

Figure 33–19. Types of joint finishes used on block walls (*courtesy of R. T. Kreh,* Masonry Skills, *Delmar Publishers Inc., 1982*)

13. If a joint other than a flush joint is desired, use a jointer to compress the mortar and create a watertight joint, figure 33–19. Tools are available to create joints that are concave, V-shaped, flush, or raked.

Walls that are above ground level may be damaged by wind pressure. This is especially true before the mortar cures. Therefore, walls higher than 8 feet should be braced on both sides at 8- to 10-foot intervals, figure 33–20. The braces are removed after the wall is strengthened by curing and by the installation of floor or roof parts.

Mortar joints must be protected from freezing for several days until they are cured. This is generally done by covering the wall with plastic or canvas when temperatures are below 32°F.

Figure 33–20. The temporary bracing of this wall will prevent wind damage until additional building parts provide adequate support.

□ STUDENT ACTIVITIES

1. Define all italicized terms used in the unit.
2. Examine samples of sand and gravel to observe the variety of particle sizes.
3. Conduct a clay and silt test on a sample of sand.
4. Prepare the forms for a small concrete project. Mix and pour the concrete. Finish and cure the concrete.
5. Visit a ready-mix concrete plant to observe the equipment and techniques.
6. Help pour concrete for a project using ready-mixed concrete.
7. Pour a footer.
8. Estimate the block needed for a wall or building.
9. Construct a small block wall.
10. Protect concrete from freezing and drying out until it is cured.
11. Construct a wooden float for use on future concrete projects.

□ SELF-EVALUATION

A. Multiple Choice. Select the best answer.
1. Concrete is a mixture of sand, gravel, water, and
   a. clay cement
   b. finishing cement
   c. finishing lime
   d. portland cement
2. Mortar is a mixture of portland cement, sand, water, and
   a. aggregate
   b. clay
   c. finishing cement
   d. finishing lime
3. The strength and durability of concrete are dependent upon the
   a. purity of water
   b. proportion of stone particles by size

c. type of cement
d. all of these

4. A concrete slab that is 6 inches deep by 10 feet wide by 20 feet long will contain
   a. 2 cubic yards of concrete
   b. 2.5 cubic yards of concrete
   c. 3.7 cubic yards of concrete
   d. 5.0 cubic yards of concrete

5. Forms for concrete slabs are usually made of
   a. 2″ × 4″
   b. 2″ × 8″
   c. 2″ × 10″
   d. none of these

6. To prevent forms from sticking to the concrete, they are treated with
   a. fat
   b. oil
   c. paint
   d. wax

7. Concrete is reinforced with
   a. air bubbles
   b. aluminum wire
   c. steel bars
   d. wood fibers

8. Concrete is cured by
   a. covering with plastic
   b. protecting from wind
   c. sprinkling with water
   d. all of these

9. A standard sized block when laid will cover an area
   a. 8 inches high by 16 inches long
   b. 4 inches high by 16 inches long
   c. 8 inches high by 12 inches long
   d. none of these

10. Courses of block are laid in a straight line by using a
    a. center point
    b. line
    c. plumb bob
    d. sighting tool

B. Matching. Match the terms in column I with those in column II.

Column I	Column II
1. sand	a. course aggregate
2. gravel	b. cement/sand/gravel
3. 1:2:3	c. curing
4. air	d. foundation for walls
5. drying	e. jointing tool
6. footing	f. fine aggregate
7. raked joint	g. entrained
8. flat end	h. corner block
9. ears	i. stretcher block
10. flush joint	j. broken block

C. Completion. Fill in the blanks with the word or words that will make the following statements correct.

1. A quart jar with 2 inches of sand for concrete mixtures should not yield more than a _____-inch layer of silt.

2. The amount of water added to a mix is important because _____.

3. A concrete mixture with uniform color and consistency and the correct proportion of ingredients is known as a _____.

4. A block wall is plumbed with a _____.

5. Concrete that is already mixed when it arrives at the job site is called _____.

APPENDICES

A. PROJECTS

B. DATA TABLES

Appendix **A**
PROJECTS

Project 1: Concrete Float

☐ **PLAN FOR CONCRETE FLOAT**

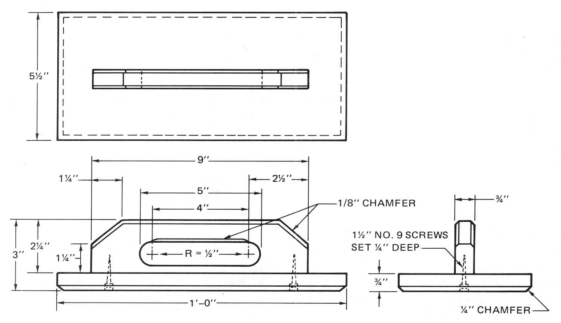

Courtesy of Department of Agricultural and Extension Education, The Pennsylvania State University, University Park, PA

BILL OF MATERIALS			
MATERIALS NEEDED	**QUANTITY**	**DIMENSIONS**	**DESCRIPTION**
Lumber (hard)	1 piece	¾″ × 5½″ × 1′0″	Base
Lumber (hard)	1 piece	¾″ × 2¼″ × 9″	Handle

□ CONSTRUCTION PROCEDURE FOR CONCRETE FLOAT

1. Lay out and cut base to the dimensions given in the plan drawing.
2. Chamfer the base edges as shown in the plan.
3. Lay out and cut the outside dimensions of the handle.
4. Locate position of 2 one-inch holes in the handle.
5. Bore the 2 one-inch holes in the handle and saw out the piece between the holes.
6. Chamfer the edges of the handle as noted in the plan.
7. Drill pilot holes for the screws.
8. Center the handle on the base; locate and drill all screw holes.
9. Fasten handle to the base using glue and two 2½″ No. 9 wood screws.
10. Sand the float and apply one coat of linseed oil.

Project 2: Push Stick

□ PLAN FOR PUSH STICK

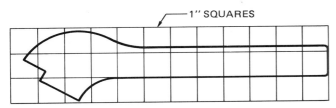

Courtesy of *T. J. Wakeman and V. L. McCoy,* The Farm Shop, *The Macmillan Company, 1960*

BILL OF MATERIALS			
MATERIALS NEEDED	**QUANTITY**	**DIMENSIONS**	**DESCRIPTION**
Lumber Unsurfaced	1 piece	1″ × 3″ × 12″	
or Surfaced	1 piece	¾″ × 3″ × 12″	
or Plywood	1 piece	½″ × 3″ × 12″	

□ CONSTRUCTION PROCEDURE FOR PUSH STICK

1. Surface the 1″ board to ¾″ thickness with a hand plane or sander.
2. Lay off 1″ squares on a piece of heavy paper or cardboard, and sketch an outline of the push stick, as shown in the drawing.
3. Cut out the pattern to use as a template.
4. Lay out the push stick by marking around the template.
5. Cut out the push stick by sawing along the outside edges of the mark with a band saw, jig saw, coping saw, or compass saw.
6. Sand and finish.

Project 3: **Miter Box**

□ PLAN FOR MITER BOX

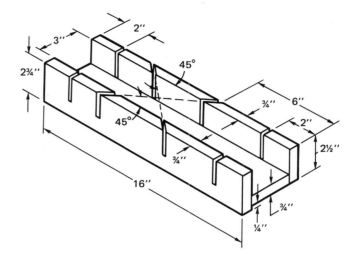

BILL OF MATERIALS			
MATERIALS NEEDED	**QUANTITY**	**DIMENSIONS**	**DESCRIPTION OR USE**
Lumber (hard)	1 piece	¾″ × 3″ × 16″	Bottom
Lumber (hard)	1 piece	¾″ × 2¾″ × 16″	Front
Lumber (hard)	1 piece	¾″ × 2½″ × 16″	Back
Wood Screws	6	No. 8 × 2″ Flat Head	Attach

□ CONSTRUCTION PROCEDURE FOR MITER BOX

1. Cut a kiln dried piece of ¾″ hard wood to 49″ long.
2. Joint one edge of the board.
3. Rip the board to 3¹⁄₁₆″ wide.
4. Joint the board to 3″ wide.
5. Saw board into three pieces 16″ long.
6. Rip pieces to ¹⁄₁₆″ wider than is shown in the materials list. Then set the jointer to remove ¹⁄₁₆″ and joint to exact width.
7. Lay off, bore, and countersink holes for wood screws. Space a hole 1″ from each end on each side. Space the other hole in the middle on each side. Fasten side pieces to bottom piece with three wood screws on each side.

(**Note:** The front side projects ¼″ below the bottom. This forms a ledge to hold the box against the edge of a table when it is being used.)

8. Lay off and saw a square cut in the side pieces 2″ from the ends, with a miter box saw. Saw both side pieces at the same time.
9. Lay off and saw 45° cuts in the side pieces 6″ from the right end, as shown in the drawing.
   These cuts serve as a guide for the saw when square or miter cuts are desired. In some instances, it may be desirable to cut other angles in the miter box to fit a particular situation. If so, the desired angle cut may be substituted for one of the 45° cuts.

Project 4: Gun Rack

□ PLAN FOR GUN RACK

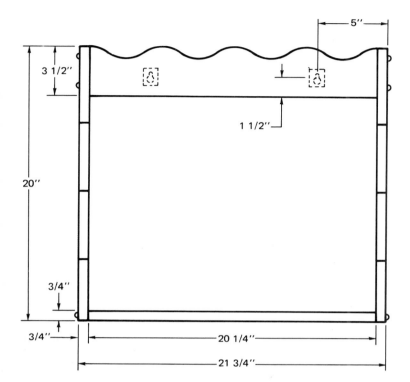

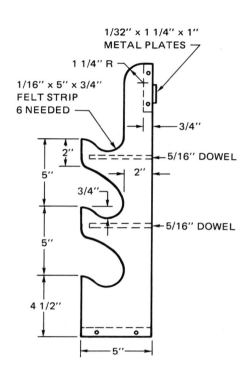

Courtesy of Agricultural Education and Agricultural Engineering, Virginia Polytechnic Institute and State University, Blacksburg, VA

BILL OF MATERIALS			
MATERIALS NEEDED	**QUANTITY**	**DIMENSIONS**	**DESCRIPTION**
Lumber, surfaced	1 piece	$\frac{3}{4}'' \times 3\frac{1}{2}'' \times 20\frac{1}{4}''$	Back
Lumber, surfaced	2 pieces	$\frac{3}{4}'' \times 5'' \times 20''$	Ends
Lumber, surfaced	1 piece	$\frac{3}{4}'' \times 5'' \times 20\frac{1}{4}''$	Bottom
Lumber, surfaced	2 pieces	$\frac{5}{16}'' \times 4\frac{1}{4}''$	Dowels
Screws	6 each	No. 8 $\times$ 1½"	Round or flat head
Metal Plates	2 each	$\frac{1}{32}'' \times 1\frac{1}{4}'' \times 1''$	Hangers

□ CONSTRUCTION PROCEDURE FOR GUN RACK

1. Cut a piece of heavy paper 3½" × 20¼". Fold the paper in half end-to-end and cut a pattern for the back.
2. Draw the pattern topline onto a piece of stock ¾" × 3½" × 20¼".
3. Use a piece of heavy paper 5" × 20" to cut a pattern for the sides.

4. Cut the topline of the back using a coping, jig, band, or bayonet saw.
5. Cut two 5" × 20" end pieces from ¾" stock.
6. Use the pattern to draw the gun supports on the end pieces.
7. Lay out the ¾" × 3½" cut on each piece to receive the back.

8. Saw out the ends.
9. Cut a ¾″ × 5″ × 20¼″ piece for the bottom.
10. Carefully drill holes and install dowel pins in the ends.
11. Carefully and thoroughly sand all surfaces, being careful not to round edges where the joints are made.

12. Glue and screw all parts together.
13. Apply a suitable finish.
14. Use stick-on felt strips in the curves where the guns are cradled.
15. Install hangers or plan to hang the rack by two nails or screws.

Project 5: Flicker/Woodpecker Houses

□ PLAN FOR FLICKER/ WOODPECKER HOUSES

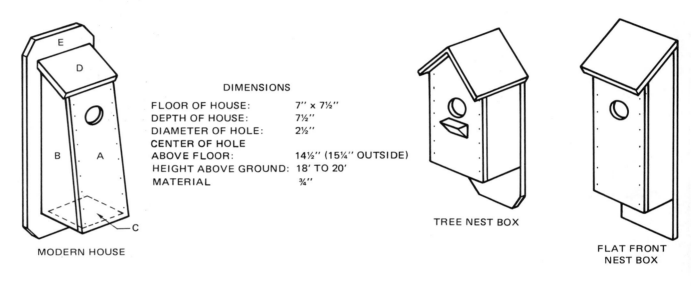

DIMENSIONS

FLOOR OF HOUSE:	7″ x 7½″
DEPTH OF HOUSE:	7½″
DIAMETER OF HOLE:	2½″
CENTER OF HOLE ABOVE FLOOR:	14½″ (15¼″ OUTSIDE)
HEIGHT ABOVE GROUND:	18′ TO 20′
MATERIAL	¾″

MODERN HOUSE

TREE NEST BOX

FLAT FRONT NEST BOX

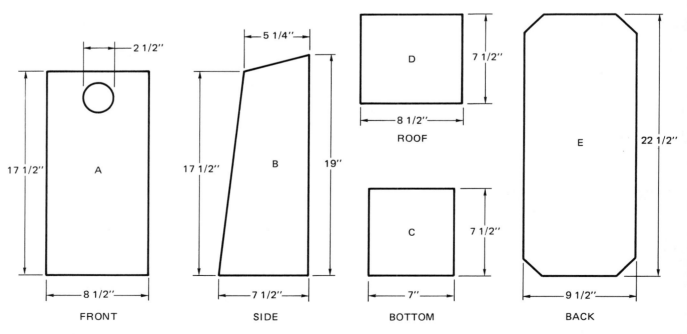

FRONT

SIDE

ROOF

BOTTOM

BACK

Courtesy of Cooperative Extension Service, Virginia Polytechnic Institute and State University, Blacksburg, VA

BILL OF MATERIALS (for modern house)			
MATERIALS NEEDED	**QUANTITY**	**DIMENSIONS**	**DESCRIPTION**
Lumber (rough and rustic)	1	¾″ × 8½″ × 17½″	Front (A)
"	2	¾″ × 7½″ × 19″	Sides (B)
"	1	¾″ × 7″ × 7½″	Bottom (C)
"	1	¾″ × 7½″ × 8½″	Roof (D)
"	1	¾″ × 9½″ × 22½″	Back (E)
Box nails or finishing nails		4d or 6d	

The flicker will nest readily in a well-placed bird-house of the proper dimensions. Roughen the interior of the nest box to assist the young woodpeckers to reach the entrance hole. Cover the bottom with sawdust so that the mother bird can shape the nest for eggs and the young birds. Material may be ½ inch instead of ¾ inch thick, in which case the width of the front and roof will be 8 inches.

□ **CONSTRUCTION PROCEDURE FOR FLICKER/WOODPECKER HOUSES**

1. Select pine, redwood, or other lumber that will withstand weather without paint. Rough lumber is generally preferred and will provide a rustic ap-pearance. Planed lumber should be used if the house is to be painted for decorative purposes, but this is not recommended for most species except Martin houses.
2. Cut all parts from a ¾″ × 9½″ × 8′ board.
3. Cut the sides to the proper angle.
4. Bevel the top edge of the front to conform to the angle of the sides.
5. Bevel the upper edge of the roof to conform to the back and sides.
6. Nail the front, sides, and bottom together.
7. Nail the back to the sides and bottom. (Center the assembly on the back for a good appearance. Leave 1″ at bottom and 2½″ at the top.)
8. Attach the roof with two hinges or nail it lightly so the roof is easily removed for periodic cleaning of the nest area.

Project 6: *Nesting and Den Boxes*

□ PLAN FOR NESTING AND DEN BOXES

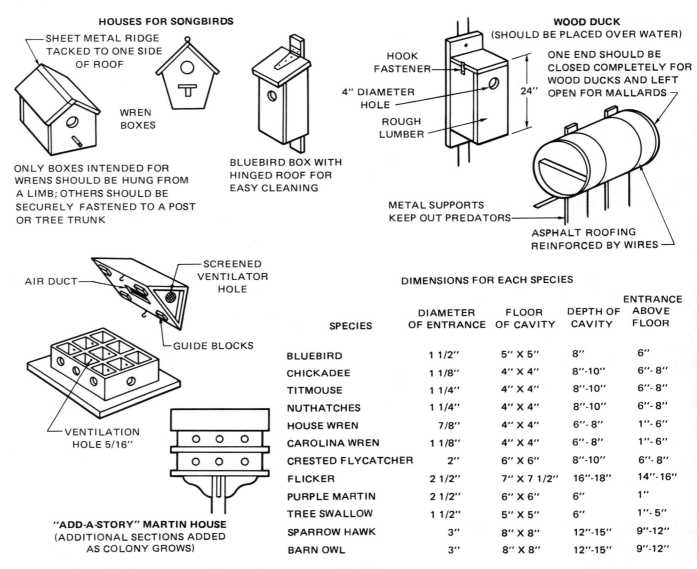

HOUSES FOR SONGBIRDS

SHEET METAL RIDGE TACKED TO ONE SIDE OF ROOF

WREN BOXES

ONLY BOXES INTENDED FOR WRENS SHOULD BE HUNG FROM A LIMB; OTHERS SHOULD BE SECURELY FASTENED TO A POST OR TREE TRUNK

BLUEBIRD BOX WITH HINGED ROOF FOR EASY CLEANING

WOOD DUCK (SHOULD BE PLACED OVER WATER)

HOOK FASTENER

4" DIAMETER HOLE

ROUGH LUMBER

24"

ONE END SHOULD BE CLOSED COMPLETELY FOR WOOD DUCKS AND LEFT OPEN FOR MALLARDS

METAL SUPPORTS KEEP OUT PREDATORS

ASPHALT ROOFING REINFORCED BY WIRES

AIR DUCT

SCREENED VENTILATOR HOLE

GUIDE BLOCKS

VENTILATION HOLE 5/16"

"ADD-A-STORY" MARTIN HOUSE (ADDITIONAL SECTIONS ADDED AS COLONY GROWS)

DIMENSIONS FOR EACH SPECIES

SPECIES	DIAMETER OF ENTRANCE	FLOOR OF CAVITY	DEPTH OF CAVITY	ENTRANCE ABOVE FLOOR
BLUEBIRD	1 1/2"	5" X 5"	8"	6"
CHICKADEE	1 1/8"	4" X 4"	8"-10"	6"- 8"
TITMOUSE	1 1/4"	4" X 4"	8"-10"	6"- 8"
NUTHATCHES	1 1/4"	4" X 4"	8"-10"	6"- 8"
HOUSE WREN	7/8"	4" X 4"	6"- 8"	1"- 6"
CAROLINA WREN	1 1/8"	4" X 4"	6"- 8"	1"- 6"
CRESTED FLYCATCHER	2"	6" X 6"	8"-10"	6"- 8"
FLICKER	2 1/2"	7" X 7 1/2"	16"-18"	14"-16"
PURPLE MARTIN	2 1/2"	6" X 6"	6"	1"
TREE SWALLOW	1 1/2"	5" X 5"	6"	1"- 5"
SPARROW HAWK	3"	8" X 8"	12"-15"	9"-12"
BARN OWL	3"	8" X 8"	12"-15"	9"-12"

Courtesy of Maryland Department of Natural Resources, Public Information Services

---

BILL OF MATERIALS

Bills of materials are not provided with these nesting and den boxes due to the variety of types pictured. Nesting and den boxes should be planned from the information provided. An appropriate bill of materials may then be developed for the plan.

□ CONSTRUCTION PROCEDURE FOR NESTING AND DEN BOXES

Construction procedures will vary with the nesting or den box being constructed. However, the procedure outlined for Project 5, Flicker/Woodpecker Houses, may be helpful.

Project 7:   Small Engine Stand—Wood

□ PLAN FOR SMALL ENGINE STAND—WOOD

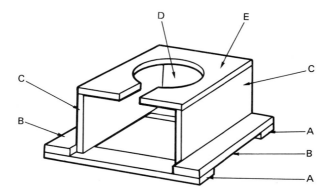

Courtesy of Department of Agricultural and Extension Education, The Pennsylvania State University, University Park, PA

BILL OF MATERIALS			
MATERIALS NEEDED	QUANTITY	DIMENSIONS	DESCRIPTION
Lumber, No. 2 White Pine, surfaced	2 pieces	¾″ × 2¼″ × 16″	–Front and back of base frame (A)
Lumber, No. 2 White Pine, surfaced	2 pieces	¾″ × 2¼″ × 12″	–Sides of base frame (B)
Lumber, No. 2 White Pine, surfaced	2 pieces	¾″ × 5½″ × 11½″	–Sides of stand (C)
Lumber, No. 2 White Pine, surfaced	1 piece	¾″ × 5½″ × 10″	–Back of stand (D)
Plywood, Good one side	1 piece	¾″ × 11½″ × 11½″	–Top of stand (E)

□ CONSTRUCTION PROCEDURE FOR SMALL ENGINE STAND—WOOD

1. All wooden joints should be made with glue and nails, or glue and screws.
2. Use pieces "A" and "B" to construct the frame base as shown in the drawing, making sure the base is square.
3. Fasten the two "C" pieces into position. One end of each "C" piece should be flush to the rear edge of the base frame; the front ends will fall short of the front edge of the base frame by ½ inch.
4. Place the "D" piece or back of the stand between the sides of the stand and flush with the back ends.
5. Determine the center of the top plywood piece "E" and draw a 7-inch diameter circle around it. Cut a 7-inch hole and a 2-inch front slot (centered on the front) before fastening the top to the stand.
6. For the front of the stand a *safety door*, not shown, should be constructed from ½-inch material. It may be fitted to the front and secured in place by small hooks and eyes or hinges.
7. All wooden parts should be painted with an enamel paint. Recommended colors are grey, green, or blue.

Project 8: Nail Box

□ PLAN FOR NAIL BOX

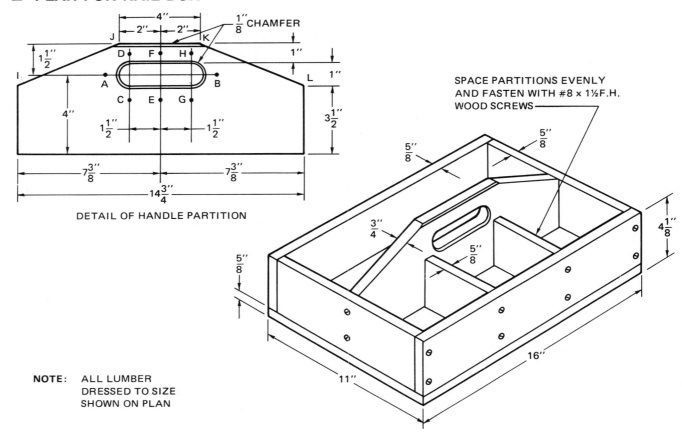

DETAIL OF HANDLE PARTITION

NOTE: ALL LUMBER
DRESSED TO SIZE
SHOWN ON PLAN

Courtesy of T. J. Wakeman and V. L. McCoy, The Farm Shop, *The Macmillan Company, 1960*

BILL OF MATERIALS			
MATERIALS NEEDED	**QUANTITY**	**DIMENSIONS**	**DESCRIPTION OR USE**
Lumber, unsurfaced*	1 piece	1″ × 6″ × 10′	
Lumber, surfaced	1 piece	¾″ × 5½″ × 14¾″	Handle partition
Lumber, surfaced	2 pieces	⅝″ × 5½″ × 16″	Bottom
Lumber, surfaced	2 pieces	⅝″ × 3½″ × 16″	Sides
Lumber, surfaced	2 pieces	⅝″ × 3½″ × 9¾″	Ends
Lumber, surfaced	2 pieces	⅝″ × 3½″ × 4½″	Partitions
Wood screws	34	1½″	No. 8 flat head

* All surfaced pieces are to be cut from this board.

□ CONSTRUCTION PROCEDURE FOR NAIL BOX

1. Surface the 1″ × 6″ × 10′–0″ board to ¾″ thickness with a thickness planer or a hand plane.
2. Cut one piece 14¾″ long for the handle partition.
3. Surface the remainder of the board to ⅝″ thickness, and cut bottom, side, end, and partition pieces to size, as shown in the bill of materials.
4. Lay off, bore, and countersink holes for screws in the side pieces. Fasten the sides to the ends with two screws at each joint starting ¾″ from each edge.
5. Fasten the bottom to the side and end pieces. Use three screws in the side pieces, two spaced 1″ from each end and one in the center. Use four screws on each end; place one screw 2″ from each corner and one 4½″ from each corner.
6. Cut the handle partition to size as shown in the bill of materials. (Check distance between end pieces before cutting handle partition to exact length.)
7. Draw lines AB, EF, CD, GH, as shown in the detail of the handle partition in the drawing.
8. Using a 1″ auger, bore holes where lines CD and GH cross line AB. Bore until the feed screw starts through the opposite side of the board; then complete the hole from the opposite side.
9. Locate points, I, J, K, and L.
10. Mark and cut lines IJ and KL.
11. Slightly round off the corners of the slot and top of handle with a wood rasp, and sand smooth.
12. Space the partitions evenly on one side of the handle, fastening each with two screws.
    Note: Partitions are left out of one side so that it may be used for small tools. If the box is to be used for nails or bolts only, partitions should be inserted in the other side.
13. Place handle and partitions in box. Square, drill holes for screws, and fasten partitions to side with two screws at each joint. Fasten handle to ends with two screws at each end.
14. Sand and finish.

Project 9: Tool Box

□ PLAN FOR TOOL BOX

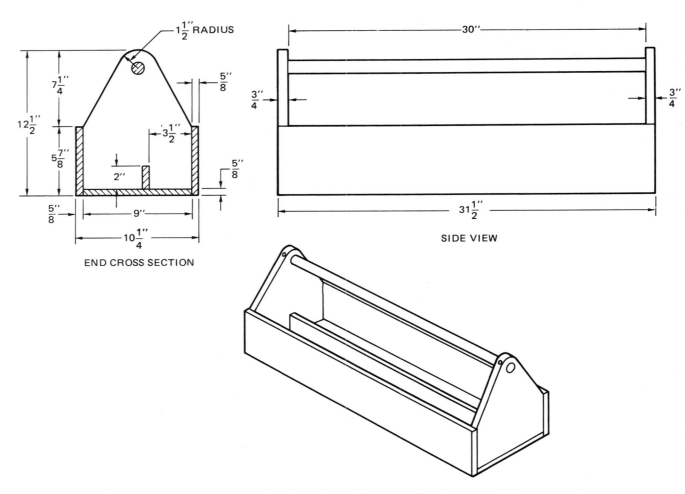

END CROSS SECTION

SIDE VIEW

Courtesy of T. J. Wakeman and V. L. McCoy, The Farm Shop, *The Macmillan Company, 1960*

BILL OF MATERIALS			
MATERIALS NEEDED	**QUANTITY**	**DIMENSIONS**	**DESCRIPTION OR USE**
Lumber, unsurfaced*	1 piece	1″ × 12″ × 8′	
Lumber, surfaced	2 pieces	¾″ × 9″ × 12½″	Ends
Lumber, surfaced	1 piece	⅝″ × 9″ × 31½″	Bottom
Lumber, surfaced	1 piece	⅝″ × 2″ × 30″	Partition
Lumber, surfaced	2 pieces	⅝″ × 5⅞″ × 31½″	Side boards
Round wood stock	1 piece	1″ diameter × 31½″	Handle (cut from broom handle or dowel)
Box of finishing nails	¼ pound	6d	
Wood screws	2	2″	No. 6 flat head

* All surfaced pieces are to be cut from this board.

□ CONSTRUCTION PROCEDURE FOR TOOL BOX

1. Surface the 1″ board to ¾″ thickness with a thickness planer.
2. Cut ends to size; shape and bore holes for handle, as shown in end cross section. Use an auger bit the same size as the ends of the handle.
3. Surface remainder of board to ⅝″ thickness with a thickness planer.
4. Cut partition, bottom, and side boards to size, as shown in the drawing.
5. Cut handle to length, and place in holes in end pieces. Glue and nail bottom, side boards, and partition in place.
6. Bore holes in end pieces and handle for the wood screws. Countersink and place screws to hold the handle in place.
7. Sand and finish.

   Note: ¾″ lumber may be substituted where ⅝″ material is specified. Flat head wood screws may be used in place of nails.

☐ PLAN FOR SADDLE RACK

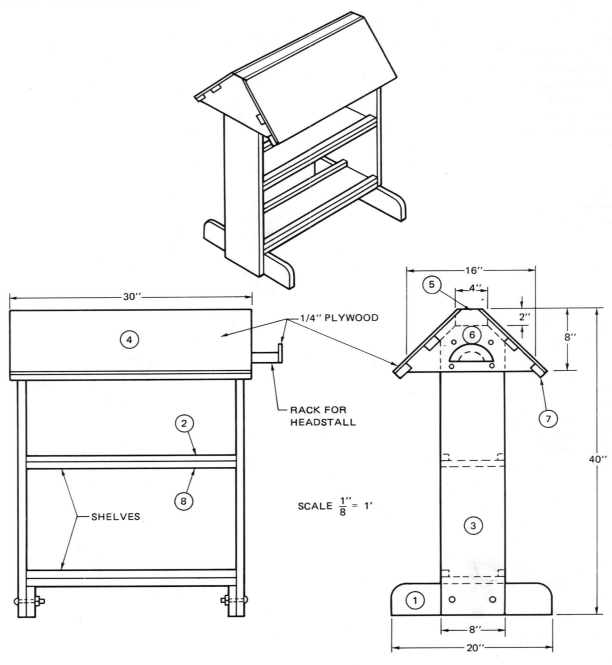

1/4" PLYWOOD

RACK FOR
HEADSTALL

SHELVES

SCALE $\frac{1''}{8}$ = 1'

30"

16"

4"

2"

8"

40"

8"

20"

Courtesy of Texas Vocational Instructional Services, Texas A & M University, College Station, TX

BILL OF MATERIALS

MATERIALS NEEDED	QUANTITY	DIMENSIONS	DESCRIPTION
Lumber	2 pieces	1″ × 4″ × 20″	Base (1)
Lumber	4 pieces	1″ × 1″ × 26″	Cleats (2)
Lumber	2 pieces	1″ × 8″ × 40″	End Supports (3)
Lumber	2 pieces	¼″ × 12″ × 30″	Top (4)
Lumber	1 piece	2″ × 4″ × 30″	Center (Ridge) (5)
Lumber	2 pieces	1″ × 8″ × 16″	Ends (6)
Lumber	4 pieces	1″ × 2″ × 30″	Ribs (7)
Lumber	2 pieces	1″ × 8″ × 26″	Shelves (8)

Note: ¾″ stock may be used in place of 1″ material.

□ CONSTRUCTION PROCEDURE FOR SADDLE RACK

1. Cut two pieces 1″ × 8″ × 16″ for the top ends. Set the boards on edge and mark the center on the top edge. Mark locations 1″ out in both directions from the centers. Draw a line from each of these points to the lower corners on each end of the boards. Saw the ends to these lines.

2. Cut the ridge to length and joint two corners so its shape matches the top of the end pieces.

3. Cut all other pieces to size as listed in the bill of materials, except the shelves and cleats. These should be cut about 1″ longer than indicated.

4. Use glue and nails or screws to assemble the top.

5. Use glue and ¼″ or 5⁄16″ carriage bolts with washers to attach the sides to the base pieces.

6. Use glue and eight ¼″ carriage bolts with washers to attach the sides to the top. Be very careful to attach the ends so the unit is straight and sits squarely on its feet on a flat surface.

7. Attach the cleats to the shelf pieces with glue and nails.

8. Measure the distance from inside to inside of the two ends at the top of the rack. Cut the length of the shelf and cleat assemblies to this dimension.

9. Install the shelf and cleat assemblies with glue and 10d finishing nails or No. 8 × 2½″ wood screws.

10. Smooth all surfaces and apply a suitable finish.

Project 11: Saw Horse—Lightweight

☐ PLAN FOR SAW HORSE— LIGHTWEIGHT

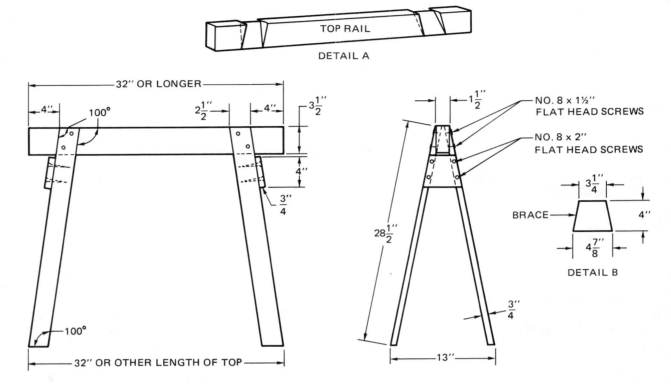

BILL OF MATERIALS			
MATERIALS NEEDED	**QUANTITY**	**DIMENSIONS**	**DESCRIPTION OR USE**
Lumber, surfaced	1 piece	1½″ × 3½″ × 32″	Top rail
*Lumber, surfaced	4 pieces	¾″ × 2½″ × 29″	Legs
Lumber, surfaced	2 pieces	¾″ × 4″ × 5″	Braces
Wood screws	8	No. 8 × 1½″	Flat head, legs
Wood screws	8	No. 8 × 2″	Flat head, braces

* Use clear lumber, preferably hardwood.

☐ CONSTRUCTION PROCEDURE FOR SAW HORSE— LIGHTWEIGHT

1. Cut the top rail from a surfaced 2 × 4.
2. Cut one leg from clear lumber so the finished size is ¾″ × 2½″ × 28½″ with ends cut at 100° angles as shown in the drawing. Tough wood such as oak or yellow pine is recommended.

3. Use the first leg as a pattern to cut the other three legs.
4. Mark "top" on one of the 1½″ edges of the top rail.
5. Mark the top of the top rail 4″ in from each end

and square a line across the 1½″ surface at each mark. (See Detail A for illustration of steps 5–8.)

6. Use a T-bevel to draw a line at 100° from each end of each line down the sides of the rail.

7. Lay the leg pattern against each line and draw a second line parallel to each. The lines for all four legs are now drawn along the sides of the top rail. These should be such that when legs are attached, the outside toe of each leg will be in line with the end of the top rail.

8. Draw a line down the center of the top of the rail between each pair of lines.

9. Use a hand saw to make multiple cuts inside the sets of parallel lines as the first step in cutting a dado for each leg. The depth of the cuts is ¾″ on the top of the rail and 0″ on the bottom of the rail.

10. Use a wood chisel to cut out the dados. Smooth the bottom of each dado with a coarse file.

11. Drill and countersink holes in the legs for wood screws.

12. Cut out two leg braces ¾″ × 4″ wide by 3¼″ and 4⅞″. See Detail B.

13. Drill and countersink for the wood screws.

14. Assemble the saw horse with glue and screws.

15. If the saw horse rocks when placed on a flat surface, place a thin block of wood on the floor against the leg. Draw a line at the top of the block and one on each side and on each edge of the leg. Repeat the process on the other three legs. Be careful that the saw horse does not rock during this procedure. Resaw the legs on the lines so established.

16. Smooth the ends of the legs on top of the rail.

17. Apply a suitable finish.

Project 12: Standard Saw Horse

□ PLAN FOR STANDARD SAW HORSE

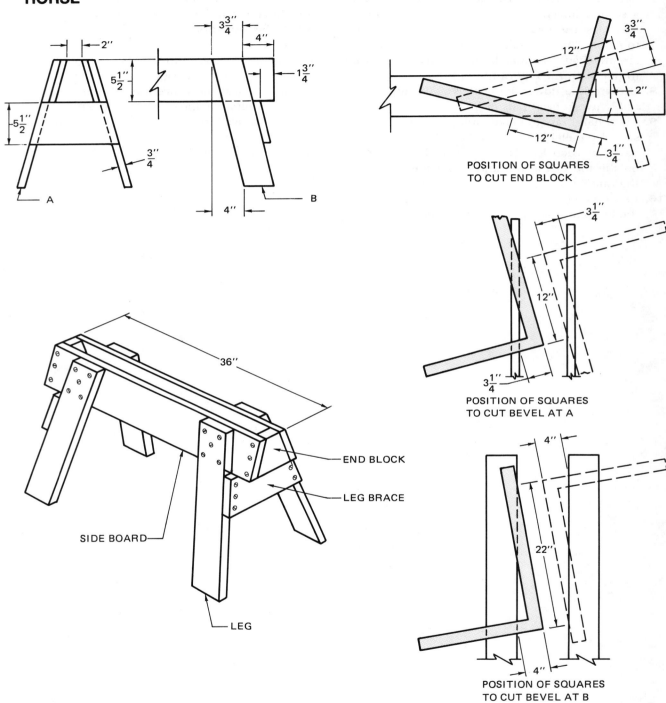

POSITION OF SQUARES TO CUT END BLOCK

POSITION OF SQUARES TO CUT BEVEL AT A

END BLOCK

LEG BRACE

SIDE BOARD

LEG

POSITION OF SQUARES TO CUT BEVEL AT B

Courtesy of T. J. Wakeman and V. L. McCoy, The Farm Shop, *The Macmillan Company, 1960*

BILL OF MATERIALS			
MATERIALS NEEDED	**QUANTITY**	**DIMENSIONS**	**DESCRIPTION OR USE**
Lumber, surfaced	2 pieces	¾″ × 5½″ × 36″	Side boards
Lumber, surfaced	2 pieces	¾″ × 5½″ × 14″	Leg braces
Lumber, surfaced	2 pieces	1¾″ × 5½″ × 6″	End blocks
Lumber, surfaced	4 pieces	¾″ × 3¾″ × 24″	Legs
Wood screws	44	1½″	No. 12 flat head

Note: 1 inch lumber may be used when ¾ inch material is specified.

□ CONSTRUCTION PROCEDURE FOR STANDARD SAW HORSE

1. Surface 1″ material for side boards, leg braces, and legs to ¾″ thickness using a thickness planer.
2. Cut side boards to length, as shown in the drawing.
3. Rip side boards to 5½″ width.
4. Set T-bevel at the same degree as the end blocks (see drawing), and use as a template for setting the jointer fence at the proper angle for beveling the top and bottom edges of the side boards.
5. Make the top and bottom bevels of the side boards parallel by jointing each bevel from the same end.
6. Surface the end blocks to 1¾″ thickness.
7. Joint one edge of each end block.
8. Lay out and mark the end blocks, as shown in the drawing. Mark the length of slope on each end block the same as the width of the side boards; then joint the bottom edge of each block to this width.
9. Fasten side boards to the end blocks with glue and No. 12 × 1½″ flat head wood screws, as shown in the drawing. Bore holes for the wood screws in the side boards and end blocks. Countersink the side boards for the screw heads.
10. Lay out and cut the legs to length, as shown in detailed drawing.
11. Bore holes, and fasten legs to side with glue and No. 12 × 1½″ flat head wood screws, as shown.
12. Set the T-bevel to the same degree as the side cuts of the legs (4″ and 22″ on the framing square), and use as a template for setting the jointer fence at the proper angle for beveling the top and bottom of the leg braces.
13. Make the top and bottom bevels of the leg braces parallel by jointing each bevel from the same end.
14. Mark leg braces to length by holding the braces tight in place and marking along the outside edges of the legs. Cut at these marks.
15. Bore holes, and fasten the leg braces to legs with glue and No. 12 × 1½″ flat head wood screws as shown in the drawing.
16. Sand and finish.

Project 13: **Workbench**

□ **PLAN FOR WORKBENCH**

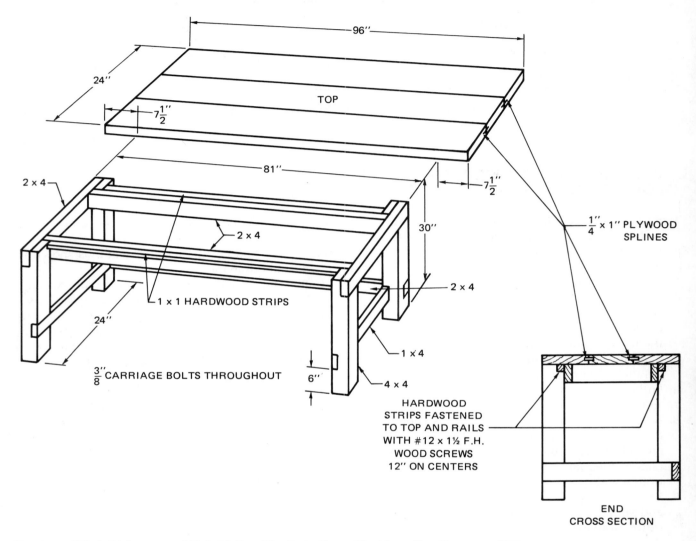

96″

24″

$7\frac{1}{2}$″

TOP

81″

$7\frac{1}{2}$″

2 x 4

2 x 4

30″

$\frac{1}{4}$″ x 1″ PLYWOOD SPLINES

2 x 4

1 x 1 HARDWOOD STRIPS

24″

1 x 4

$\frac{3}{8}$″ CARRIAGE BOLTS THROUGHOUT

6″

4 x 4

HARDWOOD STRIPS FASTENED TO TOP AND RAILS WITH #12 x 1½ F.H. WOOD SCREWS 12″ ON CENTERS

END CROSS SECTION

Courtesy of T. J. Wakeman and V. L. McCoy, The Farm Shop, *The Macmillan Company, 1960*

BILL OF MATERIALS			
MATERIALS NEEDED	QUANTITY	DIMENSIONS	DESCRIPTION OR USE
Lumber, surfaced	4 pieces	3¾″ × 3¾″ × 30″	Legs
Lumber, surfaced	2 pieces	1¾″ × 3¾″ × 6′5½″	Side rails
Lumber, surfaced	2 pieces	1¾″ × 3¾″ × 24″	End rails
Lumber, surfaced	1 piece	1¾″ × 3¾″ × 6′9″	Side braces
Lumber, surfaced	2 pieces	¾″ × 3¾″ × 22¼′	End braces
Hardwood, surfaced	2 pieces	¾″ × ¾″ × 6′1½″	Strips to fasten frame to top
Hardwood, surfaced	2 pieces	1¾″ × 7½″ × 8′0″	Top, outside pieces
Hardwood, surfaced	1 piece	1¾″ × 9½″ × 8′0″	Top, middle piece
Plywood, ¼″ thick	2 pieces	¼″ × 1″ × 8′0″	Top, splines
Carriage bolts	10	⅜″ × 4½″	
Carriage bolts	4	⅜″ × 6″	
Washers	14	⅜″	Flat
Wood screws	28	1½″	No. 12 flat head
Wood glue	3 ounces		
Paint	1 quart		For frame
Thinner	½ pint		For frame
Floor or gym seal	1 pint		For top

Note: Commercially surfaced lumber will be 1½″ × 3½″ for 2 × 4s; 1½″ × 5½″ for 2 × 6s; and 3½″ × 3½″ for 4 × 4s. Use shorter bolts if commercial lumber is used.

□ CONSTRUCTION PROCEDURE FOR WORKBENCH

1. With a thickness planer, surface the lumber for the legs, side and end rails, and side and end braces to the thickness shown in bill of materials.
2. Cut legs, side and end rails, and side and end braces to length, as shown in bill of materials.
3. Lay off and cut notches in legs for rails and leg braces. (This can be done by hand, or with a dado head on a power saw.) Remember that each leg must be made to fit a specific corner. See the drawing for the location of the notches. Be sure the notches are cut so that the end and side rails and the leg braces will fit tightly.
4. Tack the frame together. (Be sure that the nails are not placed at points where bolt holes are to be made, as described in step 5.)
5. Bore a ⅜″ hole at each joint in the frame and fasten the frame together with ⅜″ carriage bolts. Insert bolts in holes from outside, and tighten nuts.
6. Surface lumber for top pieces to 1¾″ thickness using a thickness planer. Then joint both edges of the middle piece and one edge of the outside pieces. The jointed edge must be straight and square. Clamp the top pieces together with the best sides up. If the edges of the boards do not match perfectly, repeat the process until they do.
7. Mark boards on the top side with chalk or pencil so that they may be placed in the same position for gluing.
8. Rip splines 1″ wide and 96″ long from a piece of ¼″ plywood. (The splines can be ripped from short pieces of plywood if a 96″ piece is not available. The short pieces can be placed end to end in the groove.)
9. Adjust the dado head to cut the proper size grooves for the ¼″ × 1″ splines. Establish a distance between the fence and the center of the dado that is equal to ½ the thickness of the lumber to be grooved. To be sure of the proper setting, cut a groove in two pieces of scrap wood that are the same thickness as the table top. Place the spline in the grooves of the scrap pieces to see if they are grooved correctly.
10. Cut a groove in the jointed edge of the two outside pieces and in each edge of the middle piece. The marked side of each piece must be held tightly (use a feather board) against the fence while the groove is being cut. This will make all the grooves the same distance from the top surface. If a power saw is not available, the top may be held together with glue and with ½″ × 2″ dowel pins placed about 8″ apart.

11. Glue the top together.
12. After the glue has set properly, joint one edge of the table top straight; then cut the top to 24″ wide by ripping and jointing the other edge.
13. Cut the top to 96″ in length by sawing both ends square.
14. Lay off, bore, and countersink holes in the strips to fasten the hardwood strips to the top rail. Bore one hole about 2″ from each end of each strip. Bore the other holes approximately 12″ apart.
15. Bore and countersink holes in the hardwood strips for fastening the top to the frame (base).

Space vertical holes to avoid hitting the horizontal ones.
16. Bore anchor holes in rails. Fasten strips to rails with No. 12 × 1½″ flat head wood screws.
17. Place the table top on a smooth surface with the finished side down. Place the frame over the table top, and adjust to the proper distance from both ends and sides. Fasten the strips to the table top with wood screws.
18. Finish the frame with a clear finish or with paint.
19. Sand the table top and edges. Apply two coats of floor seal or gym seal.

Project 14: Picnic Table

□ PLAN FOR PICNIC TABLE

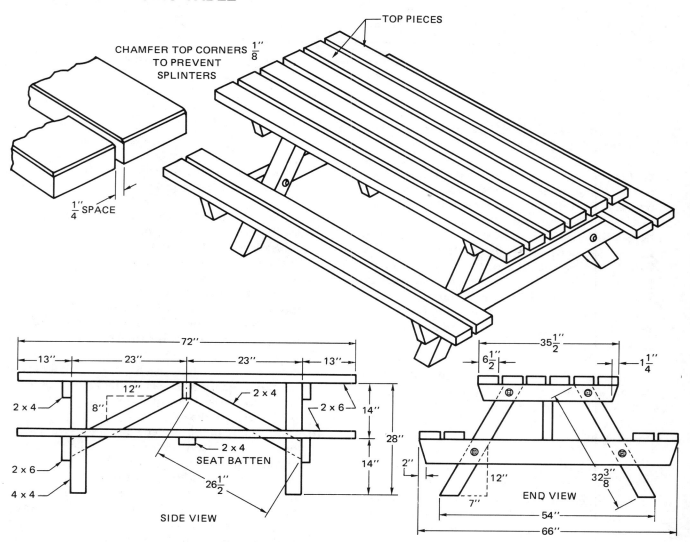

Courtesy of T. J. Wakeman and V. L. McCoy, The Farm Shop, *The Macmillan Company, 1960*

BILLS OF MATERIALS			
MATERIALS NEEDED	**QUANTITY**	**DIMENSIONS**	**DESCRIPTION OR USE**
Lumber, surfaced	10 pieces	1¾″ × 5¾″ × 72″	Top and seats
Lumber, surfaced	2 pieces	1¾″ × 5¾″ × 66″	Seat supports
Lumber, surfaced	3 pieces	1¾″ × 3¾″ × 35½″	Top supports
Lumber, surfaced	4 pieces	3¾″ × 3¾″ × 34″	Legs
Lumber, surfaced	2 pieces	1¾″ × 3¾″ × 28″	Braces
Lumber, surfaced	2 pieces	1¾″ × 3¾″ × 12″	Seat battens
Carriage bolts	8	½″ × 6½″	Ends
Washers	16	½″	Flat
Common nails	2 pounds	16d	Galvanized finish

Note: Commercially surfaced lumber will be 1½″ × 3½″ for 2 × 4s; 1½″ × 5½″ for 2 × 6s; and 3½″ × 3½″ for 4 × 4s. Use shorter bolts if commercially dressed lumber is used.

□ CONSTRUCTION PROCEDURE FOR PICNIC TABLE

1. Surface the lumber to size, as listed in the bill of materials using a thickness planer. Joint the edges smooth.
2. Cut the top and seat pieces to length, and chamfer the top corners with a hand plane (see detail drawing).
3. Cut the three top supports to length and shape. Place them on a level surface with the long edge up, and space as shown in the side view.
4. Nail the two outside top pieces to each top support with one 16d nail.
5. Square the top, and finish nailing the two outside pieces with 16d nails.
6. Space the remaining top pieces approximately ¼″ apart, and nail them in place.
7. Cut the legs and seat supports to length and angle, as shown in the end view.
8. Turn the top over, and tack the legs and seat supports in place.
9. Bore holes with a ½″ auger bit, and bolt the legs and seat supports in place with ½″ carriage bolts. Place a ½″ flat washer under the nut of each bolt.
10. Square legs with the top of the table. Cut the braces to length and shape; nail in place, as shown in the side view.
11. Cut 2″ × 4″ seat battens to length, as listed in bill of materials. Turn the table upright, and nail on seats and seat battens.
12. Apply a suitable clear, stain or paint finish for continuous exterior exposure.

Project 15: Hog-Shipping Crate

□ PLAN FOR HOG-SHIPPING CRATE

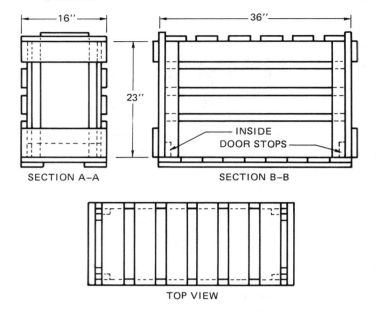

SECTION A–A

SECTION B–B

—— INSIDE
DOOR STOPS ——

TOP VIEW

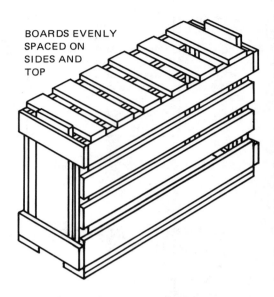

BOARDS EVENLY
SPACED ON
SIDES AND
TOP

Courtesy of T. J. Wakeman and V. L. McCoy, The Farm Shop, *The Macmillan Company, 1960*

BILL OF MATERIALS			
MATERIALS NEEDED	**QUANTITY**	**DIMENSIONS**	**DESCRIPTION OR USE**
Lumber, unsurfaced	2 pieces	1″ × 6″ × 36″	Bottom side pieces
	2 pieces	1″ × 5″ × 16″	Bottom end pieces
	2 pieces	1″ × 4″ × 36″	Runners
	9 pieces	1″ × 4″ × 16″	Flooring
	4 pieces	1″ × 4″ × 23″	Corner uprights
	6 pieces	1″ × 4″ × 36″	Top and middle side pieces
	6 pieces	1″ × 4″ × 16″	Top crosspieces
	2 pieces	1″ × 4″ × 12″	Inside doorstop
	2 pieces	1″ × 4″ × 16″	Top end pieces
	2 pieces	1″ × 10″ × 24″	Doors
Machine bolts	2	¼″ × 2½ ″	
Wing Nuts	2	¼″	
Box nails	3 pounds	8d	

□ CONSTRUCTION PROCEDURE FOR HOG-SHIPPING CRATE

1. Cut all the lumber pieces to length as shown in the bill of materials. If the crate is to be painted, surface the 1″ boards to ⅞″ thick using a thickness planer before cutting to length. Joint the edges smooth after the pieces are cut to length.

2. Nail the flooring to the two bottom side pieces (see drawing), and nail on the two runners, using 8d box or common nails.

3. Nail the two bottom end pieces to the ends of the bottom side pieces and floor. Nail the corner up-

rights to the bottom side and end pieces, as shown.

4. Nail the two top side pieces and the two end pieces to the top of the corner upright pieces. Nail on the middle side pieces.

5. Fit the doors in place, and nail the bottom inside door stops to the floor. Leave ⅛″ clearance between the door and the inside door stop.

6. Nail on the top crosspieces. Leave ⅛″ clearance between the door and the first top crosspiece.

7. Bore a ⁵⁄₁₆″ hole in the top end pieces and through the door on each end so the door can be bolted with a ¼″ machine bolt.

Note: The size of the crate may be increased for large hogs. Carriage bolts may be used in high-stress joints for added strength.

Project 16: *Wood Gate*

□ PLAN FOR WOOD GATE

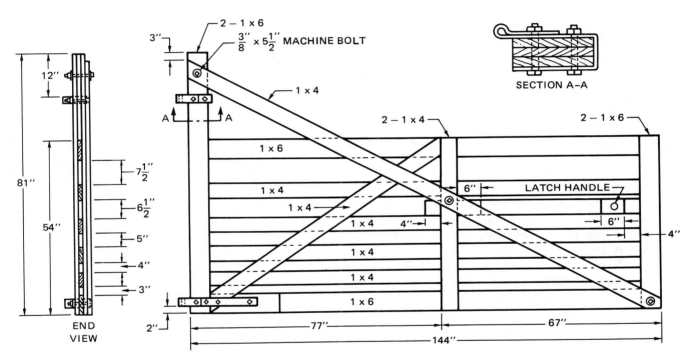

Courtesy of T. J. Wakeman and V. L. McCoy, The Farm Shop, *The Macmillan Company, 1960*

<table>
<tr><th colspan="4">BILL OF MATERIALS</th></tr>
<tr><th>MATERIALS NEEDED</th><th>QUANTITY</th><th>DIMENSIONS</th><th>DESCRIPTION OR USE</th></tr>
<tr><td>Lumber, unsurfaced*</td><td>2 pieces</td><td>1" × 6" × 81"</td><td>Hinge battens</td></tr>
<tr><td></td><td>2 pieces</td><td>1" × 6" × 54"</td><td>Front battens</td></tr>
<tr><td></td><td>1 piece</td><td>1" × 6" × 36"</td><td>Hinge filler blocks</td></tr>
<tr><td></td><td>2 pieces</td><td>1" × 4" × 54"</td><td>Center battens</td></tr>
<tr><td></td><td>1 piece</td><td>1" × 4" × 36"</td><td>Latch</td></tr>
<tr><td></td><td>2 pieces</td><td>1" × 4" × 8'0"</td><td>Short braces</td></tr>
<tr><td></td><td>2 pieces</td><td>1" × 6" × 12'0"</td><td>Top and bottom slats</td></tr>
<tr><td></td><td>4 pieces</td><td>1" × 4" × 12'0"</td><td>Inside slats</td></tr>
<tr><td></td><td>2 pieces</td><td>1" × 4" × 14'0"</td><td>Long braces</td></tr>
<tr><td>Carriage bolts</td><td>3</td><td>⅜" × 5½"</td><td></td></tr>
<tr><td>Carriage bolts</td><td>6</td><td>⅜" × 3½"</td><td></td></tr>
<tr><td>Washers</td><td>8</td><td>⅜"</td><td>Flat</td></tr>
<tr><td>Common nails</td><td>2 pounds</td><td>8d</td><td></td></tr>
<tr><td>Common nails</td><td>½ pound</td><td>6d</td><td></td></tr>
<tr><td>Hinges</td><td>2</td><td></td><td>Make or purchase</td></tr>
</table>

* Use surfaced lumber if gate is to be painted.

□ CONSTRUCTION PROCEDURE FOR WOOD GATE

1. Cut all battens and slats to length, as shown in the drawing. (If the gate is to be painted, surface all boards to ⅞" thickness.)

2. Place one front batten and one hinge batten on a flat surface. Tack the top and bottom slats in place with a 6d common nail at each joint.

3. Square gate with a framing square. Tack the inside slats at the proper spacing. The spacing may vary slightly, depending upon the width of the slats.

4. Place the two remaining end battens directly over the other two battens. Fasten permanently with 8d common nails.

5. Place the two center battens in position. Fasten permanently with 8d common nails.

6. Place the short brace for this side in position. Mark ends in line with the inside edge of the battens. Cut brace and tack it in position with 6d common nails.

7. Mark and cut the long brace and tack it in position on the brace and battens.

8. Turn the gate over. Cut braces for this side in the same manner, and fasten all joints on this side permanently with 8d common nails, except where long brace touches battens.

9. Turn the gate over. Fasten braces from this side in like manner.

10. Use ⅜ inch carriage bolts in joints where long braces cross the battens.

11. Place latch in position and nail on stop blocks. Drill hole in latch and insert a short piece of dowel or broom handle for the latch handle.

12. Nail the bottom and top hinge filler blocks on the same side of the gate that the hinges are to be placed. Fasten the hinges to the gate with carriage bolts.

13. Prime the gate and paint it with two coats of exterior paint, or finish with a wood preservative. **Note:** If additional rigidity and permanence are desired, add a $\frac{5}{16}$ inch carriage bolt at every joint in the gate.

Project 17: Salt and Mineral Box—Portable

□ PLAN FOR SALT AND MINERAL BOX—PORTABLE

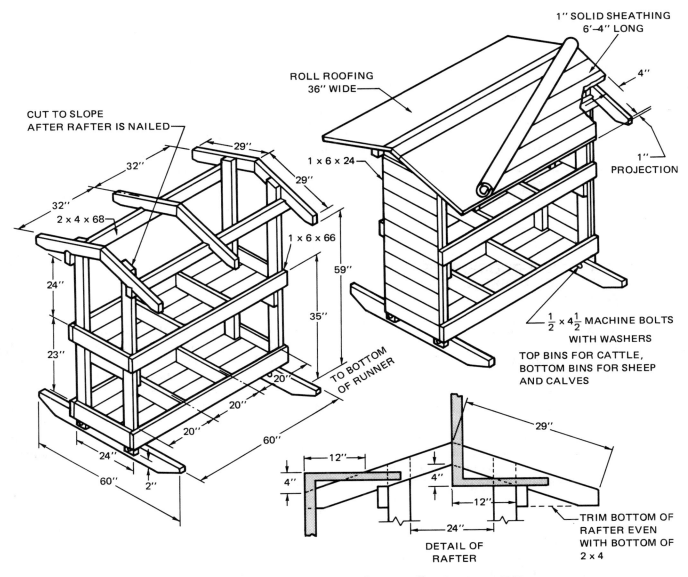

Courtesy of *T. J. Wakeman and V. L. McCoy*, The Farm Shop, *The Macmillan Company, 1960*

BILL OF MATERIALS			
MATERIALS NEEDED	**QUANTITY**	**DIMENSIONS**	**DESCRIPTION OR USE**
Lumber, unsurfaced*	2 pieces	2″ × 6″ × 60″	Runners**
	4 pieces	2″ × 4″ × 72″	Uprights
	6 pieces	2″ × 4″ × 29″	Rafters
	2 pieces	2″ × 4″ × 58″	Rafter supports
	16 pieces	1″ × 6″ × 12′0″ stock	Siding, roof sheathing, trough supports, floors, sides, pieces, partitions
Roofing nails	¼ pound	¾″	
Common nails	½ pound	16d	
Box nails	3 pounds	8d	
Machine bolts	4	½″ × 4½″	
Washers	8	½″	Flat
Mineral roofing	1 piece	13′0″ × 36″	Roll roofing

* Surface pieces if box is to be painted.
** Pressure treated material to prevent rotting is recommended.

□ CONSTRUCTION PROCEDURE FOR SALT AND MINERAL BOX— PORTABLE

1. Cut the two runners to length and shape, as shown in the drawing.
2. Cut the four uprights to length as listed in the bill of materials.
3. Lay the runners down flat, and tack uprights to the sides of the runners. See the drawing for the correct position of the uprights.
4. Square the uprights with the runners, bore holes, and bolt them to the runners with ½″ × 4½″ machine bolts. Use ½″ washers under the bolt heads and nuts. Cut the end pieces (1″ × 6″ × 24″) to length and nail to the outside of the uprights with 8d box nails.
5. Cut the rafter supports (2″ × 4″ × 68″) to length and nail to the uprights with 16d common nails.
6. Cut the two trough supports (1″ × 6″ × 24″) and nail to the inside of the uprights, as shown in the drawing. Cut flooring for the two troughs to length. Fit and nail in place with 8d box nails.
7. Square the uprights with the floor. Cut the 1″ × 6″ side pieces to length, and nail to the uprights and floor. Cut the trough partitions to length and nail to the trough sides, as shown in the drawing. Turn the salt box on its side and nail flooring to trough partitions, using 8d box nails.
8. Lay off, mark, and cut the rafters. (See the detail drawing for the position of the square in laying off the rafter.) After the first rafter is cut, it can be used as a template in marking the remaining rafters.
9. Nail the rafters in place, and saw off the ends of the uprights even with the top of the rafters.
10. Cut the 1″ × 6″ sheathing to length and nail to the rafters with 8d box nails. Cut the mineral roll roofing and nail to the sheathing with ¾″ roofing nails.
11. Apply safe paint or preservative to appropriate wood surfaces. Creosote the runners if pressure treated material is not used.

Project 18: *Cattle Feed Trough*

□ PLAN FOR CATTLE FEED TROUGH

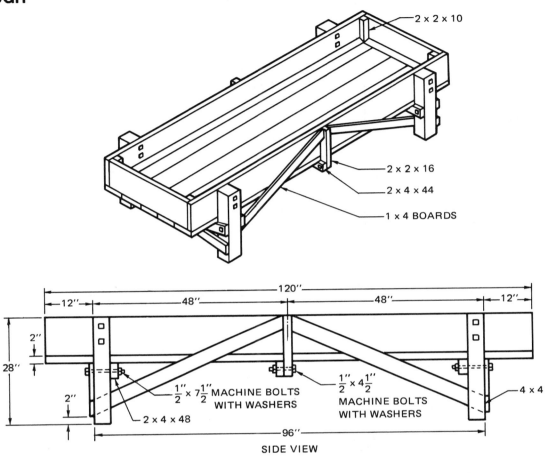

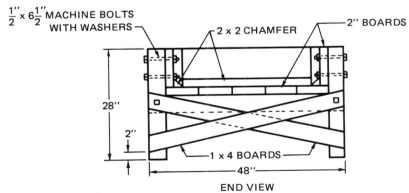

Courtesy of T. J. Wakeman and V. L. McCoy, The Farm Shop, *The Macmillan Company, 1960*

BILL OF MATERIALS			
MATERIALS NEEDED	**QUANTITY**	**DIMENSIONS**	**DESCRIPTION OR USE**
Lumber, unsurfaced	4 pieces	4″ × 4″ × 28″	Legs
	2 pieces	2″ × 4″ × 48″	End trough supports
	1 piece	2″ × 4″ × 44″	Center trough supports
	4 pieces	1″ × 4″ × 51″	End braces
	4 pieces	2″ × 8″ × 10′0″	Trough bottom*
	3 pieces	2″ × 10″ × 10′0″	Trough bottom and sides
	2 pieces	2″ × 10″ × 36″	Trough ends
	4 pieces	2″ × 2″ × 10″	Corner blocks
	2 pieces	2″ × 2″ × 16″	Center side supports
	2 pieces	2″ × 2″ × 9′5″	Chamfer strips
	2 pieces	2″ × 2″ × 33″	Chamfer strips
	4 pieces	1″ × 4″ × 54″	Side braces
Common nails	4 pounds	20d	
Common nails	1 pound	16d	
Common nails	1 pound	8d	
Machine bolts	8	½″ × 6½″	
Machine bolts	4	½″ × 7½″	
Machine bolts	6	½″ × 4½″	
Washers	36	½″	Flat
Creosote	1 gallon		

* Use surfaced or tongue and groove lumber for the bottom, if ground feed is to be fed in the trough.

□ CONSTRUCTION PROCEDURE FOR CATTLE FEED TROUGH

1. Cut the 4″ × 4″ legs to length as shown in the bill of materials.
2. Cut the end trough supports and end braces to length; square and tack them to legs with 20d common nails. See drawings. Place nails where they will not interfere with the bolt holes.
3. Bore holes, and bolt end trough supports and top ends of braces to legs with ½″ × 7½″ machine bolts.
4. Cut the pieces for the trough bottom to length; joint and nail to the trough supports with 20d common nails.
5. Cut the center trough support to length. Nail the trough bottom to the support with 20d common nails.
6. Cut the trough ends and sides to length; bore holes, and bolt sides to legs with ½″ × 6½″ machine bolts. Nail ends in place as shown in the drawings.
7. Turn trough over and nail bottom to side and end pieces with 20d common nails. Turn trough back over.
8. Cut the 2″ × 2″ center side supports to length. Bolt the side supports to the side pieces and to the center trough support with ½″ × 4½″ machine bolts.
9. Cut the four side braces to length. Square the legs with the bottom of the trough, and nail all braces to sides and legs with 8d common nails.
10. Cut the 2″ × 2″ corner blocks to length. Drill nail holes in the corner blocks and nail to end and side pieces with 16d common nails, or use 5⁄16 inch carriage bolts.
11. Cut the chamfer strips to shape and length, and nail in place with 8d common nails.
12. Apply a safe wood preservative.

Project 19: *Portable Range Feeder*

□ **PLAN FOR PORTABLE RANGE FEEDER**

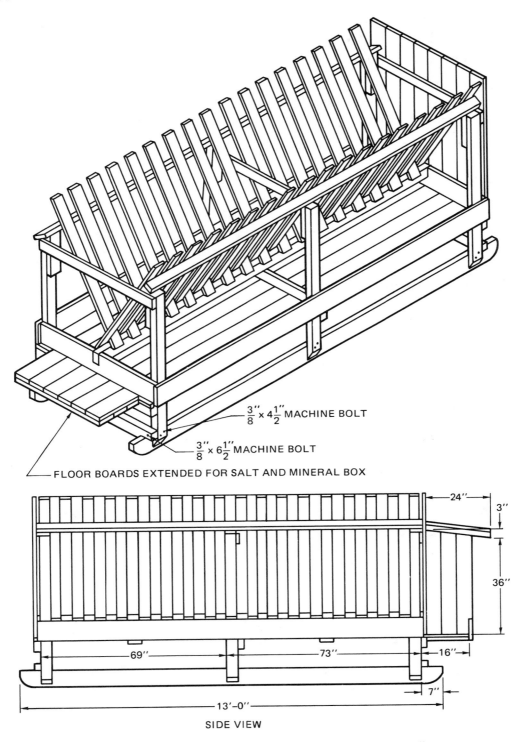

$\frac{3''}{8} \times 4\frac{1}{2}''$ MACHINE BOLT

$\frac{3''}{8} \times 6\frac{1}{2}''$ MACHINE BOLT

FLOOR BOARDS EXTENDED FOR SALT AND MINERAL BOX

24''

3''

36''

69''   73''   16''

7''

13'–0''

SIDE VIEW

Courtesy of T. J. Wakeman and V. L. McCoy, The Farm Shop, *The Macmillan Company, 1960*

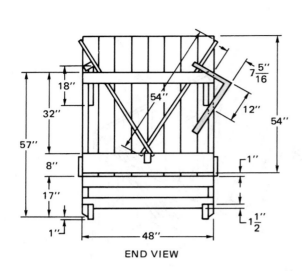

END VIEW

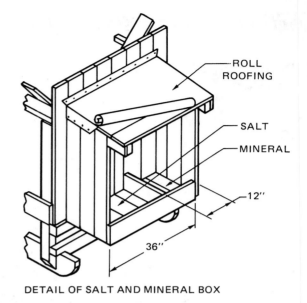

DETAIL OF SALT AND MINERAL BOX

BILL OF MATERIALS			
MATERIALS NEEDED	**QUANTITY**	**DIMENSIONS**	**DESCRIPTION OR USE**
Lumber, unsurfaced	2 pieces	2″ × 6″ × 13′0″	Runners
	6 pieces	2″ × 4″ × 60″	Uprights
	6 pieces	2″ × 4″ × 18″	Upright scabs
	3 pieces	2″ × 4″ × 48″	Trough supports
	2 pieces	2″ × 4″ × 48″	Struts between runners
	2 pieces	2″ × 4″ × 24″	Rafters
	2 pieces	2″ × 4″ × 12′0″	Top slat supports
	1 piece	2″ × 4″ × 12′0″	Bottom slat support
	3 pieces	1″ × 4″ × 48″	Top crosspieces
	1 piece	1″ × 4″ × 48″	Center crosspiece
	2 pieces	1″ × 4″ × 6″	Rafter filler block
	2 pieces	1″ × 6″ × 12′2″	Outside floor pieces
	6 pieces	1″ × 6″ × 13′6″	Inside floor pieces
	16 pieces	1″ × 6″ × 54″	End boards, vertical
	4 pieces	1″ × 6″ × 38″	Sheathing
	2 pieces	1″ × 8″ × 12′0″	Trough sides
	2 pieces	1″ × 8″ × 48″	Trough ends
	35 pieces	1″ × 4″ × 54″	Slats
	2 pieces	1″ × 8″ × 36″	Salt box subfloor
	1 piece	1″ × 8″ × 36″	Salt box end
	6 pieces	1″ × 6″ × 42″	Siding; salt box partitions
Machine bolts	12	⅜″ × 4½″	
Machine bolts	4	⅜″ × 6½″	
Washers	32	⅜″	Flat
Common nails	4 pounds	8d	
Common nails	2 pounds	20d	
Roofing nails	1 pound	¾″	
Mineral roofing	1 piece	48″ × 36″	Roll roofing
Preservative	2 gallon		

□ CONSTRUCTION PROCEDURE FOR PORTABLE RANGE FEEDER

1. Cut the runners to length and shape, as shown in the drawings.
2. Cut the uprights to length, using 7⁵⁄₁₆″ on the tongue of a framing square and 12″ on the body for marking cuts, as shown in the end view.
3. Space one upright 7″ from each end and one in the center of the runners, and tack in position. Be sure uprights are at right angles to runners. Bore holes, and bolt each upright to runners with two ⅜″ × 4½″ machine bolts.
4. Cut upright scabs, and nail to the inside top of the uprights.
5. Cut, bore holes, and bolt struts to the corner uprights with ⅜″ × 6½″ machine bolts. Be sure the notches fit right between the runners.
6. Cut trough supports, and nail to corner uprights with 20d common nails, as shown in the drawing.
7. Cut the three top crosspieces, and nail to uprights as shown.
8. Cut top slat supports 12′0″ long.
9. Place top slat supports on top of the uprights, and nail in place, as shown in the drawing. The support should project 1″ beyond the uprights at each end.
10. Cut the two outside floor pieces long enough (approximately 12′2″ long) to fit even with the outside edge of the trough supports. Notch one edge of each piece to fit around the 2″ × 4″ upright pieces, and nail to trough supports.
11. Cut the inside floor pieces long enough to extend 16″ past one end of the trough support to form the bottom for the salt box. Fit the boards tightly together, and nail in place.
12. Cut the trough ends and center crosspiece, and nail in place.
13. Cut the trough side pieces the same length as the outside bottom pieces, and nail in place.
14. Cut a notch in the center of the trough ends to fit the 2″ × 4″ bottom slat support.
15. Cut the bottom slat support, and nail in position, as shown in the drawing.
16. Cut 35 slats 54″ long.
17. Place the first slat tight against the trough end and top crosspiece, and fasten with 8d common nails. The bottom end of the slat should project 2″ below the top corner of the bottom slat support. Fasten the remaining slats to alternate sides with 8d common nails as shown in the drawing. Be sure the slats are spaced the same distance (4″) apart at the top and bottom.
18. Cut end boards 54″ long, and nail in place.
19. Cut subflooring for the salt box, and nail in place.
20. Mark the 2″ × 4″ rafters, using 3″ and 24″ on a framing square. Cut rafters, and nail in place, as shown in the detail of the salt and mineral box.
21. Cut and fasten siding to rafters and floor with 8d common nails.
22. Cut and fasten sheathing to rafters with 8d common nails.
23. Cut and nail the rafter filler blocks in position, as shown.
24. Nail roofing to sheathing.
25. Cut the salt box end, and nail to the floor and bottom edge of the side pieces.
26. Cut the salt box partition and fasten in place, as shown in the drawing.
27. Apply a suitable preservative to all surfaces.

Project 20: Tool Sharpening Gauge

□ PLAN FOR TOOL SHARPENING GAUGE

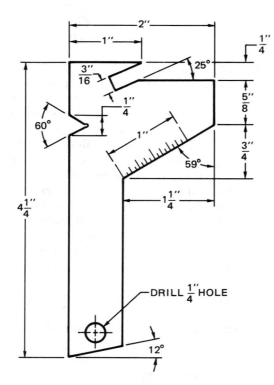

GAUGES
- PLANE IRON OR WOOD CHISEL
- COLD CHISEL OR CENTER PUNCH
- TWIST DRILL, CUTTING EDGE ANGLE AND LENGTH
- TWIST DRILL, LIP CLEARANCE

Courtesy of Department of Agricultural and Extension Education, The Pennsylvania State University, University Park, PA

BILL OF MATERIALS			
MATERIALS NEEDED	**QUANTITY**	**DIMENSIONS**	**DESCRIPTION**
Sheet metal, galvanized, 24 gauge	1	2″ × 4¼″	

□ CONSTRUCTION PROCEDURE FOR TOOL SHARPENING GAUGE

1. Measure and scribe outline on the stock using a scratch awl.
2. Cut out tool gauge with snips.
3. Use snips and a flat file to cut the 25° wood chisel slot.
4. Use a taper file to cut the 60° cold chisel vee.
5. Position and drill ¼″ hole at the narrow end of the gauge.
6. Measure and scribe a 1″ rule by ¹⁄₁₆″ graduations.
7. Cut rule indicator marks with a cold chisel or awl.
8. Smooth all edges with a flat file.

Project 21: *Feed Scoop*

□ PLAN FOR FEED SCOOP

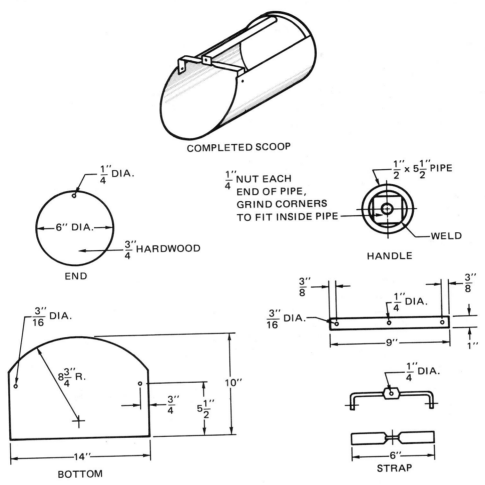

COMPLETED SCOOP

END

HANDLE

BOTTOM

STRAP

Courtesy of Department of Agricultural and Extension Education, The Pennsylvania State University, University Park, PA

BILL OF MATERIALS			
MATERIALS NEEDED	**QUANTITY**	**DIMENSIONS**	**DESCRIPTION OR USE**
Sheet metal, galvanized, 24 gauge	1	10″ × 14″	Bottom
Lumber, hardwood	1	¾″ × 6″ × 6″	End
Mild steel, hot rolled	1	⅛″ × 1″ × 9″	Strap
Steel tubing	1	½″ I.D. × 5½″	Handle
Hardware			
Stove bolt	1	¼″ × 6½″	
Nut	2	¼″	
Rivet	2	³⁄₁₆″	
Screw	6	#6 × ¾″	Pan head
Paint (nonlead)	½ pint		For end, handle, and hand bracket

□ CONSTRUCTION PROCEDURE FOR FEED SCOOP

1. Lay out and cut the sheet metal and wood end.
2. Cut and bore holes in the strap iron, ⅛″ × 1″ × 9″.
3. Cut the light steel tubing, ½″ × 5½″ long for the handle. **Note:** a ¾″ dowel may be used instead of steel tubing.
4. Bend the bottom sheet metal to fit the end.
5. Drill holes for screws and handle bolt.
6. Screw the metal to the wood end. **Note:** Screwshank or other improved nails may be used in place of screws.
7. Heat and bend the iron strap.
8. Attach the iron strap to the sheet metal using rivets.
9. Weld nuts into ends of the tubing. **Note:** Before welding, run the bolt through both nuts to check alignment.
10. Attach the handle with ¼″ × 6½″ bolt.
11. Finish and paint the handle, bracket, and wood end.

Project 22: Staple

□ PLAN FOR STAPLE

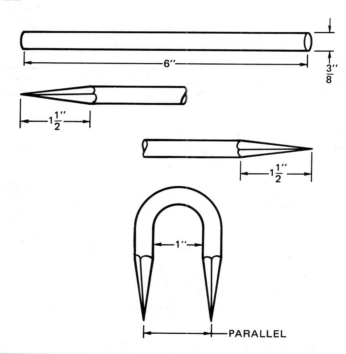

BILL OF MATERIALS			
MATERIALS NEEDED	**QUANTITY**	**DIMENSIONS**	**DESCRIPTION**
Steel, round hot or cold rolled	1	⅜″ × 6″	

□ CONSTRUCTION PROCEDURE FOR STAPLE

1. Cut a piece of ⅜″ round rod 6″ long.
2. Heat and put a long round point on one end.
3. Heat and put a long square point on the other end.

4. Bend to form a staple:
   a. One inch width between staple legs
   b. Legs of equal length

Project 23:   Cold Chisel

□ PLAN FOR COLD CHISEL

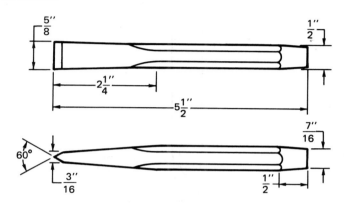

Courtesy of Department of Agricultural and Extension Education, The Pennsylvania State University, University Park, PA

BILL OF MATERIALS			
MATERIALS NEEDED	**QUANTITY**	**DIMENSIONS**	**DESCRIPTION**
Tool steel	1	½″ × 5½″	Octagonal

□ CONSTRUCTION PROCEDURE FOR COLD CHISEL

1. Heat 2¼ inches of one end of the stock to a uniform cherry red color.
2. Place one side against the anvil face. Using drawing blows, work to shape rapidly starting at the end and working back to 2¼″ taper.
3. Finish to ³⁄₁₆″ thickness and ⅝″ width at the tip. Keep area hot (from a dull red to a cherry red color) when working it.
4. Anneal by heating to cherry red and cooling slowly (12 to 24 hours) in lime or sand.
5. File and polish the forged faces. Do not grind.
6. Temper with water: (Practice on an old cold chisel)
   a. Heat 2″ to 3″ of the tip to a uniform cherry red color.
   b. Cool ¾″ to 1″ in water until drops cling to the tip when it is removed from the water.
   c. Move the tip to avoid cracks at the water line.
   d. Quickly remove scale using a steel brush or file.
   e. Observe the color changes and quench lower ¼″ on purple color. Color order is light straw, dark straw, brown, purple, dark blue and light blue.
7. Grind the cutting edge to a 60° angle. Use a tool gauge to check the angle.
8. Chamfer the opposite end to approximately ½″ by ⁷⁄₁₆″ to prevent mushrooming.

Project 24: *Drawbar Hitch Pin*

☐ PLAN FOR DRAWBAR HITCH PIN

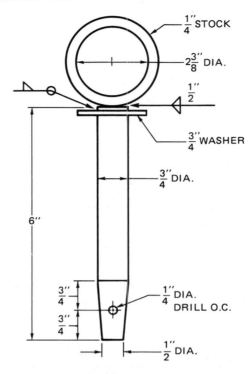

Courtesy of Department of Agricultural and Extension Education, The Pennsylvania State University, University Park, PA

BILL OF MATERIALS			
MATERIALS NEEDED	**QUANTITY**	**DIMENSIONS**	**DESCRIPTION OR USE**
Hot rolled steel, M1020	1	¼″ × 8″	Handle
C1042 cold rolled or C1045 hot rolled steel, round	1	¾″ × 6″	Pin
Washer, flat	1	¾″	
Hair pin	1	¼″ × 2″	Safety pin

☐ CONSTRUCTION PROCEDURE FOR DRAWBAR HITCH PIN

1. Bend ¼″ × 8″ round stock around 2″ pipe, using a vise as a bending aid.
2. Cut the pin to length.
3. Drill a ¼ inch hole ¾ inch from one end of the pin.
4. Shape the end as indicated in the plan.
5. Weld the washer to the other end of the pin.
6. Place handle on pin and weld.
7. Remove slag and clean with a steel brush.

Project 25: Hay Hook

☐ PLAN FOR HAY HOOK

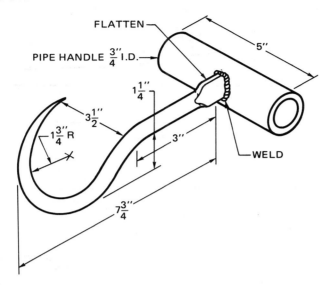

FLATTEN

PIPE HANDLE $\frac{3}{4}''$ I.D.

5''

$1\frac{1}{4}''$

$3\frac{1}{2}''$

$1\frac{3}{4}''$ R

3''

WELD

$7\frac{3}{4}''$

Courtesy of Texas Vocational Instructional Services, Texas A & M University, College Station, TX

BILL OF MATERIALS			
MATERIALS NEEDED	**QUANTITY**	**DIMENSIONS**	**DESCRIPTION OR USE**
Pipe, steel	1	¾″ ID × 5″	Handle
Rod	1	⁷⁄₁₆″ dia × 12″	Hook

☐ CONSTRUCTION PROCEDURE FOR HAY HOOK

1. Heat one end of the rod and forge it into a long slender point.
2. Finish shaping the point with a grinder.
3. Reheat the pointed end and shape it to the dimensions given in the plan drawing.
4. Heat the other end and forge a flat area about ¼″ thick and in line with the curved section.
5. Cut a piece of ¾″ ID steel pipe 5″ long.
6. Weld the hook to the center of the handle being careful to keep the hook square to the handle.
7. Grind and sand all surfaces smooth and round for a comfortable feel.

Project 26: Cant Hook

☐ PLAN FOR CANT HOOK

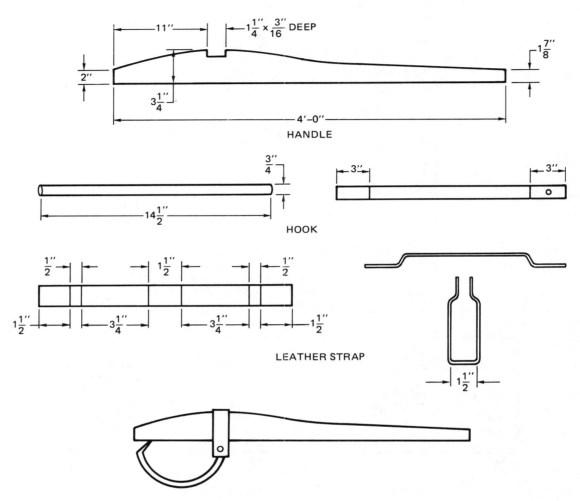

Courtesy of Douglas Hering

BILL OF MATERIALS			
MATERIALS NEEDED	**QUANTITY**	**DIMENSIONS**	**DESCRIPTION OR USE**
Lumber, oak	1 piece	2″ × 4″ × 4′	Handle
Round stock	1	¾″ dia × 14½″	Hook
Metal	1 piece	3⁄16″ × 1¼″ × 12″	Strap
Bolt, hex head	1	3⁄8″ × 1½″	
Nut, hex	2	3⁄8″	

□ CONSTRUCTION PROCEDURE FOR CANT HOOK

A. THE HANDLE

1. Plane the 48″ oak 2 × 4 to a thickness of 1½″.
2. Joint the 2 × 4 to a width of 3½″.
3. Cut one side of the 2 × 4 on a band saw to make the handle. Be sure to leave 2″ of width at the bottom, and at least 1⅞″ at the top. It is also important to leave 3¼″ of width 12 inches from the bottom to support the strap.
4. Notch the curved side of the handle to receive the strap.
5. Chamfer all edges to the desired roundness. The straight side can be jointed ¼″ at 45 degrees.

B. THE HOOK

1. Cut a piece of ¾″ round stock 14½″ long.
2. Place one end of the round stock in a gas forge and heat to a cherry red color.
3. Flatten the end of the round stock as if making a cold chisel. Stop when the end is tapered back about 3″ and is ⅜″ thick on the end. This will be the hinged end of the cant hook. Drill a ⅜″ hole after rounding the end on the grinder.
4. Heat the opposite end in the gas forge and shape it in a similar way *perpendicular* to the other end.
5. Flatten this end to ¼″, cool and sharpen like a cold chisel.
6. Reheat the sharp end until cherry red.
7. Place 1½″ of the point into a vice and bend the remainder over at an 80 degree angle forming the hook.

8. Reheat the round stock and shape the entire piece to the desired curvature.

C. THE STRAP

1. Obtain a piece of ³⁄₁₆″ × 1¼″ strap 12″ long. Mark off the dimensions with soap stone as shown.
2. Using an oxyacetylene unit, a vise and a blacksmith's hammer, bend the two ends into the shape shown (procedure optional).
3. Bend the strap into a "U" with the 1½″ dimension in the center.
4. From this point, reheating and reshaping may be necessary to fit different handles. Custom fit each strap before drilling the ⅜″ holes to receive the bolt which hinges the hook.
5. The strap should fit neatly into the ³⁄₁₆″ × 1¼″ notch in the handle.
6. Put the hook on the straight side of the handle.

D. THE ASSEMBLY

1. Assemble by putting the strap and hook in place.
2. Using a ⅜″ bolt, install a hex nut and lock washer and tighten until the desired control of the hook is achieved.
3. Saw off the excess end of the bolt and hammer it slightly to keep the nut from coming off during use.
4. If the strap is not tight on the handle, drill a ¼″ hole through the strap and handle and install a flat headed rivet or bolt to hold the strap securely to the handle.

Project 27: Round Metal Jig

□ PLAN FOR ROUND METAL JIG

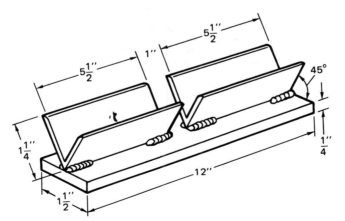

Courtesy of T. J. Wakeman and V. L. McCoy, The Farm Shop, *The Macmillan Company, 1960*

BILL OF MATERIALS			
MATERIALS NEEDED	QUANTITY	DIMENSIONS	DESCRIPTION OR USE
Angle iron	2 pieces	$\frac{3}{16}$" × 1$\frac{1}{4}$" × 1$\frac{1}{4}$" × 5$\frac{1}{2}$"	Trough
Flat iron	1 piece	$\frac{1}{4}$" × 1$\frac{1}{2}$" × 12"	Base

□ CONSTRUCTION PROCEDURE FOR ROUND METAL JIG

1. Cut the materials to length, as listed in the bill of materials.
2. Place both pieces of angle iron on top of the flat iron, as shown. Place a piece of round iron or angle iron in the V and clamp it to the flat iron to hold the two pieces of angle iron in line while welding.
3. Electric weld the angle iron to the base by welding on both sides at each end of both pieces.

Project 28: Foot Scraper

□ PLAN FOR FOOT SCRAPER

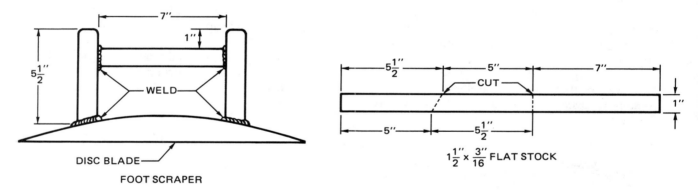

FOOT SCRAPER

1$\frac{1}{2}$" × $\frac{3}{16}$" FLAT STOCK

Courtesy of Vocational Agriculture Service, University of Illinois, Urbana, IL

BILL OF MATERIALS			
MATERIALS NEEDED	QUANTITY	DIMENSIONS	DESCRIPTION
Disc blade	1	15" diameter	Base
Band, steel	2	$\frac{3}{16}$" × 1$\frac{1}{2}$" × 5$\frac{1}{2}$"	Legs
	1	$\frac{3}{16}$" × 1$\frac{1}{2}$" × 7"	Scraper

□ CONSTRUCTION PROCEDURE FOR FOOT SCRAPER

1. Mark and cut the three pieces of flat stock with a hacksaw or torch.
2. Square the ends of the 7″ piece and round the upper ends of the upright pieces on grinder.
3. Place the scraper pieces on the edge of the welding table with the angled ends extending over the edge.
4. Clamp firmly and weld the cross piece to the uprights.

5. Fit the angled ends of the uprights to the contour of the disc blade by grinding if necessary.
6. Clean the disc blade with a wire brush or portable grinder.
7. Weld the scraper to the disc blade carefully. Use low heat. Since the disc blade is high-carbon steel, either preheat or use a low-hydrogen electrode.

Project 29: Small Engine Stand—Metal

□ PLAN FOR SMALL ENGINE STAND—METAL

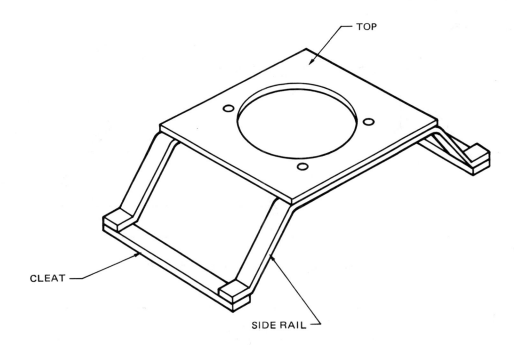

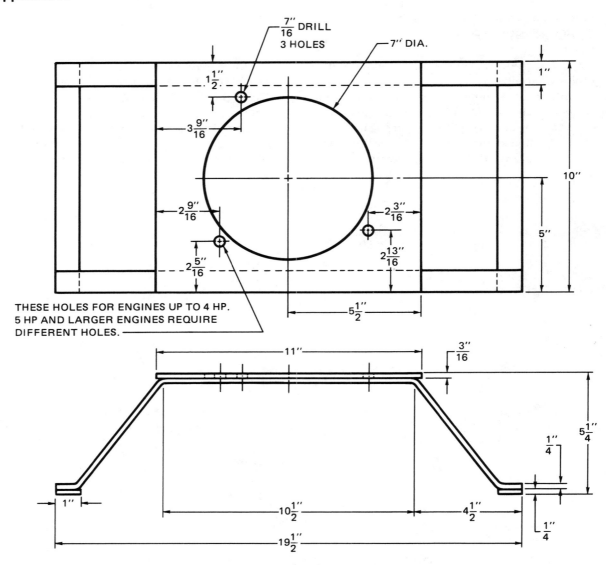

Courtesy of Agricultural Education and Agricultural Engineering, Virginia Polytechnic Institute and State University, Blacksburg, VA

BILL OF MATERIALS			
MATERIALS NEEDED	**QUANTITY**	**DIMENSIONS**	**DESCRIPTION**
Band, steel	2	¼" × 1" × 24"	Side rails
Band, steel	2	¼" × 1" × 10"	Cleats
Plate, steel	1	³⁄₁₆" × 10" × 11"	Top

□ CONSTRUCTION PROCEDURE FOR SMALL ENGINE STAND— METAL

1. Locate and mark the midpoint of the two side pieces.
2. Measure and mark points 5½" in both directions from the midpoint.
3. Measure and mark points 1" from both ends of the side pieces.
4. Bend the side pieces so they are 19½" from end to end and 4¹³⁄₁₆" from the bottom to the upper

surface. Place them side by side and be sure their shapes are identical.

5. Weld the cleats to the sides at 90° angles.
6. Lay out and center punch for all holes in the top.
7. Drill the ⁷⁄₁₆" holes.

8. Use a divider to lay out the 7" diameter hole.
9. Drill a hole inside the 7" circle to insert a metal cutting bayonet saw blade and cut out the circle; or, use a cutting torch to make the cut.
10. Weld the top to the sides.
11. Apply a suitable finish.

□ PLAN FOR SHOP STAND—ADJUSTABLE

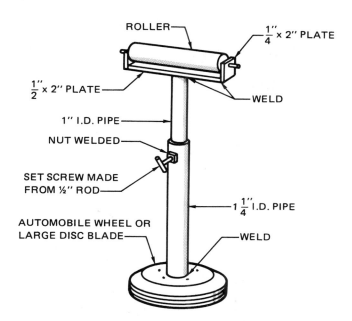

BILL OF MATERIALS			
MATERIALS NEEDED	**QUANTITY**	**DIMENSIONS**	**DESCRIPTION**
Steel automobile or truck rim	1	According to need	Base
Pipe, black steel	1	1¼" ID × *A	External post
Pipe, black steel	1	1" ID × *B	Internal post
Band, steel	1	½" × 2" × *C	Base plate
Band, steel	2	¼" × 2" × 2"	Bearing ends
Wood or metal or rubber	1	Approx. 2" dia × 12"	Roller

*A—Determined by the desired minimum height of the stand minus the height of the base and roller.
*B—Same length as A.
*C—Same length as the roller plus ¼".
Note: Standard 1" ID pipe will fit well in standard 1¼" ID pipe. 2" ID pipe will fit well in 2½" ID pipe.

☐ CONSTRUCTION PROCEDURE FOR SHOP STAND— ADJUSTABLE

1. Obtain a rim of suitable size and weight to support the height of the stand needed. A large steel disc or other heavy metal object will also work well.
2. Determine the length of pipes needed to provide the desired height of the stand.
3. Drill a ⅝" hole 1" from one end of the external post. Weld a ½" nut over the hole. Make a setscrew from ½" cold rolled round stock or use a standard bolt with a cross handle welded to it.

4. Weld the external post to the rim.
5. Obtain a standard roller or make one from 2" wood stock or 1½" ID pipe 12" long.
6. Cut the base plate ¼" longer than the roller.
7. Lay out and drill bearing holes in the bearing ends to hold the top of the roller slightly above the tops of the ends.
8. Weld the bearing ends with the roller in place.
9. Apply a suitable finish.

Project 31: Steel Post Driver

☐ PLAN FOR STEEL POST DRIVER

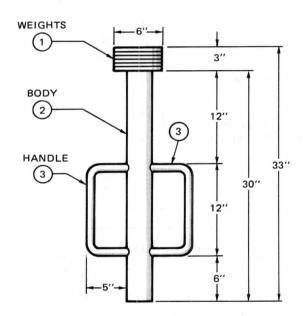

WEIGHTS ①

BODY ②

HANDLE ③

Courtesy of Texas Vocational Instructional Services, Texas A & M University, College Station, TX

BILL OF MATERIALS			
MATERIALS NEEDED	**QUANTITY**	**DIMENSIONS**	**DESCRIPTION**
Flat or plate steel	6	½" × 6" × 6" or 6" dia	Weights
Pipe	1	3" ID × 30"	Body
Round Stock	2	½" × 20"	Handles

□ CONSTRUCTION PROCEDURE FOR STEEL POST DRIVER

1. Cut six pieces for weights and weld them to form a mass 3″ thick × 6″ × 6″, or 6″ in diameter
2. Weld the weight mass securely to the body.
3. Bend the two handles so they are identical.
4. Weld the handles to the body.
5. Apply a suitable finish.

Project 32: Extension Cord Stand

□ PLAN FOR EXTENSION CORD STAND

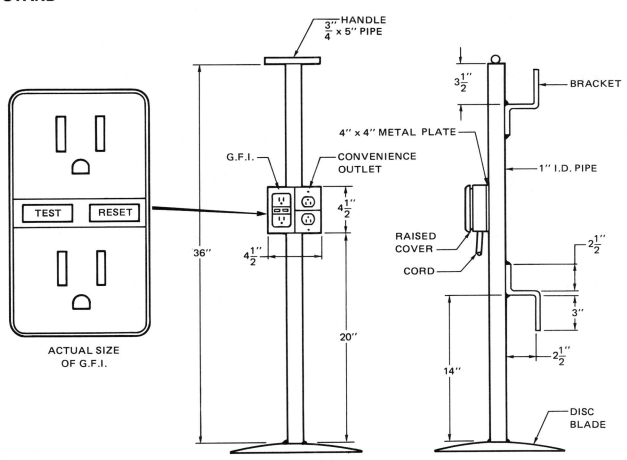

Courtesy of Department of Agricultural and Extension Education, The Pennsylvania State University, University Park, PA

BILL OF MATERIALS			
MATERIALS NEEDED	QUANTITY	DIMENSIONS	DESCRIPTION OR USE
Pipe, black iron	1	1″ ID × 36″	Upright
Pipe, black iron	1	¾″ × 5″	Handle
Band, iron	2	⅛″ × ¾″ × 8″	Brackets
Disc blade	1	12″ dia	Base
Plate, steel	1	⅛″ × 4″ × 4″	Outlet box base
Outlet box, square	1	4″ square, 1½″ deep	
*Ground fault interrupter	1		
Duplex receptacle	1		
Grounded cap, male plug	1		
Raised cover	1		
Box connector	1	¾″	
Solderless connectors	3		
Cable	100 feet		Neoprene or plastic covered portable cable, AWG 12–2 stranded conductor with ground

* Substitute a duplex receptable if the circuit is already protected by a ground fault interrupter.

□ CONSTRUCTION PROCEDURE FOR EXTENSION CORD STAND

1. Cut the handle 5″ long.
2. Cut the bracket pieces 8″ long.
3. Shape the brackets to hold the electrical cable.
4. Cut 1″ pipe to 36″ length.
5. Weld the pipe to the disc blade, using a low hydrogen electrode.
6. Weld the handle and brackets to the pipe.
7. Drill ⅛″ × 4″ × 4″ steel plate for electrical boxes; weld the plate to the pipe.
8. Bolt the electrical box to the plate.
9. Install one end of conductor into the box.
10. Install the ground fault interrupter (GFI).
11. Install the duplex receptacle so it is protected by the GFI.
12. Add the cover plate.
13. Connect the male plug to the electrical conductor.
14. Apply a suitable finish.

Project 33: Jack Stand—Automobile

☐ PLAN FOR JACK STAND— AUTOMOBILE

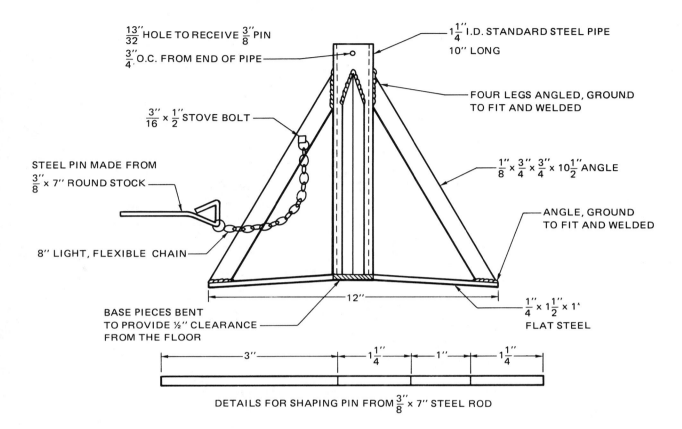

$\frac{13}{32}''$ HOLE TO RECEIVE $\frac{3}{8}''$ PIN

$\frac{3}{4}''$ O.C. FROM END OF PIPE

$1\frac{1}{4}''$ I.D. STANDARD STEEL PIPE 10'' LONG

$\frac{3}{16}'' \times \frac{1}{2}''$ STOVE BOLT

FOUR LEGS ANGLED, GROUND TO FIT AND WELDED

STEEL PIN MADE FROM $\frac{3}{8}'' \times 7''$ ROUND STOCK

$\frac{1}{8}'' \times \frac{3}{4}'' \times \frac{3}{4}'' \times 10\frac{1}{2}''$ ANGLE

ANGLE, GROUND TO FIT AND WELDED

8'' LIGHT, FLEXIBLE CHAIN

BASE PIECES BENT TO PROVIDE ½'' CLEARANCE FROM THE FLOOR

12''

$\frac{1}{4}'' \times 1\frac{1}{2}'' \times 1''$ FLAT STEEL

3'' $1\frac{1}{4}''$ 1'' $1\frac{1}{4}''$

DETAILS FOR SHAPING PIN FROM $\frac{3}{8}'' \times 7''$ STEEL ROD

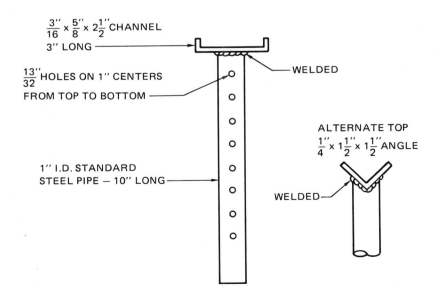

$\frac{3}{16}'' \times \frac{5}{8}'' \times 2\frac{1}{2}''$ CHANNEL 3'' LONG

WELDED

$\frac{13}{32}''$ HOLES ON 1'' CENTERS FROM TOP TO BOTTOM

ALTERNATE TOP $\frac{1}{4}'' \times 1\frac{1}{2}'' \times 1\frac{1}{2}''$ ANGLE

1'' I.D. STANDARD STEEL PIPE – 10'' LONG

WELDED

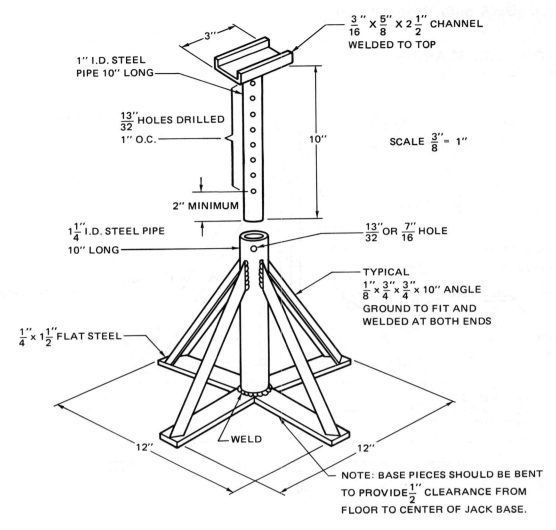

$\frac{3}{16}$" × $\frac{5}{8}$" × 2$\frac{1}{2}$" CHANNEL
WELDED TO TOP

1" I.D. STEEL
PIPE 10" LONG

$\frac{13}{32}$" HOLES DRILLED
1" O.C.

SCALE $\frac{3}{8}$" = 1"

10"

2" MINIMUM

1$\frac{1}{4}$" I.D. STEEL PIPE
10" LONG

$\frac{13}{32}$" OR $\frac{7}{16}$" HOLE

TYPICAL
$\frac{1}{8}$" × $\frac{3}{4}$" × $\frac{3}{4}$" × 10" ANGLE
GROUND TO FIT AND
WELDED AT BOTH ENDS

$\frac{1}{4}$" × 1$\frac{1}{2}$" FLAT STEEL

12"

WELD

12"

NOTE: BASE PIECES SHOULD BE BENT
TO PROVIDE $\frac{1}{2}$" CLEARANCE FROM
FLOOR TO CENTER OF JACK BASE.

BILL OF MATERIALS			
MATERIALS NEEDED	**QUANTITY**	**DIMENSIONS**	**DESCRIPTION**
Band, steel	1	$\frac{1}{4}$" × 1$\frac{1}{2}$" × 12"	Base
Band, steel	2	$\frac{1}{4}$" × 1$\frac{1}{2}$" × 5$\frac{1}{4}$"	Base
Angle, steel	4	$\frac{1}{8}$" × $\frac{3}{4}$" × $\frac{3}{4}$" × 10"	Legs
Pipe, steel	1	1$\frac{1}{4}$" ID × 10"	External post
Pipe, steel	1	1" ID × 10"	Internal post
Channel, steel	1	$\frac{3}{16}$" × $\frac{5}{8}$" × 2$\frac{1}{2}$" × 3"	Top
Round stock	1	$\frac{3}{8}$" dia × 7"	Pin, hardened
Chain	1	Light weight, 8"	Pin chain
Bolt, stove	1	$\frac{3}{16}$" × $\frac{1}{2}$"	Chain fastener

□ CONSTRUCTION PROCEDURE FOR JACK STAND— AUTOMOBILE

1. Cut all pieces to length.
2. Weld two short base pieces to the longer piece to form a 90° cross. Grind welds smooth and slightly round off the exposed ends.
3. Bend the pieces of the base slightly so there will be ½″ clearance between the floor and the center of the cross.
4. Ream the inside of the ends of the external post.
5. Center punch ¾″ from one end of the external post.
6. Scribe a mark from end to end down the center axis of the interior post.
7. Center punch at 1″ intervals on the line from end to end of the interior post.
8. Place the exterior post in a drill press vise or clamp it in a vee block and drill a $^{13}/_{32}$″ hole through both walls of the pipe.
9. Without unclamping the exterior post, slide the interior post into the exterior post and run the drill through the exterior post and on through both walls of the interior post. Drill all holes in the interior post in like fashion.

10. Weld the exterior post to the center of the base so the assembly sits with the post perpendicular to the floor.
11. Grind the legs so they fit the post and base. Tack weld the legs to the base and to the post, being careful to keep the parts and spacings equal and the post perpendicular. Weld all joints completely.
12. Weld the top to the interior post.
13. Bend a loop in the pin and attach the chain to the pin.
14. Drill a $^3/_{16}$″ hole 1½″ from the top of one leg and anchor the chain with a $^3/_{16}$″ × ½″ stove bolt.
15. Chip all welds, remove all burrs.
16. Apply a suitable finish.
    Note: All welds in this project must be high quality and a full-size pin must pass through all four pipe walls when stand is in use. For added safety, insert a hardened bolt in the first hole above the external post.

Caution:

Maximum capacity is one-half ton.

Project 34: Jack Stand—Tractor

□ PLAN FOR JACK STAND— TRACTOR

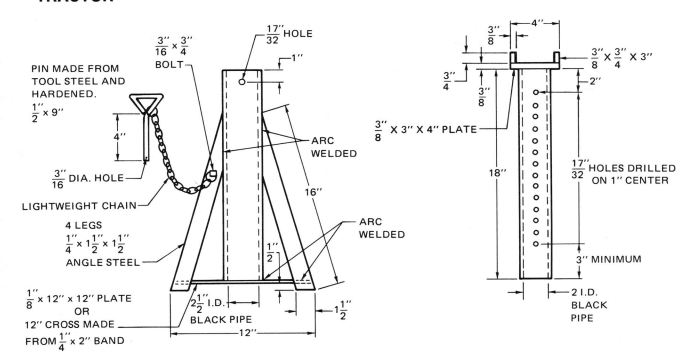

BILL OF MATERIALS			
MATERIALS NEEDED	**QUANTITY**	**DIMENSIONS**	**DESCRIPTION**
Band, steel	1	¼″ × 2″ × 15″	Base
Band, steel	2	¼″ × 2″ × 6½″	Base
Angle, steel	4	¼″ × 1½″ × 1½″ × 16″	Legs
Pipe, steel	1	2½″ ID × 18″	External post
Pipe, steel	1	2″ ID × 18″	Internal post
Plate	1	⅜″ × 3″ × 4″	Top base
Plate	2	⅜″ × ¾″ × 3″	Top rails
Round stock	1	½″ dia × 9″	Pin, hardened
Chain	1	Light weight, 12″	Pin chain
Bolt, stove	1	³⁄₁₆″ × ¾″	Chain fastener

☐ CONSTRUCTION PROCEDURE FOR JACK STAND—TRACTOR

1. Cut all pieces to length.
2. Make a cross by welding 2 pieces of ¼″ × 6½″ band to a piece of ¼″ × 15″ band to form a 90° cross. Grind the welds smooth.
3. Cut all four ends with two 45° angles so the ends will fit inside the legs.
4. Ream the inside of the ends of the external post.
5. Center punch 1″ from one end of the external post.
6. Scribe a mark from end to end down the center axis of the interior post.
7. Center punch 2″ from one end and at 1″ intervals on the line from that end to a point 3″ from the other end of the interior post.
8. Place the exterior post in a drill press vise or clamp it in a vee block and drill the ¹⁷⁄₃₂″ hole through both walls of the pipe.
9. Without unclamping the exterior post, slide the interior post into the exterior post and run the drill through the exterior post and on through both walls of the interior post. Drill all holes in the interior post in like fashion.
10. Weld the exterior post to the center of the base so the assembly sits with the post perpendicular to the floor.
11. Grind the legs so they fit the post. Tack weld the legs to the base and to the post, being careful to keep the parts and spacing equal and the post perpendicular. Weld all joints completely.
12. Weld the top on the interior post.
13. Bend a loop in the pin, harden the pin, and attach the chain to the pin.
14. Drill a ³⁄₁₆″ hole 1½″ from the top of one leg and anchor the chain with a ³⁄₁₆″ × ¾″ stove bolt.
15. Chip all welds, remove all burrs.
16. Apply a suitable finish.

Note: All welds in this project must be high quality and a full-size pin must pass through all four pipe walls when stand is in use. For added safety, insert a second ½″ hardened bolt in the first hole above the external post.

Caution:

Maximum capacity is one ton.

Project 35: Car Ramp

□ PLAN FOR CAR RAMP

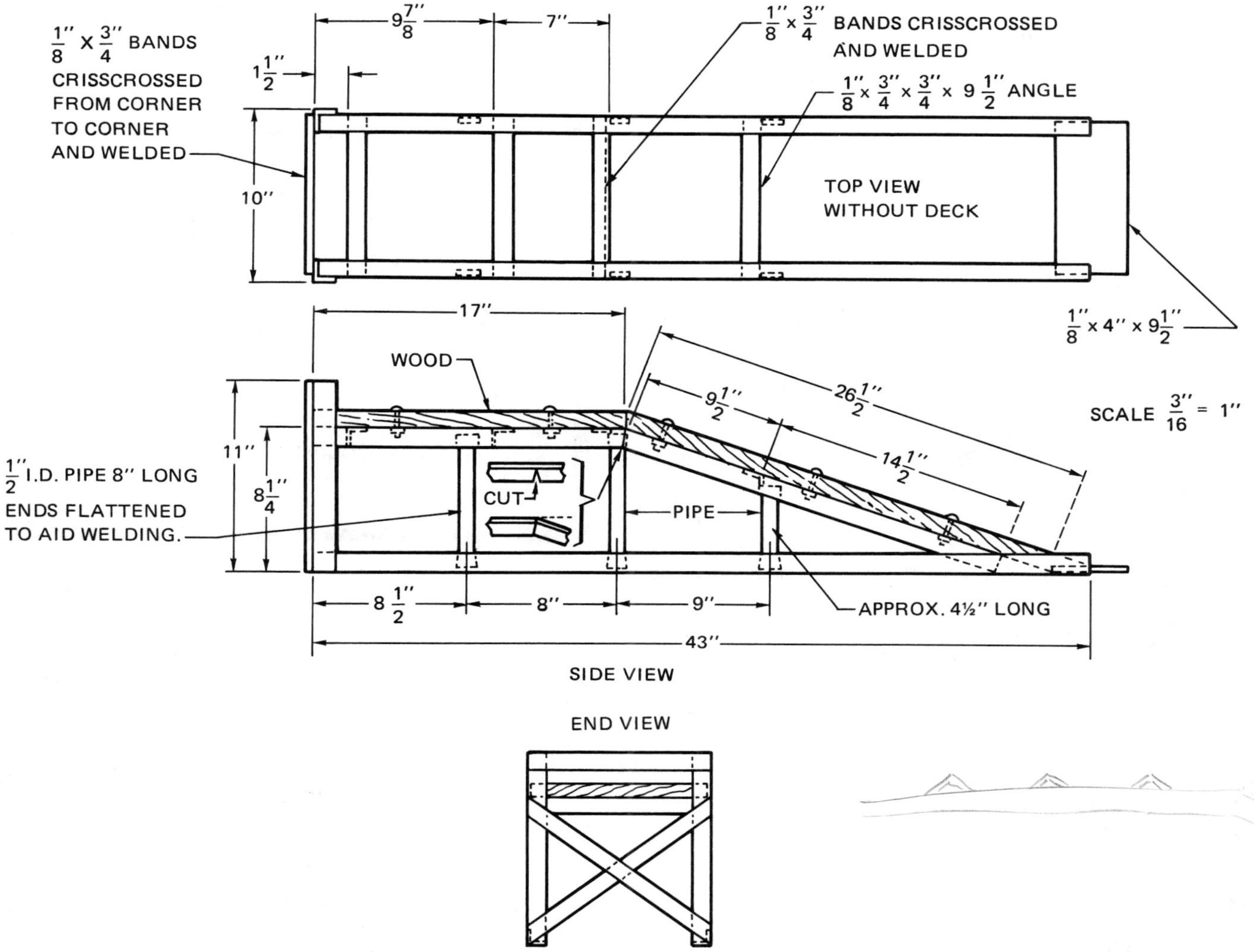

1/8" X 3/4" BANDS CRISSCROSSED FROM CORNER TO CORNER AND WELDED
9 7/8"
7"
1 1/2"
1/8" x 3/4" BANDS CRISSCROSSED AND WELDED
1/8" x 3/4" x 3/4" x 9 1/2" ANGLE
10"
TOP VIEW WITHOUT DECK
1/8" x 4" x 9 1/2"
17"
WOOD
9 1/2"
26 1/2"
SCALE 3/16" = 1"
11"
8 1/4"
14 1/2"
1/2" I.D. PIPE 8" LONG ENDS FLATTENED TO AID WELDING.
CUT
PIPE
8 1/2"
8"
9"
APPROX. 4½" LONG
43"
SIDE VIEW
END VIEW

BILL OF MATERIALS			
MATERIALS NEEDED	**QUANTITY**	**DIMENSIONS**	**DESCRIPTION**
Angle, steel	2	⅛" × ¾" × ¾" × 43"	Bottom runners
Angle, steel	2	⅛" × ¾" × ¾" × 41"	Top rails
Angle, steel	4	⅛" × ¾" × ¾" × 9½"	Cross supports
Flat (band), steel	2	⅛" × ¾" × ¾" × 11"	Corner standards
Flat (band), steel	1	⅛" × 4" × 9½"	Toe plate
Flat (band), steel	4	⅛" × ¾" × 12" (approx)	Diagonal braces
Flat (band), steel	1	⅛" × ¾" × 10"	Stop bar
*Pipe, standard, black steel	4	¼" ID × 8"	Posts
Pipe, standard black steel	2	¼" ID × 4½" (approx)	Ramp posts
**Plywood, exterior	1	¾" × 9¾" × 17" (approx)	Deck
Plywood, exterior	1	¾" × 9¾" × 26½" (approx)	Ramp deck
Bolts, carriage or stove	10	¼" × 1½"	Ramp fasteners

* Flatten ends to improve welds. Angles (⅛" × ¾" × ¾") may be used in place of pipe.

** Sawed lumber may be used if it is placed with the grain running across the ramp, or sheet steel may be welded on for decking.

☐ CONSTRUCTION PROCEDURE FOR CAR RAMP

1. Cut to length, exactly as specified, the bottom runners, top rails, cross supports, corner standards, toe plate, stop bar, and posts.
2. Assemble the parts and tack weld one side frame.

> **Caution:**
>
> **Sides must be welded so the corner standard will be on the outside of the runner and rail. When fully assembled, runners will have one flat side down and one flat side out. Rails have one flat side up and one flat side out.**

The following procedure is suggested:

a) Lay a runner and rail on a flat surface parallel to each other. One leg of each points towards the opposite piece and the other leg of each points upwards.
b) Slip one leg of a standard under the ends of the runner and rail.
c) Lay two 8″ posts in place between the runner and rail.
d) Mark the standard where the rail intersects. The top of the rail should be 8¼″ from the bottom of the runner.
e) Starting at the standard, measure out the rail 17″ and lay out and cut a notch in the rail to permit a 25° bend.
f) Place the rail in a vise and bend it to close the notch.
g) Clamp the runner, standard and rail assembly to a flat metal surface. Square the joints carefully and tack weld.
h) Place the 8″ posts in position and tack weld.
i) Tack weld the rail to the runner at the bottom of the ramp.
j) Position the 4½″ post as shown in the plan. If it's too long, shorten it as needed. Tack weld.

3. Assemble and tack weld the opposite side frame.

> **Caution:**
>
> **Make the unit match the first one except be careful to set it up so it faces the first one.**

4. Install and tack weld cross members.

> **Caution:**
>
> **Be sure the two sides are square to one another across the front and parallel from end to end.**

5. Recheck the unit to be sure it is square and parallel. Install and tack weld the toe plate, stop bar and cross braces.
6. Make permanent welds at all points.

> **Caution:**
>
> **Select a welding sequence which minimizes distortion.**

7. Chip all welds. Prime and paint the unit.
8. Measure the unit for the exact cutting of the deck pieces.
9. Cut, drill holes and install the deck pieces.

> **Caution:**
>
> **Cut angles on the ends of the ramp pieces so they fit well for maximum strength at the floor and where they meet at the top of the incline. Drill bolt holes in the frame and deck so there is room to apply nuts when bolts are installed.**

10. Prime and paint the deck parts.
    Note: All welds in this project must be high quality and fuse all parts of the joint.

> **Caution:**
>
> **Maximum capacity is one-half ton.**

Project 36: Creeper

□ PLAN FOR CREEPER

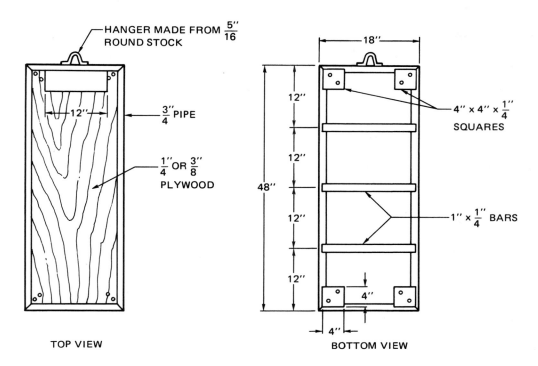

TOP VIEW

BOTTOM VIEW

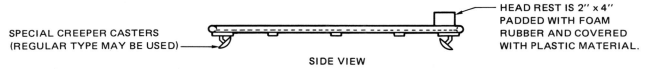

SIDE VIEW

Courtesy of Vocational Agriculture Service, University of Illinois, Urbana, IL

BILL OF MATERIALS			
MATERIALS NEEDED	**QUANTITY**	**DIMENSIONS**	**DESCRIPTION**
*Pipe, black steel	2	¾″ ID × 48″	Side rails
*Pipe, black steel	2	¾″ ID × 18″	End rails
Band, steel	3	¼″ × 1″ × 17″	Bars
Band or plate, steel	4	⅛″ or ¼″ × 4″ × 4″	Square plates
Round stock, steel	1	⁵⁄₁₆″ × 4″	Hanger
Plywood	1	¼″ or ⅜″ × 16″ × 46″	Deck
Lumber	1	2″ × 4″ × 12″	Headrest
Foam	1	As needed	Headrest
Plastic cover	1	As needed	Headrest
Bolt, carriage	8	¼″ × 1″	Deck and caster fasteners
Creeper casters	4	As needed	Casters

* Frame may also be constructed of hard wood or angle steel.

□ CONSTRUCTION PROCEDURE FOR CREEPER

1. Cut all metal pieces to size.
2. Assemble and weld the rails, bars, plates, and hanger.
3. Cut the deck to size.
4. Install the deck and casters using bolts with lock washers. Saw off excess bolt material.

5. Apply a suitable finish.
6. Construct and install the head rest using wood screws.

Project 37:   Saw Horse—Pipe

□ PLAN FOR SAW HORSE—PIPE

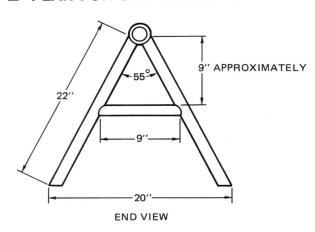

END VIEW

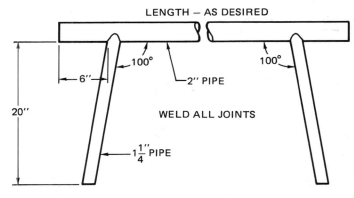

LENGTH – AS DESIRED

WELD ALL JOINTS

SIDE VIEW

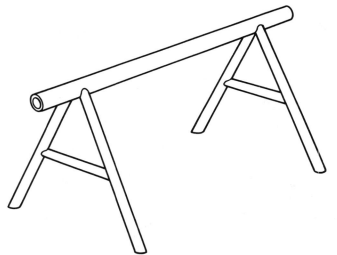

NOTE: THE DIAMETER OF PIPE MAY BE VARIED ACCORDING TO USE OF THE SAW HORSE. THE FOLLOWING MAY BE USED AS A GUIDE FOR THE LEGS:
1. LIGHT DUTY – ¾" ID PIPE
2. MEDIUM LOADS – 1" ID PIPE
3. HEAVY LOADS – 1¼" OR 1½" ID PIPE
DIAMETER OF THE TOP RAIL SHOULD BE ONE SIZE LARGER THAN THE LEGS.

BILL OF MATERIALS			
MATERIALS NEEDED	**QUANTITY**	**DIMENSIONS**	**DESCRIPTION**
Pipe, black steel	1	1¼" ID × 36"	Top rail
Pipe, black steel	4	1" ID × 20"	Legs
Pipe, black steel	2	1" ID × 9"	Braces

□ CONSTRUCTION PROCEDURE FOR SAW HORSE—PIPE

1. Cut all pipe to length.
2. Grind the ends of the legs to fit the top rail. **Note:** Welding will be easier if the end of each leg is flattened slightly where it will be fitted to the top rail.
3. Tack weld the legs to the top rail.
4. Adjust the leg positions so the saw horse stands squarely on a flat surface and all spacings are correct.

5. Weld the legs permanently.
6. Slightly flatten the ends of the braces and grind them to fit the legs.
7. Weld the braces in place.
8. Grind the bottoms of the legs so they are flat on the floor.

Project 38: Workbench Brackets

□ PLAN FOR WORKBENCH BRACKETS

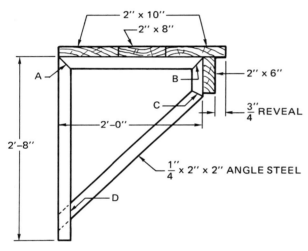

ANGLE BRACKET

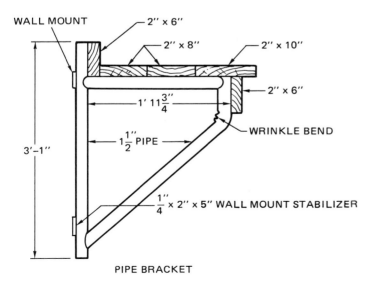

PIPE BRACKET

GENERAL

THREE BRACKETS ARE USUALLY REQUIRED FOR A 10- OR 12- FOOT BENCH. TWO MAY BE SUFFICIENT FOR AN 8- FOOT BENCH.

BOLT BRACKETS TO OR THROUGH THE WALL.

BOLT ALL WOOD PIECES TO BRACKETS, USING 3/8" CARRIAGE BOLTS. PLANKS SHOULD BE JOINTED FOR A COMBINED TOTAL BENCH WIDTH TO PROVIDE A 3/4" REVEAL ON THE FRONT. EDGES OF PLANKS MAY BE GLUED AND CLAMPED TO PROVIDE A CONTINUOUS BENCH TOP. HOLES ON TOP ARE COUNTERBORED 1 INCH AND FILLED WITH DOWEL PLUGS.

Courtesy of Vocational Agriculture Service, University of Illinois, Urbana, IL

BILL OF MATERIALS			
MATERIALS NEEDED	**QUANTITY**	**DIMENSIONS**	**DESCRIPTION**
ANGLE BRACKET (EACH): Angle, steel	2	¼″ × 2″ × 2″ × 2′8″	Legs
	1	¼″ × 2″ × 2″ × 2′0″	Horizontal support
or PIPE BRACKET (EACH): Pipe, black	2	1½″ I.D. × 3′1″	Standards
Pipe, black	1	1½″ I.D. × 1′10″	Horizontal support
Band, steel	2	¼″ × 2″ × 5″	Wall mounts

□ CONSTRUCTION PROCEDURE FOR WORKBENCH BRACKETS

Angle Iron Bracket

1. Cut 2 standards each with a 45 degree angle for the upper joint. Cut the horizontal support with a 45 degree angle on each end.
2. Cut a 30 degree wedge out at point C.
3. Square and tack weld A.
4. Heat and bend at C to create correct fits at B and D.
5. Tack weld B and D being careful to keep A and B square.
6. Weld all joints.

Pipe Bracket

1. Cut pipes to length, make wrinkle bend in one standard, and weld together.
2. Cut, drill, and weld on wall brackets.

Project 39: Workbench Bracket— Alternate Plan

□ PLAN FOR WORKBENCH BRACKET—ALTERNATE PLAN

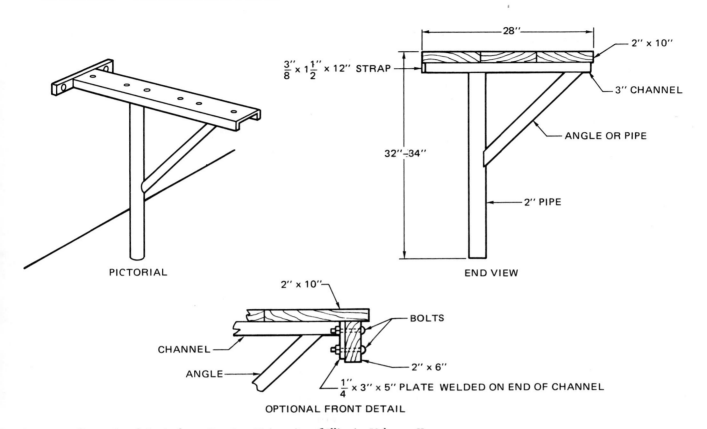

PICTORIAL

END VIEW

OPTIONAL FRONT DETAIL

Courtesy of Vocational Agriculture Service, University of Illinois, Urbana, IL

BILL OF MATERIALS			
MATERIALS NEEDED	**QUANTITY**	**DIMENSIONS**	**DESCRIPTION**
Channel, steel	1	3″ × 27″	Top
Pipe, black steel	1	2″ ID × 32″	Leg
Pipe, black steel	1	1″ ID × 15″	Brace
Band, steel	1	⅜″ × 1½″ × 12″	Bracket

□ CONSTRUCTION PROCEDURE FOR WORKBENCH BRACKET— ALTERNATE PLAN

1. Cut all pieces to length.
2. Drill ⁷⁄₁₆″ holes in the bracket.
3. Drill holes in the top as needed.

4. Round the corners of the channel that will protrude forward.
5. Weld the bracket to the top.

6. Tack weld the top to the leg and square the assembly.

7. Grind the ends of the brace to fit. Tack weld the brace.

8. Weld all joints permanently.

9. Apply a suitable finish.

Project 40:   Hand Cart

□ PLAN FOR HAND CART

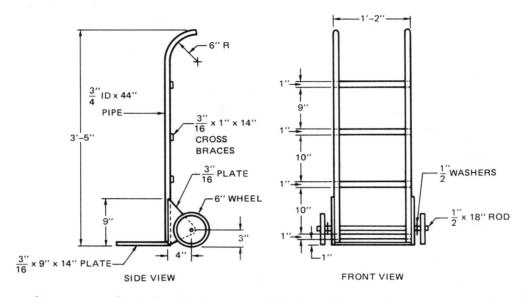

Courtesy of Department of Agricultural and Extension Education, The Pennsylvania State University, University Park, PA

BILL OF MATERIALS			
MATERIALS NEEDED	**QUANTITY**	**DIMENSIONS**	**DESCRIPTION**
Wheels, heavy duty	2	6″ × 1.50	For ½″ axle
Round, cold rolled	1	½″ × 18″ long	Axle*
Plate, steel	1	³⁄₁₆″ × 9″ × 14″	Base
Pipe, black	2	¾″ × 44″ long	Handles
Plate, steel	2	³⁄₁₆″ × 4″ × 9″	For axle supports
Band, steel	4	³⁄₁₆″ × 1″ × 14″	Cross braces
Washers	4	½″	
Cotter pins	2	⅛″ × 1″	
Metal primer	1 pint		
Enamel	1 pint		

* Axle length may vary with different wheels.

□ CONSTRUCTION PROCEDURE FOR HAND CART

1. Measure and cut the pipe to the dimension given in the bill of materials.

2. Bend the handles to a 6″ radius.

3. Cut the axle supports using an oxyacetylene torch. Grind to identical shape.

4. Drill ½″ holes with centers ½″ from the point of the axle supports. Insert the axle.

5. Cut the four cross braces and weld in place.

6. Tack weld the base in position, handles in air.

7. Position the axle supports with the wheels

mounted so that the base and wheels are level with the floor. Tack weld in place.

8. Drill the axle for the ⅛″ cotter pins to keep wheels on the cart and weld the axle in place.

9. Weld the ½″ washers to the axle on the inside of each wheel to serve as stops.

10. Complete welding all weld joints.

11. Paint the finished product with metal primer and a finish coat of enamel.

Note: If cart is to be used as a bag cart, add fenders by welding ⅛″ × 2″ × 6″ steel plate to the pipe over each wheel.

Project 41: Welding Stand—Acetylene

□ PLAN FOR WELDING STAND— ACETYLENE

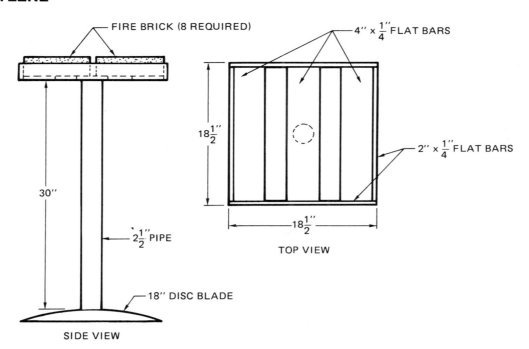

Courtesy of Vocational Agriculture Service, University of Illinois, Urbana, IL

BILL OF MATERIALS			
MATERIALS NEEDED	**QUANTITY**	**DIMENSIONS**	**DESCRIPTION**
Disc blade	1	18″ diameter	Base
Pipe, steel	1	22½″ ID ×30″	Post
Flat (band) steel	3	¼″ × 4″ × 18″	Bottom
	2	¼″ × 2″ × 18½″	Sides
	1	¼″ × 2″ × 18″	Ends

☐ CONSTRUCTION PROCEDURE FOR WELDING STAND— ACETYLENE

1. Set two rows of four fire bricks each on a flat surface. Check to see if they form a bed 18″ × 18″. If they do not, determine the changes needed in the plan to accommodate your bricks and modify the bill of materials accordingly.
2. Cut all pieces to length.
3. Tack weld the sides on the outside corners.
4. Tack weld the bottom pieces to the sides.
5. Invert the table on a flat surface and weld all joints from the underside.
6. Weld the post to the center bottom piece.
7. Weld the post to the disc base using a low-hydrogen electrode.
8. Apply a suitable finish.
9. Add the fire brick.

Project 42: Welding Stand—Arc

☐ PLAN FOR WELDING STAND— ARC

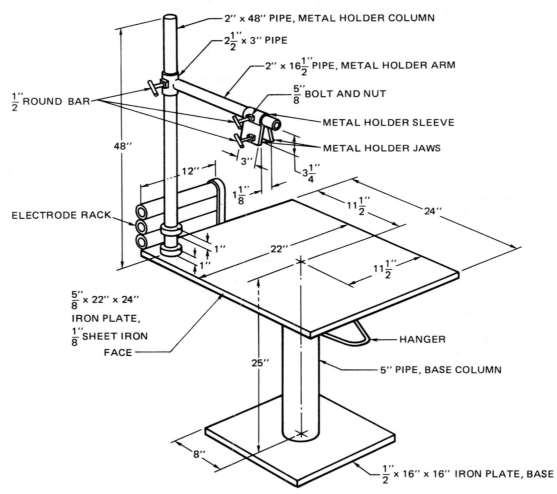

Courtesy of T. J. Wakeman and V. L. McCoy, The Farm Shop, *The Macmillan Company, 1960*

BILL OF MATERIALS			
MATERIALS NEEDED	**QUANTITY**	**DIMENSIONS**	**DESCRIPTION OR USE**
Plate iron	1 piece	½″ × 16″ × 16″	Base
	1 piece	⅝″ × 22″ × 24″	Top
	1 piece	⅛″ × 22″ × 24″	Top cover
	1 piece	¼″ × 2½″ × 7½″	Electrode rack end
	2 pieces	¼″ × 3″ × 3¼″	Clamp jaws
Black iron pipe	1 piece	5″ × 26″	Base column
	3 pieces	2″ × 11¾″	Electrode rack
	1 piece	2½″ × 2″	Rack supports
	1 piece	2″ × 48″	Metal holder column
	1 piece	2″ × 16½″	Metal holder arm
	2 pieces	2½″ × 3″	Sleeves for arm
Round iron	1 piece	½″ × 10″	Hanger
	3 pieces	½″ × 4″	Adjustment screw bars
Coupling	1	2½″ × 2″	Arm anchor
Machine bolts, nuts	3 each	⅝″ × 2½″ bolts	NC threads

□ CONSTRUCTION PROCEDURE FOR WELDING STAND—ARC

1. Center and square the 5″ pipe on the base plate, and weld them together.

Caution:

Tack weld and check for squareness before welding permanently.

2. Place the top on a flat surface; weld the other end of the 5″ pipe to the top, as shown; set the table upright.

Caution:

Tack weld and check for squareness before welding permanently.

3. Cut the 2″ coupling into two short pieces and weld the cut end of one to the table top, being careful to keep it level.

4. Thread one end of the 2″ × 48″ pipe; screw it into the coupling welded to the table top.
5. Clamp the 2″ × 11″ pipes, ends even, on a flat surface; weld 1″ at each end between the pipes; turn the pipes over; repeat the 1″ welds.
6. Place both rack supports over the 48″ column; weld the rack to them.
7. Drill a ¾″ hole in the center of one side of both 2½″ sleeves; weld a ⅝″ nut directly over the hole.
8. Shape one end of the 2″ × 16½″ pipe to fit the sleeve; weld as shown.
9. In one metal-holder jaw, drill a ¾″ hole and weld on a ⅝″ nut as in step 7.
10. Weld both jaws to the sleeve.
11. Weld a ½″ × 3¼″ round rod to the head of each bolt, as shown. Round the ends of all handles with a grinder or file.
12. Bend a ½″ rod for the electrode-holder hanger into a 2″ wide U; cover all but 1″ of each end with rubber hose; weld the ends to the underside of the table.
Note: The weldor sits at or works from the side of the table between the metal-holder column and the hanger.

Project 43: Welding Table—Combination

☐ PLAN FOR WELDING TABLE—COMBINATION

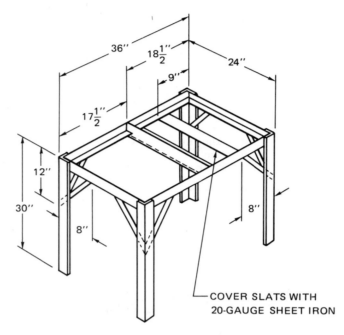

COVER SLATS WITH
20-GAUGE SHEET IRON

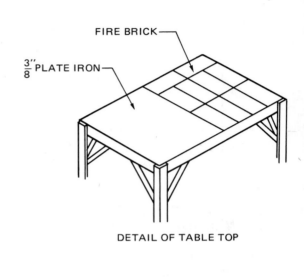

FIRE BRICK

$\frac{3}{8}''$ PLATE IRON

DETAIL OF TABLE TOP

Courtesy of T. J. Wakeman and V. L. McCoy, The Farm Shop, *The Macmillan Company, 1960*

BILL OF MATERIALS			
MATERIALS NEEDED	**QUANTITY**	**DIMENSIONS**	**DESCRIPTION OR USE**
Angle steel	4 pieces	$\frac{1}{8}'' \times 1\frac{1}{2}'' \times 1\frac{1}{2}'' \times 30''$	Legs
	2 pieces	$\frac{1}{8}'' \times 1\frac{1}{2}'' \times 2'' \times 36''$	Frame
	2 pieces	$\frac{1}{8}'' \times 1\frac{1}{2}'' \times 2'' \times 24''$	Frame
Flat steel	8 pieces	$\frac{3}{16}'' \times 1'' \times 14\frac{1}{2}''$	Braces
	2 pieces	$\frac{3}{16}'' \times 2'' \times 21''$	Bottom slats
	1 piece	$\frac{3}{16}'' \times 2'' \times 23\frac{3}{4}''$	Partition
Plate steel	1 piece	$\frac{3}{8}'' \times 17\frac{1}{2}'' \times 24''$	Top for arc welding
Sheet steel	1 piece	$18'' \times 23\frac{3}{4}''$	20 gauge
Firebrick	10	$2\frac{3}{8}'' \times 4\frac{1}{2}'' \times 9''$	

☐ CONSTRUCTION PROCEDURE FOR WELDING TABLE—COMBINATION

1. Cut all pieces to length as listed in the bill of materials. Cut the pieces for the frame at 45° angles with the 2″ outsides of the angle steel pointing upward up, as shown in the drawing.

2. Place the pieces for the frame on a smooth surface. Square and tack weld the frame on the outside and underside. Check for squareness, and then weld outside and bottom side of corners.

3. Place the partition in place, with the top edge even with the top edge of the frame, as shown in the drawing. Weld the partition to the frame, avoiding any excess weld buildup.

4. Weld one slat on the under side against the bottom edge of the partition, with the ends welded to the inside leg of the angle-iron frame. Place the other slat in the center of the space as shown in the drawing and weld on the under side to the frame.

5. Square and tack the legs to the frame.

6. Clamp the leg braces in place and weld them to the frame and legs. Finish welding the legs to the frame.

7. Place ⅜″ plate iron in position, and weld it to the top of the frame.

8. Place the sheet iron on top of the slats, as indicated in the drawing.

9. Place firebrick on top of the sheet iron.

Project 44: Gate—Metal, Small

□ PLAN FOR GATE—METAL, SMALL

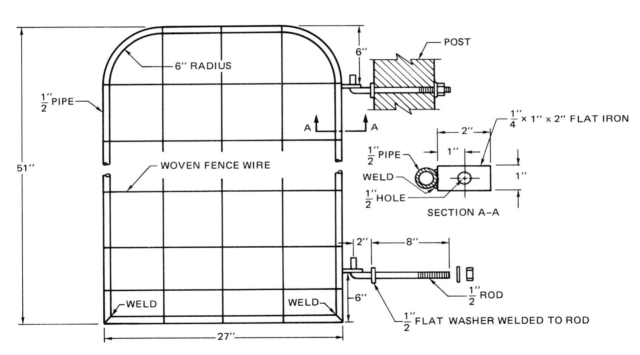

Courtesy of T. J. Wakeman and V. L. McCoy, The Farm Shop, *The Macmillan Company, 1960*

BILL OF MATERIALS			
MATERIALS NEEDED	QUANTITY	DIMENSIONS	DESCRIPTION OR USE
Pipe, steel, black	1 piece	½″ × 13′0″	Frame
Flat iron	2 pieces	¼″ × 1″ × 2″	Hinge
Round iron	2 pieces	½″ × 11″	Hinge hooks
Washers	2	½″	Flat
	2	⁷⁄₁₆″	Flat
Woven wire	1 piece	36″ × 48″	

□ CONSTRUCTION PROCEDURE FOR GATE—METAL, SMALL

1. Bend pipe for the frame as shown in the drawing, starting the first bend 45″ from one end.
2. Cut off the long end 51″ from the top of the gate at a 45° angle. Make a cut of the same angle on the opposite corner.
3. Cut the bottom piece 27″ long (or the width of the gate near the top) and weld in place.
4. Cut the flat iron for hinges to the length listed in the bill of materials. Lay off, center punch, and drill ½″ holes. Weld to the gate frame, as shown in detail section A-A.
5. Cut the hinge hooks to the length listed in the bill of materials. Make a 90° bend 1″ from one end. (The length will vary with the size of the gate post.)
6. Cut 2″ of NC threads on the other end of the hooks.
7. Drive a $\frac{7}{16}$″ washer 2″ from the bend in each hook. Weld in place, as shown in the drawing.
8. Fasten woven wire to the gate by passing the wire ends around the frame and wrapping around themselves. Use short pieces of wire to fasten the woven wire to the top and bottom of the gate.

Note: The height and width of the gate and the type of wire may be varied to suit a particular situation.

Project 45: Gate—Metal, 12-Foot or as Needed

□ PLAN FOR GATE—METAL, 12-FOOT OR AS NEEDED

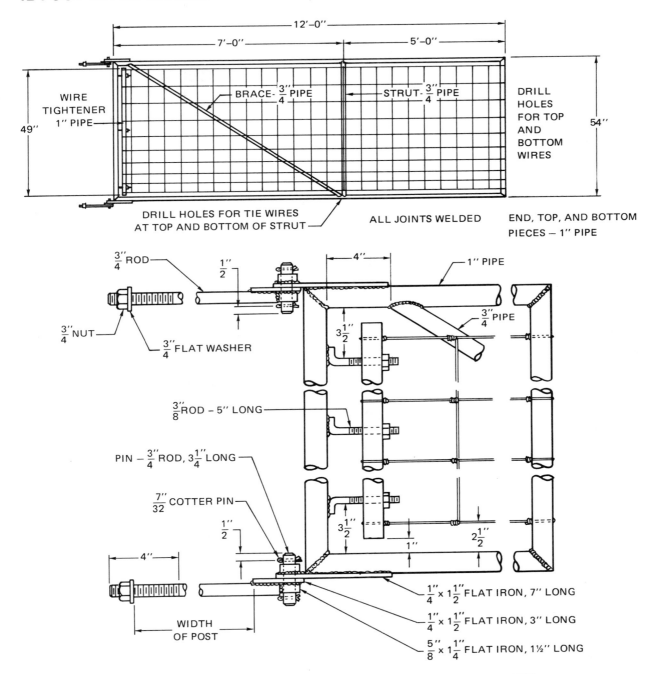

WIRE TIGHTENER 1" PIPE

49"

12'-0"

7'-0"

5'-0"

BRACE- $\frac{3}{4}$" PIPE

STRUT- $\frac{3}{4}$" PIPE

DRILL HOLES FOR TOP AND BOTTOM WIRES

54"

DRILL HOLES FOR TIE WIRES AT TOP AND BOTTOM OF STRUT

ALL JOINTS WELDED

END, TOP, AND BOTTOM PIECES — 1" PIPE

$\frac{3}{4}$" ROD

$\frac{1}{2}$"

4"

1" PIPE

$\frac{3}{4}$" NUT

$\frac{3}{4}$" FLAT WASHER

$3\frac{1}{2}$"

$\frac{3}{4}$" PIPE

$\frac{3}{8}$" ROD – 5" LONG

PIN – $\frac{3}{4}$" ROD, $3\frac{1}{4}$" LONG

$\frac{7}{32}$" COTTER PIN

$\frac{1}{2}$"

$3\frac{1}{2}$"

1"

$2\frac{1}{2}$"

4"

WIDTH OF POST

$\frac{1}{4}$" x $1\frac{1}{2}$" FLAT IRON, 7" LONG

$\frac{1}{4}$" x $1\frac{1}{2}$" FLAT IRON, 3" LONG

$\frac{5}{8}$" x $1\frac{1}{4}$" FLAT IRON, 1½" LONG

Courtesy of T. J. Wakeman and V. L. McCoy, The Farm Shop, *The Macmillan Company, 1960*

BILL OF MATERIALS			
MATERIALS NEEDED	**QUANTITY**	**DIMENSIONS**	**DESCRIPTION OR USE**
Pipe, steel, black	2 pieces	$1'' \times 12'0''$	Top and bottom
	2 pieces	$1'' \times 54''$	Ends
	1 piece	$1'' \times 49''$	Wire tightener
	1 piece	$\frac{3}{4}'' \times 51\frac{1}{2}''$	Strut
	1 piece	$\frac{3}{4}'' \times 7'9\frac{1}{4}''$	Brace
Flat iron	2 pieces	$\frac{1}{4}'' \times 1\frac{1}{2}'' \times 7''$	Hinges
	2 pieces	$\frac{1}{4}'' \times 1\frac{1}{2}'' \times 3''$	Hinges
	4 pieces	$\frac{5}{8}'' \times 1\frac{1}{4}'' \times 1\frac{1}{2}''$	Hinges
Round iron	3 pieces	$\frac{3}{8}'' \times 5''$	Tightener rods
	2 pieces	$\frac{3}{4}'' \times$ width of post $+ 2\frac{1}{2}''$	Hinge bolts
	2 pieces	$\frac{3}{4}'' \times 3\frac{1}{4}''$	Hinge pins
Nuts	3	$\frac{3}{8}''$	NC
	2	$\frac{3}{4}''$	NC
Washers	2	$\frac{3}{4}''$	Flat
Cotter pins	4	$\frac{7}{32}'' \times 1\frac{1}{2}''$	
Woven wire	1 piece	$47'' \times 13'6''$	10 strands
Wire, galvanized	2 pieces	$6''$ long	10 gauge, tie wires

□ CONSTRUCTION PROCEDURE FOR GATE—METAL, 12-FOOT OR AS NEEDED

1. Cut the pipe for the outside frame to the length as listed in the bill of materials, with corners mitered at 45°. Cut the wire tightener and strut to the length listed in the bill of materials. A hand or power hacksaw may be used to cut the pipe.

2. Center punch and drill the $\frac{1}{4}''$ holes in the wire tightener, strut, and latch end of the gate for the top and bottom wires, as shown in the detail. Locate the bottom hole $2\frac{1}{2}''$ from the bottom piece of the gate. Measure the height of the wire to determine the location of the top holes.

3. Place the ends, top, and bottom pieces together, as shown in the drawing. Place on a level surface; square and tack weld each corner on the top side.

4. Place the strut in position, and tack weld.

5. Cut the brace to length and proper angle and Tack weld.

6. Turn the gate over, and weld all joints. Place the gate in the appropriate positions and complete all welds.

7. Center punch and drill three $\frac{7}{16}''$ holes through the wire tightener for $\frac{3}{8}''$ rods, as shown in the drawing. After the first hole is drilled, place a 12" or longer $\frac{3}{8}''$ rod in the hole, and hold it in line with the drill press spindle while drilling the other two holes. This will help to drill the holes in line with each other.

8. Bend a 1" leg on one end of the three $\frac{3}{8}'' \times 5''$ tightener rods. Cut 3" of NC threads on the other end.

9. Place nuts on the three tightener rods, then insert the rods through the $\frac{7}{16}''$ holes in the tightener. Place nuts on the other side and tighten to hold the tightener rods in line while welding to the frame.

10. Lay the gate and tightener on a flat surface. Weld the legs of the tightener rods to the hinge end of the gate, as shown in the detail.

11. Remove the nuts from the tightener rods, then replace the wire tightener and the three outside nuts. Leave the nuts as near the ends of the rods as possible to allow maximum adjustment for tightening the wire.

12. Cut flat iron for hinges to the sizes listed in the bill of materials. Locate the center of the four pieces of $\frac{5}{8}''$ flat iron, and mark with a center punch for drilling.

13. Place one piece of $\frac{5}{8}''$ flat iron lengthwise $\frac{1}{4}''$ from the end of each piece of $\frac{1}{4}''$ flat iron. Be sure that the marked side is up. Clamp the two pieces together in this position, and weld all four sides.

14. Cut two pieces of $\frac{3}{4}''$ round iron to the length shown in the drawing. Weld one $\frac{3}{4}''$ rod to each of the $\frac{1}{4}'' \times 1\frac{1}{2}'' \times 3''$ pieces of flat iron. Hold

the end of the rod against the ⅝" piece welded in step 13. Cut four inches of NC threads on the other end of each rod.

15. Grind the corners of the ¼" pieces round on the reinforced ends of the hinges. Drill a ²⁵⁄₃₂" hole through the four hinge parts at the center-punched mark.

16. Cut two pieces of ¾" round iron 3¼" long for the hinge pins. Center punch and drill a ¼" hole ½" from each end.

17. Weld the 7" pieces of the hinges to the top and bottom of the gate with the reinforced side up in both cases, as shown in the drawing. A piece of ¾" round iron about 60" long should be inserted in the holes of the hinges to assist in keeping the holes in line and at the correct distance from the end of the gate while welding.

18. Attach woven wire to the latch end of the gate with the vertical stay wire parallel to the end of the gate. Place the top and bottom wires through holes in the end piece and wrap each wire around the pipe and back around itself.

19. Pull the wire tight by hand. Attach the wire to the tightener in the same manner as in step 18. Tighten nuts against the tightening bar until about one-half of the tension curve remains in the wires.

20. Tie the top and bottom wires to the strut with tie wires.

21. Apply a suitable finish.

Note: Commercial gate hinges may be used to simplify the project.

Project 46: Farrowing Crate

□ PLAN FOR FARROWING CRATE

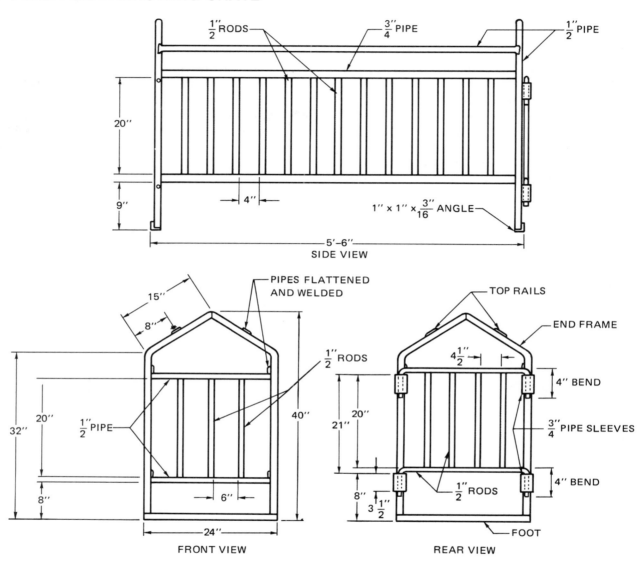

Adapted from Welding Plan No. 20, Vocational Agriculture Service, University of Illinois, Urbana, IL

BILL OF MATERIALS			
MATERIALS NEEDED	**QUANTITY**	**DIMENSIONS**	**DESCRIPTION**
Pipe, steel	2	½" ID × 5'6"	Top side rails
	4	¾" ID × 5'4"	Upper and lower side rails
Rod, steel	26	½" × 20"	Vertical slats for sides
Pipe, steel	4	½" ID × 4'	Half of end frame
	4	¾" ID × 3½"	Sleeves
	2	½" ID × 22"	Top and bottom rails of front
Rod, steel	7	½" × 20"	Vertical slats for ends
	2	½" × 32"	Top and bottom rails of rear gate
Angle, steel	2	3/16" × 1" × 1" × 24"	Feet

□ CONSTRUCTION PROCEDURE
FARROWING CRATE

1. Cut the pieces to be used in the end frame halves and feet.
2. Measure in 15" from one end and bend the four half frames to match each other.
3. Saw or grind the ridge angle on the half frames.
4. Weld the half frames at the top and weld them to the feet to form the ends.
5. Cut and weld the front rails and slats in place.
6. Cut and weld the side rails and slats in place.
7. Set up the ends and weld the side rails to the ends.
8. Make up the rear gate.
9. Cut four 3½" lengths of ¾" ID pipe for the sleeves and weld them into place.
10. Apply a suitable finish.

Project 47: Feed/Silage Cart

☐ PLAN FOR FEED/SILAGE CART

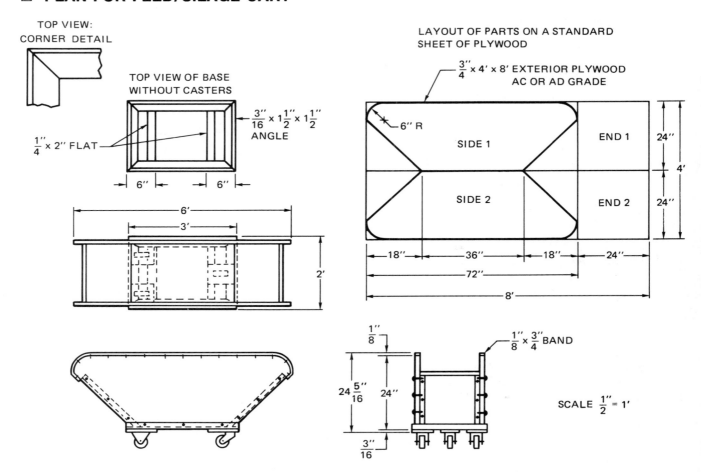

TOP VIEW:
CORNER DETAIL

TOP VIEW OF BASE
WITHOUT CASTERS

LAYOUT OF PARTS ON A STANDARD
SHEET OF PLYWOOD

BILL OF MATERIALS			
MATERIALS NEEDED	**QUANTITY**	**DIMENSIONS**	**DESCRIPTION**
Plywood, exterior AC or better recommended	2	$\frac{3}{4}'' \times 2' \times 6'$	Sides
	2	$\frac{3}{4}'' \times 2' \times 2'$	Ends
	1	$\frac{3}{4}'' \times 2' \times 3'$	Bottom
Steel, angle	2	$\frac{3}{16}'' \times 1\frac{1}{2}'' \times 1\frac{1}{2}'' \times 36''$	Sides of base
	2	$\frac{3}{16}'' \times 1\frac{1}{2}'' \times 1\frac{1}{2}'' \times 24''$	Ends of base
	4	$\frac{1}{8}'' \times 1'' \times 1'' \times 22''$	Corner angles
Steel, band	2	$\frac{1}{8}'' \times 1'' \times 90''$	Edge bands
	2	$\frac{3}{16}'' \times 2'' \times 21''$	Caster supports
Bolts, carriage, with nuts and lock washers	24	$\frac{1}{4}'' \times 1\frac{1}{4}''$	Bolt plywood to angles
Bolts, stove, flat head	6	$\frac{1}{4}'' \times 1\frac{1}{4}''$	Bolt sides to base
Bolts, stove, flat head	12	$\frac{3}{8}'' \times 1''$	Attach casters
*Caster, straight	2	6'' or 8'' dia	Rigid wheels on one end
Caster. swivel	1	6'' or 8''	Swivel wheel in center on other end
Screws, wood, flat	30	No. 6 × 1''	Fasten edge bands

* Two 6'', 8'', or 10'' wheels with supports and axle may be used in place of the straight casters. A metal support is then fabricated on the other end to level the cart on the swivel caster.

□ CONSTRUCTION PROCEDURE FOR FEED/SILAGE CART

1. Using a ¾″ × 4′ × 8′ sheet of plywood, cut the sides by cutting 2 pieces 2′ × 6′. Then lay out the other lines and finish cutting out the sides. See plan for most efficient layout.

2. Cut the ends by sawing the remaining piece in half.

3. Cut two pieces of ³⁄₁₆″ × 1½″ × 1½″ angle 36″ long and two pieces of ³⁄₁₆″ × 1½″ × 1½″ angle 24″ long with 45° angles on all ends. Weld them into a rectangular frame to form the base.

4. Cut two ¼″ × 2″ bands 21″ long.

5. Place the base on a flat surface. Weld the caster supports into place according to the size of the base of the casters.

6. Bolt the casters to the base using flat head stove bolts passing downward through countersunk holes and through the base of each caster.

7. Cut four pieces of ⅛″ × 1″ × 1″ angle 22″ long. Drill three ⁹⁄₃₂″ holes in each leg of each piece. Holes should be spaced so they are 1″ to 1½″ from each end with one in the middle. Holes must be placed so the bolts do not collide.

8. Notch the ends of the sides so they fit down into the base. Bolt the sides to the base using ¼″ × 1¼″ flat head stove bolts passing through countersunk holes in the frame and then through the wood sides. Install ¼″ flat washers and draw them slightly into the wood. Saw off any excess bolt length.

9. Using the four pieces of angle, bolt the ends to the sides at each corner. The angle steel and bolt nuts should be on the exterior of the ends and interior of the sides.

10. Cut a piece of ¾″ plywood to drop in to form the bottom. Small notches may be needed to clear the nuts at the bottom of the sides. The bottom must fit tight to prevent feed from running through. Caulk any cracks that can permit feed to pass through.

11. Cut 2 pieces of ⅛″ × ¾″ band 90″ long. Drill and countersink ½″ from each end and at 6″ intervals for No. 6 screws.

12. Shape the bands to fit the top of the sides.

13. Apply suitable finishes to all wood and metal parts.

14. Install the bands on the top of each side of the cart.

Project 48:   Hoist Frame—Portable

☐ PLAN FOR HOIST FRAME— PORTABLE

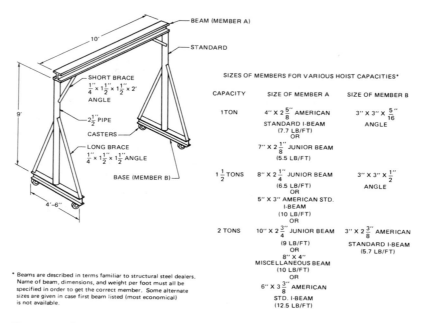

SIZES OF MEMBERS FOR VARIOUS HOIST CAPACITIES*

CAPACITY	SIZE OF MEMBER A	SIZE OF MEMBER B
1 TON	4" X 2 5/8" AMERICAN STANDARD I-BEAM (7.7 LB/FT) OR 7" X 2 1/8" JUNIOR BEAM (5.5 LB/FT)	3" X 3" X 5/16" ANGLE
1 1/2 TONS	8" X 2 1/4" JUNIOR BEAM (6.5 LB/FT) OR 5" X 3" AMERICAN STD. I-BEAM (10 LB/FT)	3" X 3" X 1/2" ANGLE
2 TONS	10" X 2 3/4" JUNIOR BEAM (9 LB/FT) OR 8" X 4" MISCELLANEOUS BEAM (10 LB/FT) OR 6" X 3 3/8" AMERICAN STD. I-BEAM (12.5 LB/FT)	3" X 2 3/8" AMERICAN STANDARD I-BEAM (5.7 LB/FT)

* Beams are described in terms familiar to structural steel dealers. Name of beam, dimensions, and weight per foot must all be specified in order to get the correct member. Some alternate sizes are given in case first beam listed (most economical) is not available.

Courtesy of Vocational Agriculture Service, University of Illinois, Urbana, IL

BILL OF MATERIALS			
MATERIALS NEEDED	**QUANTITY**	**DIMENSIONS**	**DESCRIPTION**
I beam, standard, steel	1	4" × 2⅝" × 10' (7.7 lb/ft)	Beam (member A)
Pipe, steel, standard	2	2½" × 9'	Standards
Angle, steel	2	5/16" × 3" × 3" × 4'6"	Base (member B)
Angle, steel	4	¼" × 1½" × 1½" × 6'6"	Long braces
Angle, steel	2	¼" × 1½" × 1½" × 2'	Short braces

* A 7" × 2⅛" × 10' (5.5 lb/ft) junior beam should be cheaper, if available, and may be substituted. See plan for specifications of beams for capacity greater than one ton.

☐ CONSTRUCTION PROCEDURE FOR HOIST FRAME—PORTABLE

Note: Be sure the casters used are designed to carry the combined weight of the frame, hoist, and load for which the hoist is designed.

1. Cut all pieces to length.
2. Lay the standards and member A on a level floor, square and tack weld them.

3. Check each joint between the standards and member A, and tack weld the braces into place. Weld all joints permanently.
4. Set the assembly up on the B members.
5. Check for squareness, and parallelism between the B members, then tack weld the standards to the B members.
6. Check for squareness and weld the braces between the standards and B members.

7. Weld all joints permanently.
8. Attach casters to the B members.

Caution:

Bill of materials is for a hoist with a maximum capacity of 1 ton.

Project 49: *Utility Trailer*

□ PLAN FOR UTILITY TRAILER

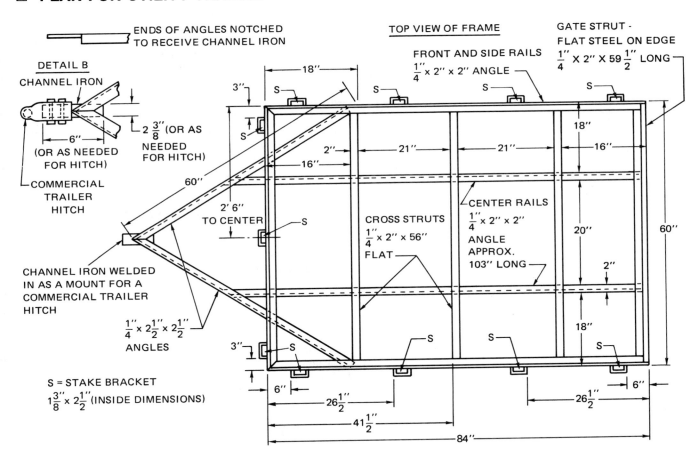

Plan by Elmer L. and Timothy B. Cooper

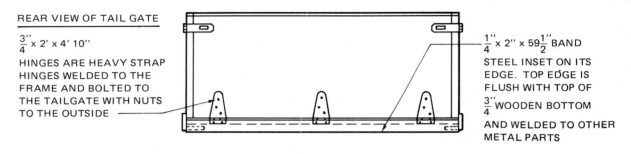

REAR VIEW OF TAIL GATE

$\frac{3''}{4}$ x 2' x 4' 10''

HINGES ARE HEAVY STRAP HINGES WELDED TO THE FRAME AND BOLTED TO THE TAILGATE WITH NUTS TO THE OUTSIDE

$\frac{1''}{4}$ x 2'' x $59\frac{1}{2}''$ BAND STEEL INSET ON ITS EDGE. TOP EDGE IS FLUSH WITH TOP OF $\frac{3''}{4}$ WOODEN BOTTOM AND WELDED TO OTHER METAL PARTS

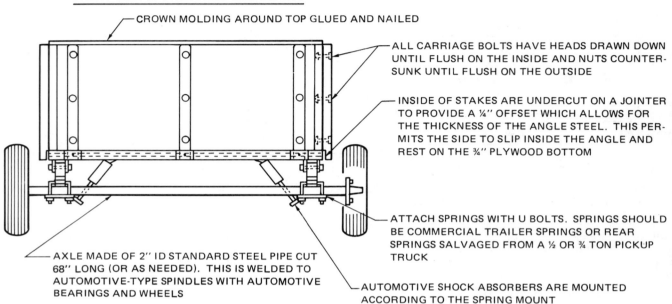

FRONT VIEW OF FRONT AND SIDE BOARDS

CROWN MOLDING AROUND TOP GLUED AND NAILED

ALL CARRIAGE BOLTS HAVE HEADS DRAWN DOWN UNTIL FLUSH ON THE INSIDE AND NUTS COUNTERSUNK UNTIL FLUSH ON THE OUTSIDE

INSIDE OF STAKES ARE UNDERCUT ON A JOINTER TO PROVIDE A ¼'' OFFSET WHICH ALLOWS FOR THE THICKNESS OF THE ANGLE STEEL. THIS PERMITS THE SIDE TO SLIP INSIDE THE ANGLE AND REST ON THE ¾'' PLYWOOD BOTTOM

ATTACH SPRINGS WITH U BOLTS. SPRINGS SHOULD BE COMMERCIAL TRAILER SPRINGS OR REAR SPRINGS SALVAGED FROM A ½ OR ¾ TON PICKUP TRUCK

AXLE MADE OF 2'' ID STANDARD STEEL PIPE CUT 68'' LONG (OR AS NEEDED). THIS IS WELDED TO AUTOMOTIVE-TYPE SPINDLES WITH AUTOMOTIVE BEARINGS AND WHEELS

AUTOMOTIVE SHOCK ABSORBERS ARE MOUNTED ACCORDING TO THE SPRING MOUNT

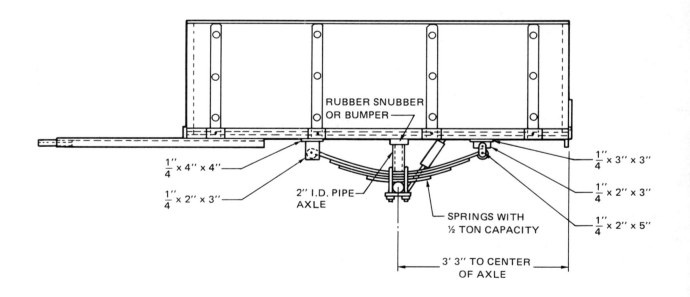

RUBBER SNUBBER OR BUMPER

$\frac{1''}{4}$ x 4'' x 4''

$\frac{1''}{4}$ x 2'' x 3''

2'' I.D. PIPE AXLE

SPRINGS WITH ½ TON CAPACITY

$\frac{1''}{4}$ x 3'' x 3''

$\frac{1''}{4}$ x 2'' x 3''

$\frac{1''}{4}$ x 2'' x 5''

3' 3'' TO CENTER OF AXLE

BILL OF MATERIALS

MATERIALS NEEDED	QUANTITY	DIMENSIONS	DESCRIPTION
Angle, steel	2	¼″ × 2″ × 2″ × 7′	Side rails
	1	¼″ × 2″ × 2″ × 5′	Front rail
	2	¼″ × 2″ × 2″ × 8′7″ (approx)	Center rails
	2	¼″ × 2½″ × 2½″ × 5′	Tongue rails
Band (flat), steel	4	¼″ × 2″ × 4′8″	Cross struts
	1	¼″ × 2″ × 4′11½″	Gate strut
	1	⅛″ or ¼″ × 2″ × 4′	Materials for pockets (or use 11 commercial pockets) 1⅜″ × 2½″ inside dimensions
	2	¼″ × 4″ × 4″	Front spring mount base
	4	¼″ × 2″ × 3″	Front spring mount
	2	¼″ × 3″ × 3″	Rear spring mount base
	4	¼″ × 2″ × 3″	Rear spring mount
	4	¼″ × 2″ × 5″	Rear spring shackle
Channel, steel	1	2⅜″ wide × 6″ long (or as needed) for the trailer hitch	Hitch mount
Trailer hitch	1	Commercial	Ball type, heavy duty
Pipe, steel, standard	1	2″ ID × 5′8″ (approx)	Axle
Springs, leaf type	2	Rated to carry ½ ton (rear truck or trailer type)	Springs with matching base mounts and U bolts
Spindles, automotive type	2	To carry ½ ton	With matching wheels and tires
Shock absorbers	2	As needed	Automotive type
Fenders	2	As needed	Commercial boat or utility trailer fenders
*Plywood, exterior type AC or better	1	¾″ × 5′ × 7′	Bottom
	2	¾″ × 2′ × 7′	Side boards
	2	¾″ × 2′ × 4′10″	Front and back boards
Lumber	11	1½″ × 2½″ × 2′1″	Stakes
Crown molding	2	½″ × ¾″ × 7′	Top of side boards
	2	½″ × ¾″ × 5′	Top of front and gate boards
T hinge, steel	3	8″ heavy duty	Back gate
Hasps, steel	4	4″ or as needed	Corner fasteners
Bolts, carriage	33	¼″ × 2½″ with hex nuts and small washers	Stake bolts
	12	¼″ × 1¼″	Hinge bolts

* Plywood may be special ordered in five foot widths, and lengths other than eight feet. Local prices should be explored to determine the most economical sizes to purchase. The trailer is designed with a five-foot width to accommodate a large lawn tractor with a four-foot mower.
All dimensions for plywood pieces must be adjusted according to the dimensions of the actual frame.

□ CONSTRUCTION PROCEDURE FOR UTILITY TRAILER

1. Cut the two side rails to length with a 45° cut in the front end. Cut ¼″ off the bottom leg in the rear.
2. Cut the front rail with 45° cuts on both ends.

Caution:

Be careful to lay out the cuts with the parts as they will be positioned on the completed trailer.

3. Weld the front and side rails together.

4. Cut the cross struts and weld them in place.

5. Cut the gate strut and weld it in place. Be careful to position the gate strut so it extends ¾" above the top edge of the back cross strut to protect the edge of the ¾" plywood bottom.

6. Make up and install the stake brackets. Drill a ³⁄₁₆" hole in the face of each bracket that is located adjacent to each corner for insertion of wood screws into the stakes.

7. Install the tongue pieces and hitch.

8. Install the center rails and weld them at all contact points.

9. Make up the spring and shock absorber mounts and shackles according to the springs being used. Place the spring mounts so the axle will be 3 to 4 inches slightly to the rear of center to provide appropriate balance in the completed trailer.

10. Make up the axle and spindle assembly to allow approximately 2" clearance between the body and tires.

11. Decide if the axle should be mounted above or below the springs to obtain the desired clearance to permit loading of the springs and to provide the desired height of the trailer to the towing vehicle.

12. Mount the axle assembly and shock absorbers.

13. Cut the plywood bottom to fit the frame and install it temporarily. Carefully lay out and drill six ¼" holes through the plywood floor and into the frame for carriage bolts to anchor the floor.

14. Cut 11 stakes for the front and sides. Cut a ¼" × 2" rabbet on the flat side of one end of each stake. The resulting end should fit into the brackets and the inside of the stake should be in line with the inside of the metal side rail. Chamfer (¼") the appropriate three edges on the opposite end of each stake to provide a finished appearance.

15. Cut and install the side boards with the stakes in their brackets so the side boards fit inside the side rails.

16. Cut and install the front board between the side boards.

17. Tack weld the T hinges to the gate strut so the hinge pins are in an absolutely straight line and parallel to the gate strut edge.

18. Cut and install the back gate. Install one bolt in each of the outside hinges. Move the gate up and down to be sure it does not bind. Install a bolt in the center hinge. Check again for binding. Weld all hinges permanently. Install all other hinge bolts.

19. Install corner hasps. The front corners may be secured with rigid corner brackets rather than hasps if desired.

20. Drill a ½ inch hole through the floor near each front corner for drainage.

21. Remove all wooden parts for painting on all sides and edges.

22. Chip all welds thoroughly. Smooth all rough areas and prepare all parts for painting.

23. Install commercial fenders if desired.

24. Prime all wood and metal parts with suitable primers.

25. Apply two coats of machinery enamel to all parts.

26. Re-install all wooden parts. Install No. 10 screws in the stakes to secure the sides.

Caution:

Trailer is designed for maximum load of ½ ton.

Project 50: Pickup Truck Racks

□ PLAN FOR PICKUP TRUCK RACKS

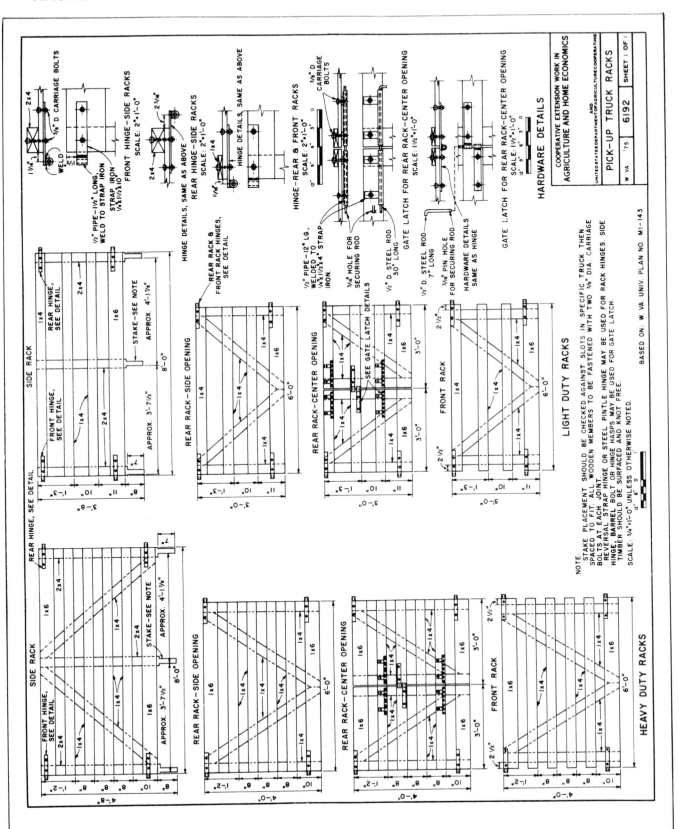

Courtesy of Department of Agricultural Engineering, University of Maryland, College Park, MD

BILL OF MATERIALS

This plan provides construction details for several types of racks. These include plans for heavy duty and light duty racks with center opening or side opening rear racks. The user must determine which weight and style of racks are preferred and then make up a bill of materials accordingly.

□ CONSTRUCTION PROCEDURE FOR PICKUP TRUCK RACKS

1. Examine the truck to determine if the truck body and pockets are such that the plans can be used as provided.
2. Select the racks from the plans that are to be used and modify them as needed to provide a custom fit.
3. Make the wooden racks.
4. Make or purchase the appropriate hardware for the corners and latches.
5. Install the hardware.
6. Apply a primer and two coats of enamel, or apply a polyurethane finish.

Project 51: Wagon Rack for Tossed Bales

☐ **PLAN FOR WAGON RACK FOR TOSSED BALES**
(*Continued on page 498*)

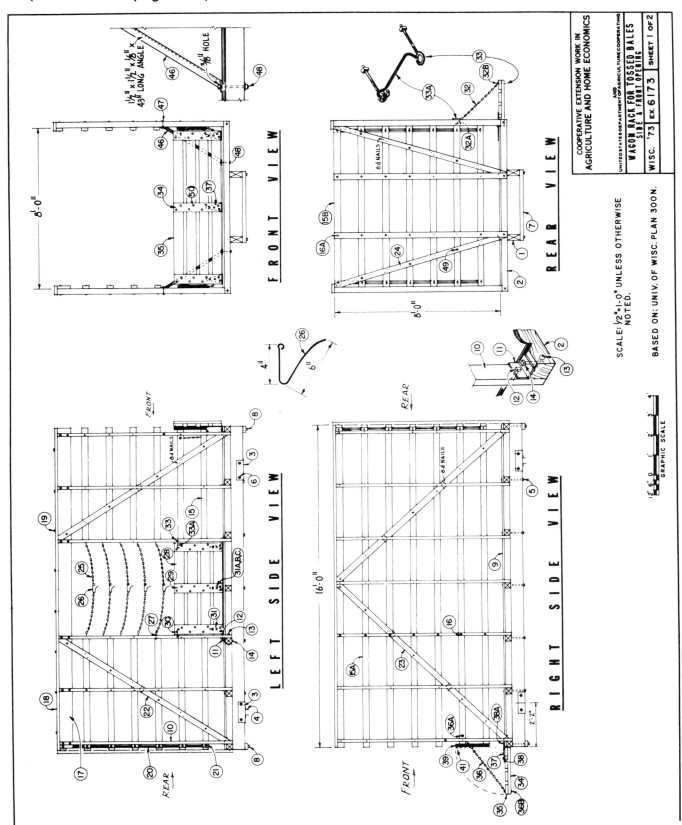

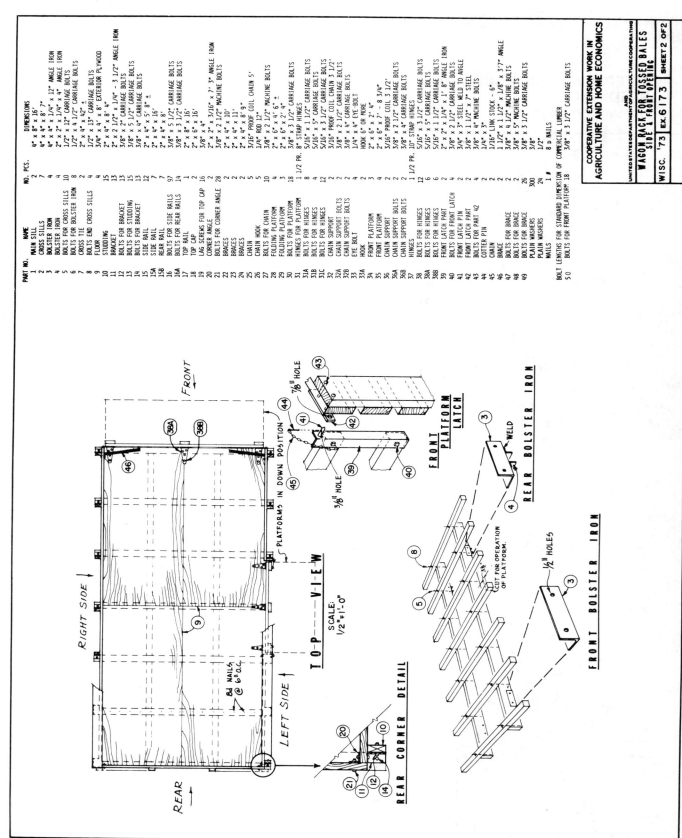

PART NO.	NAME	NO. PCS.	DIMENSIONS
1	MAIN SILL	2	4" x 8" x 16'
2	CROSS SILLS	7	4" x 4" x 8'7"
3	BOLSTER IRON	4	4" x 4" x 1/4" x 12" ANGLE IRON
4	BOLSTER IRON	4	2" x 2" x 1/4" x 4" ANGLE IRON
5	BOLTS FOR CROSS SILLS	10	1/2" x 12" CARRIAGE BOLTS
6	BOLTS FOR BOLSTER IRON	8	1/2" x 4 1/2" CARRIAGE BOLTS
7	CROSS TIE	2	2" x 4" x 42" †
8	BOLTS END CROSS SILLS	4	1/2" x 13" CARRIAGE BOLTS
9	FLOOR	4	3/4" x 4" x 8' EXTERIOR PLYWOOD
10	STUDDING	15	2" x 4" x 8' 4"
11	BRACKET	13	2" x 2 1/2" x 1/4" - 3 1/2" ANGLE IRON
12	BOLTS FOR BRACKET	13	3/8" x 2" CARRIAGE BOLTS
13	BOLTS FOR STUDDING	15	3/8" x 5 1/2" CARRIAGE BOLTS
14	BOLTS FOR BRACKET	13	3/8" x 4" CARRIAGE BOLTS
15	SIDE RAIL	12	2" x 4" x 16'
15A	SIDE RAIL	4	2" x 4" x 5' 8" †
15B	REAR RAIL	7	2" x 4" x 16'
16	BOLTS FOR SIDE RAILS	97	3/8" x 5 1/2" CARRIAGE BOLTS
16A	BOLTS FOR REAR RAILS	14	3/8" x 3 1/2" CARRIAGE BOLTS
17	TOP RAIL	1	2" x 8" x 16'
18	TOP CAP	2	2" x 6" x 16'
19	LAG SCREWS FOR TOP CAP	16	3/8" x 4"
20	CORNER ANGLE	2	2" x 2" x 3/16" x 7' 3" ANGLE IRON
21	BOLTS FOR CORNER ANGLE	28	3/8" x 2 1/2" MACHINE BOLTS
22	BRACES	2	2" x 4" x 10'
23	BRACES	2	2" x 4" x 11'
24	BRACES	2	2" x 4" x 8' 9"
25	CHAIN	5	3/16" PROOF COIL CHAIN 5'
26	CHAIN HOOK	5	1/4" ROD 12"
27	BOLTS FOR CHAIN	10	3/8" x 2 1/2" MACHINE BOLTS
28	FOLDING PLATFORM	2	2" x 6" x 4' 6" †
29	FOLDING PLATFORM	2	2" x 6" x 2' 4" †
30	BOLTS FOR PLATFORM	18	3/8" x 3 1/2" CARRIAGE BOLTS
31	HINGES FOR PLATFORM	1/2 PR.	8" STRAP HINGES
31A	BOLTS FOR HINGES	8	5/16" x 1 1/2" CARRIAGE BOLTS
31B	BOLTS FOR HINGES	5	5/16" x 5" CARRIAGE BOLTS
31C	BOLTS FOR HINGES	12	5/16" x 3 1/2" CARRIAGE BOLTS
32	CHAIN SUPPORT	2	3/16" PROOF COIL CHAIN 3 1/2'
32A	CHAIN SUPPORT BOLTS	2	3/8" x 2 1/2" CARRIAGE BOLTS
32B	CHAIN SUPPORT BOLTS	2	3/8" x 4" CARRIAGE BOLTS
33	EYE-BOLT	4	1/4" x 4" EYE-BOLT
33A	HOOK	4	HOOK 6" OR MORE
34	FRONT PLATFORM	3	2" x 6" x 2' - 3/4"
35	FRONT PLATFORM	2	2" x 6" x 7' - 8 3/4"
36	CHAIN SUPPORT	2	3/16" PROOF COIL 3 1/2'
36A	CHAIN SUPPORT BOLTS	2	3/8" x 2 1/2" CARRIAGE BOLTS
36B	CHAIN SUPPORT BOLTS	2	3/8" x 4" CARRIAGE BOLTS
37	HINGES	1/2 PR.	10" STRAP HINGES
38	BOLTS FOR HINGES	12	5/16" x 3 1/2" CARRIAGE BOLTS
38A	BOLTS FOR HINGES	6	5/16" x 5" CARRIAGE BOLTS
38B	BOLTS FOR HINGES	6	5/16" x 1 1/2" CARRIAGE BOLTS
39	FRONT LATCH PART	2	2" x 2 1/2" x 1' 8" ANGLE IRON
40	BOLTS FOR FRONT LATCH	4	3/8" x 2 1/2" CARRIAGE BOLTS
41	FRONT LATCH PIN	2	3/4" x 3" STEEL WELD TO ANGLE
42	FRONT LATCH PART	2	1/2" x 1 1/2" x 7" STEEL
43	BOLTS FOR PART 42	4	3/8" x 4" MACHINE BOLTS
44	COTTER PIN	2	1/4" x 3"
45	CHAIN	2	1/16" LINK STOCK - 6"
46	BRACE	2	1 1/2" x 1 1/2" x 1/8" x 3'7" ANGLE
47	BOLTS FOR BRACE	2	3/8" x 4 1/2" MACHINE BOLTS
48	BOLTS FOR BRACE	2	3/8" x 5" MACHINE BOLTS
49	PLAIN WASHERS	26	3/8" x 3 1/2" CARRIAGE BOLTS
	PLAIN WASHERS	300	3/8"
	NAILS	1 #	8 d NAILS

BOLT LENGTHS FOR STANDARD DIMENSION OF COMMERCIAL LUMBER

50 | BOLTS FOR FRONT PLATFORM | 18 | 3/8" x 3 1/2" CARRIAGE BOLTS

COOPERATIVE EXTENSION WORK IN
AGRICULTURE AND HOME ECONOMICS
AND
UNITED STATES DEPARTMENT OF AGRICULTURE COOPERATING

WAGON RACK FOR TOSSED BALES
SIDE & FRONT OPENING

WISC. '73 | EX. 6173 | SHEET 2 OF 2

FRONT

RIGHT SIDE

LEFT SIDE

REAR

PLATFORMS IN DOWN POSITION

3/8" HOLE

TOP - VIEW

SCALE:
1/2" = 1'-0"

8d NAILS
@ 6" O.C.

REAR CORNER DETAIL

7/8" HOLE

FRONT PLATFORM LATCH

WELD

REAR BOLSTER IRON

CUT FOR OPERATION OF PLATFORM.

1/2" HOLES

FRONT BOLSTER IRON

	BILL OF MATERIALS			
PART NO.	MATERIALS NEEDED	QUANTITY	DIMENSIONS	DESCRIPTION
1	Main sill	2	$4'' \times 8'' \times 16'$	
2	Cross sills	7	$4'' \times 4'' \times 8'7''$	
3	Bolster iron	4	$4'' \times 4'' \times \frac{1}{4}'' \times 12''$ angle iron	
4	Bolster iron	4	$2'' \times 2'' \times \frac{1}{4}'' \times 4''$ angle iron	
5	Bolts for cross sills	10	$\frac{1}{2}'' \times 12''$ carriage bolts	
6	Bolts for bolster iron	8	$\frac{1}{2}'' \times 4\frac{1}{2}''$ carriage bolts	
7	Cross tie	2	$2'' \times 4'' \times 42'' \pm$	
8	Bolts end cross sills	4	$\frac{1}{2}'' \times 13''$ carriage bolts	
9	Floor	4	$\frac{3}{4}'' \times 4' \times 8'$ exterior plywood	
10	Studding	15	$2'' \times 4'' \times 8'4''$	
11	Bracket	13	$2'' \times 2\frac{1}{2}'' \times \frac{1}{4}'' - 3\frac{1}{2}''$ angle iron	
12	Bolts for bracket	13	$\frac{3}{8}'' \times 2''$ carriage bolts	
13	Bolts for studding	15	$\frac{3}{8}'' \times 5\frac{1}{2}''$ carriage bolts	
14	Bolts for bracket	13	$\frac{3}{8}'' \times 4''$ carriage bolts	
15	Side rail	12	$2'' \times 4'' \times 5'8'' \pm$	
15A	Side rail	7	$2'' \times 4'' \times 16'$	
15B	Rear rail	7	$2'' \times 4'' \times 8'$	
16	Bolts for side rails	97	$\frac{3}{8}'' \times 5\frac{1}{2}''$ carriage bolts	
16A	Bolts for rear rails	14	$\frac{3}{8}'' \times 3\frac{1}{2}''$ carriage bolts	
17	Top rail	1	$2'' \times 8'' \times 16'$	
18	Top cap	2	$2'' \times 6'' \times 16'$	
19	Lag screws for top cap	16	$\frac{3}{8}'' \times 4''$	
20	Corner angle	2	$2'' \times 2'' \times \frac{3}{16}'' \times 7'3''$ angle iron	
21	Bolts for corner angle	28	$\frac{3}{8}'' \times 2\frac{1}{2}''$ machine bolts	
22	Braces	2	$2'' \times 4'' \times 10'$	
23	Braces	2	$2'' \times 4'' \times 11'$	
24	Braces	2	$2'' \times 4'' \times 8'9''$	
25	Chain	5	$\frac{3}{16}''$ proof coil chain $5'$	
26	Chain hook	5	$\frac{1}{4}''$ rod $12''$	
27	Bolts for chain	10	$\frac{3}{8}'' \times 2\frac{1}{2}''$ machine bolts	
28	Folding platform	4	$2'' \times 6'' \times 4'6'' \pm$	
29	Folding platform	3	$2'' \times 6'' \times 2'4''$	
30	Bolts for platform	18	$\frac{3}{8}'' \times 3\frac{1}{2}''$ carriage bolts	
31	Hinges for platform	3	$8''$ strap hinges	
31A	Bolts for hinges	8	$\frac{5}{16}'' \times 1\frac{1}{2}''$ carriage bolts	
31B	Bolts for hinges	4	$\frac{5}{16}'' \times 5''$ carriage bolts	
31C	Bolts for hinges	12	$\frac{5}{16}'' \times 3\frac{1}{2}''$ carriage bolts	
32	Chain support	2	$\frac{3}{16}''$ proof coil chain $3\frac{1}{2}'$	
32A	Chain support bolts	2	$\frac{3}{8}'' \times 2\frac{1}{2}''$ carriage bolts	
32B	Chain support bolts	2	$\frac{3}{8}'' \times 4''$ carriage bolts	
33	Eye bolt	4	$\frac{1}{4}'' \times 4''$ eye bolt	
33A	Hook	2	Hook $6''$ or more	
34	Front platform	3	$2'' \times 6'' \times 2'4''$	
35	Front platform	4	$2'' \times 6'' \times 7'8\frac{3}{4}''$	
36	Chain support	2	$\frac{3}{16}''$ proof coil $3\frac{1}{2}'$	
36A	Chain support bolts	2	$\frac{3}{8}'' \times 2\frac{1}{2}''$ carriage bolts	
36B	Chain support bolts	2	$\frac{3}{8}'' \times 4''$ carriage bolts	
37	Hinges	3	$10''$ strap hinges	
38	Bolts for hinges	12	$\frac{5}{16}'' \times 3\frac{1}{2}''$ carriage bolts	
38A	Bolts for hinges	6	$\frac{5}{16}'' \times 5''$ carriage bolts	
38B	Bolts for hinges	6	$\frac{5}{16}'' \times 1\frac{1}{2}''$ carriage bolts	
39	Front latch part	2	$2'' \times 2'' \times \frac{1}{4}'' \times 1'8''$ angle iron	
40	Bolts for front latch	4	$\frac{3}{8}'' \times 2\frac{1}{2}''$ carriage bolts	
41	Front latch pin	2	$\frac{3}{4}'' \times 3''$ steel weld to angle	
42	Front latch part	2	$\frac{3}{8}'' \times 1\frac{1}{2}'' \times 7''$ steel	
43	Bolts for part 42	4	$\frac{3}{8}'' \times 4''$ machine bolts	
44	Cotter pin	2	$\frac{1}{4}'' \times 3''$	
45	Chain	2	$\frac{1}{16}''$ link stock $- 6''$	
46	Brace	2	$1\frac{1}{2}'' \times 1\frac{1}{2}'' \times \frac{1}{8}'' \times 3'7''$ angle	
47	Bolts for brace	2	$\frac{3}{8}'' \times 4\frac{1}{2}''$ machine bolts	

			Bill of Materials (*Cont'd*)	
PART NO.	MATERIALS NEEDED	QUANTITY	DIMENSIONS	DESCRIPTION
48	Bolts for brace	2	⅜" × 5" machine bolts	
49	Bolts for brace	26	⅜" × 3½" carriage bolts	
	Plain washers	300	⅜"	
	Plain washers	24	½"	
	Nails	1 #	8d nails	
50	Bolts for front platform	18	⅜" × 3½" carriage bolts	

Note: Bolt lengths are specified for standard dimensions of commercial lumber.

□ CONSTRUCTION PROCEDURE FOR WAGON RACK FOR TOSSED BALES

1. Cut the main sills to length, install the bolster irons, and mount the sills on a suitable running gear.
2. Cut the cross sills to length.
3. Paint or use a wood preservative on the main sills and cross sills. (This procedure will delay decay of the wood due to moisture accumulation in the joints.)
4. Attach the cross sills to the main sills.
5. Saw 3½" off one end of the center cross sill as shown if a center platform is to be installed.
6. Cut the side studding (uprights) to length and bolt them to the cross sills.
7. Nail the floor boards in place and saw them off flush.
8. Cut the side rails to length and bolt them to the studding.
9. Install the top caps.
10. Install the diagonal bracing.
11. Cut the two rear studs to length and bolt them in place.
12. Bolt the rear rails in place.
13. Install the rear diagonal braces.
14. Install the corner angles between the rear and side rails.
15. Make and install the front platform.
16. Install the front angle braces.
17. Make and install the side platform.
18. Install chains and all remaining hardware.
19. Apply paint or other suitable finish.

Appendix B
DATA TABLES

TABLE 1   WEIGHTS AND MEASURES

LIQUID/DRY MEASURE	LINEAR MEASURE
16 fluid ounces or 1 pound (water) ....1 pint	12 inches ....1 foot
2 pints ....1 quart	3 feet ....1 yard
4 quarts ....1 gallon	5½ yards (16½ feet) ....1 rod
8 quarts ....1 peck	320 rods ....1 mile
4 pecks ....1 bushel	1760 yards (5,280 feet) ....1 mile
1 bushel ....1¼ cu ft	**SQUARE MEASURE**
2 barrels ....1 hogshead	144 square inches.............................1 square foot
31½ gallons (US) ....1 barrel	9 square feet ....1 square yard
7.48 gallons (US) ....1 cu ft	30¼ square yards ....1 square rod
1 gallon (US) ....231 cu in	160 square rods ....1 acre (43,560 sq ft)
1 inch of water	640 acres ....1 square mile
per acre...............................27,154 gallons	1 square mile ....1 section
1 inch of water per hectare ....67,885 gallons	**CUBIC MEASURE**
(1 hectare = 2.5 acres)	1728 cubic inches ....1 cu ft
WEIGHT MEASURE	27 cubic feet ....1 cu yd
Gram ....15.432 Grains	128 cu ft (8′ × 4′ × 4′)...........................1 cord
Gram ....0353 Ounce	1′ × 1′ × 1″ ....1 bd ft
Kilogram ....2.2046 Lb	**LAND MEASURE**
Kilogram ....0011 Ton (Short)	To find number of acres:
Met. Ton ....1.1025 Ton (Short)	Divide no. of sq ft by 43,560
Grain ....064 Gram	Divide no. of sq yds by 4,840
Ounce ....28.35 Grams	Divide no. of sq rods by 160
16 ounces ....1 pound	Divide no. of sq chains by 10
Lb ....453.5 Grams	**HEIGHT OF TREE OR BUILDING**
100 pounds ....1 hundredweight	To estimate the height of a tree or building:
20 hundredweight ....1 ton	1. Measure the height (H_1) of a nearby object which is vertical.
Ton (Short) ....907.18 Kilgms.	2. Measure the length of the shadow (S_1) cast by that object.
Ton (Short) ....907 Met. Ton	3. Measure the length of the shadow (S_2) cast by the tree or building.
Ton (Short) ....2,000 Lb	4. Length of shadow (S_2) of tree or building times the height (H_1) of the object divided by length of shadow (S_1) of the object equals the height (H_2) of the tree or building ($H_2 = S_2 \times H_1 \div S_1$)
1 gallon of water weighs 8.34 pounds	
1 gallon of milk weighs 8.6 pounds	
TEMPERATURE	
°C. = (°F. − 32) × 5/9	
°F. = (9/5 °C) + 32	

TABLE 2 METRIC CONVERSIONS AND MEASUREMENTS

COMPARISON OF THE INTERNATIONAL METRIC SYSTEM AND THE ENGLISH SYSTEM OF MEASUREMENT

1 Centimeter	= .3937 inches	1 Kilometer	= 1000 meters	1 Gallon	= 3.785 liters
1 Inch	= 2.54 centimeter	1 Kilometer	= .62137 miles	1 Gram	= 15.43 grains
1 Foot	= 30.48 centimeter	1 Sq. Centimeter	= .155 Sq. inches	1 Ounce	= 28.35 grams
1 Meter	= 39.37 inches	1 Sq. Decimeter	= 100 cu. centimeters	1 Kilogram	= 1000 grams
1 Meter	= 100 centimeters	1 Cu. Centimeter	= .061 cu. inches	1 Kilogram	= 2.205 lbs.
1 Meter	= 1.094 yards	1 Cu. Decimeter	= 1000 Cu. centimeter	1 Pound	= 7000 grains
1 Meter	= 1000 millimeters	1 Cu. Meter	= 100 liters	1 Pound	= .4536 kilograms
1 Millimeter	= .001 meter	1 Fluid Ounce	= 29.54 milliliters	1 Kilogram	= 1000 milliliters
1 Yard	= .9144 meter	1 Liter	= 1000 cu. centimeters	1 Kilogram	= 1 liter
1 Mile	= 1609.344 meters	1 Liter	= 1.057 quarts		

APPROXIMATE CONVERSION OF COMMON UNITS

U.S. to Metric	*Metric to U.S.*

LENGTH

U.S. to Metric	Metric to U.S.
1 inch = 25.0 millimeters (mm)	1 millimeter (mm) = 0.04 inch
1 foot = 0.3 meter (m)	1 meter (m) = 3.3 feet
1 yard = 0.9 meter	1 meter = 1.1 yards
1 mile = 1.6 kilometers (km)	1 kilometer (km) = 0.6 mile

AREA

U.S. to Metric	Metric to U.S.
1 sq. inch = 6.5 sq. centimeters (cm²)	1 sq. centimeter (cm²) = 0.16 sq. inch
1 sq. foot = 0.09 sq. meter (m²)	1 sq. meter (m²) = 11.0 sq. feet
1 sq. yard = 0.8 sq. meter	1 sq. meter = 1.2 sq. yards
1 acre = 0.4 hectare*	1 hectare = 2.5 acres
1 sq. mile = 2.6 sq. kilometers	1 sq. kilometer = 0.39 sq. mile
*1 hectare equals 10,000 sq. meters	

MASS

U.S. to Metric	Metric to U.S.
1 grain = 64.8 milligrams (mg)	1 milligram (mg) = 0.015 grain
1 ounce (dry) = 28.0 grams (g)	1 gram (g) = 0.035 ounce
1 pound = 0.45 kilogram (kg)	1 kilogram (kg) = 2.2 pounds
1 short ton = 9.071 kilograms	1 metric ton = 1.102 tons (short)

VOLUME

U.S. to Metric	Metric to U.S.
1 cubic inch = 16.0 cubic centimeters (cm³)	1 cubic centimeter (cm³) = 0.06 cubic inch
1 cubic foot = 0.03 cubic meter (m³)	1 cubic meter (m³) = 35.0 cubic feet
1 cubic yard = 0.76 cubic meter	1 cubic meter = 1.3 cubic yards
1 teaspoon = 5.0 milliliters (ml)	1 milliliter (ml) = 0.2 teaspoon
1 tablespoon = 15.0 milliliters	1 milliliter = 0.07 tablespoon
1 fl. ounce = 30.0 milliliters	1 milliliter = 0.03 ounce
1 cup = 0.24 liter (l)**	1 liter (l) = 4.2 cups
1 pint = 0.47 liter	1 liter = 2.1 pints
1 quart (liq.) = 0.95 liter	1 liter = 1.1 quarts
1 gallon (liq.) = 0.004 cubic meter	1 cubic meter = 264.0 gallons
1 peck = 0.009 cubic meter	1 cubic meter = 113.0 pecks
1 bushel = 0.04 cubic meter	1 cubic meter = 28.0 bushels
**1 liter equals 1 cubic decimeter (dm³)	

Courtesy of The Maryland State Department of Education, Division of Vocational Technical Education, Baltimore, MD

TABLE 3 APPROXIMATE DILUTION RATIOS AND PROPORTIONS

DILUTION OF LIQUID PESTICIDES AT VARIOUS CONCENTRATIONS				
Dilution	1 Gal.	3 Gal.	5 Gal.	15 Gal.
1–100	2 tbs. + 2 tsp.	½ cup	¾ cup + 2 tsp.	2 cups + 6½ tbs.
1–200	4 tsp.	¼ cup	6½ tbs.	1 cup + 3⅓ tbs.
1–400	2 tsp.	2 tbs.	3 tbs.	½ cup + 1 tbs.
1–800	1 tsp.	1 tbs.	1 tbs. + 2 tsp.	4 tbs. + 2 tsp.
1–1000	¾ tsp.	2¼ tsp.	1 tbs. + 2 tsp.	4 tbs.

NOTE: 1 gal. = 16 cups or 256 tbs. or 768 tsp.
1 cup = 16 tbs.; 1 tbs. = 3 tsp.

WATER	QUANTITY OF MATERIAL					
100 gals.	1 lb.	2 lb.	3 lb.	4 lb.	5 lb.	6 lb.
25 gals.	4 oz.	8 oz.	12 oz.	1 lb.	1¼ lb.	1½ lb.
5 gals.	3 tbs.	1½ oz.	2½ oz.	3¼ oz.	4 oz.	5 oz.
1 gal.	1 tsp.	2 tsp.	1 tbs.	4 tsp.	5 tsp.	2 tbs.

Equivalent quantities of liquid materials (emulsion concentrates, etc.) for various quantities of water based on pints per 100 gallons.

WATER	QUANTITY OF MATERIAL					
100 gals.	½ pint	1 pint	2 pints	3 pints	4 pints	5 pints
25 gals.	2 fl. oz.	4 fl. oz.	8 fl. oz.	12 fl. oz.	1 pint	1¼ pint
5 gals.	1 tbs.	1 fl. oz.	2 fl. oz.	2½ fl. oz.	3 fl. oz.	4 fl. oz.
1 gal.	½ tsp.	1 tsp.	2 tsp.	3 tsp.	4 tsp.	5 tsp.

TABLE 4 CUBIC AIR CONTENT OF A GREENHOUSE

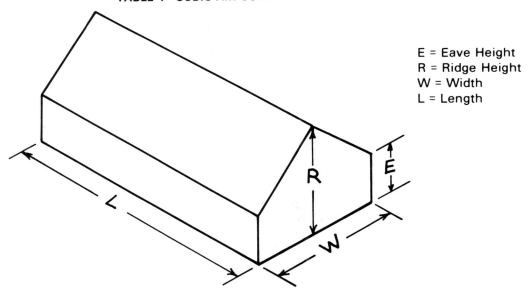

E = Eave Height
R = Ridge Height
W = Width
L = Length

FORMULA: $\dfrac{E + R}{2}$ x W x L = Cubic Content

Courtesy of The Maryland State Department of Education, Division of Vocational Technical Education, Baltimore, MD

TABLE 5 WEIGHTS PER BUSHEL OF COMMODITIES

COMMODITY	WEIGHT PER BUSHEL (POUNDS)
FIELD CROPS	
Barley	48
Cottonseed	30
Corn, ear	70
Corn, shelled	56
Oats	32
Sorghum seed	56
Soybeans	60
Wheat	60
LEGUMES	
Lespedeza, common	25
Lespedeza, Korean	40
Lespedeza, Kobe	25
Clover, crimson	60
Clover, white	60
Alfalfa	60
Vetch	60
GRASSES	
Bermuda	40
Dallis	15
Bahia	15–17
Fescue	10–30
Orchard	14
Rye	53
FRUITS	
Apples	48
Peaches	48
Pears	50
VEGETABLES	
Potatoes, Irish	56
Potatoes, Sweet	50–55
Cabbage	50
Cucumbers	48
Okra	32–36
Green beans	30
English peas	30
Dried peas	60
Turnips	54

TABLE 6 MOISTURE CONTENT OF GRAINS FOR LONG-TERM STORAGE

CROP	MAXIMUM MOISTURE CONTENT
Wheat	12%
Oats	13%
Barley	13%
Grain sorghums	12%
Shelled corn	13%
Soybeans	11%
Rice	12%

TABLE 7   QUICK REFERENCE FRACTION TO DECIMAL EQUIVALENTS

Fraction	Decimal
1/64	.015625
1/32	.03125
3/64	.046875
1/16	.0625
5/64	.078125
3/32	.09375
7/64	.109375
1/8	.125
9/64	.140625
5/32	.15625
11/64	.171875
3/16	.1875
13/64	.203125
7/32	.21875
15/64	.234375
1/4	.250
17/64	.265625
9/32	.28125
19/64	.296875
5/16	.3125
21/64	.328125
11/32	.34375
23/64	.359375
3/8	.375
25/64	.390625
13/32	.40625
27/64	.421875
7/16	.4375
29/64	.453125
15/32	.46875
31/64	.484375
1/2	.500
33/64	.515625
17/32	.53125
35/64	.546875
9/16	.5625
37/64	.578125
19/32	.59375
39/64	.609375
5/8	.625
41/64	.640625
21/32	.65625
43/64	.671875
11/16	.6875
45/64	.703125
23/32	.71875
47/64	.734375
3/4	.750
49/64	.765625
25/32	.78125
51/64	.796875
13/16	.8125
53/64	.828125
27/32	.84375
55/64	.859375
7/8	.875
57/64	.890625
29/32	.90625
59/64	.921875
15/16	.9375
61/64	.953125
31/32	.96875
63/64	.984375
1	1.000

TABLE 8 NAIL DATA

PENNY	LENGTH (IN.)	GAUGE NO. OF SHANK			APPROX. NO. PER LB.		
		Common	Box	Finish	Common	Box	Finish
2d	1	15	15½	16½	830	1010	1351
3d	1¼	14	14½	15½	528	635	807
4d	1½	12½	14	15	316	473	584
6d	2	11½	12½	13	168	236	309
8d	2½	10¼	11½	12½	106	145	189
10d	3	9	10½	11½	69	94	121
16d	3½	8	10	11	49	71	90
20d	4	6	9	10	31	52	62
30d	4½	5			24		
40d	5	4			18		
60d	6	2			11		

TABLE 9 WOOD SCREW INFORMATION

WOOD SCREW SIZE	SHANK HOLE SIZE	PILOT HOLE SIZE		AUGER BIT SIZES BY 16TH FOR COUNTERBORE HOLE
	All Woods	Hardwood	Softwood	
	Nearest Fractional Size Drill	Nearest Fractional Size Drill	Nearest Fractional Size Drill	
0	1/16	1/32	—	—
1	5/64	1/32	1/32	—
2	3/32	3/64	1/32	3
3	7/64	1/16	3/64	4
4	7/64	1/16	3/64	4
5	1/8	5/64	1/16	4
6	9/64	5/64	1/16	5
7	5/32	3/32	1/16	5
8	11/64	3/32	5/64	6
9	3/16	7/64	5/64	6
10	3/16	7/64	3/32	6
11	13/64	1/8	3/32	7
12	7/32	1/8	7/64	7

Standard screw lengths in inches are ³⁄₁₆, ¼, ⅜, ½, ⅝, ¾, ⅞, 1, 1¼, 1½, 1¾, 2, 2¼, 2½, 2¾, 3, 3½, and up.

TABLE 10 GRADES OF HARDWOOD LUMBER*

Grades	Clear cut ** (%)	Minimum Dimensions (W × L)	Principal Farm Uses
Firsts	91.7	6″ × 8′	Flooring, trim, furniture, implement & tool handles
Seconds	83.3	6″ × 8′	
Selects	91.7	4″ × 6′	
No. 1 Common	66.7	3″ × 4′	Permanent structures (barns, pens, feeders, troughs, fences)
No. 2 Common	50.0	3″ × 4′	
Sound & Wormy	33.3	3″ × 4′	Temporary structures (sheds, fences, feeders, crates)
No. 3A Common	33.3	3″ × 4′	
No. 3B Common	25.0	3″ × 4′	

* Adopted by National Hardwood Lumber Association.
** Percentage of face surface that is free of all defects.

Courtesy of T. J. Wakeman and V. L. McCoy, The Farm Shop, *The Macmillan Company, 1960*

TABLE 11 LUMBER DIMENSIONS

TYPE OF LUMBER	ROUGH SIZE (IN.) (SAWMILL CUT)	ACTUAL SIZE (IN.) (S 4 S)*	BD. FT. PER FT. OF LENGTH	AMOUNT TO ADD TO COVER**
	1 × 4	¾ × 3½	⅓	⅒
Square	1 × 6	¾ × 5½	½	¹⁄₁₆
Edge	1 × 8	¾ × 7¼	⅔	¹⁄₁₆
Boards	1 × 10	¾ × 9¼	⅚	¹⁄₂₀
	1 × 12	¾ × 11¼	1	¹⁄₂₄
	2 × 4	1½ × 3½	⅔	——
	2 × 6	1½ × 5½	1	——
	2 × 8	1½ × 7¼	1⅓	——
Framing	2 × 10	1½ × 9¼	1⅔	——
	2 × 12	1½ × 11¼	2	——
	4 × 4	3½ × 3½	1⅓	——
	6 × 6	5½ × 5½	3	——

* Surfaced on all four sides.
** Amount to add to total surface to allow for surfacing and matching of lumber. (Does not include carpenter's waste.)

TABLE 12 WOOD MEASURE

A board foot of lumber = A Unit 1′ long × 1′ wide × 1″ thick or its equivalent
1 U.S. cord of wood = A Pile 4′ wide × 4′ high × 8′ long or its equivalent
1 cord foot of wood = 4′ wide × 4′ high × 1′ long or its equivalent
Board feet = Number of pieces of wood × thickness in inches × width in inches × length in feet ÷ 12

TABLE 13 POSITIONS OF FRAMING SQUARE FOR SOME COMMON ANGLES

Position of		For Angle at	
Body	Tongue	Body	Tongue
12″	1¹⁄₁₆″	5°	85°
12″	2⅛″	10°	80°
12″	3³⁄₁₆″	15°	75°
12″	4⅜″	20°	70°
12″	5⁹⁄₁₆″	25°	65°
12″	6¹⁵⁄₁₆″	30°	60°
12″	8⅜″	35°	55°
12″	10¹⁄₁₆″	40°	50°
12″	12″	45°	45°

Courtesy of T. J. Wakeman and V. L. McCoy, The Farm Shop, *The Macmillan Company, 1960*

TABLE 14 RECOMMENDED R.P.M. FOR WOOD TURNING

Stock Diam.	Rounding Stock	General Cutting	Finishing
1″	900–1400	2400–3000	3000–4400
2″	800–1200	2000–2700	2700–4000
3″	700–1050	1800–2400	2400–3500
4″	600– 900	1600–2100	2100–3000
5″	550– 750	1400–1800	1850–2600
6″	500– 650	1200–1550	1600–2200
7″	450– 550	1000–1300	1350–1650
8″	400– 500	800–1050	1100–1400
9″	350– 450	600– 700	750–1150
10″	300– 400	400– 500	600– 950

Courtesy of T. J. Wakeman and V. L. McCoy, The Farm Shop, *The Macmillan Company, 1960*

TABLE 15 ABRASIVE PAPER GRADES

GRIT	EQUIV. "O" SERIES	GENERAL DESCRIPTION	REMARKS
600	——	Super Fine	Range of abrasive papers used for wet sanding
500	——		
400	10/0		
360	——		
320	9/0		
280	8/0	Very Fine	Used for dry sanding all finishing undercoats.
240	7/0		
220	6/0		
180	5/0	Fine	For final sanding of bare wood. Good for smoothing old paint
150	4/0		
120	3/0	Medium	Use for general wood sanding. Good for first smoothing of old paint, plaster patches
100	2/0		
80	1/0		
60	½	Coarse	For rough-wood sanding
50	1		
40	1½		
36	2	Very Coarse	Too coarse for pad sanders. Heavy machines and high speed recommended
30	2½		
24	3		

TABLE 16 RECOMMENDED R.P.M. OF HIGH-SPEED DRILLS IN VARIOUS METALS*

Drill Size**	Wrought Iron, Low-Carbon Steel	Medium-Carbon Steel	High-Carbon Tool Steel	Cast Iron	Aluminum & Brass
$\frac{1}{16}$	5000–6700	4000–4900	3000–3600	4200–6100	12,000–18,000
$\frac{1}{8}$	2500–3350	2100–2450	1500–1800	2100–3000	6000–9000
$\frac{3}{16}$	1600–2200	1400–1600	1000–1200	1400–2000	4000–6000
$\frac{1}{4}$	1200–1700	1000–1200	750–900	1100–1500	3000–4500
$\frac{5}{16}$	1000–1300	850–950	600–725	850–1200	2400–3600
$\frac{3}{8}$	800–1100	700–800	500–600	700–1000	2000–3000
$\frac{7}{16}$	700–950	600–700	425–525	600–875	1700–2500
$\frac{1}{2}$	600–850	450–600	375–450	525–750	1500–2250
$\frac{9}{16}$	550–750	425–550	340–400	475–675	1325–2000
$\frac{5}{8}$	500–650	400–500	300–350	425–600	1200–1800
$\frac{11}{16}$	450–600	375–450	275–325	375–550	1100–1650
$\frac{3}{4}$	400–550	350–400	250–300	350–500	1000–1500
$\frac{13}{16}$	375–500	325–375	235–275	325–460	900–1375
$\frac{7}{8}$	350–450	300–350	225–250	300–425	850–1275
$\frac{15}{16}$	325–425	275–325	210–235	275–400	800–1200
1	300–400	250–300	200–225	250–375	750–1125

*Reduce rpm one-half for carbon drills.
**For number and letter drills, use speed of nearest fractional-sized drill.

Courtesy of T. J. Wakeman and V. L. McCoy, The Farm Shop, *The Macmillan Company, 1960*

TABLE 17 DATA FOR SCREW EXTRACTORS

Screw or Bolt (In.)	Twist Drill (In.)	Extractor (No.)	Pipe (In.)
$\frac{3}{16}-\frac{1}{4}$	$\frac{5}{64}$	1	
$\frac{1}{4}-\frac{5}{16}$	$\frac{7}{64}$	2	
$\frac{5}{16}-\frac{7}{16}$	$\frac{5}{32}$	3	
$\frac{7}{16}-\frac{9}{16}$	$\frac{1}{4}$	4	$\frac{1}{8}$
$\frac{9}{16}-\frac{3}{4}$	$\frac{17}{64}$	5	$\frac{1}{4}$
$\frac{3}{4}-1$	$\frac{13}{32}$	6	$\frac{3}{8}$
$1-1\frac{3}{8}$	$\frac{17}{32}$	7	$\frac{1}{2}$
$1\frac{3}{8}-1\frac{3}{4}$	$\frac{13}{16}$	8	$\frac{3}{4}$
$1\frac{3}{4}-2\frac{1}{8}$	$1\frac{1}{16}$	9	1
$2\frac{1}{8}-2\frac{1}{2}$	$1\frac{5}{16}$	10	$1\frac{1}{4}$
$2\frac{1}{2}-3$	$1\frac{9}{16}$	11	$1\frac{1}{2}$
$3-3\frac{1}{2}$	$1\frac{15}{16}$	12	2

Courtesy of T. J. Wakeman and V. L. McCoy, The Farm Shop, The Macmillan Company, 1960

TABLE 18 SHEET METAL RIVETING DATA

Metal Gauge	Rivets			Drill, Punch	Rivet Set
	Weight	Diam.	Length		
28	14 oz.	.109″	$\frac{3}{16}$″	$\frac{7}{64}$″	#8
26	1 lb.	.112″	$\frac{13}{64}$″	$\frac{1}{8}$″	#7
24	2 lb.	.144″	$\frac{17}{64}$″	$\frac{9}{64}$″	#5
22	2½ lb.	.148″	$\frac{9}{32}$″	$\frac{5}{32}$″	#4
20	3 lb.	.160″	$\frac{5}{16}$″	$\frac{11}{64}$″	#4

Courtesy of T. J. Wakeman and V. L. McCoy, The Farm Shop, The Macmillan Company, 1960

TABLE 19 SELF-TAPPING SHEET METAL SCREW DATA (POINTED TYPE)

Screw No.	Screw Lengths*	Metal Gauge	Drill Size No.	Drill Size Inches
4	$\frac{1}{4}-\frac{3}{4}$	29–22	42	$\frac{3}{32}$
6	$\frac{3}{8}-1$	29–20	38	$\frac{7}{64}$
8	$\frac{3}{8}-1\frac{1}{2}$	26–18	32	$\frac{1}{8}$
10	$\frac{3}{8}-2$	24–18	29	$\frac{9}{64}$
12	$\frac{1}{2}-2$	24–18	25	$\frac{5}{32}$
14	$\frac{1}{2}-2$	24–18	12	$\frac{3}{16}$

* Standard screw lengths in inches are: $\frac{1}{4}$, $\frac{3}{8}$, $\frac{1}{2}$, $\frac{5}{8}$, $\frac{3}{4}$, $\frac{7}{8}$, 1, $1\frac{1}{4}$, $1\frac{1}{2}$, and 2. This column shows range of screw lengths available—example, for No. 4, lengths are $\frac{1}{4}$, $\frac{3}{8}$, $\frac{1}{2}$, $\frac{5}{8}$, and $\frac{3}{4}$.

Courtesy of T. J. Wakeman and V. L. McCoy, The Farm Shop, The Macmillan Company, 1960

TABLE 20 FLUXES FOR SOFT SOLDERING

METAL	FLUX	CHEMICAL NAME
Aluminum	Special Compounds	
Brass, copper	Cut acid, rosin, sal ammoniac	Zinc chloride, colophony, ammonium chloride
Galvanized iron	Muriatic acid	Hydrochloric acid
Iron, steel	Cut acid, sal ammoniac	Zinc chloride, ammonium chloride
Tin	Rosin, cut acid	Colophony, zinc chloride

TABLE 21 MELTING POINTS OF SOFT SOLDERS

Tin	Lead	Liqui- fies at *
63%	37%	360°F
50	50	415
45	55	437
40	60	459
33	67	621

*First melting point for each type is 360°F.

Courtesy of T. J. Wakeman and V. L. McCoy, The Farm Shop, The Macmillan Company, 1960

TABLE 23 HACK SAW BLADES

Teeth per In.	Stock	Material
14	1″ & over	cast iron, machine steel, brass, copper, aluminum, bronze, slate
18	¼″ to 1″	annealed toolsteel, high-speed steel, rails, bronze, copper, aluminum
24	⅛″ to ¼″	iron, steel, drill rod, brass & copper tubing, wrought-iron pipe, conduit, trim
32	⅛″ & less	same as for 24 teeth per in.

Courtesy of T. J. Wakeman and V. L. McCoy, The Farm Shop, The Macmillan Company, 1960

TABLE 22 RECOMMENDED COLORS FOR TEMPERING VARIOUS TOOLS*

Degrees Fahrenheit	Color	Kind of Tool
430	Yellow	Scrapers, lathe cutting tools, hammers
470	Straw	Punches, dies, hack-saw blades, drills, taps, knives, reamers
500	Brown	Axes, wood chisels, drifts, shears
530	Purple	Cold chisels, center punches, rivet sets
560	Blue	Screw drivers, springs, gears, picks, saws
700	Grey	Soft (annealed), must harden again

*Because of the difference in the quality of steel used in tool manufacture, the color recommendations will not always apply. E.g., it may be necessary to cool a cold chisel when at a straw or blue color instead of purple to get the desired temper.

Courtesy of T. J. Wakeman and V. L. McCoy, The Farm Shop, The Macmillan Company, 1960

TABLE 24 FUEL-OIL MIXTURES

MIX.	U.S. GALLONS		IMPERIAL GALLONS		METRIC	
	FUEL	OIL	FUEL	OIL	PETROL	OIL
16:1	1 Gallon	8 oz.	1 Gallon	10 oz.	4 Liters	.250 Liter
	3 Gallons	24 oz.	3 Gallons	30 oz.	12 Liters	.750 Liter
	5 Gallons	40 oz.	5 Gallons	50 oz.	20 Liters	1.250 Liter
	6 Gallons	48 oz.	6 Gallons	60 oz.	24 Liters	1.500 Liter
24:1	1 Gallon	5.33 oz.	1 Gallon	6.4 oz.	4 Liters	.160 Liter
	3 Gallons	16 oz.	3 Gallons	19.2 oz.	12 Liters	.470 Liter
	5 Gallons	26.66 oz.	5 Gallons	32 oz.	20 Liters	.790 Liter
	6 Gallons	32 oz.	6 Gallons	38.4 oz.	24 Liters	.940 Liter
32:1	1 Gallon	4 oz.	1 Gallon	5 oz.	4 Liters	.125 Liter
	3 Gallons	12 oz.	3 Gallons	15 oz.	12 Liters	.375 Liter
	5 Gallons	20 oz.	5 Gallons	25 oz.	20 Liters	.625 Liter
	6 Gallons	24 oz.	6 Gallons	30 oz.	24 Liters	.750 Liter
50:1	1 Gallon	2.5 oz.	1 Gallon	3 oz.	4 Liters	.080 Liter
	3 Gallons	8 oz.	3 Gallons	9 oz.	12 Liters	.240 Liter
	5 Gallons	13 oz.	5 Gallons	15 oz.	20 Liters	.400 Liter
	6 Gallons	15.5 oz.	6 Gallons	18.5 oz.	24 Liters	.480 Liter

Courtesy of Tecumseh Products Company

TABLE 25 ELECTRODES FOR FARM WELDING

TYPE OF WORK	ELECTRODE AWS No.	ELECTRODE NEMA Color	CURRENT & POLARITY (S-straight) (R-reversed)	WELDING POSITION	PENE-TRATION	CHARACTERISTICS	USES
MILD STEEL	E6010	None	DC-R	All	Deep	Digging affects it; leaves rough, rippled surface; slow burn off; a strong weld with much spatter.	All mild-steel welding; 1/8″ very good for holes and cutting; best farm electrode.
	E6011	Blue spot	All	All	Deep		
	E6012	White spot	All	All	Medium	Melted metal is gummy.	Fill in gaps and poor fit ups; medium arc keeps sagging metal from closing the circuit.
	E6013	Brown spot	All	All	Shallow		Very good for down welds on thin sheet metal.
	E6013	Brown spot	All	All	Shallow	General-purpose, easy to use; less burn through owing to shallower penetration.	All mild steel; for extra strength, lay a heavy first bead or two light beads.
	E6014	Brown spot	All	All	Shallow	Cross between E6013 and E6024; easy to use; fast burn off.	Second best for farm; same uses as above.
	E6024	Yellow spot	All	Flat, Horizontal	Medium	Fast deposit; slag removes itself if amperage is right and edges not pinned by poor electrode motion.	All downhand welds, in fillets and horizontal; 3/32″ good as spray rod on vertical down welds.
	E6027	Silver spot	All	Flat, Horizontal	Medium	Less undercut than E6024; better weld; slag not as easily removed; less penetration on rusty metal.	Same as E6024, but see Characteristics (to the left).
LOW-ALLOY METAL	E7016	(See Character.)	AC, DC-R	All	Varies	Requires clean surface*; NEMA code: green group, orange spot, blue end.	All difficult welds; rather hard for beginners to use.
NON-MACHIN-ABLE CAST IRON	No AWS; non-machin-able cast iron	Orange end	AC, DC-S	All	Medium	Requires clean surface*; low heat needed reduces cracking of weld or work: *do not permit work to become dark red.*	For shielded arc welds if not to be machined; hold close arc, but coating must not touch molten metal; intermittent beads not over 3″ long; peen lightly after each bead; cool and clean before next bead.
CAST IRON	No AWS; cast i.	Orange spot	AC, DC-R	All	Medium	Same as above, except machinable.	All cast iron, machinery gears, housings, parts, mold boards, etc.
STAIN-LESS STEEL	No AWS; s. steel 308-16	Yellow group & end	All	All	Medium	Requires short arc, electrode held 15° in direction of travel.	High-to-medium carbon-alloy steel and most nonferrous metals; auto bumpers, mold boards, etc.
HARD-SUR-FACING	No AWS numbers or NEMA color. Consult manufac-urers' catalogs according to uses.		AC, DC-R	All	Light		Metal-to-metal wear: building-up gives moderate hardness to resist shock and abrasion.
			AC, DC-S	All	Light	For all hardsurfacing: requires clean surface*; not for joining parts; build up worn parts with a strength rod, then hardsurface; for wear resistance, lay straight or weavy bead not over 3/4″ wide; remove slag before new bead.	Metal in rocky soil: build-up for resistance to impact and severe abrasion; for mild or carbon steel, low-alloy or high-manganese steel.
			AC, DC-S	All	Light		Metal in sandy soil: to resist any kind of abrasion, mild impact; for carbon, alloy, or manganese steel.
			AC, Arc torch	All	Light		Knife edges: fine-grain alloy powder applied with carbon arc gives smooth, abrasion-resistance surface.

* This means that surfaces must be cleaned of rust, grease, oil, moisture, and other foreign matter by wire brushing or grinding.

Courtesy of T. J. Wakeman and V. L. McCoy, The Farm Shop, *The Macmillan Company, 1960*
Note: See Figure 25–19 for additional information on electrode selection.

TABLE 26   ELECTRICAL WIRE SIZES FOR COPPER AT 115–120 VOLTS

COPPER up to 200 Amperes, *115- 120 Volts,* Single Phase, Based on 2% Voltage Drop

Length of Run in Feet — Compare size shown below with size shown to left of double line. Use the larger size.

Load in Amps	Types R, T, TW	Types RH, RHW, THW	Overhead in Air Bare & Covered Conductors	30	40	50	60	75	100	125	150	175	200	225	250	275	300	350	400	450	500	550	600	650	700
5	12	12	10	12	12	12	12	12	12	12	10	10	10	10	8	8	8	8	6	6	6	6	4	4	4
7	12	12	10	12	12	12	12	12	12	10	10	8	8	8	8	6	6	6	6	4	4	4	4	4	3
10	12	12	10	12	12	12	12	10	10	8	8	8	6	6	6	6	4	4	4	3	3	2	2	2	
15	12	12	10	12	12	10	10	10	8	6	6	6	4	4	4	4	3	2	2	1	1	0	0	00	00
20	12	12	10	12	10	10	8	8	6	6	4	4	4	4	3	3	2	2	1	1	0	0	00	00	00
25	10	10	10	10	10	8	8	6	6	4	4	4	3	3	2	2	1	1	0	0	00	00	000	000	000
30	10	10	10	10	8	8	8	6	4	4	4	3	2	2	1	1	1	0	00	00	000	000	000	4/0	4/0
35	8	8	10	10	8	8	6	6	4	4	3	2	2	1	1	0	0	00	00	000	000	4/0	4/0	4/0	250
40	8	8	10	8	8	6	6	4	4	3	2	2	1	1	0	0	00	00	000	000	4/0	4/0	250	250	300
45	6	8	10	8	8	6	6	4	4	3	2	1	1	0	0	00	00	000	000	4/0	4/0	250	250	300	300
50	6	6	10	8	6	6	4	4	3	2	1	1	0	0	00	00	000	000	4/0	4/0	250	250	300	300	350
60	4	6	8	8	6	4	4	4	2	1	1	0	00	00	000	000	000	4/0	250	250	300	300	350	400	400
70	4	4	8	6	6	4	4	3	2	1	0	00	00	000	000	4/0	4/0	250	300	300	350	400	400	500	500
80	2	4	6	6	4	4	3	2	1	0	00	00	000	000	4/0	4/0	250	300	300	350	400	400	500	500	600
90	2	3	6	6	4	4	3	2	1	0	000	000	000	4/0	4/0	250	250	300	350	400	500	500	500	600	600
100	1	3	6	4	4	3	2	1	0	00	000	000	4/0	4/0	250	250	300	350	400	500	500	600	600	600	700
115	0	2	4	4	4	3	2	1	0	00	000	4/0	4/0	250	300	300	350	400	500	500	600	600	700	700	750
130	00	1	4	4	3	2	1	0	00	000	4/0	4/0	250	300	300	350	400	500	500	600	600	700	750	800	900
150	000	0	2	4	2	1	1	0	000	4/0	4/0	250	300	350	350	400	500	500	600	700	700	800	900	900	1M
175	4/0	00	2	3	2	1	0	00	000	4/0	250	300	350	400	400	500	500	600	700	750	800	900	1M		
200	250	000	1	2	1	0	00	000	4/0	250	300	350	400	500	500	500	600	700	750	900	1M				

Source: Agricultural Wiring Handbook, *National Food and Energy Council,* 1984

TABLE 27   ELECTRICAL WIRE SIZES FOR ALUMINUM AT 115–120 VOLTS

ALUMINUM up to 200 Amperes, *115- 120 Volts,* Single Phase, Based on 2% Voltage Drop

Length of Run in Feet — Compare size shown below with size shown to left of double line. Use the larger size.

Load in Amps	Types R, T, TW	Types RH, RHW, THW	Overhead in Air Bare & Covered Conductors Single Triplex	30	40	50	60	75	100	125	150	175	200	225	250	275	300	350	400	450	500	550	600	650	700
5	12	12	10	12	12	12	12	12	10	10	8	8	6	6	6	6	4	4	4	4	3	3	2	2	1
7	12	12	10	12	12	12	12	10	10	8	8	6	6	6	6	4	4	4	4	3	3	2	2	2	1
10	12	12	10	12	12	10	10	8	8	6	6	4	4	4	4	3	3	2	2	1	1	0	0	0	0
15	12	12	10	12	10	10	8	8	6	4	4	4	3	3	2	2	2	1	0	0	00	00	000	000	
20	10	10	10	10	8	8	6	6	4	4	3	3	2	2	1	1	0	0	00	00	000	000	4/0	4/0	4/0
25	10	10	10	8	8	6	6	4	4	3	2	2	1	1	0	0	00	00	000	000	4/0	4/0	250	250	300
30	8	8	10	8	6	6	4	4	3	2	2	1	0	0	00	00	000	4/0	4/0	250	250	300	300	350	
35	6	8	10	8	6	6	4	4	3	2	1	0	0	00	00	000	000	4/0	4/0	250	300	300	350	350	400
40	6	8	10	6	6	4	4	3	2	1	0	0	00	00	000	000	4/0	4/0	250	300	300	350	350	400	500
45	4	6	10	6	6	4	4	3	2	1	0	00	00	000	000	4/0	4/0	250	300	300	350	400	400	500	500
50	4	6	• 8	6	4	4	3	2	1	0	00	00	000	000	4/0	4/0	250	300	300	350	400	400	500	500	600
60	2	4	6	6	4	3	3	2	0	00	00	000	4/0	4/0	250	250	300	350	350	400	500	500	500	600	700
70	2	2	6	4	4	3	2	1	0	00	000	4/0	4/0	250	300	300	350	400	500	500	600	600	700	700	750
80	1	2	6	4	3	2	1	0	00	000	4/0	4/0	250	300	300	350	350	500	500	600	600	700	750	800	900
90	0	2	4	4	3	2	1	0	00	000	4/0	250	300	300	350	350	400	500	500	600	600	700	800	900	1M
100	0	1	4	3	2	1	0	00	000	4/0	250	300	300	350	400	400	500	500	600	600	700	750	900	1M	
115	00	0	2	3	1	1	0	00	000	4/0	300	300	350	400	500	500	600	600	700	800	900	1M			
130	000	00	2	2	1	0	00	000	4/0	250	300	350	400	500	500	600	600	700	800	900	1M				
150	4/0	000	1	2	0	00	000	250	300	350	400	500	500	600	600	700	800	900	1M						
175	300	4/0	0	1	0	00	000	4/0	300	350	400	500	600	600	700	750	800	900							
200	350	250	00	0	00	000	4/0	250	300	400	500	600	600	700	750	900	900								

Source: Agricultural Wiring Handbook, *National Food and Energy Council,* 1984

TABLE 28 ELECTRICAL WIRE SIZES FOR COPPER AT 230–240 VOLTS

COPPER up to 400 Amperes, *230-240 Volts*, Single Phase, Based on 2% Voltage Drop

Compare size shown below with size shown to left of double line. Use the larger size.

Load in Amps	Types R, T, TW	Types RH, RHW, THW	Overhead in Air Bare & Covered Conductors	50	60	75	100	125	150	175	200	225	250	275	300	350	400	450	500	550	600	650	700	750	800
5	12	12	10	12	12	12	12	12	12	12	12	12	12	12	12	10	10	10	10	8	8	8	8	6	6
7	12	12	10	12	12	12	12	12	12	12	12	10	10	10	10	8	8	8	8	6	6	6	6	6	6
10	12	12	10	12	12	12	12	10	10	10	10	10	8	8	8	8	6	6	6	6	4	4	4	4	4
15	12	12	10	12	12	12	10	10	10	8	8	8	6	6	6	6	6	4	4	4	4	4	3	3	2
20	12	12	10	12	12	10	10	8	8	8	6	6	6	6	4	4	4	4	3	3	2	2	2	1	1
25	10	10	10	12	10	10	8	8	6	6	6	6	4	4	4	4	3	3	2	2	1	1	1	0	0
30	10	10	10	10	10	10	8	6	6	6	4	4	4	4	3	2	2	1	1	1	0	0	00		
35	8	8	10	10	10	8	8	6	6	4	4	4	4	3	3	2	2	1	1	0	0	0	00	00	00
40	8	8	10	10	8	8	6	6	4	4	4	3	3	2	2	1	1	0	0	00	00	00	000	000	
45	6	8	10	10	8	8	6	6	4	4	4	3	3	2	2	1	1	0	0	00	00	000	000	000	
50	6	6	10	8	8	6	6	4	4	4	3	3	2	2	1	1	0	0	00	00	000	000	000	4/0	4/0
60	4	6	8	8	8	6	4	4	4	3	2	2	1	1	0	0	00	00	000	000	4/0	4/0	4/0	250	
70	4	4	8	8	6	6	4	4	3	2	2	1	1	0	0	00	00	000	000	4/0	4/0	4/0	250	250	300
80	2	4	6	6	6	4	4	3	2	2	1	1	0	0	00	00	000	000	4/0	4/0	250	250	300	300	300
90	2	3	6	6	6	4	4	3	2	1	1	0	0	00	00	000	000	4/0	4/0	250	250	300	300	350	350
100	1	3	6	6	4	4	3	2	1	1	0	0	00	00	000	000	4/0	4/0	250	250	300	300	350	350	400
115	0	2	4	6	4	4	3	2	1	0	0	00	00	000	000	4/0	4/0	250	300	300	350	350	400	400	500
130	00	1	4	4	4	3	2	1	0	00	00	000	000	4/0	4/0	250	300	300	350	400	400	500	500	500	
150	000	0	2	4	4	3	1	0	0	00	000	000	4/0	4/0	4/0	250	300	350	350	400	500	500	500	600	600
175	4/0	00	2	4	3	2	1	0	00	000	000	4/0	4/0	250	250	300	350	400	400	500	500	600	600	600	700
200	250	000	1	3	2	1	0	00	000	000	4/0	4/0	250	300	350	350	400	500	500	600	600	700	700	700	750
225	300	4/0	0	3	2	1	0	00	000	4/0	4/0	250	300	300	350	400	500	500	600	600	700	700	750	800	900
250	350	250	00	2	1	0	00	000	4/0	4/0	250	300	300	350	350	400	500	600	600	700	700	750	800	900	1M
275	400	300	00	2	1	0	00	000	4/0	250	300	350	350	400	500	500	600	600	700	700	800	900	900	1M	
300	500	350	000	1	1	0	000	4/0	4/0	250	300	350	350	400	500	500	600	700	700	800	900	900	1M		
325	600	400	4/0	1	0	00	000	4/0	250	300	300	350	400	500	500	600	600	700	750	900	900	1M			
350	600	500	4/0	1	0	00	000	4/0	250	300	350	400	400	500	500	600	700	750	800	900	1M				
375	700	500	250	0	0	00	4/0	250	300	350	400	500	500	500	600	600	700	800	900	1M					
400	750	600	250	0	00	000	4/0	250	300	350	400	500	500	500	600	700	750	900	1M						

Conductors in overhead spans must be at least No. 10 for spans up to 50 feet and No. 8 for longer spans. See NEC, Sec. 225-6 (a).
See Note 3 of NEC "Notes to Tables 310-16 through 310-19".

Source: Agricultural Wiring Handbook, *National Food and Energy Council, 1984*

TABLE 29 ELECTRICAL WIRE SIZES FOR ALUMINUM AT 230–240 VOLTS

ALUMINUM up to 400 Amperes, *230-240 Volts*, Single Phase, Based on 2% Voltage Drop

Compare size shown below with size shown to left of double line. Use the larger size.

Load in Amps	Types R, T, TW	Types RH, RHW, THW	Overhead Bare & Covered Single	Overhead Bare & Covered Triplex	50	60	75	100	125	150	175	200	225	250	275	300	350	400	450	500	550	600	650	700	750	800
5	12	12	10		12	12	12	12	12	12	12	10	10	10	10	8	8	8	6	6	6	6	6	4	4	
7	12	12	10		12	12	12	12	12	10	10	10	8	8	8	6	6	6	6	4	4	4	4	4	4	
10	12	12	10		12	12	12	10	10	8	8	8	8	6	6	6	6	4	4	4	3	3	3	2	2	
15	12	12	10		12	12	10	8	8	8	6	6	6	4	4	4	4	3	3	2	2	1	1	1	0	
20	10	10	10		10	10	8	8	6	6	6	4	4	4	4	3	3	2	2	1	1	0	0	0	00	
25	10	10	10		10	8	8	6	6	4	4	4	3	3	2	2	1	1	0	0	00	00	00	000	000	
30	8	8	10		8	8	8	6	4	4	4	3	3	2	2	2	1	0	0	00	00	00	000	000	000	4/0
35	6	8	10		8	6	6	4	4	3	3	2	2	1	1	0	0	00	00	000	000	4/0	4/0	4/0		
40	6	8	10		8	6	6	4	4	3	3	2	1	1	0	0	00	00	000	000	4/0	4/0	4/0	250	250	
45	4	6	10		8	6	6	4	4	3	2	2	1	1	0	0	00	00	000	000	4/0	4/0	250	250	300	
50	4	6	8		6	6	4	4	3	2	2	1	1	0	0	00	00	000	000	4/0	4/0	250	250	300	300	300
60	2	4	6	6	6	6	4	3	2	1	0	0	00	00	00	000	4/0	4/0	250	250	300	300	350	350	350	
70	2	2(a)	6	4	6	4	4	3	2	1	0	0	00	00	000	000	4/0	4/0	250	300	300	350	350	400	500	
80	1	2(a)	6	4	4	4	3	2	1	0	00	00	000	000	4/0	4/0	250	300	300	350	400	500	500	500		
90	0	2(a)	4	2	4	4	3	2	1	0	00	00	000	4/0	4/0	250	300	300	350	400	500	500	500	600		
100	0	1(a)	4	2	4	3	2	1	0	00	00	000	000	4/0	4/0	250	300	300	350	400	500	500	600	600		
115	00	0(a)	2	1	4	3	2	1	0	00	000	000	4/0	250	300	300	350	400	500	500	600	600	700	700		
130	000	00(a)	2	0	3	2	1	0	00	000	000	4/0	250	250	300	300	350	400	500	500	600	700	700	750	800	
150	4/0	000(a)	1	00	2	2	1	00	000	000	4/0	250	250	300	300	350	500	500	600	600	700	750	800	900	900	
175	300	4/0(a)	0	000	2	1	0	00	000	4/0	250	300	300	350	400	400	500	600	600	700	750	900	900	1M		
200	350	250	00	4/0	1	0	00	000	4/0	250	300	300	350	400	400	500	600	600	700	750	900	900	1M			
225	400	300	000		1	0	00	000	4/0	250	300	350	400	500	500	500	600	700	750	900	1M	1M				
250	500	350	000		0	00	000	4/0	250	300	350	400	500	500	500	600	700	750	900	1M						
275	600	500	4/0		0	00	000	4/0	250	300	300	400	400	500	600	600	750	900	1M							
300	700	500	250		00	00	000	250	300	350	400	500	500	600	600	700	800	900	1M							
325	800	600	300		00	000	4/0	250	300	400	500	500	600	600	700	750	900	1M								
350	900	700	300		00	000	4/0	300	350	400	500	600	600	700	750	800	900									
375	1M	700	350		000	000	4/0	300	350	500	500	600	700	700	800	900	1M									
400		900	350		000	4/0	250	300	400	500	600	600	700	750	900	900										

Conductors in overhead spans must be at least No. 10 for spans up to 50 feet and No. 8 for longer spans. See NEC, Sec. 225-6 (a).
See Note 3 of NEC "Notes to Tables 310-16 through 310-19".

Source: Agricultural Wiring Handbook, *National Food and Energy Council, 1984*

TABLE 30 AVERAGE DAILY WATER REQUIREMENTS OF FARM ANIMALS

Horse . . . . 10-12 gallons

Beef Cattle . . . . 8-12 gallons

Dairy Cows (dry) . . . . 8-12 gallons

Dairy Cows . . . . 12-20 gallons

Hogs . . . . 4-11 gallons

Sheep . . . . 1-2 gallons

Layers (100) . . . . 5 gallons

Broilers (100) . . . . 1-10 gallons

Courtesy of The Maryland State Department of Education, Division of Vocational Technical Education, Baltimore, MD

TABLE 31 STEEL AND WROUGHT-IRON PIPE DATA*

Nom. Pipe Size	Outside Diam.	Standard Inside Diam.	Standard Weight Lb. Per Ft.	Extra Strong Inside Diam.	Extra Strong Weight Lb. Per Ft.	Threads Per Inch	Tap Drill Size	Hole Size **	Length of Thread	Dist. Pipe Goes into Fitting
$\frac{1}{8}$	0.405	0.269	0.24	0.215	0.31	27	$\frac{5}{16}$	$\frac{7}{16}$	$\frac{13}{32}$	$\frac{1}{4}$
$\frac{1}{4}$	0.540	0.364	0.42	0.302	0.53	18	$\frac{7}{32}$	$\frac{9}{16}$	$\frac{5}{8}$	$\frac{3}{8}$
$\frac{3}{8}$	0.675	0.493	0.57	0.423	0.74	18	$\frac{19}{32}$	$\frac{11}{16}$	$\frac{5}{8}$	$\frac{3}{8}$
$\frac{1}{2}$	0.840	0.622	0.85	0.546	1.09	14	$\frac{23}{32}$	$\frac{7}{8}$	$\frac{13}{16}$	$\frac{1}{2}$
$\frac{3}{4}$	1.050	0.824	1.13	0.742	1.47	14	$\frac{15}{16}$	$1\frac{1}{16}$	$\frac{7}{8}$	$\frac{1}{2}$
1	1.315	1.049	1.68	0.957	2.17	$11\frac{1}{2}$	$1\frac{3}{16}$	$1\frac{7}{16}$	$1\frac{1}{32}$	$\frac{9}{16}$
$1\frac{1}{4}$	1.660	1.380	2.27	1.278	3.00	$11\frac{1}{2}$	$1\frac{5}{16}$	$1\frac{11}{16}$	$1\frac{1}{16}$	$\frac{5}{8}$
$1\frac{1}{2}$	1.900	1.610	2.72	1.500	3.63	$11\frac{1}{2}$	$1\frac{25}{32}$	$1\frac{15}{16}$	$1\frac{1}{16}$	$\frac{11}{16}$
2	2.375	2.067	3.65	1.937	5.04	$11\frac{1}{2}$	$2\frac{3}{16}$	$2\frac{7}{16}$	$1\frac{1}{8}$	$\frac{3}{4}$
$2\frac{1}{2}$	2.875	2.469	5.79	2.323	7.66	8	$2\frac{5}{8}$	$2\frac{15}{16}$	$1\frac{5}{8}$	$\frac{7}{8}$
3	3.500	3.068	7.58	2.900	10.25	8	$3\frac{1}{4}$	$3\frac{9}{16}$	$1\frac{11}{16}$	$1\frac{5}{16}$

*All measurements are stated in fractional or decimal inches.
**Size of hole to drill to insert pipe through walls, floors, and other materials.

Courtesy of T. J. Wakeman and V. L. McCoy, The Farm Shop, *The Macmillan Company, 1960*

TABLE 32 STANDARD MALLEABLE FITTINGS

Nominal Size (in.)	General Dimensions (inches)					
	A	B	C	D	E	F
$\frac{1}{8}$	$1\frac{1}{16}$	$1\frac{1}{16}$	$1\frac{5}{16}$	—	1	$1\frac{3}{16}$
$\frac{1}{4}$	$1\frac{3}{16}$	$\frac{3}{4}$	$1\frac{1}{16}$	1	$1\frac{3}{16}$	$1\frac{5}{16}$
$\frac{3}{8}$	$1\frac{5}{16}$	$1\frac{3}{16}$	$1\frac{3}{16}$	$1\frac{1}{8}$	$1\frac{7}{16}$	1
$\frac{1}{2}$	$1\frac{1}{8}$	$\frac{7}{8}$	$1\frac{5}{16}$	$1\frac{1}{4}$	$1\frac{5}{8}$	$1\frac{1}{8}$
$\frac{3}{4}$	$1\frac{5}{16}$	1	$1\frac{1}{2}$	$1\frac{1}{2}$	$1\frac{7}{8}$	$1\frac{5}{16}$
1	$1\frac{1}{2}$	$1\frac{1}{8}$	$1\frac{11}{16}$	$1\frac{11}{16}$	$2\frac{1}{8}$	$1\frac{7}{16}$
$1\frac{1}{4}$	$1\frac{3}{4}$	$1\frac{5}{16}$	$1\frac{15}{16}$	$2\frac{1}{16}$	$2\frac{7}{16}$	$1\frac{11}{16}$
$1\frac{1}{2}$	$1\frac{15}{16}$	$1\frac{7}{16}$	$2\frac{1}{8}$	$2\frac{5}{16}$	$2\frac{11}{16}$	$1\frac{7}{8}$
2	$2\frac{1}{4}$	$1\frac{11}{16}$	$2\frac{1}{2}$	$2\frac{13}{16}$	$3\frac{1}{4}$	$2\frac{1}{4}$
$2\frac{1}{2}$	$2\frac{11}{16}$	$1\frac{15}{16}$	$2\frac{7}{8}$	$3\frac{1}{4}$	$3\frac{7}{8}$	—
3	$3\frac{1}{16}$	$2\frac{3}{16}$	$3\frac{3}{16}$	$3\frac{11}{16}$	$4\frac{1}{2}$	—

Courtesy of *T. J. Wakeman and V. L. McCoy*, The Farm Shop, *The Macmillan Company, 1960*

TABLE 33 RECOMMENDED MIXTURES FOR CONCRETE

Uses of Concrete	Water (gal.) for 1 Sack Cement—Sand				Trial Batch 1 Sack Cement plus		Space Trial Batch Fills (ft³)	Materials for 1 yd³**		
	Dry	Damp	Wet	Very Wet	Sand (ft³)	Gravel (ft³)		Cement (sack)	Sand (ft³)	Gravel (ft³)
Acid-, alkali-resistant; dairy, creamery floors	5*	$4\frac{3}{4}$	$4\frac{1}{2}$	$4\frac{1}{4}$	$1\frac{3}{4}$	2	3.38	8	14	16
Medium wear; reinforced; water-tight; floors, tanks, etc.	6*	$5\frac{1}{2}$	5	$4\frac{1}{4}$	$2\frac{1}{4}$	3	4.32	$6\frac{1}{4}$	14	19
Medium wear, indoor, underground, no water pressure	7*	$6\frac{1}{4}$	$5\frac{1}{2}$	$4\frac{3}{4}$	$2\frac{3}{4}$	4	5.4	5	14	20

* Called 5-, 6-, or 7-gallon pastes, indicating that there are 5, 6, or 7 gallons of water to each sack of cement, including the amount of water in the sand.
** To calculate materials for 1 cu. ft., divide these figures by 27.

Courtesy of *T. J. Wakeman and V. L. McCoy*, The Farm Shop, *The Macmillan Company, 1960*

TABLE 34 APPROXIMATE MATERIALS REQUIRED FOR ONE CUBIC YARD OF CONCRETE

Proportions of Concrete or Mortar			Quantities of Materials		
Cement	Sand	Gravel	Sacks Cement	Cu. Yds. Sand	Cu. Yds. Gravel
1	1.5	—	15.5	0.86	—
1	2.0	—	12.8	.95	—
1	2.5	—	11.0	1.02	—
1	3.0	—	9.6	1.07	—
1	1.5	3	7.6	.42	0.85
1	2.0	2	8.2	.60	.60
1	2.0	3	7.0	.52	.78
1	2.0	4	6.0	.44	.89
1	2.5	3.5	5.9	.55	.77
1	2.5	4	5.6	.52	.83
1	2.5	5	5.0	.46	.92
1	3.0	5	4.6	.51	.85
1	3.0	6	4.2	.47	.94

TABLE 35 AREAS COVERED BY ONE CUBIC YARD OF MIXED CONCRETE

Depth, in.	Sq. ft.	Depth, in.	Sq. ft.	Depth, in.	Sq. ft.
1	324	4 3/4	68	8 1/2	38
1 1/4	259	5	65	8 3/4	37
1 1/2	216	5 1/4	62	9	36
1 3/4	185	5 1/2	59	9 1/4	35
2	162	5 3/4	56	9 1/2	34
2 1/4	144	6	54	9 3/4	33
2 1/2	130	6 1/4	52	10	32.5
2 3/4	118	6 1/2	50	10 1/4	31.5
3	108	6 3/4	48	10 1/2	31
3 1/4	100	7	46	10 3/4	30
3 1/2	93	7 1/4	45	11	29.5
3 3/4	86	7 1/2	43	11 1/4	29
4	81	7 3/4	42	11 1/2	28
4 1/4	76	8	40	11 3/4	27.5
4 1/2	72	8 1/4	39	12	27

TABLE 36 PIGMENTS FOR COLORING CONCRETE

Color Desired	Pigment to Use
Blues	Cobalt oxide
Browns	Brown oxide of iron
Buffs	Synthetic yellow oxide of iron
Greens	Chromium oxide
Reds	Red oxide of iron
Grays & slate effects	Black iron oxide or carbon black (never use common lamp black)

Courtesy of T. J. Wakeman and V. L. McCoy, The Farm Shop, *The Macmillan Company, 1960*

GLOSSARY

abuse–to use wrongly, make bad use of, or misuse.

acrylonitrite-butadiene-styrene (ABS)–a type of plastic pipe used for sewerage and underground applications.

adaptor–a fitting used to connect pipes of different types.

adhesive–sticky substance used to bind two materials.

adjust–to set a part or parts to function as designed.

aerator–device that mixes air with water as it comes from a faucet.

aerosol–high-pressure container with a valve and spray nozzle.

agribusiness–broad range of activities associated with agriculture.

agricultural mechanics–selection, operation, maintenance, service, sale, and use of power units, machinery, equipment, structures, and utilities in agriculture.

agriculture–enterprises involving the production of plants and animals, along with supplies, services, mechanics, products, processing, and marketing related to those enterprises.

air atomization–paint split into tiny droplets using compressed air.

air-dried lumber–sawed lumber separated with wooden strips and protected from rain and snow for six months or more.

air-entrained concrete–ready-mixed concrete with tiny bubbles of air trapped throughout the mixture to strengthen it.

air tool–a tool powered by compressed air.

alkyd–oil-base paint.

allen screw–a screw with a six-sided hole in the head.

alloy–mixture of two or more metals.

alternating current (AC)–current that reverses its direction 60 times per second.

alternator–device that produces alternating electrical current but usually equipped with diodes to change current to direct current.

aluminum–tough, light, and durable metal; an element associated with high-quality paint.

American Welding Society (AWS)–an organization that supports education in welding processes and that developed a system of numerical classification of electrodes.

ampere (A) (amp)–a measure of the rate of flow of current in a conductor.

anneal–to cool steel slowly so as to make it soft and malleable.

annual rings–pattern in a tree caused by the hardening and disuse of tubes in the woody parts.

anvil–heavy steel object used to help bend, cut, and shape metal.

apparatus–equipment necessary to carry out a function.

arc–the discharge of electricity through an air space.

arc welder–a machine that produces current for welding.

arc welding–*see* shielded metal arc welding.

armature–the rotating part of a motor.

armored cable–flexible metal sheath with individual wires inside.

asphalt–black paint pigment with excellent hiding power.

backfire–a loud snap or popping noise in a gas torch which generally blows out the flame.

back saw–a saw that has very fine teeth and a stiff metal back.

back-stepping–series of short welds each made in the backward direction while the bead progresses forward.

balance–weight equally distributed on all sides of center.

band saw–power tool with saw teeth on a continuous blade or band.

base metal–main piece of metal or main component.

bead–continuous and uniform line of filler metal.

belt sander–power tool with a moving sanding belt.

bench stone–sharpening stone designed to rest on a bench.

bench yoke vise–a tool useful for holding pipe.

bevel–a sloping edge.

bill of materials–a listing of materials with specifications that are needed in a project.

bit brace–a device with a large crank-type handle used to turn wood cutting bits.

black pipe–steel pipe painted black that has little resistance to rusting.

blade guide–metal blocks or wheels that support the blade of a band saw.

bleeding the lines–removing gas pressure from all lines and equipment.

blind cut–cut made by piercing a hole with a sabre saw blade or by slowly lowering a portable hand saw blade into the material.

blue–the safety color used for signs of warning or caution.

board foot–amount of wood equal to a board 1 inch thick, 1 foot wide, and 1 foot long, or 144 cubic inches in volume.

border–heavy, solid, black line drawn close to the outer edges of paper used for drawing plans.

border line–*See* border.

bore–to make or drill a hole.

bottom dead center (BDC)–piston at its lowest point (point nearest the crankshaft).

branch circuit–electrical wiring, switches, and outlets that extend out from a service entrance panel.

brass–mixture of copper and zinc.

braze welding–bonding with metals and alloys that melt at or above 840° F where capillary action does not occur.

brazing–bonding with metals and alloys that melt at or above 840° F where capillary action occurs.

break line–solid, zigzag line used to show that the illustration stops but the object does not.

brick set–wide chisel used for breaking masonry units.

broom finish–rough surface formed by pushing a broom over concrete before it hardens to improve footing or traction.

bull float–smooth board attached to a long handle used to smooth out newly poured concrete.

bushing–a fitting used to connect pipe to pipe fittings of larger sizes. Also a device used to fill the space between a shaft and a larger hole in a blade or wheel.

business–work done for profit.

butt joint–joint formed by placing two pieces end to end in line or at a 90° angle.

butt weld–bead laid between two pieces of metal set edge to edge or end to end.

cadmium–metal used for plating of fasteners and other steel products for rust-resistance.

calcium–a low-quality paint pigment.

caliper–an instrument used to measure the diameter or thickness of an object.

cam–a raised area on a shaft.

cap–a fitting used to close a pipe end.

capillary action–rising of the surface of a liquid at the point of contact with a solid.

cap screw–a hex-head screw that is threaded over its entire length and is generally two inches or shorter.

carbon arc torch–a device which holds two carbon sticks and produces a flame from the energy of an electric welder.

carbonizing flame–a flame with an excess of acetylene.

cardiopulmonary resuscitation (CPR)–a first aid technique to provide oxygen to the body and circulate blood when breathing and heart beat stop.

carriage bolt–a threaded fastener with a round head over square shoulders.

casting–a mold that holds molten metal.

cast iron soil pipe–iron pipe with thick walls used mostly for sewerage systems.

caulk–material that stretches, compresses, and rebounds to maintain a tight seal between materials as they expand and contract.

cement paste–a paste made by mixing portland cement and clean water in exact proportions.

center line–a long- short-long line used to indicate the center of a round object on a diagram or plan.

center punch–steel punch with a sharp point.

CFM–cubic feet per minute.

chain vise–a tool useful for holding pipe.

chain wrench–wrench that utilizes a chain to grip and turn pipe.

chalking–a process whereby paint gradually washes off in the rain to stay bright.

chalk line–a cotton cord with chalk applied used to create long, straight lines.

chamfer–cutting down of a corner between the edge and face of a board.

chipping hammer–a hammer with a sharp edge and/ or point used to remove slag from a welding bead.

chlorinated polyvinyl chloride (CPVC)–plastic pipe suitable for hot water lines.

chuck–a device with jaws which open and close to receive and hold bits or other objects.

circuit–an electrical source and wires connected to a light, heater, or motor.

circuit breaker–a switch that trips and breaks the circuit when more than a specified amount of current passes through it.

clay–the smallest group of soil particles.

clinch–bending a nail over and driving the flattened end down into the material.

clockwise–the direction that hands turn on a clock.

cloth–fabric material used in some measuring tapes.

coarse aggregate–gravel; larger particles of stone used in concrete.

cold chisel–a piece of tool steel shaped, tempered, and sharpened to cut mild steel when driven with a hammer.

combination square–type of square that combines many tools in one.

combustion–burning.

commutator–the part of a motor that changes the flow of current at appropriate times.

compatible–going together with no undesirable reactions.

compress–to reduce in volume by pressure.

compressed air–air pumped under high pressure which may be carried by pipe, tubing, or reinforced hose.

compression fitting–a fitting that grips copper or steel tubing by compressing a special collar with a tapered, threaded nut.

compression ratio–the relationship between the volume of an engine cylinder at the beginning and end of the compression stroke.

compression stroke–movement of an engine piston to squeeze or compress the air-fuel mixture.

concave–hollow or curved in.

concrete–a mixture of stone aggregates, sand, portland cement, and water that hardens as it dries.

conductor–any material that will permit electrons (electricity) to move through it.

conduit–metal tube with individual insulated wires inside.

construction joint–place where one pouring of concrete stops and another starts.

contaminant–any material that does not belong in a substance.

continuity–connectedness.

continuity tester–device used to determine if electricity can flow between two points.

continuous duty–tool that can be used all the time for a 6- or 8-hour day.

control joint–planned break which permits concrete to expand and contract without cracking.

convex–curved out.

coolant–liquid used to cool parts or assemblies.

coping saw–a saw that has a very thin and narrow blade supported by a spring steel frame.

cordless–tool containing a rechargeable battery pack to drive the unit when not plugged into an electrical outlet.

core–the hollow space in a masonry block.

corner pole–a straight piece of wood or metal held plumb by diagonal supports and used to support a line when laying block or brick.

corrosion–reaction of metal to liquids and gases that causes them to deteriorate or break down.

counterclockwise–opposite of the way hands move around a clock.

countersink–tapered hole for a screw head.

coupling–a fitting used to connect two pieces of similar pipe.

course–a row of masonry units.

crack a cylinder–turn gas on and off quickly to blow dust from the opening.

craftsman–skilled worker.

crankshaft–a shaft with an offset projection that converts circular motion to reciprocal motion, or vise versa.

crater–a low spot in metal where the force of a flame has pushed out molten metal.

creosote–popular wood preservative which is black in color.

crosscut–to cut across the grain of wood.

crosscut saw–saw with teeth cut and filed to a point, used to cut across the grain of wood.

crown–the part of a tool which receives the blow of a hammer.

cuprinol–popular wood preservative used when plants will be near the wood.

curing–proper drying of concrete.

cutting tool–tool used to cut, chop, saw, or otherwise remove material.

cycle–all the events that take place as an engine takes in air and fuel, compresses the air-fuel mixture, burns the fuel, and expels the burned gasses.

cylinder–long round tank with extremely thick walls built to hold gases under great pressure.

dado–square or rectangular groove in a board.

dado head–special blade that is adjustable to cut kerfs from ⅛ to ¾ inch in one pass.

dado joint–rectangular groove cut in a board with the end or edge of another board inserted.

decibel (dB)–standard unit of sound.

diameter–distance across the center of a circle or round object.

diaphragm–a flat piece of rubber that creates a seal between moving and nonmoving parts.

die–an instrument used to cut threads onto a rod or bolt.

die stock–handle used to turn a die.

digging tool–device used to move soil or rock.

dimension–measurement of length, width, or thickness.

dimension line–solid line with arrowheads at the ends to indicate the length, width, or height of an object or part.

direct current (DC)–current that flows in one direction continuously.

disc sander–tool with sanding grit on a revolving plate.

dividers–two sharp steel legs connected on one end that are used to make arcs and circles.

double cut file–file that has teeth laid out in two directions.

double extra heavy–steel pipe with walls thicker than standard and extra heavy pipe.

double insulated–motor with electrical parts insulated or separated from the user by special insulation inside the motor and by its insulating plastic motor housing.

dovetail joint–a joint formed by interlocking parts of two pieces.

dowel–a round piece of wood.

drawfiling–a filing procedure done by placing the file at a 90° angle to the metal.

drawing–a picture or likeness made with a pencil, pen, chalk, crayon, or other instrument.

draw the temper–modify the temper of steel to render it softer.

dress–to remove material and leave a smooth surface.

drill press–stationary tool used for making holes in metal and other materials.

driving tool–tool used to move another tool or object.

drop cloth–material used to protect floors, furniture, shrubbery, and the like from falling paint.

dual–two.

ductile–able to be bent slightly without breaking.

duplex receptacle–double receptacle wired so that both outlets are on the same circuit.

duty cycle–the proportion of time a motor or welder can run without over heating.

dysfunctional–does not work the way it should.

ear–short extensions on the ends of masonry block.

efficient–able to produce with a minimum of time, energy, and expense.

elbow (ell)–a fitting used where a single pipe line changes direction.

electricity–form of energy that can produce light, heat, magnetism, and chemical changes.

electrode–a metal welding rod coated with flux and used with an electric welder.

electrode holder–a spring-loaded device with insulated handles used to grip welding electrodes.

electromagnet–a core of magnetic material surrounded by a coil of wire through which an electric current is passed to magnetize the core.

employment–work done for which one is paid by the hour, day, week, month, or year.

enamel–paint with a gloss or semigloss finish.

end marking–color on the end of an electrode.

end nailing–nailing through the thickness of one piece and into the end of another piece.

epoxy–a synthetic paint material with exceptional adhesion and wear-resistant qualities.

exhaust–burned gases removed by the motion of a piston.

exhaust stroke–movement of a piston which expels burned gases from a cylinder.

extension line–solid line showing the exact area specified by a dimension.

exterior–paint able to withstand moisture and outside weather conditions.

extinguish–to put out a fire by cooling, smothering, or removing fuel.

extractor–device that removes moisture and oil from compressed air.

extra heavy–pipe with thicker than standard walls.

eye–a piece of metal bent into a small circle.

face–surface that is intended for use.

farmstead layout–arrangement of buildings on a farm.

fastener–device used to hold two or more pieces of material together.

faucet–device that controls the flow of water from a pipe or container.

faucet washer–rubberlike part that creates a seal when pressed against a metal seat.

feather–to sand so that a chipped edge is tapered.

ferrous–metal that comes from iron ore.

ferrule–a metal collar fitted on a handle to prevent splitting of the wood.

field–magnetic force around an electric wire or iron core.

field use–location where repairs and other operations are done.

50–50 solder–solder composed of 50 percent tin and 50 percent lead.

filament–special metal element in a vacuum that produces light when electricity flows through it.

file–flat, round, half-round, square, or three-sided piece of metal with fine teeth.

filler rod–metal in the form of a long, thin rod used to add to or fill joints when brazing and welding.

fillet weld–a weld placed in a joint created by a 90° angle.

fine aggregate–sand and other small particles of stone.

fingering–dividing and clustering of the bristles of a paint brush.

finish–chemical layer that protects the surface of a material.

finishing lime–powder made by grinding and treating limestone for use in masonry materials.

finishing sander–tool with a small sanding pad driven in a forward-backward or circular pattern.

fire–flame; to make a spark jump across an air gap.

fire triangle–the three conditions—fuel, oxygen, and heat—that must be present to produce a fire.

fitting–a part used to connect pieces of pipe or to connect other objects to pipe.

fixture–a base or housing for a light bulb, fan motor, or other electrical device.

flammable–capable of burning easily.

flaring block–tool used to hold tubing when flaring.

flaring tool–tool used to expand the opening in tubing to a bell shape.

flashback–burning inside an oxyfuel torch that causes a squealing or hissing noise.

flat finish–dull or without shine.

flat head screw–screw that has a flat head with a tapered underside designed to fit down into the material being secured.

flat nailing–two flat pieces nailed to each other.

float valve assembly–a device used to control liquid flow into a reservoir such as a flush tank in a toilet.

fluorescent light–light tube that glows as a result of electricity flowing through gases.

flush plate–thin device fastened across a joint to provide support.

flush valve–a device used to control water used in the flushing process.

flux–material that removes tarnish or corrosion, prevents corrosion from developing, and acts as an agent to help solder spread over metal.

folding rule–rigid rule of 2 to 8 feet in length which folds into a compact unit.

footer (footing)–a continuous slab of concrete that provides a solid, level foundation for block and other masonry.

form–a metal or wooden structure that contains and shapes concrete until it hardens.

formulate–to put together according to a formula.

four-cycle engine–*see* four-stroke-cycle engine.

four-stroke-cycle engine–an engine with four strokes per cycle.

framing square (carpenter's square)–a flat square with a body and tongue.

frost line–the maximum depth that the soil freezes in a given locality.

fuel–any material that will burn.

full scale–a drawing the same size as the object it represents.

fuse–plug or cartridge containing a strip of metal that melts when more than a specified amount of current passes through it.

fusion–joining by melting

fusion welding–joining parts by melting them together.

galvanized–coated with zinc for rust resistance.

gas–any fluid substance that can expand without limit.

gas forge–a gas-burning unit used to heat pieces of metal and temper tools.

gas furnace–gas-burning unit used to heat objects.

gauge–a device used to determine the dimension of materials or space; a device used to measure and indicate pressure in a hose, pipe, or tank.

generator–a device that produces direct current.

glazing compound–special puttylike material used to install window glass.

gloss–paint with a shiny finish.

gradations–numbers and lines stamped or painted on measuring devices.

grain–lines on lumber caused by the annual rings in a tree.

graph paper–paper laid out in squares of equal size.

gravel–particles of stone larger than sand, also called coarse aggregate.

green–the safety color used to indicate the presence of safety equipment, safety areas, first aid, and medical practice.

grinding wheel–abrasive cutting particles formed into a wheel by a bonding agent.

ground clamp–connector used to attach a cable, wire or object to a ground source.

ground (gee) clip–a device by which ground wires are attached to electrical boxes.

ground-fault interrupter (GFI)–a device that cuts off the electricity if even very tiny amounts of current leave the normal circuit.

grounding–making an electrical connection between a piece of equipment and the earth.

hacksaw–a device that holds a blade designed for cutting metal.

hammer–striking action.

hand drill–a manually operated device with gears that drive a bit faster than the handle turns.

hands-on–actually doing something rather than reading or hearing others tell about it.

hand saw–a saw used to cut across boards or to rip boards and panels.

hand stone–a sharpening stone designed to be held in the hand.

hand tool–a tool operated by hand to do work.

hardware–fasteners; objects made from metal.

heat–type of energy which causes the temperature to rise.

hidden line–a series of dashes which indicates the presence of unseen edges.

hiding power–the ability of a material to create color and mask out the presence of colors over which it is spread.

high-speed drill–a twist drill made and tempered specially to drill metal.

hinge–an object that pivots and permits a door or other object to swing back and forth or up and down.

holding tool–a tool used to grip wood, metal, plastic, and other materials.

hollow core solder–solder with flux inside.

hollow core block–masonry block with two or three holes per block.

hollow ground–a saw blade with teeth wider at the points than at the base.

horizontal–flat or level.

horizontal shaft engine–an engine with a crankshaft that lies crossways for normal operation.

hose–flexible line that carries gases or liquids.

ID–inside diameter.

ignition–a spark igniting an air-fuel mixture.

impurity–any material other than the base metal.

inclined plane–a surface at an angle to another surface; one of the six simple machines.

insulator–material that provides great resistance to the flow of electricity.

intake stroke–engine process of taking fuel and air into the combustion chamber.

interior–paint that will not hold up if exposed to weather.

internal combustion engine–device that burns fuel inside a cylinder to create a force that drives a piston.

iron–ferrous; an element associated with high-quality paint.

item–separate thing.

jigsaw–a saw that cuts very short curves by reciprocal action.

jobbers-length drill–a drill of longer than standard length.

joint–the place where two pieces come together.

jointer–a machine with rotating knives used to straighten and smooth edges of boards; a masonry tool used to shape joints.

K–classification of thick wall copper pipe or tubing.

kerf–opening in a board made by a saw; opening in steel made by oxyfuel cutting.

keyhole saw–a saw designed for starting in holes.

kiln-dried–lumber that has been dried by heat in a special oven called a kiln.

kilo–one thousand.

kilowatthour–the use of 1000 watts for one hour.

knockout–partially punched impression in electrical boxes.

L–classification of medium wall copper pipe or tubing.

labor–to struggle or work hard to keep running.

lag screws (lag bolts)–screws with coarse threads designed for use in structural timber or lead anchors.

laminate–to fasten two or more flat pieces together with an adhesive.

lap joint–joint formed by fastening one member face-to-face on another member of an assembly.

latex–water-base paint.

laying block–the process of mixing mortar, applying it to masonry block, and placing the block to create walls.

layout–a plan, map, or pattern for future operations.

layout tool–tool used to measure or mark wood, metal, and other materials.

lead–heavy metal element that stays in the body once it is ingested; one component of solder.

leader line–solid line with an arrow used with an explanatory note to point to a specific feature.

leaner–a greater proportion of air and a lesser proportion of fuel in an air-fuel mixture.

level–a device used to determine if an object has the same height at two or more points. *See* spirit level.

line–small diameter material stretched tightly between two or more points.

linear–in a straight line.

loading the brush–the process of dipping bristles into paint.

M–classification of thin wall copper pipe or tubing.

machine bolt–fastener with a square or hex head on one end and threads on the last inch or so on the other end.

magnesium–a low-quality paint pigment.

magnetic field–area around an iron core or electric wire influenced by the presence of magnetism.

magnetic flux–lines of magnetic force which occur in patterns between poles of a magnet.

magnetism–a force that attracts or repels iron or steel.

maintenance–doing the tasks that keep a machine in good condition.

malleable–workable.

malleable cast iron–combination of cast iron core and ductible metal.

manifold–device with openings to serve two or more units.

mask–to cover so paint will not touch.

masonry–anything constructed of brick, stone, tile, or concrete units held in place with portland cement.

masonry units–block made from concrete or cinders.

mechanic–a person specifically trained to perform mechanical tasks.

mechanical–of or having to do with a machine, mechanism, or machinery.

mechanics–the branch of physics dealing with motion and the action of forces on bodies or fluids.

meter (m)–a metric unit of linear measure equalling 39.37 inches.

meter–a device that measures electricity, gas, or liquid that passes through a system.

metric system–a decimal system of measures and weights.

milliampere–thousandth of an ampere.

millimeter (mm)–thousandth of a meter.

miter–an angle.

miter box–a device used to cut molding and other narrow boards at angles.

miter gauge–adjustable, sliding device to guide stock into a saw at the desired angle.

miter joint–joint formed by cutting the ends of two pieces at a 45° angle.

mix–the ratio of materials in concrete or mortar.

molding head–a device that holds knives to shape wood into moldings of various types.

momentum–turning force of the flywheel and other moving parts that carries an engine through non-power strokes.

mortar–a mixture of portland cement, finishing lime, water, and sand.

mortar bed–a layer of mortar.

mushroom–to spread out over an edge.

mushroomed–a spread or pushed-over condition caused by being struck repeatedly.

nail–fastener that is driven into the material it holds.

nail set–punchlike tool with a cupped end.

nap–soft, woolly, threadlike surface.

National Electrical Manufacturers Association (NEMA)–the group that developed the system of color coding electrodes.

neat's-foot oil–light yellow oil obtained by boiling the feet and shinbones of cattle and used to soften and preserve leather.

neoprene–tough rubberlike material that is resistant to moisture and rotting.

neutral flame–flame with a balance of acetylene and oxygen.

neutral wires–conductors with white insulation that carry current from an applicance back to the source.

new wood–wood that has never been painted or sealed.

new work–original construction, including new wood.

nipple–a short piece of steel pipe threaded at both ends.

noise intensity–energy in the sound waves.

nominal–the size of material as used in the name for trade and building purposes.

nonmetallic sheathed cable–cable composed of copper or aluminum wires covered with paper, rubber, or vinyl for insulation and protection.

north pole–end of a magnet opposite the south pole.

nut–threaded device with shoulders.

oakum–ropelike material that makes a watertight seal when compacted.

object line–solid line in a drawing that shows visible edges of an object.

occupation–business, employment, or trade.

occupational cluster–group of related jobs.

occupational division–group of occupations or jobs within a cluster that require similar skills.

OD–outside diameter.

off-the-farm agricultural jobs–those jobs requiring agricultural skills, but not regarded as farming or ranching.

ohm–a measure of the resistance of a material to the flow of electrical current.

Ohm's Law–the relationship between electric current (I), electromotive force (E), and resistance (R): E = IR.

oil-base–paint containing some type of oil as the vehicle.

old work–wood that was previously painted or finished.

open circuit–an electrical circuit in which there is at least one place where current cannot flow.

orange–safety color used to designate machine hazards such as edges and openings.

orbital–circular or egg-shaped pattern.

orbital sander–a finishing sander that travels in a circular pattern.

O ring–piece of rubber shaped like an *O* which fits in a groove on a shaft and prevents liquids from passing.

oval head screw–screw with a head that extends both above and below the surface of the material being held.

oxidation–combining with oxygen.

oxide–the product resulting from oxidation of metal.

oxidizing flame–flame with an excess of oxygen; hottest type of oxyfuel flame.

oxyacetylene–oxygen and acetylene combined.

oxyfuel–the combination of nearly pure oxygen and a combustible gas to produce a flame.

oxyfuel cutting–process in which steel is heated to the point where it burns and is removed in a thin line called a kerf.

oxygen–gas in the atmosphere which is necessary to support combustion.

packing–soft, slippery, wear-resistant material used to seal space between a moving and nonmoving part.

pad–metal for practice welding.

paint–substance consisting of pigments and a vehicle.

paint film–material left after paint has dried.

paint roller–a cylinder that turns on a handle and is used to apply paint.

pan head screw–screw that has a head that looks like a frying pan turned upside down.

parallel–two edges or lines the same distance apart at all points.

pass–one bead of metal or other material.

pattern–a model or guide for something to be made.

penny–a unit of measure used to designate the length of most nails.

pentachlorophenol–a popular wood preservative sold in a nearly clear vehicle.

permanent magnet–a piece of iron or steel that holds its magnetism.

perpendicular–at a 90° angle to some object.

phillips head screw–a screw with screwdriver slots in the shape of a plus sign.

phillips screwdriver–a screwdriver with the tip shaped like a plus sign (+).

pictorial drawing–a drawing that shows three views in one.

pierce–to make a hole by pushing through.

pigment–a solid coloring substance suspended in a liquid, or vehicle.

pilot hole–a small hole drilled in material to guide the center point of larger drills; hole drilled to receive the threaded part of a screw.

pipe–rigid tubelike material.

pipe dope–substance applied to threaded fittings to prevent leakage at pipe connections.

pipe fitting–*See* fitting.

pipe stand–work area for cutting and threading pipe.

piston–a sliding cylinder fitting within a cylindrical vessel.

pivot–to turn or swing on.

plane–a tool that shaves off small amounts of wood and leaves the surface smooth.

plane iron–the cutting part of a plane.

planer–a machine with turning knives that dress the sides of boards to a uniform thickness.

plan reading–reading or interpreting scale drawings.

plastic–term used for a group of synthetic materials made from chemicals and molded into objects.

play the flame–to alternately move a flame into and out of an area to carefully control temperatures.

pliable–able to move without separating if pushed or pulled.

plow bolt–a bolt with a square tapered head that is flush with the surface when installed.

plug–a device used to close a pipe fitting.

plug tap–a device used to cut pipe threads in a hole.

plumb–vertical to the axis of the earth or in line with the pull of gravity.

plumb bob–a round, tapered piece of metal attached to a plumb line.

plumbing–installing and repairing water pipes and fixtures.

plumb line–a string with a plumb bob attached.

poles–opposite ends of a magnet.

policy–firm rules that govern management.

polyethlene (PE)–plastic pipe used extensively for cold water lines.

polyurethane–a clear, durable, water-resistant finish.

polyvinyl chloride (PVC)–a relatively new type of plastic pipe suitable for interior plumbing.

poppet valve–a valve that controls the flow of air and gases by moving up and down.

pop rivet–a tubular rivet that is enlarged by pulling a ball-ended stem in it until the stem breaks leaving the rivet secured.

port–a special hole in the cylinder wall of a two-cycle engine to permit gases to flow in or out of the cylinder.

portable circular saw (power handsaw)–a lightweight, motor-driven, round-bladed saw.

portland cement–dry powder made by burning limestone and clay followed by grinding and mixing.

positive (hot) wires–conductors with black, red, or blue insulation that carry current to an appliance.

power handsaw–*See* portable circular saw.

power machine–a tool driven by electric motor, hydraulics, air, gas engine, or some force other than or in addition to human power.

power stroke–the engine process in which burning fuel expands rapidly but evenly to drive the piston down.

power tool–a tool operated by some source other than human power.

precleaner–a device that removes large particles from air entering an air cleaner.

pressure feed system––a spray painting system that delivers paint under pressure to the mixing nozzle of the spray gun.

primer–a special paint used to seal bare wood and prepare metal surfaces for high-quality top coats.

procedure–a sequence of steps for doing something; a particular course of action.

profit–income made from the sale of goods or services.

project–special activity planned and conducted with the purpose of learning.

propellant–the gas that propels paint toward the object being painted.

protractor–an instrument for drawing or measuring angles.

puddle–a small pool of liquid metal.

purge the lines–remove undesirable gases.

purple–the safety color used to designate radiation hazards.

push drill (automatic drill)–a drill with a spiral shaft that turns a chuck when pushed against an object.

push stick–a wooden device with a notch in the end to push or guide stock on the table of a power tool.

putty–soft material containing oils that keep it pliable over a long period of time, used to seal joints or fill holes.

rabbet–a cut or groove at the end of a board made to receive another board and form a joint.

rabbet joint–a joint formed by setting the end or edge of one board into a groove in the end or edge of another board.

radial arm saw–a power circular saw which rolls along a horizontal arm.

rasp–a file with very coarse teeth.

ratio–proportion of one component to another by weight or volume.

receptacle–a device for receiving an electrical plug.

reciprocal–back and forth.

reciprocate–return; move back and forth.

reciprocating saw–a saw with a stiff blade that moves back and forth.

recondition–to do what is needed to put back into good condition.

red–the safety color used to designate areas or items of danger or emergency.

reducer–a fitting used to connect pipes of different sizes.

reed valve–a flat, flexible plate that permits air or liquid to pass in one direction but seals when the flow reverses.

regulator–a device that keeps pressure at a set level or controls the rate of flow of a gas or liquid.

reinforced concrete–concrete slabs or structures that are strengthened with embedded steel rods or wire mesh.

renewable natural resources–resources provided by nature that can replace or renew themselves.

repair–to replace a faulty part or make it work correctly.

repel–to push away.

represent–to stand for or to be a sign or symbol of.

resistance–any tendency of a material to prevent electrical flow.

reverse (positive) polarity (RP)–DC current flowing in the opposite direction from straight polarity; to reverse the direction of current.

reversible–capable of running backward as well as forward.

revolution–one complete turn of 360°.

richer–a mixture with an increased proportion of fuel to air.

rig–a piece of apparatus assembled to conduct an operation.

right triangle–a three-sided figure with one 90° angle.

rip–to cut the long way on a board or with the grain.

rip fence–a guide that helps keep work in a straight line with a saw blade.

rip saw–a saw with teeth filled to a knifelike edge and used to cut with the grain.

rivet–to spread or shape by hammering; a fastening device held in place by spreading one or both ends.

roller cover–a hollow, fabric-covered cylinder that slides onto the handle of a paint roller.

root–the deepest point in a weld.

root pass–the first welding pass that is made in a joint.

rot–to decay or break down into other substances.

rough lumber–lumber as it comes from the sawmill.

rounded up or down–going to the next higher or lower whole number.

round head screw–screw that extends in an even curve above the surface of the material being held.

RPM–revolutions per minute.

rust–reddish brown or orange coating that results when iron reacts with air and moisture.

sabre saw–a reciprocal saw used primarily for cutting curves or holes in wood, metal, cardboard, and similar materials.

saddle soap–a product used to clean, soften, and preserve leather.

safe–free from harm or danger.

safety–freedom from accidents.

sag–shifting of a large area of paint downward.

sal ammoniac–a cube or block of special flux for cleaning soldering coppers.

sand–small particles of stone.

satin–a wood or metal finish with low sheen or shine.

saturate–to add a substance until the excess starts to run out.

sawhorse–trestle; a wood or metal bar with legs, used for temporary support of materials.

scale–an instrument with all increments shortened according to proportion; numbers and gradations on measuring tools; a rigid steel or wooden measuring device; the size of a plan compared with that of the object it represents.

scale drawing–a drawing that represents an object in exact proportion although the object is larger or smaller than the drawing itself.

scope–size and complexity.

score–to scratch.

scratch awl–a sharply pointed tool with a wooden or plastic handle used to mark metal.

screeding–striking off excess concrete to create a smooth and level surface.

screw–a fastener with threads that bite into material as it is turned.

screwdriver—a turning tool with a straight tip, phillips tip, or special tip.

screw plate–a set of taps, dies, and handles used for making threads.

scriber–a very small, metal, sharp-tipped marker.

seal–to apply a coating that fills or blocks the pores so that no material can pass through the surface.

sealer–a coating used to fill pores and prevent material from passing through a surface

seat–a stationary or nonmoving part that is designed to seal when a moving part is pressed against it.

semigloss–paint with a slight shine or gloss.

serrated–notched.

service drop–an assembly of electrical wires, connectors, and fasteners used to transmit electricity from a transformer to a service entrance panel.

service entrance panel–a box with fuses or circuit breakers where electricity enters a building.

set–to drive the head of a nail below the surface.

shank–the nonthreaded part of a nail or screw.

shank hole–the hole provided for the shank of a screw.

shear–to cut by action of opposed cutting edges.

shears–large scissorlike tools for cutting sheet metal and fabrics.

sheet metal screw–a screw that has threads wide enough to permit thin metal to fit between the ridges of the threads.

shellac–a natural material that comes from an insect and is used to seal wood.

shielded metal arc welding (stick welding)–welding with electrical power as a source of heat and rods covered with flux which form a gaseous shield around the molten metal until it solidifies.

shock–the body's reaction to an electric current.

short circuit–a condition that occurs when electricity flows back to its source too rapidly and blows fuses, burns wires, and drains batteries.

shroud–to cover; a cover.

silicon–a low-quality paint pigment.

silt–a substance composed of intermediate sized soil particles.

single-pole switch–the one switch that controls one or more lights and/or outlets.

siphon system–a spray painting system that creates a vacuum at the nozzle of the gun and utilizes atmospheric pressure to push the paint to the nozzle.

60-cycle current–electricity that alternates direction of flow 60 times per second.

sketch–a rough drawing of an idea, object, or procedure.

slag–the product formed when burning steel combines with oxygen.

slag box–a container of water or sand placed to catch hot slag and metal from the cutting process.

sliding T bevel–a device used to draw angles on boards or metal.

slotted head screw–a screw with one straight slot across the head.

slow moving vehicle (SMV) emblem–reflective emblem consisting of an orange triangle with a red strip on each of the three sides.

snips–a large scissorlike tool for cutting sheet metal and fabrics.

soapstone–a soft, gray rock that shows up well when marked on most metals.

solder–mixture of tin and lead.

soldering–bonding with metals and alloys that melt at temperatures below 840° F.

soldering copper–a tool consisting of a handle, steel shank, and copper tip used to heat metal for soldering.

solderless connector–*See* wire nut.

solidify–to harden or change from a liquid to a solid.

south pole–end of a magnet opposite the north pole.

spar varnish–special finish for wood exposed to high moisture conditions.

species–plants or animals with the same permanent characteristics.

speed indicator–a device used to measure revolutions per minute (RPM) of a turning shaft or part.

spirit level–a tool containing alcohol in a sealed, curved tube with a small air space or bubble, used to determine whether an object has the same height at two or more points.

spot marking–color on the surface of the wire of an electrode.

spot welder, spot welding–a device and process used when sheet metal must be fastened in many places near the edges.

spray gun–a device that releases paint in the form of a fine spray.

square–a device used to draw angles for cutting and to check the cuts for accuracy.

square head–a head with four equal sides.

standard pipe–steel pipe sold or used unless extra heavy or double extra heavy is specified.

staple–a piece of wire with both ends sharpened and bent to form two legs of equal length.

stationary–having a fixed position.

steam cleaner–a portable machine that uses water, a pump, and a burner to produce steam.

steel screws–screws made of steel with a blued, galvanized, cadmium, nickel, chromium, or brass finish.

stick welding–*See* shielded metal arc welding.

stock–a piece of material such as wood or metal.

stove bolt–a round head bolt with a straight screwdriver slot, threaded its entire length.

straight cut file–a file that has straight teeth going in one direction only.

straight (negative) polarity (SP)–DC current flowing in one direction, the opposite of reverse polarity.

strands–small wires placed with others to form bundles to improve flexibility and conductivity.

stringer bead–weld bead produced without weaving.

stroke–the movement of a piston from top to bottom or from bottom to top.

supervised occupational experience program (SOEP)–activities of the student outside the agricultural class or laboratory done to develop agricultural skills.

sweating–process of soldering a piece of copper pipe into a fitting.

switch–a device used to stop the flow of electricity.

switching wires (traveler wires)–pair of wires attached to the light-colored screws on three-way or four-way switches.

table saw (bench saw)–a stationary circular saw with either a tilting arbor or a tilting table.

tacking–making a small weld to hold metal parts temporarily.

tack rag–a rag dampened with solvent.

tacky–sticky.

tang–tapered end on a file.

tap–a hardened, brittle, fluted tool used to cut threads into holes in metal.

tape–flexible measuring device that rolls onto a spool.

tap wrench–a device used to turn a tap.

tee–a fitting used where a pipe line splits in two directions.

teflon tape–tape used on threaded fittings to prevent water or gas leakage at pipe connections.

temper–to heat a piece of tool steel followed by controlled cooling so as to control the degree of hardness.

tensile strength–the amount of tension or pull a weld can withstand.

threads–grooves of even shape and taper that wrap continuously around a shank or hole.

three-view drawing–a drawing that shows three views or sides of an object in three separate representations.

three-way switch–a switch that permits a light or receptacle to be controlled from two different locations by a pair of switches.

tilting arbor–a motor, belt, pulley, shaft, and blade assembly that tilts as a unit in a power saw.

tilting table–a table that can be set at various angles.

tin–one component of solder.

tinning–bonding filler material to a base metal.

tip cleaners–rods with rough edges designed to remove soot, dirt, or metal residue from the hole in the tip of a torch.

titanium dioxide–a high-quality pigment used in many paints.

title block–the section of a drawing reserved for information about the drawing in general.

toe nail–to drive a nail at an angle near the end of one piece and into the face of another piece.

tongue and groove (T&G)–board with a tonguelike edge on one side and a groove on the other which permits boards to be locked together.

tool–any instrument used in doing work.

tool fitting–to clean, reshape, repair, or resharpen a tool.

tool steel–steel with a specific carbon content that allows the tool to be annealed and tempered.

top dead center (TDC)–position of a piston when at its highest point (furthest from the crankshaft).

torch–an assembly that mixes gases and discharges them to support a controllable flame.

torque–a twisting force; to twist.

trade–specific kinds of work or businesses, especially those that require skilled mechanical work.

transformer–a device that converts high voltage from high-power lines to 230 volts for home and farm installations; device used to step current up or down.

traveler wires–*See* switching wires.

trestle–*See* sawhorse.

trim and shutter paint–paint formulated to stay bright without chalking.

troubleshooting–determining what causes a malfunction in a machine or process.

try square–a tool used to try or test the accuracy of cuts that have been made.

tubing–pipe that is flexible enough to bend.

tubing cutter–a tool used for cutting and smoothing pipe or tubing.

turbine–a high-speed rotary engine driven by water, steam, or other gases.

turning tool–a tool used to turn nuts, bolts, or screws.

two-cycle engine (two-stroke-cycle engine)–an engine with two strokes per cycle.

undercoater–special paint used to prepare surfaces for high-quality top coats.

union–a fitting that can be easily opened between two pieces of similar threaded pipe.

unique–having no like or equal.

valve–a device that controls the flow of water or gas.

variable speed motor–a motor whose speed can be controlled by the operator.

varnish–clear, tough, water-resistant finish.

vehicle–a device for carrying something.

vertical down–a weld made by moving downward across the metal.

vertical shaft engine–an engine whose crankshaft is vertical for normal operation.

vertical up–a weld made by moving upward on the metal.

viscosimeter–an instrument used to measure the rate of flow of a liquid.

viscosity–tendency to flow.

vista green–special shade of green used as a focal color.

voltage drop–loss of voltage as electricity travels through a wire.

volt-ohm-milliampere meter (VOM)–a popular combination meter used in electronics.

volt (V), voltage–a measure of electrical pressure.

warp–to bend or twist out of shape.

washed sand–sand flushed with water to remove clay or silt.

washer–a device with a hole used to increase the holding power or prevent loosening of a bolt, nut, or screw.

water-base–paint containing water as the vehicle.

waterproof–material that is sealed against the entrance of water.

watthour–the use of one watt of electricity for one hour.

watt (W)–a measure of energy available or work that can be done using one ampere at one volt.

weaving–moving an electrode sideways to create a wider bead.

wedge–a piece of wood or metal that is thick on one end and tapers down to a thin edge on the other; to pack tightly.

weld–to join by fusion; the seam created by fusion.

weldor–a person who welds.

whet–to sharpen by rubbing on a stone.

white–the color used to mark traffic areas; color of insulation on neutral wires.

white and black stripes–stripes used as traffic markings.

wire nut–an insulated solderless connector used to connect the ends of electrical wires.

wood–the hard, compact, fibrous material that comes from the stems and branches of trees.

wood filler–a thick material used to fill the pores of open-grained woods.

wood preservatives–liquids used on wood to prevent rotting and insect damage.

wood screws–screws with threads designed to bite into wood fibers and draw down when turned.

workable mix–the consistency of wet concrete when the various ingredients are mixed together correctly.

Y–a fitting used where a pipe line splits in two directions at an angle of less than 90°.

yellow–the safety color meaning caution used to identify parts of machines such as wheels, levers, and knobs that control or adjust the machine.

zinc–element associated with high-quality paint; material used for galvanizing steel.

INDEX